In this series original contributions of mathematical research in all fields are contained, such as

- research monographs
- collections of papers to a single topic
- reports on congresses of exceptional interest for mathematical research.

This series is aimed at promoting quick information and communication between mathematicians of the various special branches.

In diese Reihe werden Originalbeiträge zu allen Gebieten der mathematischen Forschung aufgenommen wie

- Forschungsmonographien
- Sammlungen von Arbeiten zu einem speziellen Thema
- Berichte von Tagungen, die für die mathematische Forschung besonders aktuell sind.

Die Reihe soll die schnelle Information und gute Kommunikation zwischen den Mathematikern der verschiedenen Fachgebiete fördern.

Manuscripts
in English and German comprising at least 100 pages and not more than 500 pages can be admitted to this series. With respect to a quick publication the manuscripts are reproduced photomechanically. Authors who are interested in this series please turn directly to the 'Akademie-Verlag'. Here you will get more detailed information about the form of the manuscripts and the modalities of publication.

Manuskripte
in englischer und deutscher Sprache, die mindestens 100 Seiten und nicht mehr als 500 Seiten umfassen, können in diese Reihe aufgenommen werden. Im Interesse einer schnellen Publikation werden die Manuskripte auf fotomechanischem Weg reproduziert. Autoren, die an der Veröffentlichung entsprechender Arbeiten in dieser Reihe interessiert sind, wenden sich bitte direkt an den Akademie-Verlag. Sie erhalten dort genauere Informationen über die Gestaltung der Manuskripte und die Modalitäten der Veröffentlichung.

P. Burmeister

A Model Theoretic Oriented Approach to Partial Algebras

Mathematical Research · Mathematische Forschung

Wissenschaftliche Beiträge
herausgegeben von der
Akademie der Wissenschaften der DDR
Institut für Mathematik

Band 32

A Model Theoretic Oriented Approach to Partial Algebras

by

P. Burmeister

A Model Theoretic Oriented Approach to Partial Algebras

Introduction to Theory and Application of Partial Algebras-Part I

by

Peter Burmeister

Akademie-Verlag Berlin 1986

Autor:

Prof. Dr. Peter Burmeister
Technische Hochschule Darmstadt
Fachbereich Mathematik
D 6100 Darmstadt
Schloßgartenstraße 7

Die Titel dieser Schriftenreihe werden vom Originalmanuskript der Autoren reproduziert.

ISSN 0138-3019

ISBN 3-05-500176-1

Erschienen im Akademie-Verlag Berlin, DDR-1086 Berlin, Leipziger Str. 3-4

Lizenznummer: 202 · 100/407/86
Printed in the German Democratic Republic
Gesamtherstellung: VEB Kongreß- und Werbedruck, 9273 Oberlungwitz
Lektor: Dr. Reinhard Höppner
LSV 1025
Bestellnummer: 763 597 8 (2182/32)
04000

PREFACE

This book is the first of two parts about partial algebras. The second part [Re 84] (Structural Induction on Partial Algebras) has been written by H.REICHEL and provides several applications to computer science. Moreover, the theory presented there is more specially tailored to the interests of computer science and contains some deeper treatment of some special partial algebras, so called hierarchical partial algebras.

In comparison to the second part the general theory of partial algebras in this first one is developed to a broad extent (as far as it already exists), although not all features which have already been studied can be presented within the scope of this book (see section 22 for some more details). Thus choices have been made according to the interests of the author and of some colleagues with whome he has a close collaboration. In this connection especially H.Andréka, I.Németi and I.Sain as well as B.Wojdyło and A.Pasztor should be mentioned whose encouragement to a great extent finally caused the author to write this book.

Actually the idea of a book on universal algebra which also treats partial algebras as fundamental structures had already been pursued at the end of the sixties by the now deceased teacher of the author, Jürgen Schmidt, who, however, never finished this project. One reason might have been that in the early seventies there seemed to be almost no larger interest or possibility of applications for partial algebras. This, however, has changed during the last years, when the necessity of a more rigorous theoretical foundation of computer science and the growing engagement in abstract (algebraic) data types -- as they are e.g. used in programming languages like PASCAL or ADA -- caused among computer scientists also a growing interest in universal algebra. As a next step some of them also turned their interest towards partial algebras, since many operations occurring in computer science are only partial: Operations and the evaluation of functions or expressions used or defined in programs sometimes are not everywhere defined -- either because of the delimitations given by the used computer or because of the definition of the expressions. Moreover, the abstract theory of algorithms shares many features with the one of partial algebras (see [Ti 81]), and this led H.THIELE in [Th 66] to develop a logic for partial algebras which is basically the one we use here. -- Further applications -- or hints to corresponding literature (see e.g. [AN 79f] as such a source) -- can also be found in the literature mentioned in subsection 22.2.
A quite old source of interest in the study of partial operations is the one concerning partial recursive functions (see e.g. [Kl 52], [Mal 74] and [Hoe 72], also [Ku 77]). Another one may come from quantum mechanics, e.g. quantum logic, in theoretical physics, where the fact that not all phenomena connected with the objects of interest can be observed and described at the same time gives rise to partial operations (see [KoSp 68]).

Despite a possible growing interest in features of partial algebras only few people have investigated or really used these structures so far, since most possible users had great reservations against these structures whose treatment seemed to be so much more complicated than the one of total algebras and for which they did not see much of a theory around. As a matter of fact almost every concept from the theory of total algebras often splits into more than three corresponding concepts for partial algebras, and for a long time it seemed hard to distinguish some rules behind it -- but this has changed. On the other hand one has either the possibility of completing a partial algebra in a more or less satisfactory way, or one can consider partial algebras as special relational systems whose theory provides some tools in order to handle such structures -- but not always the most adequate ones.

Nevertheless there has been developed -- in particular during the last decade -- a theory of partial algebras which may really be a starting point for the application of these structures. It is based on category and model theoretical methods and provides some insight into the reasons for the fact that there really is such a wealth of concepts around for partial algebras -- one reason for this richness being the fact that the theory of relational systems can totally be embedded into the theory of partial algebras. But the theory of partial algebras is much more appropriate to the task of dealing with relations which in fact are the graphs of functions than the theory of relational systems is. Thus we think it to be worth-while to write a book like this one which introduces into the theory of partial algebras and shows up their special features in comparison to total algebras. One further reason is the fact that out of the theory of partial algebras one may get a better understanding of many features of so-called many-sorted or heterogeneous algebras which are those structures mostly used in computer science. -- One may say that this is too large a detour for the understanding of heterogeneous algebras, but as long as one has not yet decided on how to treat errors, even by using many-sorted algebras one cannot avoid partiality as the common example of the abstract data type "STACKS" (see the introduction and later sections) shows. And partial algebras include all such structures, especially when one uses the approach via (what we shall call) partial $\underline{S}$-algebras which is only sort of a formal but very useful further generalization, all of whose final applications cannot yet be foreseen.

The present book has another feature which makes it different from almost any other presentation of universal algebra so far, namely the fact that also infinitary operations are allowed. Although this requires at some places some more technical details, we hope that this fact does not burden the understandability too much, in particular since we can avoid almost everywhere the treatment of ordinal numbers (following J.Schmidt). Nevertheless on some occasions one has to deal with cardinality restrictions, where in the finitary case the occurring bounds usually can be replaced by the cardinality $\#\omega_0$ of the set of all natural numbers. One reason for allowing infinite arities is the possible application of infinitary structures in connection with so-called ω-posets, which are used in some part of computer science (cf. D.LEHMANN, A.PASZTOR [LePa 80] and A.PASZTOR [Pa 83], [Pa 84]). Infinitary operations also arise in possible applications in topology (see G.A.EDGAR [Ed 73]), or in the theory of ordered sets.

The objection might be raised that in this book we do not provide enough really important applications for the theory presented. There might be some point to this, since this is the more theory oriented part of a two-volume project, and for some more special part of the theory of partial algebras the second volume of this project (written by H.REICHEL) provides some really important examples; however, the special theory presented there cannot cover all their features. Therefore the general theory provided here shows how one can proceed in situations not covered by the more special theory. Moreover we hope that the existence of a theory for partial algebras like this one will encourage possible users to produce new applications. Although we do not know what a possible user will really need, we hope that in this book he will find enough information in order to develop by himself the special part of the theory necessary for him. And thus we touch a further intention of this book -- together with its second part --, namely to encourage other mathematicians or computer scientists to use these two books as a platform from which they can start a further development of the theory.

Such a first treatment and representation of a very general theory like this one on partial algebras will probably not be all around as smooth as one would like it to be, and we beg the reader in advance to pardon this. The first chapter of this book has to a great extent already been the contents of lectures of J.Schmidt's at Bonn (Germany) in 1965/66 (see [Sch 66b]) and Athens (Georgia, USA) in 1967/68 (see [Sch 68]), as well as of the author's at Bonn in 1971 and at Darmstadt for several times, while most of the material in the subsequent chapters has not yet been presented in this form and framework before; and it contains at some places research results not yet presented anywhere else. Only a rough overview of the guiding ideas has been given in [B 82] (see also [AN 76] for a considerable portion of the logic and category oriented aspects), where especially the model theoretical aspects of the theory have been presented (without proofs). However the whole presentation is new as far as it is based on the study of what we call partial $\underline{S}$-algebras. Therefore we might not have found the best way of representation right at the beginning. On the other hand we wanted this book to appear as quickly as possible because of the other reasons mentioned above. -- If not every one who contributed to the theory of partial algebras presented here has received due credit, we apologize for that: it is not due to bad will but only to lack of space (e.g. as far as the schools of J.Słomiński and H.-J.Hoehnke are concerned), to some forgetfulness or even to ignorance on part of the author. -- Let us also mention that we do not treat relations separately, since they can easily be encoded as partial operations without any loss of algebraic features; however, observe that in this context the class of all relational systems of some given type is not an equational but only an implicational subvariety of the class of all partial algebras of the corresponding type.

Finally I want to express my gratitude to all those who have supported this project. In the first place I have to mention my family for its patience and understanding, and in particular my wife Hildegund who also did a lot of proof-reading and filled in most of the special symbols. Moreover I thank Hans-Jürgen Hoehnke, Hajnal Andréka, István Németi, Ildiko Sain, Ana Pasztor and Manuel Siegmund-Schultze for valuable remarks and discussions, and also -- together with Peter Winkler, Peter Alles and Wolfgang Thumser -- for proof-reading all or parts of the book. I should not forget to thank all those who helped me getting started on text processing on a personal computer. -- I also want to thank the Akademie-Verlag for publishing this book, e.g. Dr. R. Höppner for his patience -- and his slight pressure.

Darmstadt, July 1985 Peter Burmeister

CONTENTS

PREFACE 5

INTRODUCTION 11

CHAPTER I -- BASIC DEFINITIONS AND PROPERTIES 17

1 The concept of homogeneous and heterogeneous partial algebras 17

1.1 Partial algebras 18
1.2 Heterogeneus partial algebras 21
1.3 Cardinal and ordinal dimension 23

2 Homomorphisms, congruence relations and factor algebras 24

2.1 Homomorphisms 24
2.2 Isomorphisms 25
2.3 Endomorphisms and automorphisms 27
2.4 Full and closed homomorphisms 28
2.5 Congruence relations 30
2.6 Factor algebras 33
2.7 The Homomorphism Theorem 35
2.8 The congruence lattice of a full homomorphic image 37
2.9 The analogue of the Second Isomorphism Theorem of E. Noether 38
2.10 Mono- and epimorphisms 39
2.11 Partial S-algebras and heterogeneous partial algebras 40
2.12 Special observations concerning partial S-algebras 42

3 Substructures, subalgebras, closed subsets, generation 44

3.1 Substructures of partial algebras 45
3.2 Comparison of substructures 48
3.3 The global process of generation 49
3.4 "Local" generation (generation "from below") 49
3.5 The principle of algebraic induction 51
3.6 Some further results concerning generation, epimorphisms 52
3.7 The analogue of E. Noether's First Isomorphism Theorem 54

4 Direct and reduced products, coproducts, direct limits and related constructions 55

4.1 (Direct) Products (homogeneous case) 55
4.2 Pullbacks and products of partial S-algebras 59
4.3 ("Global") Coproducts and multiple pushouts 61
4.4 Direct limits (= directed colimits) 64
4.5 Reduced products and ultraproducts 69
4.6 Some properties of the operators **H**,.. **S**,.. and **P**,.. 71

5 Universal solutions, free partial algebras, Peano algebras, terms and term operations 75

5.1 The general concepts of universal solution and freeness 75
5.2 Partial Peano algebras 77
5.3 Free completions 80
5.4 The Generalized Recursion Theorem 82
5.5 The algebraic quasi-ordering on partial algebras 83
5.6 Tree representation of terms 85
5.7 (Global) Term operations and polynomials 86
5.8 The Epimorphism Theorem 89
5.9 The initial closed congruence relations of a free completion 91
5.10 On the cardinalities of M-generated partial algebras 92
5.11 On the existence of **K**-universal solutions 93

6 The equational theory of partial algebras 95

6.1 Existence-equations 95
6.2 Description of the model classes (E-varieties) 98
6.3 More facts about freeness, fully invariant congruences 102
6.4 Characterization of E-equational theories 104

CHAPTER II -- A BASIS FOR A TWO-VALUED MODEL THEORY FOR PARTIAL ALGEBRAS 111

7 A "first order" language for (infinitary) partial ($\underline{S}$-) algebras 111

7.1 The syntax 112
7.2 The semantics 115
7.3 Coincidence and Substitution Theorem 116

8 Elementary implications, universal Horn formulas 119

8.1 Some basic kinds of universal Horn formulas 119
8.2 Algebraic encoding of universal Horn formulas 123
8.3 Preservation properties of algebraic operators -- first visit 125
8.4 The diagonal-fill-in property -- motivation 127

9 Preservation (reflection) of formulas by mappings 128

9.1 Preservation and reflection 129
9.2 Examples for the reflection of formulas by homomorphisms 132

10 Factorization systems of partial algebras 137

10.1 The Galois connection induced by the unique-diagonal-fill-in-property . 138
10.2 Factorization systems 140
10.3 Examples of factorization systems for partial algebras 146

CHAPTER III -- BIRKHOFF TYPE RESULTS FOR ELEMENTARY IMPLICATIONS 152

11 Direct limits revisited 152

11.1 Morphisms between directed systems 152
11.2 A device to measure "smallness" 155
11.3 Direct limits and the operator "Λ" 156
11.4 Sections, retractions and direct limits 158
11.5 Some further concepts connected with direct limits 159
11.6 Proof of Theorem 8.3.4 160

12 A Meta Birkhoff Type Theorem 162

12.1 $\mathbf{HSP}^{r,\mathbf{k}}$-cones 162
12.2 A Meta Birkhoff Theorem 163
12.3 "Compatible" classes **H** and **S** in **Alg**($\underline{S}$) 171

13 Construction of examples for the Meta Birkhoff Theorem 174

13.1 **k**-dual algebroidalness of $\Lambda^{op}(S)$ for some classes S of **Alg**($\underline{S}$)-morphisms 175
13.2 **H**-projective partial $\underline{S}$-algebras 177
13.3 Examples in **Alg**($\underline{S}$) for the Meta Birkhoff Theorem 180
13.4 Special examples 183

14 Closed classes of epimorphisms (elementary implications) 186

14.1 Preparatory results 186
14.2 The general theorem 188
14.3 Closed sets of E-, ECE- and QE-equations and of $\mathbf{H}_f\mathbf{S}_c\mathbf{P}^{r,\mathbf{k}}$-implications . 190
14.4 Closed sets of ECH- and Horn-formulas 196

CHAPTER IV -- SPECIAL TOPICS 198

15 Algebraic operations 199

15.1 Superposition 199
15.2 Algebraic partial operations of the first and second kind 203
15.3 Algebraic partial operations of the third kind 207
15.4 Algebraic total operations 210
15.5 More about algebraic partial operations and terms; algebraic partial functions 213
15.6 Algebraic partial operations and homomorphisms 215

16 Semilattices of (weak) completions 216

16.1 The semilattice of (weak) minimal completions 217
16.2 The complete lattice of normal minimal completions 219

17 Independence, free partial algebras and primitive classes 222

17.1 Some properties of independence and independence classes 222
17.2 Relatively free completions of partial algebras 223
17.3 Relatively free completions of relatively free partial algebras 228
17.4 Characterization of heterogeneous relatively free partial algebras 231
17.5 Generators and relations 233
17.6 Initial objects and representation by canonical terms 235

18 Dependence of operations on arguments and element basis property 239

18.1 General concepts 239
18.2 Dependence on variables and algebraic independence 242
18.3 A counterexample for the infinitary homogeneous case 246

19 Congruence relations revisited 247

19.1 Closed congruence relations of finitary partial algebras 247
19.2 P-reflecting homomorphisms and P-closed congruence relations 252

20 Subdirect representations of partial algebras 255

20.1 The general concept and remarks 255
20.2 Classes of full and surjective homomorphisms -- arbitrary monomorphisms 256
20.3 The cases of full respectively closed monomorphisms 260
20.4 Subdirectly irreducible mono-unary partial algebras 264

21 More facts about homogeneous E-varieties 266

21.1 Initial homomorphic extensions 267
21.2 On the lattice of initial closed congruence relations -- homogeneous case 271
21.3 Relativized algebraic partial operations -- homogeneous case 274

APPENDIX

22 Survey on some additional concepts and results 277

22.1 Other concepts of homomorphisms 277
22.2 Further topics and hints to applications 279

23 Some definitions and facts from set theory 281

23.1 Some remarks on notations 281
23.2 Some remarks on an axiomatic approach 284
23.3 Ordinals 286
23.4 Cardinals 289

24 Ordered sets, algebraic lattices and Galois connections 291

24.1 Partially ordered sets lattices 291
24.2 **k**-directed posets, **k**-inductivity, **k**-algebraic lattices 293
24.3 Closure systems and closure operators 294
24.4 Galois connections 297

REFERENCES 299

LIST OF SYMBOLS 306

INDEX 310

TABLES: 1.1 Cardinal and ordinal dimension 23
8.1 Preservation of formulas by algebraic operators 126
9.1 Homomorphisms defined by reflection of formulas 135
10.1 Factorization systems in **Alg**($\underline{S}$) 150
12.1 Compatibility of the operators **H** and **S** 174
13.1 Pairs (**H**,**S**) satisfying the assumptions of the Meta Birkhoff Theorem . 180
13.2 Shape of the implications 181
20.1 (**E**,**M**)-subdirect representation theorems 264

INTRODUCTION

Algebra has a long history which started with the investigation of rules for very special calculations with natural numbers more than three thousand years ago. Later rational numbers and calculations with them were added, irrational numbers were discovered about two thousand years ago, and only a few centuries ago it was accepted that complex numbers form a suitable extension of real numbers. During all those centuries algebra was the mathematical theory which mainly dealt with the solvability of equations in special ranges of numbers and with the description of such solutions. And the extension of the "world of numbers" at each stage was caused by the desire to add solutions of the equations under consideration -- mostly linear equations or polynomial eqations of low degree -- to the existing and accepted domain (i.e. to complete a given partial algebra). Moreover, one wanted to transfer the rules of calculation as far as possible also to the extended domain. Finally with the introduction and wide acceptance of complex numbers a stage was reached, where every linear or polynomial equation was -- theoretically -- solvable within this domain, and other questions arose, in particular the one of which polynomial equations could be solved by expressions using n-th roots and rational operations only. This led -- among others -- to the introduction and investigation of abstract groups, fields and rings, independently from a special concept of "numbers". Thus quite different problems than before intrigued algebraists: construction, investigation and description -- in particular classification -- of axiomatically defined algebraic structures. While, at the beginning, this mainly concerned structures close to groups, rings, fields, modules and vector spaces, possibly also boolean algebras or lattices, at some stage abstraction went so far as to investigate sets with (mostly finitary) operations in general (up to isomorphism), this was the "hour of birth" of universal algebra, which is usually closely related to the paper [Bi 35] of Garrett BIRKHOFF.

At the beginning universal algebra was pushed forward just to find a general background for the different algebraic structures arising within mathematics itself partly supported by the hope that such a theory could be useful in connection with problems within special algebraic theories. But it has turned out that some of these hopes could not be fulfilled; universal algebra rather **provides** problems for classical algebra than that it **solves** problems for classical algebra. It proves to be useful in particular when new algebraic structures arise, since it provides all the known results common to all algebraic structures (in general or under certain restrictions like being equationally or implicationally defined). Thus it has proved to be quite useful for some aspects of theoretical computer science, where computers, programming languages, programs or abstract data types can be interpreted as, be described by or be better understood when considered as algebraic structures.

Now this is a point, where partial algebraic structures might come into the picture, since often the operations occurring in computer science are not defined on the whole set of elements involved. One has tried to overcome this fact in several ways. Let us consider some examples in order to illustrate this:

EXAMPLE 1: A Mealy-type automaton is usually algebraically described as a system $\underline{A} = (S,I,O;\lambda,\rho)$ which consists of three non-empty sets, namely I -- the set of all input elements --, O -- the set of all output elements -- and S -- the set of all possible (internal) states of the automaton. Furthermore one has two functions: $\lambda:S \times I \to S$, the so-called transition function, and $\rho:S \times I \to O$, the so-called output function (an input $i(\in I)$ which is applied to $\underline{A}$ while in a state $s(\in S)$ causes $\underline{A}$ to switch into the state $\lambda(s,i)$ $(\in S)$ and yields an output $\rho(s,i)$ $(\in O)$).

This is a prototype of a so-called many-sorted algebra. Considered as a partial algebra it would correspond to $(S \dot{\cup} I \dot{\cup} O;\lambda',\rho')$ -- "$\dot{\cup}$" designating disjoint union --, where λ and λ' as well as ρ and ρ' in each case have the same graph, but λ' as well as ρ' are now binary partial operations on the carrier set $S \dot{\cup} I \dot{\cup} O$.
Thus one method is to subdivide the carrier set of an algebraic structure into subsets, such that one can specify for each argument of any given operation, from which subset it may be taken. This has led to the study of so-called many-sorted or heterogeneous algebras. But even this may not help directly in every case, as the following example will show.

EXAMPLE 2: A data type STACK OF INTEGERS is usually described -- except for error values -- as the so-called "initial object" in the class of all two-sorted (partial) algebras $\underline{A} = (S,I;\theta,SUCC,PRED,PUSH,TOP,POP,E)$, such that I and S are non-empty sets containing θ (zero) and E (the empty stack!) as distinguished elements, respectively (i.e. θ and E are so-called nullary constants), and the operations behave as follows:

SUCC, PRED:$I \to I$, satisfying (1) $SUCC(PRED(i)) = i$
and (2) $PRED(SUCC(i)) = i$ for every $i \in I$,

PUSH:$S \times I \to S$

POP:$S \to S$, satisfying (3) $POP(PUSH(s,i)) = s$,

TOP:$S \to I$, satisfying (4) $TOP(PUSH(s,i)) = i$ $(s \in S, i \in I)$.

The intuitive meaning of these operations is the following: If I is really the set of integers, then $SUCC(i) = i+1$ and $PRED(i) = i-1$, moreover PUSH(s,i) puts integer i as a new entry on top of the stack s, while TOP(s) reads the item at the top of the stack s, if there is any such item, and is undefined, if $s = E$, i.e. when s is the empty stack; similarly POP(s) throws away the item at the top of the stack s if s is non-empty $(s \neq E)$ and is undefined if s is empty $(s = E)$. θ and E are so-called nullary constants which in this connection are required to be defined, what we shall formally express later by the axioms

$$\theta \stackrel{e}{=} \theta \quad \text{and} \quad E \stackrel{e}{=} E .$$

At this point we should already say something about the use of equations as axioms in connection with partial algebras, which will more formally be introduced in Chapter I, section 6 (see also section 7): There has long been a great uncertainty among algebraists on how to generalize the concept of an identity from total algebras to partial ones; the basic concept which we shall use here means that for all possible replacements of the variables occurring on any side of an identity both "sides" of the identity have to "exist" (i.e. the corresponding expressions (terms) have to be evaluable in the partial algebra under consideration), and the values have to be equal. Because of these hidden existence statements in this kind of semantics we also call equations for partial algebras "existence equations" (more or less for didactical reasons) and stress this by writing "$\stackrel{e}{=}$" instead of "=" -- this corresponds to the habit of logicians of using a different equality sign in the object language and in the meta-language, respectively. We would like to stress that we shall use the usual equality symbol in the meta-language in the same way as existence-equality of the object language: For the given expressions both sides have to exist and have to take the same value! -- With this starting point at hand almost all known interpretations of identities for partial algebras can be expressed by validity of usually relatively simple open first order formulas (in general elementary implications or conjunctions of them). Actually this starting point is the key for a successful introduction of a powerful two-valued model theory for partial algebras, as we shall formally do in section 7.

Returning to Example 2 we realize that so far we have specified a really partial two-sorted algebra. In order to make it a total algebra we have to introduce for each sort at least one "error-value" in order to be able to assign a value to POP(E) and TOP(E), respectively. If, however, we would like to keep the above axioms valid also in the completion, we will end up with an infinite set of different error-values, although this is not what one usually wants in the concrete case, showing (or knowing in advance) that the "error-values" are not original, but only some necessity for the final concrete implementation. Here the theory of partial algebras allows a wide range of possibilities of completion, and in connection with a specific implementation one has to choose the most appropriate one.

Nevertheless ("total") many-sorted algebras may help in so far as one often introduces a new sort BOOLEAN which applies to exactly two elements TRUE and FALSE, and every partial (many-sorted) operation is replaced by a function which maps its graph to TRUE and everything else which would be possible because of "correct sorts" to FALSE. This method imposes strong -- and sometimes not wanted -- restrictions on the admissible homomorphisms i.e. in connection with the comparison of structures. On the other hand this method can also be applied in connection with relations, where it just yields the characteristic function of the relation under consideration (again with the indicated heavy restrictions on homomorphisms). For a detailed discussion of such completions in computer science we refer to [GoTWa 78] and [Go 78] from the ADJ-group.

We think that it might usually be easier first to handle partiality in its own right and then to propose adequate completions, for instance to use in Example 2 above a so-called "heterogeneous normal one-point completion" (i.e. wherever necessary, to a set of a specific sort exactly one new point is added and all undefined items of operations with output in this sort are mapped onto this new element, the so-called "error-element" of the sort under consideration). Other completions are conceivable, too (see Chapter IV). Thus we present in this book an introduction into the theory of partial algebras, which -- as we hope -- may serve as a basis for further investigations using partial algebras or for further research on partial algebras.

There is one well-known example of -- homogeneous or heterogeneous -- partial algebras, where usually nobody thinks of completing them: <u>small categories</u>. Let us briefly sketch the homogeneous description -- the two-sorted one is presented in detail in section 1 in

1.2.1.(iii) (see also 8.1.1 for further items):

EXAMPLE 3: As carrier set of a description of a given **small category** $\underline{C}$ as a (homogeneous) partial algebra one chooses the set $M := Mor(\underline{C})$ of all $\underline{C}$-morphisms (the reader may e.g. think of the set of all linear mappings between finite powers of a given field $\underline{Q}$ considered as vector spaces over $\underline{Q}$ -- or even more concretely of the set of all finite matrices with components in $\underline{Q}$). Fundamental operation is a (binary) partial multiplication "·" (the composition of morphisms), which is defined for two elements f, g of M (i.e. for two morphisms) iff (if and only if) the codomain (target) of f is equal to the domain of g -- since in this approach objects are not included, they are replaced by their identity morphisms. In order to have a possibility within the partial algebraic structure to distinguish somewhere the domain and the codomain of an (abstract) morphism one also introduces two unary total operations -- here denoted by Dom* and Cod* ("*" in order of distinguishing these operations from the two-sorted ones which have "objects" as values), which map a morphism f onto the identity morphism Dom*(f) respectively Cod*(f) of the domain respectively of the codomain of f. Thus small categories can be considered as partial algebras (M; · , Dom*, Cod*) with one binary and two unary (partial) operations such that one has for arbitrary f, g, h in M (observe that in a composition of morphisms the morphism on the left is applied first):

(i) Dom* and Cod* are unary total operations, · is a binary partial operation, and $f \cdot g$ is defined iff $Cod^*(f) = Dom^*(g)$.

(ii) The multiplications $Dom^*(f) \cdot f$ and $f \cdot Cod^*(f)$ are always defined and each has value f.

(iii) $Dom^*(f \cdot g) = Dom^*(f)$ and $Cod^*(f \cdot g) = Cod^*(g)$, whenever $f \cdot g$ is defined.

(iv) If $f \cdot g$ and $g \cdot h$ are defined, then $(f \cdot g) \cdot h$ and $f \cdot (g \cdot h)$ are defined, and they are equal: $(f \cdot g) \cdot h = f \cdot (g \cdot h)$ (weak associativity).

The class **Cat*** of all partial algebras (M; · , Dom*, Cod*) satisfying (i) through (iv) above is then identical with the class of all abstract small categories described by their set of morphisms only.

We shall see that this class of all small categories is not axiomatizable by equations (since a "weak homomorphic image" of a category need not be a category), however by elementary implications of a very simple structure -- namely by such elementary implications which behave like equations whenever the model under consideration is a total algebra (see 8.1.1 and the concept of ECE-equations -- i.e. of existentially conditioned existence-equations -- in 8.1.2).

Category theory has developed its own special concepts, but it should be observed that e.g. functors between (small) categories are nothing else but the usual homomorphisms between the corresponding partial algebras.

Let us add a further example of partial algebras:

EXAMPLE 4: In theoretical computer science so-called ω-posets play an important rôle. These are partially ordered sets (posets) $(P;\leq)$ in which each countable chain monotonously indexed by natural numbers ($\in \omega = \omega_0$) has a supremum: i.e. if $p_0, p_1, p_2, \ldots, p_n, \ldots$ ($n \in \omega$) are elements of P such that $p_n \leq p_{n+1}$ for each $n \in \omega$, then the supremum $\sup(p_n | n \in \omega)$ of this "ω-chain" exists. -- The morphisms are the mappings preserving order and the partial supremum "sup". -- D.LEHMANN and A.PASZTOR have observed in [LePa 80] that ω-posets can be considered as infinitary partial

algebras:
The partial order relation "$\le$" is presented by a binary partial operation, say "*", such that $x*y$ is defined iff $x \le y$; and if $x \le y$, then $x*y:= x$ (say). Moreover, one has an ω-ary partial operation, say Sup, such that $\mathrm{Sup}(x_n|n \in \omega)$ is defined iff $x_n \le x_{n+1}$ for all $n \in \omega$, and if $\mathrm{Sup}(x_n|n \in \omega)$ is defined, then its value is the supremum of the argument chain.
The axioms for ω-posets are basically of the same nature (besides containing infinite conjunctions) as those for small categories (namely ECE-equations, see 13.4.4)
Thus all that is said in this book on classes definable by ECE-equations is applicable to the classes of (small) categories respectively ω-posets, and no new structure theory is needed. More material on varieties of ω-posets and ω-continuous algebras (i.e. ω-posets which have "$\le$- and Sup-compatible" operations, see 13.4.5) can be found in [Pa 83] and [Pa 84].

Example 4 already suggests that the theory of partial algebras as presented in this book includes the theory of (partial) algebras with relations, since each n-ary relation, say R, on some set A -- considered as set of n-tuples with elements from A -- ($n > 0$) can be "encoded" by a partial n-ary operation, say ϕ_R, on A, whose domain is R, and which maps each n-tuple from R onto, say, its first component. -- Again we obtain a variety defined by ECE-equations.
We do not claim that it be always better to treat relational systems as partial algebras; we only want to stress that -- although we do not treat relations separately -- most of the material contained in this book can also be applied without any additional effort to partial algebras with relations.

However, we should also mention that a reader who is mainly interested in the algebraic theory and model theory of fields -- another standard example for partial algebras, since the formation of a multiplicative inverse is a partial operation -- might be disappointed by this book, since our model theoretic oriented results in Chapters II and III apply at most to Horn varieties (i.e. classes axiomatizable by universal Horn formulas), and the class of all fields is not definable by universal Horn formulas. But the model theoretic language which we use for partial algebras at least allows us to axiomatize the class of all fields by universal formulas -- the existential quantifiers usually needed are hidden in our concept of (existence-) equations (see sections 6, 7 and 14).

After these introductory examples which may already provide a slight idea of some intentions of this book, let us say a few words in general about the main concepts -- a larger survey of the basic ideas and concepts guiding us here has been presented in the survey article [B 82] --:
We try to present a theory of partial algebras which fits neatly between the theories of relational systems (and two-valued model theory) and of total algebras, respectively. This has been the main guideline in particular for the choice of the basic concept of a homomorphism: it should in each case coincide with the corresponding concepts of the two theories. Then most of the properties usually encountered with homomorphisms between partial algebras have a model theoretic description or interpretation (see Chapter II).

Modest tools from category theory are also applied, when we thought that it would help a better understanding, and in order to relate the results presented here to more category oriented papers in theoretical computer science: we hope that our -- in parts -- parallel use of category theoretical and algebraic language and the "translations" between these two languages will help to apply universal algebra, in particular partial algebras, as well as the abstract results from category theory in concrete situations. Because of limitations in space, time, etc., this book cannot cover all the aspects of the theory of partial algebras, but we hope that it can serve as a good introduction.

Since each chapter is preceded by a brief survey of the contents in its sections, we only mention here the main intentions of each chapter:
Chapter I provides basic algebraic (and related category theoretical) concepts with respect to partial algebras, such as are usually contained in an introductory text on universal algebra, and ends with the E-equational theory. -- In Chapter II an introduction to a two-valued ("extended") first order logical language ("extended" in order to include the infinitary and many-sorted cases) is presented and an investigation of its applications to a "structuring" of the properties of homomorphisms is added. -- In Chapter III we prove a meta Birkhoff-type theorem for elementary implications and universal Horn formulas (due to H.Andréka, I.Németi and I.Sain) and study to a broad extent its impact on the theory of partial algebras. -- In Chapter IV several topics are presented, most of them concerning concepts related to (relative) freeness. It should be observed that, as long as we "move in front of a category theoretical background," many-sortedness does not cause much trouble, but when we "dive" in Chapter IV into intrinsic more set theoretic properties of (constructions of special kinds of) partial algebras, then one encounters a lot of complications in part explicable by the fact that for many-sorted partial algebras existence-equations are -- on a more elementary level, where the different sorts are expressed by unary relations and axioms for them -- actually elementary implications and not equations (of this level). This may explain -- besides further instances given in the text --, why -- in our opinion -- basic concepts from model theory have to be used on a very early stage in order to understand the behaviour of partial algebras and of concepts related to them.
In an Appendix basic definitions and facts from set theory and order theory are collected as an introduction to the main text. Moreover, some hints on further aspects of the theory (or of other theories) of partial algebras are given (section 22).

CHAPTER I

BASIC DEFINITIONS AND PROPERTIES

In this first chapter we present a great part of the universal algebraic properties of partial algebras. Starting with basic definitions and concepts and their main properties we develop the theory of partial algebras up to their equational theory. As already indicated in the introduction, heterogeneous partial algebras are included in the considerations, but based on the investigations of homogeneous partial algebras: In a category theoretical language one would say that the category of all heterogeneous partial algebras of some given similarity type is considered as a comma category **Alg**$(\tau)\downarrow\underline{S}$ of the corresponding category of homogeneous partial algebras, where τ is the corresponding homogeneous type, and $\underline{S}$ is a homogeneous partial algebra encoding the heterogeneous similarity type, the so-called sort algebra (cf. subsection 2.11).

In **section 1** homogeneous and heterogeneous partial algebras are defined, and some concepts connected with the similarity type are discussed.

In **section 2** we introduce homomorphisms as structure preserving mappings from the start- into the target-object, study their relationship with congruence relations, and prove the Homomorphism Theorem and its consequences (e.g. Isomorphism Theorems). Besides injectivity, surjectivity and bijectivity, also closedness and fullness are recognized as important properties of homomorphisms between partial algebras. As a by-product we get the description of heterogeneous partial and total algebras indicated above.

In **section 3** closed subsets and the process of generation of partial algebras by sets are investigated -- also in connection with the properties of homomorphisms mentioned above.

Section 4 is dedicated to the construction of new partial algebras from given ones, e.g. as direct products, coproducts, reduced products or direct limits.

In **section 5** universal solutions, e.g. absolutely free completions, terms and term operations are introduced and investigated, and a general Recursion Theorem as well as a homomorphism theorem for epimorphisms and an existence theorem for universal solutions are presented. Moreover, upper bounds for the cardinalities of partial algebras depending on the cardinal dimension and the cardinality of a generating subset are given.

Finally, **section 6** contains the introduction of the concept of (E-) equations and the Birkhoff-type (E-) equational theory for partial algebras.

1 The concept of homogeneous and heterogeneous partial algebras

In this section the concepts of homogeneous and heterogeneous partial algebras of a given similarity type, as already intuitively used in the introduction, are made precise. For set theoretical prerequisites see section 23 in the appendix.

1.1 Partial algebras

Since they form the fundament of (partial) operations and algebraic structures, we first consider

1.1.1 DEFINITION (of sequences). Let A and K be any sets. A K-ary sequence (sequence of type or arity K) (with elements) in A is any mapping from the set K into the set A. We shall use different notations for sequences -- as suitable --, e.g. $\underline{a}:K \to A$, $(a_k)_{k\in K}$, $(a_k | k \in K)$, $(k\underline{a} | k \in K)$ or $((k)\underline{a} | k \in K)$; if K is some natural number, say n, then we also use the notation $(a_0,\ldots,a_{n-1})$ or sometimes $(a_1,\ldots,a_n)$, since we shall usually identify the set A^n of all mappings from $\{0,\ldots,n-1\}$ into A with the set $A \times A \times\ldots\times A$ (n factors) of all n-tuples of elements of A.
In general we denote by A^K the set of all K-ary sequences in A.
#A always designates the cardinality of set A (cf. section 23.4 in the Appendix).

1.1.2 REMARK. Notice that A^1 or $A^{\{k\}}$ for any one-element set is usually identified with the set A itself, if no confusion can arise, and that $A^0 = A^\emptyset$ consists of exactly one element, namely the empty sequence $\emptyset$: $A^0 = \{\emptyset\}$. Especially we have:
(i) $A^K = \emptyset$ iff $A = \emptyset$ and $K \neq \emptyset$.
(ii) $\#A^K \leq 1$ iff $\#A \leq 1$ or $K = \emptyset$.
(iii) $\#A^K = 1$ iff $\#A = 1$ or $K = \emptyset$. ∎

1.1.3 DEFINITION (of partial operations) and REMARKS. As before let A and K be any sets. A partial operation of type (arity) K on A (a K-ary partial operation on A) is any partial mapping ϕ out of A^K into A: ϕ:dom $\phi \to A$, where dom $\phi \subseteq A^K$ is its domain.
If, however, dom $\phi = A^K$, then ϕ is called a (total, full or complete) K-ary operation on A, while in the case of dom $\phi = \emptyset$ ϕ is called the empty or discrete K-ary partial operation on A.
Observe that for any given arity K we usually have exactly one discrete partial operation of type K on A, while we usually have quite a lot of total K-ary operations on A.
If $K = \emptyset$, then any K-ary partial operation ϕ is called a nullary partial operation on A, and if its domain is non-empty, then it is often called a (nullary) constant in A. -- Notice that the total nullary operations on A are in one-to-one correspondence with the elements of A, and they are usually identified with their unique value.
We would also like to mention that every K-ary partial operation ϕ on A can be considered in a natural way as a $(K \cup \{K\})$-ary relation on A, i.e. as a subset of $A^{K\cup\{K\}}$, namely by considering graph(ϕ):= $\{(\underline{a},\underline{a}\phi) | \underline{a} \in \text{dom } \phi\}$, the graph of ϕ, having a distinguished index, namely K. -- Notice, that in the general context we shall usually apply functions from the right, and that we also compose them in that way, while concrete functions and operators are often applied from the left, and for operators the corresponding way of composition is then used; for binary concrete operations infix notation is often used, too. -- On the other hand, every K-ary relation R on A for a non-empty set K, i.e. every subset $R \subseteq A^K$, can be considered as a K-ary partial operation on A by distinguishing some element, say k_0 in K, and defining $\phi_R:R \to A$ by $\underline{a}\phi_R := (k_0)\underline{a}$ iff $\underline{a} \in R$, i.e. ϕ_R is a partial K-ary projection of A onto the k_0-th component.
If the reader follows all the subsequent definitions and statements carefully, he may easily realize that everything that is said about partial operations and partial alge-

bras equally applies to relations and relational systems, only that something might either become trivial (as for instance everything concerning "generation" (cf. section 3)), while other items will become a little bit more complicated (as for instance some descriptions by first order axioms will need some additional axioms).
While mappings f from a set A into a set B are usually noted as $f:A \to B$, the fact that ϕ is a K-ary **partial** operation on A and not necessarily everywhere defined on A^K will sometimes be indicated by the notation $\phi:A^K \dashrightarrow A$ -- "$\dashrightarrow$" will always indicate a partial mapping.

1.1.4 DEFINITION (of partial algebras). Throughout this book Ω will designate an arbitrary, but fixed set, which we shall call the set of operation symbols; the elements of Ω, i.e. the operation symbols themselves will usually be denoted by ϕ, ϕ', ... or ϕ_1, ϕ_2, ... (cf. 1.1.5.(iii) below).
Moreover, τ will designate a sequence of empty or non-empty sets (often of natural or ordinal numbers) indexed by the elements of Ω, i.e. $\tau := (K_\phi)_{\phi\in\Omega}$, the similarity type (or briefly: type) under consideration. When all sets K_ϕ ($\phi \in \Omega$) are finite, then τ is said to be a finitary type, otherwise we speak of an infinitary type.
Let A be any -- empty or non-empty -- set. A (homogeneous) partial algebra of (similarity) type τ with carrier set A is an ordered pair $(A,(\phi^{\underline{A}})_{\phi\in\Omega})$ $(=:\underline{A})$, such that for each $\phi \in \Omega$, $\phi^{\underline{A}}$ is a K_ϕ-ary partial operation on A : $\phi^{\underline{A}}:A^{K_\phi} \dashrightarrow A$. $\phi^{\underline{A}}$ is called the fundamental partial operation of $\underline{A}$ corresponding to the operation symbol ϕ ($\in \Omega$).
If each fundamental operation of a partial algebra $\underline{A}$ is total, then $\underline{A}$ is called a total (or complete (universal)) algebra -- since we shall always deal with universal or partial universal algebras, we shall usually omit the adjective "universal", since no confusion with "algebras", i.e. "hypercomplex systems" in the classical sense, is to be feared. If, in the contrary, each fundamental partial operation is discrete (empty), then $\underline{A}$ will be called the discrete partial algebra of type τ with carrier set A, and we shall often identify it with the carrier set A itself -- sometimes, in order to stress discreteness, we shall also write $\underline{A}_d$ for the discrete partial algebra of type τ on A.
By **Alg**(τ) we shall denote the class (species) of all partial algebras of some given similarity type τ, while **TAlg**(τ) will designate its subclass of all total algebras of type τ. -- Partial algebras of the same similarity type are called similar; if they are of finitary type, they are called finitary, otherwise infinitary.

1.1.5 REMARKS. (i): If not explicitly stated differently, partial algebras considered at the same time -- e.g. in connection with homomorphisms (cf. section 2) or classes of partial algebras -- are always assumed to be similar.
(ii): The symbol "_" is used in connection with sets in order to indicate that they carry the structure of a partial algebra. In " $\underline{A} = (A,(\phi^{\underline{A}})_{\phi\in\Omega})$ " the upper index "$\underline{A}$" on the right hand side is just indicating that $\phi^{\underline{A}}$ is a realization of the operation symbol ϕ in the partial algebra $\underline{A}$. It has become quite customary in universal algebra to use a notation like this one.
(iii): If anybody is scared by the fact that we allow arbitrary sets as arities for fundamental operations, he is free to think of K_ϕ as an ordinal (which we assume to be the set of all smaller ordinals ($0 := \emptyset$), cf. section 23.3), and if he does not need infinitary operations at all, he may think of K_ϕ to be any natural number, and he may

skip all statements referring to the fact that arities might be infinite.

(iv): What we have introduced above is the notation which we shall use in definitions and general statements. However, when we consider special examples, where only finitely many operations are given, then we shall often present Ω as $\Omega = \{\phi_1,\ldots,\phi_n\}$ or even $\Omega = (\phi_1,\ldots,\phi_n)$, τ as $\tau = (K_1,\ldots,K_n)$, and any partial algebra $\underline{A}$ as $(A;\phi_1^{\underline{A}},\ldots\phi_n^{\underline{A}})$ in agreement with the common notation. In very special cases, such as groups, rings, and derived structures, we shall also omit as usual the upper index "$\underline{A}$". E.g. we denote (multiplicative) groups as total algebras $(G;\cdot,e,^{-1})$ of type $(2,0,1)$, or fields as partial algebras $(F;+,0,-,\cdot,1,^{-1})$ of type $(2,0,1,2,0,1)$, where all operations apart from "$^{-1}$" are total.

It should be observed, too, that e.g. $(A;+,\cdot)$ and $(A;\cdot,+)$ (where "+" and "$\cdot$" are different, but in both cases the same, operations) have to be distinguished.

When we deal with binary operations, we shall often use the "infix notation", writing the operator between the two arguments, especially when dealing with "classical" algebraic structures like semigroups, groups, rings, fields, categories, etc. -- The dot for multiplication will often be omitted.

1.1.6 PRESENTATION of EXAMPLES. Usually we shall describe a concrete partial algebra by its carrier set and the graphs of its fundamental partial operations. But sometimes we shall rather sketch finitary partial algebraic structures graphically, indicating the elements of the carrier set by small circles. And the application of an operation to a sequence is indicated by either an arrow, when the operation is **unary** (i.e. the arity is a one-element set!), or by lines -- usually labeled with the position in the sequence, when this matters -- starting at the arguments and leading to a symbol (small rectangle or circle different from the one indicating the elements) representing the partial operation under consideration. From this symbol there starts an arrow leading to the value of the operation with respect to the given sequence. Different unary partial operations will usually be distinguished by a different "form" of the arrows: continuous line, broken line or dotted line -- usually we shall not need more than three different unary partial operations in order to illustrate some feature; non-empty fundamental constants will be distinguished by writing the corresponding operation at the circle representing the value. Thus such examples may look like the following one for a similarity type, say $\tau := (2,2,1,1,1,0,0)$, where $\Omega := (\phi_1,\ldots,\phi_7)$, the indices corresponding to the place in the type-sequence. A partial algebra $\underline{A}$ of this type may then be described as follows (observe that we identify one-element sequences with the elements of the carrier set!) -- the graphical description is given in Figure 1.1:

$A := \{a,b,c,d,e,f,g\}$,

$\mathrm{graph}(\phi_1^{\underline{A}}) := \{((a,a),d),\ ((a,b),c),\ ((b,a),c)\}$,

$\mathrm{graph}(\phi_2^{\underline{A}}) := \{((f,e),d)\}$,

$\mathrm{graph}(\phi_3^{\underline{A}}) := \{(e,b),\ (f,c)\}$,

$\mathrm{graph}(\phi_4^{\underline{A}}) := \mathrm{graph}(\phi_5^{\underline{A}}) := \{(e,g),\ (g,f),\ (f,e)\}$,

$\phi_6^{\underline{A}} := a$, $\phi_7^{\underline{A}} := b$.

Symbols: ϕ_1 : □ , ϕ_2 : ○ , ϕ_3 :−−, ϕ_4 :⋯⋯, ϕ_5 :〰〰.

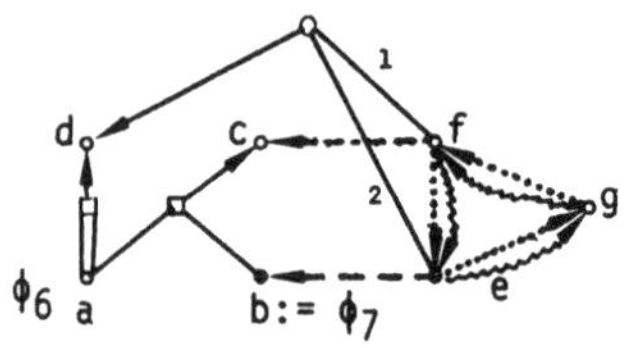

Figure 1.1.

1.2 Heterogeneous partial algebras

1.2.1 REMARKS and EXAMPLES. **(i):** Mainly in Computer Science, but to some extent also in Mathematics there arise algebraic structures which can be considered as partial algebras, where the domains of the partial fundamental operations can easily be defined by subdividing the carrier set, say A, into a family, say $(A_s)_{s\in S}$, of -- usually disjoint -- subsets such that all arguments in a fixed position of a fundamental operation belong to exactly one of these subsets, and all values of a specific operation have to belong to a well-defined subset.

(ii): Probably the most well-known example from Mathematics -- although usually described a little bit differently -- are **vector spaces** $\underline{V}$, where the carrier set is divided into two disjoint sets, namely the sets V_s of **scalars** and V_v of **vectors**, respectively, and V_s carries the structure of a field, while V_v carries the structure of an abelian group, and there is a multiplication, say $°:V_s \times V_v \to V_v$ satisfying the well-known axioms for the multiplication of vectors with scalars.

(iii): Another example are small categories $\underline{\mathbf{C}}$, where $\mathbf{C}$ is also subdivided into two disjoint subsets, usually denoted by $Ob(\underline{\mathbf{C}})$ and $Mor(\underline{\mathbf{C}})$, the subsets of objects and morphisms, respectively -- cf. the different, but equivalent definition in the introduction, where we only used one sort of elements, namely the one of morphisms. Small categories can then be defined as partial algebras $\underline{\mathbf{C}}$ of type (2,1,1,1), corresponding to the set $\{\cdot, \mathrm{Dom}, \mathrm{Cod}, \mathbf{1}\}$ -- in the same order as in the type --, such that

$\cdot:Mor(\underline{\mathbf{C}}) \times Mor(\underline{\mathbf{C}}) \overset{\cdot}{\to} Mor(\underline{\mathbf{C}})$	is a partial operation, and
$\mathrm{Dom}, \mathrm{Cod}:Mor(\underline{\mathbf{C}}) \to Ob(\underline{\mathbf{C}})$	as well as
$\mathbf{1}:Ob(\underline{\mathbf{C}}) \to Mor(\underline{\mathbf{C}})$	are total operations.

"$\cdot$" designates the composition of morphisms, while "Dom" and "Cod" assign to each morphism its domain (start object) and its codomain (target object), and "$\mathbf{1}$" assigns to each object its canonical isomorphism (if the category is realized by sets with mappings, this is the identity automorphism, and then "id" is used instead of $\mathbf{1}$), such that the usual axioms for categories are satisfied (cf. HERRLICH & STRECKER [HS 73]) for all $m,m',m'' \in Mor(\underline{\mathbf{C}})$ and for all $A \in Ob(\underline{\mathbf{C}})$:

(C1) $m\cdot m'$ is defined iff $\mathrm{Cod}\, m = \mathrm{Dom}\, m'$.

(C2) If $m\cdot m'$ and $m'\cdot m''$ are defined, then $(m\cdot m')\cdot m''$ and $m\cdot(m'\cdot m'')$ are defined and equal. ("Associativity" of composition)

(C3) If $m\cdot m'$ is defined, then $\mathrm{Dom}(m\cdot m') = \mathrm{Dom}(m)$ and $\mathrm{Cod}(m\cdot m') = \mathrm{Cod}(m')$.

(C4) $\mathbf{1}_A$ -- the value of $\mathbf{1}$ on A -- satisfies:

(a) $\mathrm{Dom}(\mathbf{1}_A) = \mathrm{Cod}(\mathbf{1}_A) = A$;

(b) if $\mathbf{1}_A\cdot m$ is defined, then $\mathbf{1}_A\cdot m = m$;

(c) if $m\cdot\mathbf{1}_A$ is defined, then $m\cdot\mathbf{1}_A = m$.

-- Notice that abstract categories in general are defined similarly, with the only difference that $Ob(\underline{\mathbf{C}})$ and $Mor(\underline{\mathbf{C}})$ are allowed to be proper classes, and the additional restriction, that for any two objects A and B the class

$Mor(A,B):= \{m \in Mor(\underline{\mathbf{C}}) \mid \mathrm{Dom}(m) = A \text{ and } \mathrm{Cod}(m) = B\}$ is a proper set.

(iv): The two examples mentioned above and those of the introduction suggest that it might be helpful in some cases to have an appropriate language to handle such situations, although they can also be treated within the framework of partial algebras in general.

Although we shall present in section 2 a different approach to heterogeneous or many-sorted partial algebras, which relates them more closely to the homogeneous case, we briefly present in this subsection the straightforward generalization to the partial case of their usual definition (cf. P.I.HIGGINS [Hg 63], H.LUGOWSKI [Lu 76] or PART II). The reader who has difficulties in understanding the following definition of heterogeneous partial algebras should skip this subsection and take the characterization in subsection 2.11 as a definition.

1.2.2 DEFINITION (of heterogeneous partial algebras). Throughout this book S will designate an arbitrary but fixed **non-empty** set, which will be called the *set of sorts*, and the elements of S will be denoted by s, s', s", ... or s_1, s_2, ..., and will be called *sorts*.
As in 1.1.4, Ω designates the set of fundamental operation symbols, and $\tau = (K_\varphi)_{\varphi\in\Omega}$ is the family of arities under consideration. For a heterogeneous similarity type we need two additional mappings, say $\text{INPUT}:\Omega \to \bigcup\{S^{K_\varphi} \mid \varphi \in \Omega\}$ and $\text{OUTPUT}:\Omega \to S$,

such that for each $\varphi \in \Omega$ we have $(\varphi)\text{INPUT} \in S^{K_\varphi}$

-- notice that (φ)INPUT is itself a K_φ-ary sequence. A *heterogeneous* (or *many-sorted*) *similarity type* is then a triple $\Sigma := (\tau,\text{INPUT},\text{OUTPUT})$ -- notice that, as in the homogeneous case, Ω is implicitely fixed by this definition.
Now, a *heterogeneous* (or *many-sorted*) *partial algebra* is an ordered pair of families $(A^` := (A_s)_{s\in S}, (\varphi^{\underline{A}^`})_{\varphi\in\Omega})$ (=:$\underline{A}^`$), where for each $\varphi \in \Omega$, $\varphi^{\underline{A}} : \text{dom}\,\varphi^{\underline{A}^`} \to A_{(\varphi)\text{OUTPUT}}$, with

(*) $$\text{dom}\,\varphi^{\underline{A}^`} \subseteq \{\underline{a} \mid (k)\underline{a} \in A_{(k)((\varphi)\text{INPUT})},\ k \in K_\varphi\} .$$

$\varphi^{\underline{A}^`}$ is called a *heterogeneous fundamental partial operation* of $\underline{A}^`$. For each $s \in S$, A_s is called the *phylum of sort* s of $\underline{A}^`$.
When we have equality in (*) for all $\varphi \in \Omega$, then $\underline{A}^`$ is called a (*total*) *heterogeneous* or *many-sorted algebra* (in some literature a *Σ-algebra*).
All other notions are defined as in the homogeneous case.

1.2.3 REMARKS. (i): Let us look at the above definition for the finitary case, and let us assume that all arities are natural numbers, e.g. let us consider one total n-ary heterogeneous operation, say φ, on $(A_s)_{s\in S}$. Then $(\varphi)\text{INPUT} =: (s_0,\ldots,s_{n-1})$ and $(\varphi)\text{OUTPUT} =: s_\varphi$ such that $\varphi^{\underline{A}^`}$ is a mapping from $A_{s_0} \times A_{s_1} \times \ldots \times A_{s_{n-1}}$ into A_{s_φ}.
If φ is allowed to be a partial operation of the above type, its domain has to be a subset of the above cartesian product, while all the other data remain the same.
(ii): We shall discuss some more examples of heterogeneous partial algebras in section 2, thus we do not give further ones at this place.
(iii): At this point "`" is used to indicate that the carrier is a family of sets rather than a set, and that the types of the (partial) operations are restricted with respect to this family. Starting from subsection 2.11 we shall mostly omit this "reminder", since we shall treat the homogeneous and the heterogeneous case in a parallel way.
(iv): As far as the presentation of examples is concerned, we apply the same methods as discussed in 1.1.6. The different phyla are usually surrounded by a closed curve, and sometimes the whole structure is surrounded by an additional curve, especially, when several structures are considered at the same time. Often a sketch of this kind will substitute the formal set theoretical description.

1.3 Cardinal and ordinal dimension

In what follows we shall often need some cardinal and ordinal numbers connected with a given similarity type τ. In the heterogeneous case the cardinality $\#S$ of the set of sorts plays an additional rôle. We recall from the Appendix that the successor cardinal of a cardinal number n is denoted by n^+. Although we shall first need them in section 3, we introduce them already here in order to compare them, and and to avoid these definitions to be too much scattered around.

1.3.1 DEFINITION (of ordinal dimension). The <u>ordinal dimension</u> $\delta := \delta(\tau)$ of a given similarity type τ is the least **regular** ordinal number ζ (cf. section 23 in the Appendix) such that $\#\zeta > \#K_\phi$ for each $\phi \in \Omega$.
-- The ordinal dimension will mainly be needed in connection with the "local" description of generation (from below) of subalgebras and other substructures (cf. section 3).

1.3.2 DEFINITION (of rank and cardinal dimension). The <u>cardinal dimension</u> $\mathbf{d} := \mathbf{d}(\tau)$ of a given similarity type τ is the least additively inaccessible (i.e. regular) cardinal number $\mathbf{c}$ (cf. the Appendix) such that $\mathbf{c} > \#K_\phi$ for each $\phi \in \Omega$.
Often we shall need the <u>infinite cardinal dimension</u> $\mathbf{d}^* := \max\{\#\omega_0, \mathbf{d}\}$, and in connection with heterogeneous partial algebras we sometimes also need the <u>heterogeneous infinite cardinal dimension</u> $\mathbf{d}^{\&} := \max\{\mathbf{d}^*, (\#S)^+\}$, where S is the corresponding set of sorts.
Another useful cardinal derived from the similarity type is the <u>rank</u> $\mathbf{r} := \sup\{\mathbf{c} \mid \mathbf{c} < \mathbf{d}\}$.

1.3.3 REMARKS and EXAMPLES. (i): The cardinal dimension will mainly be used in connection with inductivity properties of subalgebra lattices and direct limits, especially as an upper bound for other cardinalities, while the rank is usually more precise, when a lower bound for cardinalities is needed.
(ii): In Table 1.1 we compare the values of the above cardinal and ordinal numbers. As an additional tool we use $\mathbf{s} := \sup\{\#K_\phi \mid \phi \in \Omega\}$. As a rule one can state that $\mathbf{d} = \#\delta$ except for the case when $\mathbf{d} = 2$, since then $\delta = \omega_0$. We do not prove the statements included there, since they easily follow from the definitions above and the properties of cardinal and ordinal numbers. ∎

Properties of τ $(\mathbf{s} := \sup\{\#K_\phi \mid \phi \in \Omega\})$	δ	$\mathbf{d}$	$\mathbf{d}^*$	$\mathbf{r}$
$\Omega = \emptyset$	0	0	$\#\omega_0$	0
$\Omega \neq \emptyset$: $\mathbf{s} = 0$	1	1	$\#\omega_0$	0
$\mathbf{s} = 1$	ω_0	2	$\#\omega_0$	1
$1 < \mathbf{s} \leq \#\omega_0$, all K_ϕ finite	ω_0	$\#\omega_0$	$\#\omega_0$	$\#\omega_0$
$\mathbf{s} = \#\omega_0$, at least one K_ϕ infinite	ω_1	$\#\omega_1$	$\#\omega_1$	$\#\omega_0$
$\mathbf{s} = \#\omega_\alpha > \#\omega_0$ additively accessible	$\omega_{\alpha+1}$	$\#\omega_{\alpha+1}$	$\#\omega_{\alpha+1}$	$\#\omega_\alpha$
$\mathbf{s} = \#\omega_\alpha$ additively inaccessible, $\mathbf{s} = \#K_\phi$ for some $\phi \in \Omega$	$\omega_{\alpha+1}$	$\#\omega_{\alpha+1}$	$\#\omega_{\alpha+1}$	$\#\omega_\alpha$
$\mathbf{s} = \#\omega_\alpha$ additively inaccessible, $\mathbf{s} > \#K_\phi$ for all $\phi \in \Omega$	ω_α	$\#\omega_\alpha$	$\#\omega_\alpha$	$\#\omega_\alpha$

Table 1.1.

2 Homomorphisms, congruence relations and factor algebras

Next to the concept of partial algebras the one of homomorphisms between similar partial algebras, i.e. of structure preserving mappings between them, is the most important one. We think that among all the possibilities which have been introduced so far somewhere in the literature, the concept of a homomorphism which we shall use here is in some way the most natural one, judged from its close relationship to the corresponding concepts for total universal algebras as well as for relational systems, and from its connections to model theory, which will be elaborated especially in Chapter II.

Although we shall treat for quite a while only homogeneous partial algebras, we already include in the next definition the concept of homomorphisms between heterogeneous partial algebras, since it will give a better understanding of what we shall do in subsection 2.11 below (however, 2.1.1.(ii) may also be skipped).

2.1 Homomorphisms

2.1.1 DEFINITION (of homomorphisms).

(i) (homogeneous case): Let $\underline{A}$ and $\underline{B}$ be two (homogeneous) partial algebras of the same similarity type $\tau = (K_\phi)_{\phi\in\Omega}$, and let $f: A \to B$ be any mapping from A into B. f -- or more precisely the triplet $(\underline{A}, \text{graph}(f), \underline{B})$, although we shall use this notation very rarely -- is a homomorphism from $\underline{A}$ into $\underline{B}$ -- and we shall indicate this fact by writing $f: \underline{A} \to \underline{B}$ -- iff

(*) for all $\phi \in \Omega$, $\underline{a} \in A^{K_\phi}$ and $b \in A$ the following holds:

$$\text{if } \underline{a}\phi^{\underline{A}} = b, \text{ then } (\underline{a}f)\phi^{\underline{B}} = (b)f,$$

where the implication is to be read in the following way: "Whenever $\underline{a}\phi^{\underline{A}}$ exists and has value b, then $(\underline{a}f)\phi^{\underline{B}}$ exists and has value $(b)f$ " -- this value is also often written as bf. -- By Hom(**Alg**(τ)) we denote the class of all homomorphisms between partial algebras of type τ.

(ii) (heterogeneous case): Let $\underline{A}$\` and $\underline{B}$\` be heterogeneous partial algebras of a heterogeneous similarity type (τ,INPUT,OUTPUT) -- we use the same notation as in 1.2. Then a homomorphism $f: \underline{A}` \to \underline{B}`$ is a family $(f_s: A_s \to B_s)_{s\in S}$ of mappings such that for each $\phi \in \Omega$, $\underline{a} \in \text{dom } \phi^{\underline{A}`}$, $b \in A_{(\phi)\text{OUTPUT}}$ and $\underline{a}f_{(\phi)\text{INPUT}} := (((k)\underline{a})f_{(k)((\phi)\text{INPUT})} \mid k \in K_\phi)$:

$$\text{if } \underline{a}\phi^{\underline{A}`} = b, \text{ then } (\underline{a}f_{(\phi)\text{INPUT}})\phi^{\underline{B}`} = (b)f_{(\phi)\text{OUTPUT}}.$$

If $(\phi)\text{INPUT} = (s_1,\ldots,s_n)$ and $(\phi)\text{OUTPUT} = s$, then this is illustrated in the **commutative diagram** of Figure 2.1.

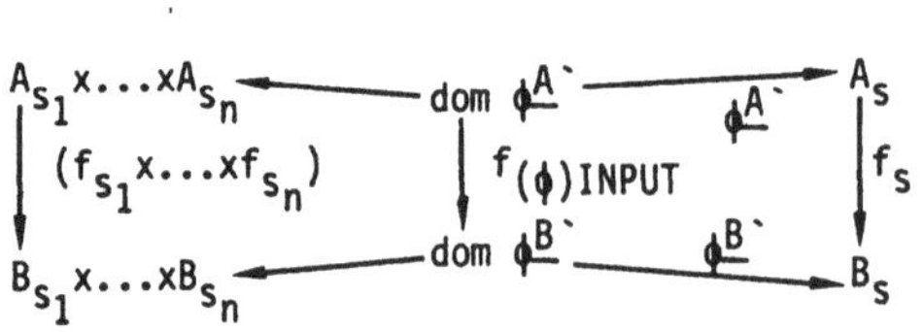

Figure 2.1.

2.1.2 REMARKS. (i): Notice that in the case of total algebras the implication 2.1.1.(*) reduces to the well-known condition " $(\underline{a}\phi^{\underline{A}})f = (\underline{a}f)\phi^{\underline{B}}$ ", and that in the case of partial algebras there is some sort of "completeness requirement" to the structure of $\underline{B}$ with respect to the structure of $\underline{A}$, but not vice versa. This is also expressed in the equivalent description of (*) for the graphs of the fundamental operations:

(*)' For all $\phi \in \Omega$ one has: $[\mathrm{graph}(\phi^{\underline{A}})]f^{(K_\phi \cup \{K_\phi\})} \subseteq \mathrm{graph}(\phi^{\underline{B}})$.

-- Notice that "$f^{(K_\phi \cup \{K_\phi\})}$" is the mapping induced by f by componentwise application. The simple proof of the equivalence between (*) and (*)' is left to the reader as an exercise; the clue is the equality $(\underline{a}f,(\underline{a}f)\phi^{\underline{B}}) = (\underline{a}f,(\underline{a}\phi^{\underline{A}})f)$. ∎

(ii): Notice that all homomorphisms between total algebras in the usual sense are examples for Definition 2.1.1.(i). In addition each identity mapping from a discrete partial algebra on some set A onto any partial algebra on A is a homomorphism in the sense of 2.1.1.(i). More concrete examples can be found e.g. in the Figures of the following subsections.

(iii): As far as heterogeneous partial algebras are concerned we only want to point out at this place that homomorphisms between them treat different phyla as if they were disjoint. This will provide us with another definition -- which is from a category theoretical point of view equivalent with the usual one -- of heterogeneous partial algebras in subsection 2.11. Until then we shall only deal with homogeneous partial algebras in order that the fundamental concepts will not be too much burdened with additional technicalities, which -- at the end -- will prove to be unnecessary. However, the reader may consider it as an exercise to extend the concepts and results of the following subsections to heterogeneous partial algebras, before we do this at the end of section 2 in connection with the alternative representation. -- It is easy to realize that linear mappings between vector spaces -- and more generally all homomomorphisms between similar total heterogeneous algebras -- are homomorphisms in the sense of 2.1.1.(ii), and that 2.1.2.(ii) also applies to the heterogeneous case.

2.1.3 PROPOSITION. Let $\underline{A}, \underline{B}, \underline{C} \in \mathbf{Alg}(\tau)$.

(i) The identity mapping $id_A:A \to A$ is always a homomorphism from $\underline{A}$ into itself: $id_A:\underline{A} \to \underline{A}$ (sometimes also denoted by 1_A).

(ii) If $f:\underline{A} \to \underline{B}$ and $g:\underline{B} \to \underline{C}$ are homomorphisms, then their <u>composition</u> $f \cdot g$ (mostly also written as fg, where $(a)(fg) := (af)g$ $(a \in A)$) is also a homomorphism $f \cdot g:\underline{A} \to \underline{C}$.

(iii) If $f:A \to B$ is any mapping, then f is always a homomorphism from the discrete partial algebra $\underline{A}_d$ into $\underline{B}$, i.e. $f:\underline{A}_d \to \underline{B}$ -- also written as $f:A \to \underline{B}$.

PROOF. Use condition 2.1.1.(*). ∎∎∎

2.1.4 REMARK. Properties 2.1.3.(i) and 2.1.3.(ii) together with the well-known properties of mappings in general tell us that (for every similarity type τ) $\mathbf{Alg}(\tau)$ together with all homomorphisms between its objects forms a (concrete) category (cf. [HS 73] for basic definitions on categories).

2.2 Isomorphisms

2.2.1. In any category $\underline{\mathbf{C}}$ a $\underline{\mathbf{C}}$-morphism $f:A \to B$ is an <u>isomorphism</u> iff there exists a morphism $g:B \to A$ such that $f \cdot g = 1_A$ and $g \cdot f = 1_B$; g is then called an <u>inverse</u> of f and is uniquely determined by f. Two $\underline{\mathbf{C}}$-objects A and B are called <u>isomorphic</u> -- in symbols: $A \cong B$ --, iff there exists an isomorphism, say $i:A \to B$, between them. Iso($\underline{\mathbf{C}}$) designates the class of all isomorphisms in $\underline{\mathbf{C}}$.

Because of 2.1.4 we can adopt the same definition and notation for isomorphisms between partial algebras, and we get the following criterion:

2.2.2 PROPOSITION. Let $\underline{A}, \underline{B} \in \mathbf{Alg}(\tau)$, and let $f:\underline{A} \to \underline{B}$ be any homomorphism. Then the following statements are equivalent:

(i) f is an isomorphism between $\underline{A}$ and $\underline{B}$.

(ii) f is bijective, and the inverse mapping $f^{-1}:B \to A$ is also a homomorphism $f^{-1}:\underline{B} \to \underline{A}$.

(iii) f is bijective, and for all $\phi \in \Omega$: $[\mathrm{graph}(\phi^{\underline{A}})]f^{(K_\phi \cup \{K_\phi\})} = \mathrm{graph}(\phi^{\underline{B}})$.

(iv) f is bijective, and for all $\phi \in \Omega$ and for all $\underline{a} \in A^{K_\phi}$ the following holds: $\underline{a} \in \mathrm{dom}\, \phi^{\underline{A}}$ iff $\underline{a}f \in \mathrm{dom}\, \phi^{\underline{B}}$ (i.e. $[\mathrm{dom}\, \phi^{\underline{A}}]f^{(K_\phi)} = \mathrm{dom}\, \phi^{\underline{B}}$) .

PROOF. One has to check the definitions of homomorphism and isomorphism above, and to observe the following two facts: $fg = id_A$ and $gf = id_B$ iff $g = f^{-1}$ (whence (i) $\Leftrightarrow$ (ii)), and the equivalence between 2.1.1.(*) and 2.1.2.(*)', which establishes the other equivalences. ∎∎∎

2.2.3 REMARKS. (i): It should be noticed that from the equivalence of (i) and (iv) in 2.2.2 it easily follows in the case when both $\underline{A}$ and $\underline{B}$ are total algebras, that every bijective homomorphism from $\underline{A}$ onto $\underline{B}$ is an isomorphism, while this it not true for arbitrary partial algebras; e.g. $id_A:A \to \underline{A}$ is always a bijective homomorphism, but it is an isomorphism iff $\underline{A}$ is discrete, too.

(ii): But even a bijective homomorphism from some partial algebra onto itself need not be an isomorphism, as the following example shows (as an exercise one should decide, whether it is necessary to choose the carrier set to be infinite!):

Let $\mathbf{N}$ be the set of all natural numbers including 0, and let $2\mathbf{N}$ and $2\mathbf{N}+1$ designate the subsets of all even (respectively odd) numbers. Consider the partial algebra $(\mathbf{N};\phi)$ of type (1), where $\mathrm{graph}(\phi) := \{(n,n) \mid n \in 2\mathbf{N}\}$. Moreover, let $f:\mathbf{N} \to \mathbf{N}$ be given by

$$(n)f := \begin{cases} n+2 \text{ , if } n \in 2\mathbf{N} \text{ ,} \\ 0 \quad\;\; \text{ , if } n = 1 \text{ ,} \\ n-2 \text{ , else} \quad\quad . \end{cases}$$

This situation is illustrated in Figure 2.2, which also shows that f is bijective, but no isomorphism, since $(0)f^{-1} \notin \mathrm{dom}\, \phi$.

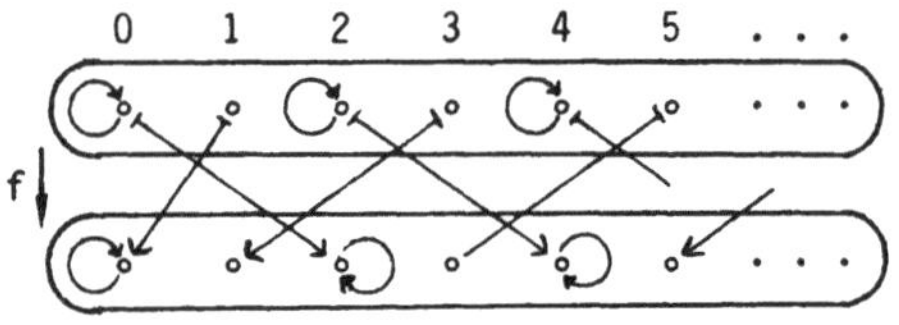

Figure 2.2.

The statement in 2.2.3.(i) about total algebras is strengthened in

2.2.4 PROPOSITION. Let $\underline{A}$ be any **non-empty** partial algebra of type τ. Then the following statements are equivalent:

(i) $\underline{A}$ is total.

(ii) Every bijective homomorphism $f:\underline{A} \to \underline{B}$ into any similar partial algebra $\underline{B}$ is an isomorphism (and $\underline{B}$ is then a total algebra, too).

PROOF. Let $\underline{A}$ be total, and let $f:\underline{A} \to \underline{B}$ be any bijective homomorphism. Then 2.2.2.(iv) easily implies that $\underline{B}$ is total, and that f^{-1} is also a homomorphism, thus f is an isomorphism. ∎

Conversely, assume that $\underline{A}$ is not total. Since $A \neq \emptyset$, we may choose some fixed element c in A and define for every ϕ in Ω a total operation ϕ^* on A by

$$\underline{a}\phi^* := \begin{cases} \underline{a}\phi^{\underline{A}}, & \text{if } \underline{a} \in \operatorname{dom} \phi^{\underline{A}} \\ c, & \text{else} \end{cases} \qquad \text{for any } \underline{a} \in A^{K_\phi}.$$

$\underline{A}^* := (A, (\phi^*)_{\phi\in\Omega})$ is called an <u>inner</u> <u>one-point</u> <u>completion</u> of $\underline{A}$, and obviously $id_A : \underline{A} \to \underline{A}^*$ is a bijective homomorphism, which is not an isomorphism. ■■■

2.2.5 PROPOSITION. Let $\underline{A}$, $\underline{B}$ and $\underline{C}$ be similar partial algebras and $f:\underline{A} \to \underline{B}$, $g:\underline{B} \to \underline{C}$ any homomorphisms.

(i) $id_A:\underline{A} \to \underline{A}$ is always an isomorphism.

(ii) If f is an isomorphism, then $f^{-1}:\underline{B} \to \underline{A}$ is an isomorphism, too.

(iii) If f and g are isomorphisms, then $fg:\underline{A} \to \underline{C}$ is an isomorphism, too. ■■■

2.2.6 COROLLARY. For every similarity type τ the isomorphism relation "$\cong$" on **Alg**(τ) has the properties of an equivalence relation: reflexivity, symmetry and transitivity. The equivalence classes of "$\cong$" are proper classes, and they are called <u>isomorphism types</u>. ■■■

2.3 Endomorphisms and automorphisms

2.3.1 DEFINITION. Let $\underline{A}$ be any partial algebra. The homomorphisms from $\underline{A}$ into itself are called <u>endomorphisms</u> of $\underline{A}$, and those endomorphisms of $\underline{A}$, which are at the same time isomorphisms, are called <u>automorphisms</u> of $\underline{A}$. By End($\underline{A}$) and Aut($\underline{A}$) we denote the sets of all endomorphisms (respectively automorphisms) of $\underline{A}$.

2.3.2 PROPOSITION. For every partial algebra $\underline{A}$ one has:

(i) $\underline{End}(\underline{A}) := (End(\underline{A}); \cdot, id_A)$ is a monoid;

(ii) $\underline{Aut}(\underline{A}) := (Aut(\underline{A}); \cdot, id_A, ^{-1})$ is a group

(where "$\cdot$" is the composition of homomorphisms).

PROOF. This is an immediate consequence of previous statements and the fact that the composition of mappings -- and therefore of homomorphisms -- is associative, whenever it is defined (and here it is always defined). ■■■

2.3.3 REMARK. It should be noticed that in general -- as the example in 2.2.3.(ii) shows -- a bijective endomorphism need not be an automorphism; but it is a simple set theoretical argument to show that every bijective endomorphism of a partial algebra with a finite carrier set is an automorphism. From 2.2.4 one may easily infer that every bijective endomorphism of a total algebra is an automorphism.

Sometimes it might be useful to know the following result:

2.3.4 PROPOSITION. (CAYLEY, BIRKHOFF) Let $\underline{M} := (M; \cdot, e)$ be any monoid, and define for every $m \in M$ the mappings $l_m : M \to M$, $x \mapsto m \cdot x$, and $r_m : M \to M$, $x \mapsto x \cdot m$ $(x \in M)$.

(i) $(End((M, (l_m)_{m\in M})); \cdot, id_M) = (\{r_m | m \in M\}; \cdot, r_e) \cong \underline{M}$,

i.e. every monoid is isomorphic to the endomorphism monoid of some unary total algebra.

(ii) If $\underline{M}$ is a group, then $\underline{M}$ is isomorphic to the automorphism group of $(M,(1_m)_{m\in M})$ in the way described in (i). ■

Notice that in the above result the similarity type depends on #M. In [PuTr 80] by A.PULTR and V.TRNKOVA much stronger results of similar nature can be found.

2.4 Full and closed homomorphisms

2.4.1. The definition of a homomorphism between partial algebras as given in 2.1.1 is a very weak one, since it contains only a statement about "richness" of the structure in the codomain depending on the structure on its domain, but not vice versa. This fact -- and more precisely some model theoretic aspects of 2.1.1 which we shall discuss in detail in Chapter II -- allows to introduce many more properties of homomorphisms between partial algebras than can be done in the case of total algebras. We shall discuss this in more detail in later chapters. In this introductory chapter we restrict considerations to the following two properties, which both are trivially satisfied by all homomorphisms between total algebras:

2.4.2 DEFINITION. Let $f:\underline{A}\to\underline{B}$ be any homomorphism between partial algebras of the same type $\tau = (K_\phi)_{\phi\in\Omega}$.

(i) f is said to be a <u>full homomorphism</u> from $\underline{A}$ into $\underline{B}$ iff one has:

(fh) For every $\phi\in\Omega$, and for every $\underline{a}\in A^{K_\phi}$:
If $\underline{a}f \in \mathrm{dom}\,\phi^{\underline{B}}$ and $(\underline{a}f)\phi^{\underline{B}} \in [A]f$, then there exists $\underline{a}' \in \mathrm{dom}\,\phi^{\underline{A}}$ such that $\underline{a}f = \underline{a}'f$.

(ii) f is said to be a <u>closed homomorphism</u> from $\underline{A}$ into $\underline{B}$ iff one has:

(ch) For every $\phi\in\Omega$ and for every $\underline{a}\in A^{K_\phi}$ one has:
If $\underline{a}f\in\mathrm{dom}\,\phi^{\underline{B}}$, then $\underline{a}\in\mathrm{dom}\,\phi^{\underline{A}}$.

2.4.3 REMARKS. (i): Notice that in G.GRÄTZER's book [G 68] those homomorphisms, which we call "closed", are called "strong", and this terminology is also used in many papers on partial algebras. But in papers of J.SCHMIDT and of the author this property has already been called "closed", while "strong" was used for what GRÄTZER and also we here in this book call "full". Confusion is even greater, since "strong" has been used in papers of J.SŁOMIŃSKI and B.WOJDYŁO in a third meaning (cf. [BWo 82] for further details). This is the main reason that we try to avoid "strong" in connection with homomorphisms throughout this book.

(ii): Intuitively spoken a homomorphism is **full**, iff the restriction to the exact image set of the structure in the codomain is the one "induced" by the structure of the preimage via the homomorphism (see the "Homomorphism Theorem" below and the concept of relative subalgebras in section 3). Similarly, a homomorphism is **closed**, iff the domains of the operations in the preimage are exactly the inverse images of the domains of the corresponding operations in the codomain.

2.4.4 PROPOSITION. Let $f:\underline{A}\to\underline{B}$ be any homomorphism between similar partial algebras.

(i) f is **full**, iff for every $\phi\in\Omega$:

$$[\mathrm{graph}(\phi^{\underline{A}})]f^{(K_\phi\cup\{K_\phi\})} = \mathrm{graph}(\phi^{\underline{B}}) \cap ([A]f)^{K_\phi\cup\{K_\phi\}}.$$

(ii) f is **closed**, iff for every $\phi \in \Omega$: $[\mathrm{dom}\,\phi^{\underline{B}}](f^{(K_\phi)})^{-1} = \mathrm{dom}\,\phi^{\underline{A}}$.

(iii) If f is closed, then it is full, but not vice versa.

PROOF. (i) and (ii) follow immediately from the properties (fh) and (ch), respectively, and it is easily seen that (cf) implies (fh), whence we have the first statement in (iii). Finally observe that the homomorphism f indicated in Figure 2.3 is full, but not closed, while g is not even full. ■■■

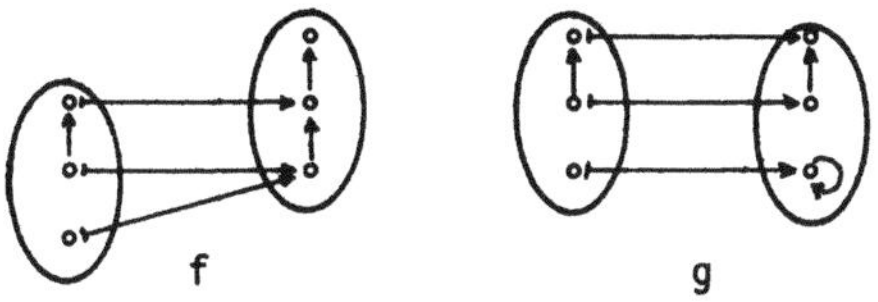

full, not closed   not full

Figure 2.3.

2.4.5 PROPOSITION.

(i) The composition of closed homomorphisms is closed.

(ii) The composition of full homomorphisms is not necessarily full.

(iii) Let $f:\underline{A} \to \underline{B}$ and $g:\underline{B} \to \underline{C}$ be homomorphisms, then one has:

- (a) If f and g are full, and if g is injective or if f is surjective, then fg is full.
- (b) If fg is full and g is injective, then f is full.
- (c) If fg is full and f is surjective, then g is full.
- (d) If fg is full and surjective, then g is full and surjective.
- (e) If fg is closed, then f is closed.
- (f) If fg is closed and f is surjective, then g is closed.

PROOF. In what follows let $f:\underline{A} \to \underline{B}$ and $g:\underline{B} \to \underline{C}$ be arbitrary homomorphisms.

Ad (i): Let f and g be closed, and assume $\underline{a}fg \in \mathrm{dom}\,\phi^{\underline{C}}$ for some $\underline{a} \in A^{K_\phi}$ and some $\phi \in \Omega$. Then $\underline{a}f \in \mathrm{dom}\,\phi^{\underline{B}}$, since g is closed, whence $\underline{a} \in \mathrm{dom}\,\phi^{\underline{A}}$, since f is closed. ■

Ad (ii): Consider Figure 2.4. ■

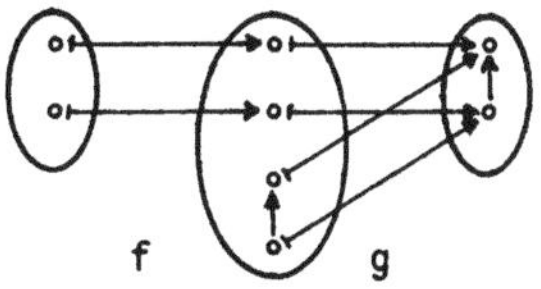

f and g are full, fg is not full

Figure 2.4

Ad (iii): The simple proofs are all quite similar and will, moreover, follow from more general results in 9.1.7 and 9.2.16.(iv) in connection with reflection and weak reflection of formulas by mappings, whence we leave them at this point as exercises. ■■■

In connection with the properties of homomorphisms discussed above we also get new criteria for isomorphisms, when we apply 2.4.4 to the results of 2.2.2:

2.4.6 PROPOSITION. Let $f:\underline{A} \to \underline{B}$ be a homomorphism between similar partial algebras. Then the following statements are equivalent:

(i) f is an isomorphism.

(ii) f is bijective and closed.

(iii) f is bijective and full. ■■■

2.4.7 PROPOSITION. Any homomorphism starting from a total algebra is closed.

This is an immediate consequence of the Definition 2.4.2.(ii) of closedness. ■■■

2.5 Congruence relations

2.5.1. In group theory the surjective homomorphisms starting from a given group $\underline{G}$ are characterized (up to unique isomorphism) by the normal subgroups of $\underline{G}$. For rings and Boolean algebras this is done by so-called ideals, but in each case these are special subobjects, while already in the cases of semigroups, monoids or lattices one cannot find subobjects which do this job. But it is also well known in the cases of groups, rings or Boolean algebras that the distinguished subobjects mentioned above induce equivalence relations on the carrier sets such that these relations are compatible with the fundamental operations. This fact can be generalized to partial algebras.

2.5.2 DEFINITION. Let $\underline{A}$ be any partial algebra of some type $(K_\phi)_{\phi\in\Omega}$. A relation $R \subseteq A \times A$ is called a congruence relation (or briefly a congruence) on $\underline{A}$, iff it satisfies (Cong 1) and (Cong 2) below:

(Cong 1) R is an equivalence relation on A.

(Cong 2) For every $\phi \in \Omega$ and for any two sequences $\underline{a},\underline{b} \in \mathrm{dom}\,\phi^{\underline{A}}$ one has:
If $((k)\underline{a},(k)\underline{b}) \in R$ for each $k \in K_\phi$, then $(\underline{a}\phi^{\underline{A}},\underline{b}\phi^{\underline{A}}) \in R$.

R will be called a closed congruence relation (or a closed congruence) on $\underline{A}$, iff R satisfies in addition:

(Cong 3c) For every $\phi \in \Omega$ and for any sequences $\underline{a},\underline{b} \in A^{K_\phi}$ one has:
If $\underline{a} \in \mathrm{dom}\,\phi^{\underline{A}}$, and $((k)\underline{a},(k)\underline{b}) \in R$ for all $k \in K_\phi$, then $\underline{b} \in \mathrm{dom}\,\phi^{\underline{A}}$.

By $\mathrm{Cong}(\underline{A})$ we denote the set of all congruence relations on $\underline{A}$, while $\mathrm{Cong}_c(\underline{A})$ designates the set of all closed congruence relations on $\underline{A}$. Moreover, we introduce: $\underline{\mathrm{Cong}}(\underline{A}) := (\mathrm{Cong}(\underline{A});\subseteq)$, and $\underline{\mathrm{Cong}}_c(\underline{A}) := (\mathrm{Cong}_c(\underline{A});\subseteq)$, the corresponding partially ordered sets with respect to set theoretical inclusion.

The most immediate consequences of this definition are collected in the

2.5.3 COROLLARY. For every partial algebra $\underline{A}$ and for every homomorphism $f:\underline{A} \to \underline{B}$ one has:

(i) The identity relation $\Delta := \Delta_A := \{(a,a) \mid a \in A\}$ and the universal relation $\nabla := \nabla_A := A \times A$ are always congruence relations on $\underline{A}$. Δ is always closed, while ∇ is closed iff every $\phi^{\underline{A}}$ is either total or discrete (some may be total while others may be discrete!).

(ii) The relation $R_f := \{(a,a') \mid a,a' \in A, af = a'f\}$ is always a congruence relation on $\underline{A}$. It is called the congruence relation induced by f or the kernel of f, and it is also denoted by ker f .

(iii) If f is closed, then ker f is closed, too.

PROOF. Ad (i): The first statements in (i) are obvious from the definition. Assume that each fundamental operation of $\underline{A}$ is either total or discrete, then obviously (Cong 3c) is satisfied by ∇. Now assume that ∇ is closed, and let $\underline{a},\underline{b} \in A^{K_\phi}$ such that $\underline{a} \in \mathrm{dom}\phi^{\underline{A}}$; then $\underline{a}$ and $\underline{b}$ are componentwise equivalent with respect to ∇, and $\underline{b} \in \mathrm{dom}\,\phi^{\underline{A}}$ by (Cong 3c) showing that $\phi^{\underline{A}}$ is total whenever it is not discrete and ∇ is closed. ∎

Ad (ii): It is well known that R_f is always an equivalence relation. Let $\underline{a},\underline{a}' \in \mathrm{dom}\,\phi^{\underline{A}}$ such that $\underline{a}f = \underline{a}'f$, then $(\underline{a}\phi^{\underline{A}})f = (\underline{a}f)\phi^{\underline{B}} = (\underline{a}'f)\phi^{\underline{B}} = (\underline{a}'\phi^{\underline{A}})f$, showing that also (Cong 2) is satisfied. ∎

Ad (iii): If, finally, f is a closed homomorphism, $\underline{a} \in \text{dom}\, \phi^{\underline{A}}$, and $\underline{b}f = \underline{a}f$ ($\in \text{dom}\, \phi^{\underline{B}}$), then also $\underline{b} \in \text{dom}\, \phi^{\underline{A}}$ by the closedness of f. ■■■

Recalling that **d** designates the cardinal dimension, and that $\mathbf{d^*} = \max\{\#\omega_0, \mathbf{d}\}$ we get:

2.5.4 PROPOSITION and NOTATION. Let $\underline{A}$ be any partial algebra. Then:

(i) The set theoretical intersection of arbitrarily many congruence relations on $\underline{A}$ is always a congruence on $\underline{A}$, thus $\text{Cong}(\underline{A})$ is a closure system on $A \times A$; and the corresponding closure operator (cf. section 24 in the Appendix) is denoted by $\mathbf{Cong}_{\underline{A}}$, i.e. for every binary relation M on A one has
$\mathbf{Cong}_{\underline{A}}(M) = \bigcap\{R \mid R \in \text{Cong}(\underline{A}), \text{ and } M \subseteq R\}$.
$\mathbf{Cong}_{\underline{A}}(M)$ is called the congruence (relation) of $\underline{A}$ generated by M.

(ii) The closure system $\text{Cong}(\underline{A})$ is $\mathbf{d^*}$-inductive, and $\mathbf{Cong}_{\underline{A}}$ is a $\mathbf{d^*}$-algebraic closure operator. Hence, for each $M \subseteq A \times A$:
$\mathbf{Cong}_{\underline{A}}(M) = \bigcup\{\mathbf{Cong}_{\underline{A}}(N) \mid N \subseteq M \text{ and } \#N < \mathbf{d^*}\}$.

(iii) $\text{Cong}_c(\underline{A})$ forms an initial segment of $\text{Cong}(\underline{A})$ (i.e. whenever $R,S \in \text{Cong}(\underline{A})$, $R \subseteq S$, and S is closed, then R is closed). Moreover, $\text{Cong}_c(\underline{A})$ is $\mathbf{d^*}$-inductive.

PROOF. It is routine to check that the intersection of any family of congruences of a partial algebra $\underline{A}$ is itself a congruence, where the intersection of an empty family is defined to be ∇_A (see subsection 2.12 for the intersection of the empty family of congruences of heterogeneous partial algebras). Thus (i) easily follows from the general facts about closure systems and closure relations (cf. section 24). ■

Ad (ii): Let $\mathbf{B} \subseteq \text{Cong}(\underline{A})$ be any $\mathbf{d^*}$-directed system of congruences of $\underline{A}$; we have to show that $R^* := \bigcup\mathbf{B}$ is again a congruence on $\underline{A}$: Obviously, R^* is reflexive and symmetric, and it is also transitive, since $(a,b),(b,c) \in R^*$ imply $(a,b) \in R$ and $(b,c) \in R'$ for some $R,R' \in \mathbf{B}$. By the $\mathbf{d^*}$-directedness of $\mathbf{B}$ there is T in $\mathbf{B}$ containing both R and R'; therefore $(a,c) \in T \subseteq R^*$. -- Now assume $\underline{a},\underline{b} \in \text{dom}\, \phi^{\underline{A}}$ and $((k)\underline{a},(k)\underline{b}) \in R^*$ for all $k \in K_\phi$; thus for each $k \in K_\phi$ we have $((k)\underline{a},(k)\underline{b}) \in R_k \in \mathbf{B}$. Since $\#K_\phi < \mathbf{d^*}$, and since $\mathbf{B}$ is $\mathbf{d^*}$-directed, there is $R' \in \mathbf{B}$ containing all R_k, and hence all $((k)\underline{a},(k)\underline{b})$. Since R' is a congruence, we get $(\underline{a}\phi^{\underline{A}}, \underline{b}\phi^{\underline{A}}) \in R' \subseteq R^*$, showing that R^* is a congruence relation on $\underline{A}$. If, in addition, all congruences of $\mathbf{B}$ are closed, then R^* is closed, too: Namely, assume that $\underline{a},\underline{b} \in A^{K_\phi}$ are componentwise R^*-equivalent as above, and that $\underline{a} \in \text{dom}\, \phi^{\underline{A}}$; then we can find -- as above -- an $R' \in \mathbf{B}$ such that $((k)\underline{a},(k)\underline{b}) \in R'$ for every $k \in K_\phi$ (notice that $\mathbf{B}$ is non-empty, since it is $\mathbf{d^*}$-directed, and the empty subset has an upper bound in $\mathbf{B}$). Since R' is closed by assumption, we may infer that $\underline{b} \in \text{dom}\, \phi^{\underline{A}}$, i.e. R^* satisfies (Cong 3c). -- Finally let us observe that $\{\mathbf{Cong}_{\underline{A}}(N) \mid N \subseteq M, \#N < \mathbf{d^*}\}$ is $\mathbf{d^*}$-directed, since $\mathbf{d^*}$ is additively inaccessible, and therefore the union of less than $\mathbf{d^*}$ many sets of cardinality less than $\mathbf{d^*}$ again has cardinality less than $\mathbf{d^*}$. Thus the union of the above set of "$\mathbf{d^*}$-small generated" congruences is again a congruence contained in $\mathbf{Cong}_{\underline{A}}(M)$, but containing M; therefore equality has to hold. ■

Ad (iii): That $\text{Cong}_c(\underline{A})$ is $\mathbf{d^*}$-inductive has already been shown in connection with (ii); and it is easy to realize that any congruence contained in a closed congruence satisfies (Cong 3c) and is therefore closed, too. ■■■

In the **finitary** case we can say more in connection with closed congruences:

2.5.5 LEMMA. If the similarity type under consideration is **finitary**, then the supremum of any two **closed** congruences on $\underline{A}$ is again **closed**, moreover it is the equivalence theoretical supremum of the two congruences under consideration, i.e.:
For any two **closed** congruences $R,R' \in \mathrm{Cong}_c(\underline{A})$ one has

$$(*) \qquad \mathbf{Cong}_{\underline{A}}(R \cup R') = \mathbf{U}\{(R \circ R')^n \mid n < \omega_0\} = \sup_e(R,R') =: R \vee_e R' \in \mathrm{Cong}_c(\underline{A}) ,$$

where $R \circ R' := \{(r,s) \mid \text{there exists } t \in A \text{ such that } (r,t) \in R \text{ and } (t,s) \in R'\}$ is the relational product of R and R', $(R \circ R')^0 := \Delta_A$ and $(R \circ R')^{n+1} := (R \circ R')^n \circ (R \circ R')$.
Moreover, there exists a largest closed congruence $R_c(\underline{A})$ on $\underline{A}$, namely the set theoretical union of all closed congruences on A is again a closed congruence on $\underline{A}$.

2.5.6. Notice that the assumptions of Lemma 2.5.5 are in particular satisfied for every **finitary total** algebra $\underline{A}$ and **all congruences** on $\underline{A}$.

PROOF of 2.5.5. Without loss of generality (w.l.o.g.) let all K_ϕ be natural numbers, say k_ϕ, let $T := \mathbf{U}\{(R \circ R')^n \mid n < \omega_0\}$, and let $V := \mathbf{Cong}_{\underline{A}}(R \cup R')$. Then -- by transitivity -- T is contained in every equivalence relation containing both R and R', and therefore especially in V. If we can show that T is itself a congruence, then $T = V$: Obviously, T is an equivalence relation, since so are R and R'. Thus there remains to show (Cong 2) and (Cong 3c): By the definition of T there exist for any $(a,b) \in T$ elements $z_0 := a$, $z_1, \ldots, z_{2n-1}$, $z_{2n} := b$, such that for $0 \leq i < n$ one has: $(z_{2i}, z_{2i+1}) \in R$ and $(z_{2i+1}, z_{2i+2}) \in R'$. We shall call such a sequence an **R-R'-chain of length** n **between** a and b (such a sequence can always be assumed to have 2n+1 elements, since we may repeat elements because of reflexivity!). Let us now consider any $\phi \in \Omega$, and let us set $k := k_\phi$. Moreover, let $\underline{a} = (a_1, \ldots, a_k)$, $\underline{b} = (b_1, \ldots, b_k) \in A^k$, such that $\underline{a} \in \mathrm{dom}\, \phi^{\underline{A}}$. We claim that $\underline{b} \in \mathrm{dom}\, \phi^{\underline{A}}$ and $(\underline{a}\phi^{\underline{A}}, \underline{b}\phi^{\underline{A}}) \in T$:
First let us observe that we may choose the R-R'-chains between a_i and b_i all of the same length $n := \max\{n_1, \ldots, n_k\}$, where n_i is the original length of such a chain $(1 \leq i \leq k)$, by repeating elements if necessary. Thus, let $z_0^i := a_i$, $z_1^i, \ldots, z_{2n-1}^i$, $z_{2n}^i := b_i$ be the corresponding R-R'-chains, and let $\underline{z}_r := (z_r^1, \ldots, z_r^k)$ $(0 \leq r \leq 2n)$. By the closedness of R and R' and by (Cong 2) it is not difficult to prove by induction on r that $\underline{z}_r \in \mathrm{dom}\, \phi^{\underline{A}}$ (since $\underline{z}_0 = \underline{a} \in \mathrm{dom}\, \phi^{\underline{A}}$), and that $(\underline{z}_{2i}\phi^{\underline{A}}, \underline{z}_{2i+1}\phi^{\underline{A}}) \in R$ and $(\underline{z}_{2i+1}\phi^{\underline{A}}, \underline{z}_{2i+2}\phi^{\underline{A}}) \in R'$; thus $(\underline{a}\phi^{\underline{A}}, \underline{b}\phi^{\underline{A}}) \in T$, since $\underline{z}_0 = \underline{a}$ and $\underline{z}_{2n} = \underline{b}$; moreover, $\underline{b} \in \mathrm{dom}\, \phi^{\underline{A}}$. -- As a consequence of what we have just shown we get that $\mathrm{Cong}_c(\underline{A})$ is $(\#\omega_0\text{-})$ directed, since it is non-empty and contains with any two congruences also their supremum. By inductivity its union is a closed congruence; this was to be shown. ■■■

2.5.7 REMARKS and EXAMPLES. (i): Closedness of **both** congruences involved in 2.5.5.(*) is necessary. For instance one might consider the five-element partial algebra of Figure 2.5, where the closed curves indicate the congruence classes of a closed congruence, while the broken ones indicate the classes of a congruence which is not closed, and the dotted curves indicate the classes of their equivalence theoretical supremum, while the congruence theoretical supremum is ∇.

—— :R --- :R' ··· : $R \vee_e R'$
$\mathbf{Cong}_{\underline{A}}(R \cup R') = \nabla$

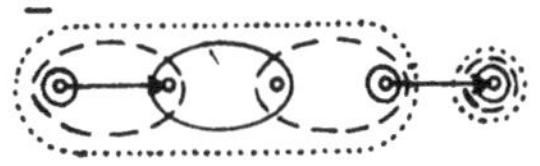

Figure 2.5

(ii): Notice that in the **infinitary case** there exists in general no greatest closed congruence relation. As an **example** consider the partial algebra $(\mathbf{N};\phi)$ ($\mathbf{N}$ designating the set of all natural numbers) of similarity type $(\mathbf{N})$ such that

dom $\phi := \{\underline{a} := (a_0,\dots,a_n,\dots) \mid$ infinitely many elements of $\mathbf{N}$ occur infinitely often in $\underline{a}\}$;

$\underline{a}\phi := 0$, whenever $\underline{a} \in$ dom ϕ .

Let $R_n := \Delta \cup \{(k,l) \mid 1 \leq k,l \leq n\}$ for all $n \in \mathbf{N}$.

Then each R_n is a closed congruence relation, since from a sequence belonging to dom ϕ only finitely many values can be equivalent modulo R_n to different elements in the corresponding component of a second sequence which therefore has to have infinitely many elements of $\mathbf{N}$ occurring infinitely often, while R_n trivially satisfies (Cong 1) and (Cong 2). But the union (= supremum) of all the R_n is ∇, which is not closed (cf. 2.5.3.(i)). ∎

(iii): As an application of the above result one may consider a partial MEALY-automaton $(I \mathbin{\dot\cup} S \mathbin{\dot\cup} O;\mu,\rho) =: \underline{A}$ of type (2,2), where $\mu: I \times S \to S$ and $\rho: I \times S \to O$. Then a <u>reduction</u> of $\underline{A}$ is a **closed** homomorphism $f:\underline{A} \to \underline{B}$ onto a similar automaton $\underline{B}$ such that the restriction of f to $I \cup O$ is the identity mapping (the <u>input sets</u> (I) (respectively the <u>output sets</u> (O)) of $\underline{A}$ and $\underline{B}$ are assumed to be the same, and only the <u>set of states</u> $S =: S_{\underline{A}}$ and $S_{\underline{B}}$ may differ). Then it is easy to check that there exists a "smallest" reduct of $\underline{A}$ corresponding to the largest closed congruence relation of $\underline{A}$, which has only one-element congruence classes for the elements of $I \cup O$. Such a largest closed congruence relation exists because of the following lemma:

2.5.8 LEMMA. Let $\underline{A}$ be a **finitary** partial algebra and M any subset of A. Let **B** be a **chain** of (closed) congruence relations which have only one-element congruence classes for each element of M (i.e. none of them identifies two different elements of M). Then **∪B** is a (closed) congruence of $\underline{A}$ with the same property with respect to M. Especially there always exists (in the finitary case!) a largest (closed) congruence on $\underline{A}$ not identifying any two different elements of M.

PROOF. This is an immediate consequence of 2.5.5.(*) and of the inductivity of <u>Cong</u>($\underline{A}$) respectively <u>Cong</u>$_c$($\underline{A}$). ∎∎∎

2.6 Factor algebras

2.6.1. In 2.5.3 we have already realized that the equivalence relation R_f induced by a homomorphism $f:\underline{A} \to \underline{B}$ is always a congruence relation on $\underline{A}$. One aim of the following construction is to show the converse: that every congruence relation on $\underline{A}$ is induced by a (full) homomorphism (and every closed congruence relation by a closed homomorphism, which in each case can also be assumed to be surjective).

2.6.2 CONSTRUCTION and DEFINITION (of factor algebras). Let $\underline{A}$ be any partial algebra of type $(K_\phi)_{\phi\in\Omega}$, and let R be any congruence on $\underline{A}$. By $[x]R$ we denote the congruence class of $x \in A$ with respect to R, and we define as usual:

$A/R := \{[x]R \mid x \in A\}$, the set of all congruence classes of elements of A, and

$nat_R: A \to A/R$, $x \mapsto [x]R$, the <u>natural projection</u> from A onto A/R.

In order to get an algebraic structure on A/R, the <u>quotient set</u> of A with respect to R, we define for each $\phi \in \Omega$:

$\mathrm{dom}\, \phi^{\underline{A}/R} := \{([x_k]R \mid k \in K_\phi) \mid (x_k \mid k \in K_\phi) \in A^{K_\phi}$, and there is $(y_k \mid k \in K_\phi) \in \mathrm{dom}\, \phi^{\underline{A}}$ such that $(x_k, y_k) \in R$ for each $k \in K_\phi\}$;

and if $([x_k]R \mid k \in K_\phi) \in \mathrm{dom}\, \phi^{\underline{A}/R}$, and if $(y_k \mid k \in K_\phi) \in \mathrm{dom}\, \phi^{\underline{A}}$ satisfies $(x_k, y_k) \in R$ for each $k \in K_\phi$, then we define $([x_k]R \mid k \in K_\phi)\phi^{\underline{A}/R} := [(y_k \mid k \in K_\phi)\phi^{\underline{A}}]R$.

Notice that this definition is independent from the choice of the sequence $\underline{y}$, namely, let $\underline{z} := (z_k \mid k \in K_\phi) \in \mathrm{dom}\, \phi^{\underline{A}}$ be a sequence which is also componentwise equivalent to $\underline{x}$, then -- by the transitivity of R -- we have $(y_k, z_k) \in R$ for each $k \in K_\phi$, and -- by (Cong 2) -- $(\underline{y}\phi^{\underline{A}}, \underline{z}\phi^{\underline{A}}) \in R$, i.e. the two representative sequences lead to the same congruence class. Thus -- for each $\phi \in \Omega$ -- $\phi^{\underline{A}/R}$ is a well-defined K_ϕ-ary partial operation on A/R such that nat_R becomes a full homomorphism $nat_R : \underline{A} \to \underline{A}/R :=$ $:= (A/R, (\phi^{\underline{A}/R})_{\phi\in\Omega})$. $\underline{A}/R$ is called the <u>factor algebra</u> or <u>quotient algebra</u> of $\underline{A}$ with respect to the congruence R -- notice that we shall omit here the adjective "partial" even though $\underline{A}/R$ will usually also be partial when $\underline{A}$ is partial. We reformulate the above observations in

2.6.3 PROPOSITION. Let $\underline{A}$ be any partial algebra, and let R be any congruence on $\underline{A}$. Then $\underline{A}/R$ as defined in 2.6.2 is a partial algebra on A/R such that the natural projection nat_R becomes a full and surjective homomorphism $nat_R : \underline{A} \to \underline{A}/R$ and R is the congruence relation on $\underline{A}$ induced by nat_R: $R = \ker nat_R$. -- We shall see later that the structure on $\underline{A}/R$ is the weakest one which allows nat_R to be a homomorphism.
Moreover, R is closed iff nat_R is closed.

PROOF. From the construction in 2.6.2 most of the statements are obvious, one might even say that one makes the structure of $\underline{A}/R$ in such a way that nat_R becomes a full and surjective homomorphism. By recalling 2.5.2.(Cong 3c) and the definition of closed homomorphisms in 2.4.2.(ii) also the last statement becomes obvious in connection with 2.6.2. ■■■

2.6.4 COROLLARY. The quotient algebra of any partial algebra $\underline{A}$ is always of the same similarity type as $\underline{A}$. And if a fundamental operation, say $\phi^{\underline{A}}$, of $\underline{A}$ is total (discrete), then $\phi^{\underline{A}/R}$ is total (discrete) for each quotient algebra $\underline{A}/R$ of $\underline{A}$, too. ■■■

2.6.5 REMARK. It should be observed that the homomorphisms between partial algebras and the congruence relations of any given partial algebra depend very much on the chosen similarity type. Because of the equivalence between full and surjective homomorphisms and congruence relations -- which will be made even more evident in the next subsection -- it suffices to consider this dependence for congruence relations. As an example we consider the four element lattice (cf. its HASSE diagram in Figure 2.6) both as a lattice $(M_2; \vee, \wedge)$ of type (2,2) and as a meet semilattice $(M_2; \wedge)$.

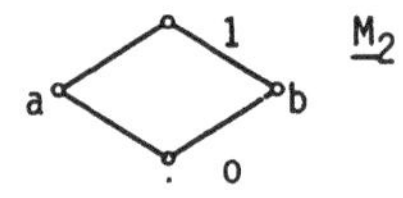

Figure 2.6.

Let the equivalence relation R on M_2 have the equivalence classes {1} and {o,a,b}; then it is easy to check that R is a congruence relation of the meet semilattice but not of the lattice.
That every semigroup homomorphism between groups is already a group homomorphism, and that every semigroup congruence of a group is already a group congruence are important exceptions which cannot be generalized. E.g. a non-surjective semigroup homomorphism

between two semigroups of type (2) which both have a (left as well as right) neutral element need not carry the neutral element of the start semigroup onto the one of the target semigroup, i.e. it need not be a monoid homomorphism.

2.7 The Homomorphism Theorem

The core of diagram completion theorems, among which the Homomorphism Theorem is one of the most important ones, is the following set theoretical result.

2.7.1 DIAGRAM COMPLETION LEMMA. Let $f:A \to B$ and $g:A \to C$ be arbitrary mappings such that f is **surjective**, and let R_f and R_g be the equivalence relations on A induced by f and g, respectively. Then there exists a mapping $h:B \to C$ such that $g = fh$, iff $R_f \subseteq R_g$ (see Figure 2.7).

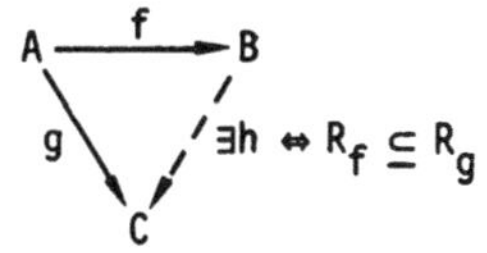

Figure 2.7.

Moreover, if h exists, then:

(a) h is uniquely determined by f and g.

(b) h is **injective**, iff $R_f = R_g$.

(c) h is **surjective**, iff g is surjective.

PROOF. Let us first assume that $h:B \to C$ exists and satisfies $g = fh$, and let $(a,b) \in R_f$; then $ag = afh = bfh = bg$, i.e. $(a,b) \in R_g$. Conversely, assume $R_f \subseteq R_g$; then -- in order to achieve $g = fh$ -- we have to define:

(*) $\qquad (b)h := (a)g$, whenever $(a)f = b$, for all $b \in B$.

Since f is surjective, h assigns to each $b \in B$ at least one value. Let us now assume that $af = a'f = b$; then $(a,a') \in R_f$, and because of $R_f \subseteq R_g$: $(a,a') \in R_g$, i.e. $ag = a'g$; therefore bh is uniquely determined by (*). ∎

For the rest of the proof let us assume that $R_f \subseteq R_g$ is satisfied, i.e. that h exists. (a) is then an immediate consequence of the surjectivity of f, and for the same reason both directions of (c) hold. In order to prove (b) let us assume $R_f = R_g$. Then $bh = b'h$, $b = af$ and $b' = a'f$ imply $ag = a'g$, and therefore $b = b'$. Conversely, let h be injective and $(a,a') \in R_g$; then $afh = ag = a'g = a'fh$, and $af = a'f$ because of the injectivity of h; hence $(a,a') \in R_f$. ∎∎∎

2.7.2 COROLLARY. Let, in addition to the assumptions of the above Lemma, $\underline{A}$, $\underline{B}$ and $\underline{C}$ be similar partial algebras, and let f and g be homomorphisms between them such that f is **full** (besides being surjective). Then h is a homomorphism, whenever it exists. Moreover, if h exists, then:

(d) h is a **full** homomorphism, iff g is a full homomorphism.

(e) h is an **isomorphism**, iff g is a surjective and full homomorphism, and $R_f = R_g$.

(f) If g is **closed**, then h is closed, too; if f is closed, then h is closed iff g is closed.

PROOF. Let the assumptions of the Corollary be satisfied. Let $\underline{b} \in \operatorname{dom} \phi^{\underline{B}}$; since f is surjective and full, there is $\underline{a} \in \operatorname{dom} \phi^{\underline{A}}$ such that $\underline{af} = \underline{b}$. Thus $\underline{bh} = \underline{afh} = \underline{ag}$; therefore $\underline{bh} \in \operatorname{dom} \phi^{\underline{C}}$, since g is a homomorphism. Therefore: $\underline{b}\phi^{\underline{B}}h = \underline{af}\phi^{\underline{B}}h = \underline{a}\phi^{\underline{A}}fh =$ $= \underline{a}\phi^{\underline{A}}g = \underline{ag}\phi^{\underline{C}} = \underline{afh}\phi^{\underline{C}} = \underline{bh}\phi^{\underline{C}}$, showing that h is a homomorphism, too. ∎

Now, (d) is a consequence of 2.4.5.(iii).(a) and (c) and the surjectivity of f; (e) follows from 2.4.6.(iii) using 2.7.1 above; and (f) is a consequence of 2.4.5.(iii).(f) and 2.4.5.(i). ■

2.7.3 REMARKS. (i): The results above show that the rôle of normal subgroups in group theory or ideals in the theory of rings or Boolean lattices is played in general by congruence relations. Thus -- in universal algebra -- the **kernel** of a homomorphism f between similar partial algebras is indeed the induced congruence relation R_f; therefore it really deserves this notation as kernel: $R_f = \ker f$.

(ii): It is usually less known -- since almost trivial -- that 2.7.1 and 2.7.2 have **dual statements**:

(a): Let $f:A \to B$ and $g:C \to B$ be mappings such that f is **injective.** Then there exists a (unique!) mapping $h:C \to A$ such that $hf = g$, iff $[C]g \subseteq [A]f$. -- h is **surjective,** iff $[C]g = [A]f$, and it is **injective,** iff g is injective.

(b): If $\underline{A}$, $\underline{B}$ and $\underline{C}$ are partial algebras such that f and g above are homomorphisms -- f being full and injective --, then h is a homomorphism whenever it exists. Moreover, h is full iff g is full; if g is closed, then h is closed; if f is closed, then h is closed iff g is closed; h is an isomorphism iff g is injective and full, and $[A]f = [C]g$.

The easy proofs are left as exercises. ■

The results of 2.6 and 2.7 so far combine to the

2.7.4 FIRST HOMOMORPHISM THEOREM. Let $g:\underline{A} \to \underline{C}$ be any homomorphism between similar partial algebras $\underline{A}$ and $\underline{C}$, and set $R := \ker g$.

(i) Then there exists a unique injective homomorphism $i:\underline{A}/R \to \underline{C}$ such that $g = nat_R \cdot i$.

(ii) i (from (i)) is an isomorphism iff g is surjective and full.

(iii) If hj is any factorization of g, where $h:\underline{A} \to \underline{D}$ is a full and surjective homomorphism and $j:\underline{D} \to \underline{C}$ is an injective homomorphism (and $g = hj$), then there exists a unique isomorphism $k:\underline{A}/R \to \underline{D}$ such that $h = nat_R \cdot k$ and $kj = i$.

This can also be expressed by saying that the diagrams in Figure 2.8 are in all parts commutative.

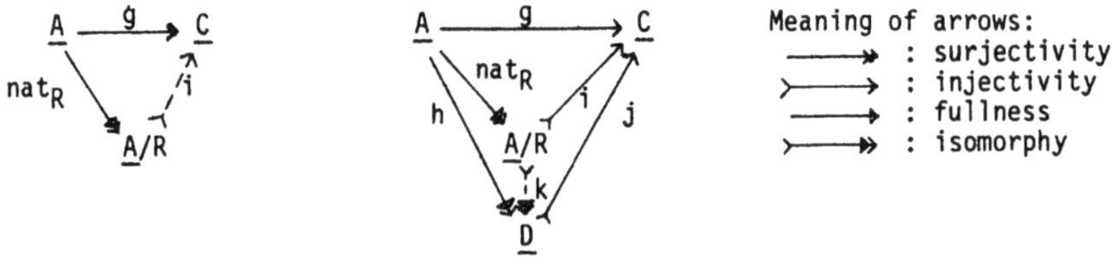

Figure 2.8. ■

2.7.5 REMARK. It should be observed that 2.7.4.(iii) means that the class of all full and surjective homomorphisms and the class of all injective homomorphisms form a "factorization system" (cf. Chapter II, section 10) in the category of all partial algebras and their homomorphisms, since it can easily be checked that both classes are closed with respect to composition and contain all isomorphisms.

2.7.6 DEFINITON. Let $\underline{A}$ and $\underline{B}$ be similar partial algebras, then we introduce the following notation:

$\underline{B}$ is called a (weak) homomorphic image of $\underline{A}$, iff there exists a surjective homomorphism from $\underline{A}$ onto $\underline{B}$.
$\underline{B}$ is called a full (or strong) homomorphic image of $\underline{A}$, iff there exists a full and surjective homomorphism from $\underline{A}$ onto $\underline{B}$.
$\underline{B}$ is called a closed homomorphic image of $\underline{A}$ iff there exists a closed and surjective homomorphism from $\underline{A}$ onto $\underline{B}$.

2.7.7 REMARKS. (i): The First Homomorphism Theorem tells us that the total information about a full homomorphic image $\underline{B}$ of some partial algebra $\underline{A}$ with respect to a full and surjective homomorphism, say $f:\underline{A} \to \underline{B}$ is contained in the ordered pair $(\underline{A}, \ker f)$, i.e. within the structure of $\underline{A}$. Hence there is up to unique isomorphism only a **set** of representatives of the classes of "equivalent" full homomorphic images of $\underline{A}$ ("equivalence" meaning that they belong to the same congruence relation). But it should also be observed that two full homomorphic images of $\underline{A}$ may be isomorphic without the existence of an isomorphism carrying one of the full and surjective homomorphisms into the other one: Take e.g. the void type (i.e. sets), and consider for a set A with at least two elements the two projections $pr_1:A \times A \to A$ $((a,b) \mapsto a)$ and $pr_2:A \times A \to A$ $((a,b) \mapsto b)$, then we have two full homomorphic images of $A \times A$, which are not equivalent in the above sense, but which are isomorphic (via id_A; we could have done the same, say, for a total structure on $\underline{A}$ and for the product $\underline{A} \times \underline{A}$ (see section 4)).
(ii): It should be observed that in the case of total algebras the three kinds of homomorphic images defined above coincide; therefore in this case each homomorphic image of a total algebra is totally described within $\underline{A}$ by its kernel. We shall discuss later how to describe other kinds of homomorphic images of partial algebras in close connection with $\underline{A}$ and the corresponding homomorphism. From 2.7.2 we may already infer that the closed homomorphic images of $\underline{A}$ are (up to isomorphism) in one-to-one correspondence with the closed congruence relations of $\underline{A}$.
(iii): It should have become clear so far from the above results, why we had to require the surjective homomorphism f in 2.7.2 to be **full**, since only then we can guarantee that an existing diagram completion of the set theoretical case is **always** a homomorphism.

2.8 The congruence lattice of a full homomorphic image

First we state a simple but useful fact:

2.8.1 LEMMA. Let $f:\underline{A} \to \underline{B}$ and $g:\underline{B} \to \underline{C}$ be homomorphisms, then
$$\ker fg = [\ker g](f^{(2)})^{-1} . \blacksquare$$

2.8.2 PROPOSITION. Let $f:\underline{A} \to \underline{B}$ be a homomorphism, $R_1 \in \mathrm{Cong}(\underline{A})$, $R_2 \in \mathrm{Cong}(\underline{B})$.
(i) If f is full and surjective, and $\ker f \subseteq R_1$, then $[R_1]f^{(2)} \in \mathrm{Cong}(\underline{B})$.
(ii) $[R_2](f^{(2)})^{-1} \in \mathrm{Cong}(\underline{A})$ and $\ker f \subseteq [R_2](f^{(2)})^{-1}$.
(iii) If f is closed and surjective, and if R_1 and R_2 are closed congruence relations, respectively, and if $\ker f \subseteq R_1$, then $[R_1]f^{(2)}$ and $[R_2](f^{(2)})^{-1}$ are closed congruence relations, too.

PROOF. Ad (i): From the Diagram Completion Lemma 2.7.1 and its Corollary we conclude that there exists a homomorphism $g:\underline{B} \to \underline{A}/R_1$ such that $fg = nat_{R_1}$. Obviously,

$[R_1]f^{(2)} \subseteq \ker g$, but because of the surjectivity of f also the other inclusion holds, showing that $[R_1]f^{(2)} = \ker g$ is a congruence on $\underline{B}$. ■

Ad (ii): Notice that $\ker f = [\Delta_B](f^{(2)})^{-1} \subseteq [R_2](f^{(2)})^{-1} = \ker(f\cdot nat_{R_2})$, where the last equality follows from 2.8.1 and shows that $[R_2](f^{(2)})^{-1} \in Cong(\underline{A})$. ■

Ad (iii): If f is closed and surjective, and if R_1 is closed and contains $\ker f$, then the homomorphism g in the proof of (i) is closed according to 2.4.5.(iii).(f); therefore $[R_1]f^{(2)}$ is a closed congruence according to (i) and 2.5.3.(iii). Also from 2.5.3 we conclude that $[R_2](f^{(2)})^{-1}$ is closed, since $f\cdot nat_{R_2}$ is closed (cf. 2.7.2.(f) and 2.4.5.(i)). ■■■

2.8.3 COROLLARY. Let $f:\underline{A} \to \underline{B}$ be a full and surjective homomorphism.

(i) $f^{(2)}$ induces an isomorphism between the interval $([\ker f, \nabla_A];\subseteq)$ of $\underline{Cong}(\underline{A})$ and all of $\underline{Cong}(\underline{B})$ $(R \mapsto [R]f^{(2)})$.

(ii) If, in addition, f is closed, then $f^{(2)}$ induces, too, an isomorphism between the "principal end" $([\ker f,\uparrow];\subseteq)$ of $\underline{Cong_c}(\underline{A})$ and all of $\underline{Cong_c}(\underline{B})$ on the other hand. (See Figure 2.9.)

PROOF. It remains only to show that the assignment $R \mapsto [R]f^{(2)}$ is injective, but this follows easily from the assumption that $\ker f \subseteq R$. ■■■

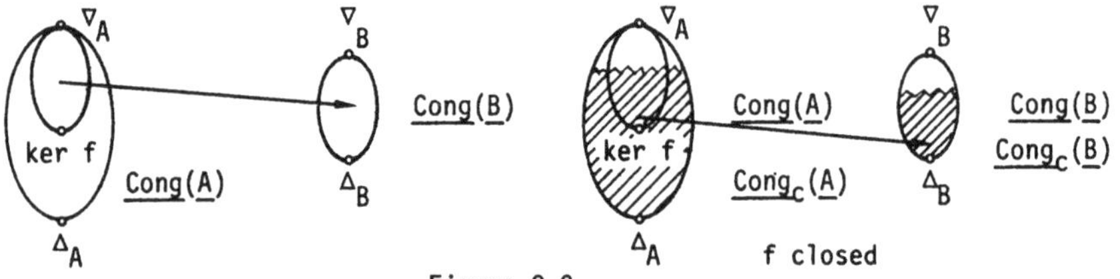

Figure 2.9.

2.8.4 REMARK. We would like to observe that in the case, when f above is closed, but not necessarily surjective, then the preimage of a closed congruence is also always closed. ■

2.9 The analogue of the Second Isomorphism Theorem of E.NOETHER

2.9.1 NOTATION. If $\underline{A}$ is any partial algebra, and R and R' are congruences on $\underline{A}$ such that $R \subseteq R'$, then we denote by R'/R the congruence $[R']nat_R^{(2)}$ of the factor algebra $\underline{A}/R$, i.e. the image of R' in $\underline{A}/R$ (see 2.8.2.(i)).

2.9.2 THEOREM. Let $\underline{A}$ be any partial algebra, and let $R,R' \in Cong(\underline{A})$ such that $R \subseteq R'$. Then one has

$$\underline{A}/R' \cong (\underline{A}/R)/(R'/R) ,$$

where the isomorphism is given by the assignment $[a]R' \mapsto [[a]R](R'/R)$ $(a \in A)$.

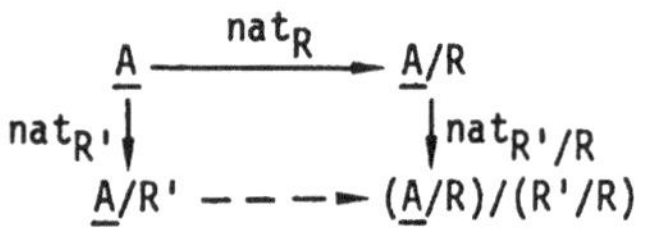

Figure 2.10.

PROOF. Consider the diagram in Figure 2.10 with $R' = \ker nat_{R'} = \ker (nat_R\cdot nat_{R'/R})$, where $nat_{R'}$, nat_R and $nat_{R'/R}$ -- and therefore also $nat_R\cdot nat_{R'/R}$ -- are full and surjective homomorphisms. Thus the required isomorphism exists according to 2.7.2.(e). ■■■

2.9.3. The Second Isomorphism Theorem of group theory (cf. van der WAERDEN [vdW 66], p. 150) is an immediate consequence of the above Theorem, since there is a one-to-one correspondence between the normal subgroups and the congruence relations of a group. The analogue of the First Isomorphism Theorem of group theory can only be formulated in the next section, when substructures are around.
Observe that in 2.9.2 for closed R' the congruences R and R'/R are closed, too.

2.10 Mono- and epimorphisms

2.10.1. In classical algebra -- and abstracted from it in category theory -- two kinds of morphisms play an important rôle: the monomorphisms and their duals, the epimorphisms. We have seen that in the theory of partial algebras there are additional properties -- which are, as we shall see in Chapter II, of a more model theoretical nature, since they are connected with the preservation or reflection of special formulas --, but mono- and epimorphisms are sort of a "basis" for those additional properties, too.

2.10.2 DEFINITION. Let $\underline{\mathbf{C}}$ be any category. A morphism $f:A \to B$ between any two $\underline{\mathbf{C}}$-objects A and B is called a <u>monomorphism</u>, iff for any two morphisms $g,h:C \to A$ one has that $gf = hf$ implies $g = h$. Dually, f is called an <u>epimorphism</u>, iff for any two morphisms $k,l:B \to D$ $fk = fl$ implies $k = l$. Mono($\underline{\mathbf{C}}$) and Epi($\underline{\mathbf{C}}$) designate the corresponding classes of all monomorphisms and of all epimorphisms in $\underline{\mathbf{C}}$, respectively.

Since we have not yet discussed the process of generation, we can not yet characterize epimorphisms in the category of partial algebras of some given type, but it is already easy to realize that each surjective homomorphism is an epimorphism. And we have:

2.10.3 PROPOSITION. In the category **Alg**(τ) of all partial algebras of similarity type τ with the homomorphisms as morphisms, the monomorphisms are exactly the injective homomorphisms. Moreover we have -- since this holds in arbitrary categories -- that the composition of monomorphisms is a monomorphism, and the composition of epimorphisms is an epimorphism, each isomorphism is a mono- as well as an epimorphism, but -- in contrast to species of total algebras and to the classes of "hierarchical equoids" (cf. Part II by H.REICHEL), in general a <u>bimorphism</u>, i.e. a monomorphism which is at the same time an epimorphism, need not be an isomorphism.

PROOF. That every injective homomorphism is a monomorphism follows already from the fact that every injective mapping has the cancellation property (for all other mappings) required for monomorphisms. Conversely, let $f:\underline{A} \to \underline{B}$ be a homomorphism which is not injective, and let $a,b \in A$ be different elements such that $af = bf$ (such elements then exist!). Let $\underline{C} := (\{c\};\emptyset)$ be a discrete one-element partial algebra of type τ, and let $g,h:\underline{C} \to \underline{A}$, $cg := a$, $ch := b$. Obviously, g and h are homomorphisms from $\underline{C}$ into $\underline{A}$ satisfying $gf = hf$, but $g \neq h$; therefore f is not a monomorphism. -- Let $\underline{A}$ be a total algebra on a non-empty carrier set, and of a non-trivial similarity type, then $id_A:A \to \underline{A}$ is a bimorphism which is not an isomorphism. ■■■

2.10.4 REMARKS. **(i):** The above argument can be repeated in every full subcategory of **Alg**(τ), which has at least one "free object with at least one generator" (cf. sections 5, 6 and 17).
(ii): Notice that in the book [Pi 68] of R.S.PIERCE "epimorphisms" are defined as full and surjective homomorphisms -- described in our terminology.

2.11 Partial $\underline{S}$-algebras and heterogeneous partial algebras

2.11.1 REMARKS and OBSERVATIONS. **(i):** As we have already mentioned in the Introduction and in 1.2 most applications of universal algebra in computer science are concerned with so-called heterogeneous or many-sorted algebras, and some of the structures encountered there are partial. Thus we do not want to exclude them from this text. On the other hand we want to burden this presentation as little as possible with the relatively technical description of many-sorted partial algebras. For this purpose we want to introduce a different, but equivalent description of heterogeneous partial algebras, which seems to us to be more closely related to the homogeneous case and allows to a large extent to treat the homogeneous and the heterogeneous case in a parallel way. This way of presentation was suggested to us by remarks and ways of representation in the habilitation theses of I.KUPKA in [Ku 77] and H.REICHEL in [Re 79].
(ii): The starting point is the fact that the phyla of a many-sorted partial algebra are usually treated as if they were disjoint, as it is also the fact in Definition 2.1.1.(ii). Thus we assume that they really **are disjoint** -- or at least we can make them disjoint w.l.o.g. --; therefore we can view at the carrier of a heterogeneous partial algebra as a set A together with a mapping $v_A: A \to S$, where S is the set of sorts under consideration, and a **phylum** A_s is nothing else but $(s)v_A^{-1}$ $(s \in S)$.
(iii): Moreover, let us endow S with the structure of a homogeneous partial algebra $\underline{S}$ by defining $\operatorname{graph}(\phi^{\underline{S}}) := \{((\phi)\mathrm{INPUT},(\phi)\mathrm{OUTPUT})\}$, for each $\phi \in \Omega$.
$\underline{S}$ is called the <u>sort algebra</u> of the heterogeneous similarity type under consideration. Notice that $\underline{S}$ is total iff S has exactly one element or the similarity type τ contains at most nullary constants, and in the former case we actually deal with homogeneous partial algebras.
(iv): Now, let $\underline{A}`$ be any heterogeneous partial algebra of the similarity type under consideration, let A be the disjoint union of the phyla $A_s$, and let us consider dom $\phi^{\underline{A}`}$ as a subset of A^{K_ϕ} and $\phi^{\underline{A}`}$ as a mapping from dom $\phi^{\underline{A}`}$ into A (for each $\phi \in \Omega$) -- and we now denote this operation by $\phi^{\underline{A}}$ in order to distinguish it from the heterogeneous one. Then one can realize that v_A is a homomorphism from $(A,(\phi^{\underline{A}})_{\phi\in\Omega})$ into $\underline{S}$. ∎

These observations motivate the following

2.11.2 DEFINITION. Let $\underline{S}$ be any partial algebra of type τ with **non-empty** carrier set S. Then a <u>partial $\underline{S}$-algebra $\underline{A}`$</u> is a pair $(\underline{A},v_A)$, where $\underline{A}$ is any homogeneous partial algebra, and v_A is a homomorphism from $\underline{A}$ into $\underline{S}$, the <u>sort mapping</u> or <u>sort homomorphism</u> of $\underline{A}$. $\underline{S}$ is called the <u>corresponding sort algebra</u>.
$\underline{S}$ is called a <u>heterogeneous sort algebra</u> -- and $\underline{A}` := (\underline{A},v_A)$ a <u>heterogeneous (partial) $\underline{S}$-algebra</u> --, iff $\#\operatorname{dom}\phi^{\underline{S}} = 1$ for each $\phi \in \Omega$.

A homomorphism $f:\underline{A}^{`} \to \underline{B}^{`}$ between partial $\underline{S}$-algebras $\underline{A}^{`}$ and $\underline{B}^{`}$ -- also called an $\underline{S}$-homomorphism -- is a homomorphism $f:\underline{A} \to \underline{B}$ satisfying $fv_B = v_A$.
The class of all partial $\underline{S}$-algebras -- and also the corresponding category including all $\underline{S}$-homomorphisms -- is denoted by **Alg**($\underline{S}$).

2.11.3 REMARKS. (i): Notice that in terms of category theory the category **Alg**($\underline{S}$) defined in 2.11.2 is nothing else but the **comma category** **Alg**$(\tau)\downarrow\underline{S}$ (cf. HERRLICH & STRECKER [HS 73] or MacLANE [Ml 71]).

(ii): In what follows we shall only consider heterogeneous partial $\underline{S}$-algebras, but we want to stress at this point that most of the results with category theoretical background will also be true for arbitrary partial $\underline{S}$-algebras. And the approach via $\underline{S}$-algebras opens a possibility of generalization which is not so easily seen for the usual concept of heterogeneous algebras.

(iii): In the following sections we shall treat homogeneous and heterogeneous partial $\underline{S}$-algebras in a parallel way, often without mentioning $\underline{S}$ explicitly. We shall even usually also omit the symbol "`", but the reader should be aware that every set A will be connected with a sort mapping v_A into $\underline{S}$, and for a subset $M \subseteq A$ its sort mapping is the restriction of v_A to M: $v_M := v_A|M$.

(iv): Below we shall show that the concepts of heterogeneous partial algebras and of heterogeneous partial $\underline{S}$-algebras are equivalent; therefore the use of the same notation ($\underline{A}^{`}$) for both kinds of structures is justified, but notice that in the case of partial $\underline{S}$-algebras the fundamental operations of $\underline{A}^{`}$ are still denoted by $\phi^{\underline{A}}$ and not by $\phi^{\underline{A}^{`}}$. However, a reader who is not familiar with heterogeneous algebras should **skip the rest of this subsection** and accept 2.11.2 as the basic definition.

2.11.4 THEOREM. Let (τ,INPUT,OUTPUT) be any heterogeneous similarity type, let $\underline{S}$ be the partial algebra of type τ on the set S of sorts such that graph($\phi^{\underline{S}}$):= := {((ϕ)INPUT,(ϕ)OUTPUT)} for each $\phi \in \Omega$. Then the category **Alg**(τ,INPUT,OUTPUT) of all heterogeneous partial algebras of type (τ,INPUT,OUTPUT) is canonically isomorphic to the category **Alg**($\underline{S}$) as defined in 2.11.2 (and vice versa).
Moreover, the subcategory **TAlg**(τ,INPUT,OUTPUT) of all total heterogeneous algebras of type (τ,INPUT,OUTPUT) corresponds in this connection exactly to the subcategory **TAlg**($\underline{S}$) of **Alg**($\underline{S}$) of those pairs ($\underline{A}$,v_A) where $v_A:\underline{A} \to \underline{S}$ is a closed homomorphism (we shall call them total $\underline{S}$-algebras) -- and the homomorphisms being just the same as in **Alg**($\underline{S}$).

PROOF. We have already shown in 2.11.1 how we can associate with each heterogeneous partial algebra $\underline{A}^{`}$ of type (τ,INPUT,OUTPUT) a partial $\underline{S}$-algebra ($\underline{A}$,v_A). It is easy to realize that these assignments yield a "homomorphism between categories" (i.e. a functor, cf. 2.12.7 below), which is injective.
Conversely, let $\underline{S}$ be a heterogeneous sort algebra, and let ($\underline{A}$,v_A) be any partial $\underline{S}$-algebra, and define, for each $s \in S$, $A_s := (s)v_A^{-1}$. Then it is easy to realize that $((A_s)_{s\in S},(\phi^{\underline{A}})_{\phi\in\Omega})$ is a heterogeneous partial algebra of type (τ,INPUT,OUTPUT), where we define {((ϕ)INPUT,(ϕ)OUTPUT)}:= graph($\phi^{\underline{S}}$) ; moreover, if we had started with a heterogeneous partial algebra, say $\underline{A}^{`}$, then the composition of the two constructions above leads to a heterogeneous partial algebra canonically isomorphic to the one we have started with, and the same is true for the composition of the above constructions in the other way.

Now, let $f`:\underline{A}` \to \underline{B}`$ be any homomorphism between heterogeneous partial algebras of type $(\tau,\mathrm{INPUT},\mathrm{OUTPUT})$, and let the graph of f be the disjoint union of the family $(\mathrm{graph}(f_s) | s \in S)$, and let $(\underline{A},v_A)$ and $(\underline{B},v_B)$ be the two homogeneous partial algebras together with their homomorphisms into $\underline{S}$ corresponding to $\underline{A}`$ and $\underline{B}`$, respectively. Then one sees immediately that $fv_B = v_A$. -- Conversely, let $(\underline{A},v_A:\underline{A} \to \underline{S})$ and $(\underline{B},v_B:\underline{B} \to \underline{S})$ be given, where $\underline{A}$ and $\underline{B}$ are homogeneous partial algebras, let $f:\underline{A} \to \underline{B}$ be a homomorphism satisfying $fv_B = v_A$, and let $f`:= (f_s:A_s \to B_s)_{s\in S}$ be the family of restrictions of f to the preimages of the elements of S -- the equality "$fv_B = v_A$" implies that the image of $A_s$ is indeed contained in $B_s$ $(s \in S)$. Then one easily computes that $f`$ is a homomorphism between $\underline{A}`$ and $\underline{B}`$, the two heterogeneous partial algebras corresponding to $(\underline{A},v_A)$ and $(\underline{B},v_B)$, respectively. In both directions these constructions of homomorphisms are compatible with composition of homomorphisms, where $f`\cdot g` = (f_s|s \in S)\cdot(g_s|s \in S):= (f_s\cdot g_s|s \in S)$ for any two composable homomorphisms $f`$ and $g`$ between heterogeneous partial algebras. ∎

It still remains to prove the statement about total heterogeneous partial algebras, but if v_A is closed, then, for each $\phi \in \Omega$, dom $\phi^{\underline{A}}$ has to be the cartesian product of $(A_{(k)((\phi)\mathrm{INPUT})} | k \in K_\phi)$ -- where $A_s = (s)v_A^{-1}$ $(s \in S)$ --, but this just means that the corresponding heterogeneous partial algebra is total. The converse is proved similarly. ∎∎∎

2.11.5 COROLLARY. Let $f:(\underline{A},v_A) \to (\underline{B},v_B)$ be a homomorphism in **Alg**($\underline{S}$), where the first pair represents a total heterogeneous algebra. Then $f:\underline{A} \to \underline{B}$ is closed.

PROOF. This is an immediate consequence of the fact that $v_A = fv_B$ is closed; therefore 2.4.5.(iii).(e) implies that f is closed. ∎∎∎

2.12 Special observations concerning partial $\underline{S}$-algebras

In the previous subsection we have defined (heterogeneous) partial $\underline{S}$-algebras and have shown that this concept is equivalent to the one of heterogeneous partial algebras. Now we extend the concepts introduced so far for homogeneous partial algebras to the case of partial $\underline{S}$-algebras (in this subsection we still use the "reminder" "`"):

2.12.1 DEFINITION. Let $\underline{S}$ be any heterogeneous sort algebra -- throughout the rest of this book $\underline{S}$ will have this meaning, even if we do not stress it explicitly -- and let $\underline{A}`$ be any partial $\underline{S}$-algebra.

(i): An $\underline{S}$-homomorphism $f:\underline{A}` \to \underline{B}`$ is <u>full</u> or <u>closed</u>, respectively, iff the ordinary homomorphism $f:\underline{A} \to \underline{B}$ is full, respectively closed.

(ii): A relation $R \subseteq A \times A$ is an <u>$\underline{S}$-congruence</u> of $\underline{A}`$, iff it is the kernel of an $\underline{S}$-homomorphism starting from $\underline{A}`$, and it is called a <u>closed $\underline{S}$-congruence</u> of $\underline{A}`$, iff it is the kernel of a closed $\underline{S}$-homomorphism starting from $\underline{A}`$.

(iii) Let $A`:= (A,v_A)$ and $B`:= (B,v_B)$ be any S-sets. Then we define $A` \times B`:= \mathbf{U}(A_s \times B_s|s \in S)$, which later on will also be denoted by $A \times B$, since in the case of S-sets this is the correct product (see section 4). -- Let us also mention that in what follows we shall understand by "A^B" the set $A^{`B`}$ of all S-mappings $f:B` \to A`$, i.e. satisfying $fv_A = v_B$. In particular, for each $\phi \in \Omega$,

$kv_{K_\phi} := (k)((\phi)\text{INPUT})$ in $\underline{S}$ $(k \in K_\phi)$.

(iv): By $\text{Cong}(\underline{A})$ and $\text{Cong}_c(\underline{A})$ we denote from now on the sets of all $\underline{S}$-congruences respectively closed $\underline{S}$-congruences of $\underline{A}$\`. The corresponding ordered sets are also denoted as in the homogeneous case; only in special cases we use "$\text{Cong}(\underline{A}^{\backprime})$" or "$\text{Cong}_c(\underline{A}^{\backprime})$".

(v): $\underline{\text{End}}(\underline{A}^{\backprime})$ and $\underline{\text{Aut}}(\underline{A}^{\backprime})$, the endomorphism monoid and the automorphism group of $\underline{A}$\` are defined as in the homogeneous case and will also be denoted as $\underline{\text{End}}(\underline{A})$, respectively $\underline{\text{Aut}}(\underline{A})$. -- By $\text{Hom}(\mathbf{Alg}(\underline{S}))$ we denote the class of all $\underline{S}$-homomorphisms.

2.12.2 PROPOSITION. Let $\underline{A}$\` be any partial $\underline{S}$-algebra. Then

(i) $\text{Cong}(\underline{A}^{\backprime})$ is the set of all congruence relations of $\underline{A}$ -- the underlying homogeneous partial algebra --, which are contained in $\ker v_A = A^{\backprime} \times A^{\backprime}$, i.e. $\ker v_A$ ($=:\nabla_A$ from now on!) is the largest element of the complete lattice $\underline{\text{Cong}}(\underline{A}^{\backprime}) := (\text{Cong}(\underline{A}^{\backprime}); \subseteq)$.

(ii) $\underline{\text{Cong}_c}(\underline{A}^{\backprime})$ is the initial segment of all those closed congruences of $\underline{A}$\` contained in $\ker v_A$. If type τ is finitary, then there also exists a largest element $R_c(\underline{A}^{\backprime})$ in $\underline{\text{Cong}_c}(\underline{A}^{\backprime})$, namely $R_c(\underline{A}^{\backprime}) = R_c(\underline{A}) \cap \ker v_A$, where $R_c(\underline{A})$ is the largest closed congruence relation of $\underline{A}$.

PROOF. Ad (i): Let $\underline{A}$\` and $\underline{B}$\` be any partial $\underline{S}$-algebras, $f:\underline{A} \to \underline{B}$ any full and surjective homomorphism, and $R := \ker f$. Since f is an $\underline{S}$-homomorphism, iff $fv_B = v_A$, we have $R \in \text{Cong}(\underline{A}^{\backprime})$, iff $R \subseteq \ker v_A$ - as can be easily derived from the Diagram Completion Lemma 2.7.1. ∎

Ad (ii): This is an immediate consequence of (i) and the definition of closed $\underline{S}$-homomorphisms. ∎∎∎

2.12.3 OBSERVATION. The above result and the observations in 2.11 show that all the preceding results on homomorphisms and congruences carry over to $\underline{S}$-homomorphisms, also as far as full and closed ($\underline{S}$-) homomorphisms and congruences are concerned, as long as one observes the restriction given by $\ker v_A$ in connection with congruences. For instance we always have to remember that now ∇_A means $\nabla_{A^{\backprime}} := \ker v_A = A^{\backprime} \times A^{\backprime}$.

As an example we reformulate Corollary 2.8.3.(i) as

2.12.4 PROPOSITION. Let $f:\underline{A}^{\backprime} \to \underline{B}^{\backprime}$ be a full and surjective $\underline{S}$-homomorphism. Then $f^{(2)}$ induces an isomorphism between the interval $([\ker f, \nabla_{A^{\backprime}}]; \subseteq)$ of $\underline{\text{Cong}}(\underline{A}^{\backprime})$ and all of $\underline{\text{Cong}}(\underline{B}^{\backprime})$ $(R \mapsto [R]f^{(2)})$. ∎∎∎

Also for closed $\underline{S}$-congruences the corresponding result is true. Finally let us observe:

2.12.5 PROPOSITION and NOTATION. Let $\underline{A}$\` be a partial $\underline{S}$-algebra and $R \subseteq A \times A$ any relation. Then there exists a smallest $\underline{S}$-congruence, say R', of $\underline{A}$\` containing R, iff $R \subseteq \nabla_{A^{\backprime}}$; and if R' exists, then it is given by $R' := \mathbf{Cong}_{\underline{A}}(R)$ and will be denoted by $\mathbf{Cong}_{\underline{A}^{\backprime}}(R)$. Each relation $R \subseteq \nabla_{A^{\backprime}}$ will be called an $\underline{S}$-relation on A\`. ∎∎∎

2.12.6. Let us repeat that in what follows "partial algebra $\underline{A}$ " will always mean "partial $\underline{S}$-algebra $\underline{A}$\`" for some arbitrary but fixed heterogeneous sort algebra $\underline{S}$ of the (homogeneous) type τ under consideration, even when $\underline{S}$ will not be explicitely mentioned (or only at the beginning of a section), except when we stress that it is homogeneous (then $\underline{S}$ is the one-element total algebra). Moreover, mappings and homomorphisms will

always have to be compatible with the sort mappings.
It should be observed, too, that we allow empty carrier sets of homogeneous and empty phyla of heterogeneous partial algebras; therefore the sort homomorphisms v_A need not be surjective. Especially in the heterogeneous case we would otherwise get into trouble with the existence of subalgebras generated by subsets which do not contain elements from each phylum.

2.12.7 EXAMPLE. In order to get a better feeling for all these concepts, let us consider for the class (category) **Cat** of all "small" categories $\underline{\mathbf{C}}$ (i.e. $Ob(\underline{\mathbf{C}})$ and $Mor(\underline{\mathbf{C}})$ are always sets) as class of heterogeneous partial algebras the concept of homomorphisms, which here are usually called <u>functors</u>: We have in this case (cf. 1.2.1.(iii)) $S := \{mor, ob\}$, $graph(\cdot^{\underline{S}}) := \{((mor,mor),mor)\}$, $graph(Dom^{\underline{S}}) := \{((mor),ob)\} =: \; =: graph(Cod^{\underline{S}})$, $graph(1^{\underline{S}}) := \{((ob),mor)\}$. For every $\underline{\mathbf{C}} \in \mathbf{Cat}$ we then have $Ob(\underline{\mathbf{C}}) := (ob)v_{\mathbf{C}}^{-1}$, $Mor(\underline{\mathbf{C}}) := (mor)v_{\mathbf{C}}^{-1}$. And if $\underline{\mathbf{C}}$ and $\underline{\mathbf{C}}'$ are any two small categories, then a <u>functor</u> $f:\underline{\mathbf{C}} \to \underline{\mathbf{C}}'$, i.e. an $\underline{S}$-homomorphism from $\underline{\mathbf{C}}$ into $\underline{\mathbf{C}}'$, is any mapping $f:\mathbf{C} \to \mathbf{C}'$ such that

(a) $[Ob(\underline{\mathbf{C}})]f \subseteq Ob(\underline{\mathbf{C}}')$ and $[Mor(\underline{\mathbf{C}})]f \subseteq Mor(\underline{\mathbf{C}}')$;

(b) $(m \cdot^{\underline{\mathbf{C}}} m')f = (m)f \cdot^{\underline{\mathbf{C}}'} (m')f$ for all composable $m, m' \in Mor(\underline{\mathbf{C}})$;

(c) $(Dom^{\underline{\mathbf{C}}}(m))f = Dom^{\underline{\mathbf{C}}'}((m)f)$ and $(Cod^{\underline{\mathbf{C}}}(m))f = Cod^{\underline{\mathbf{C}}'}((m)f)$ $(m \in Mor(\underline{\mathbf{C}}))$;

(d) $(1^{\underline{\mathbf{C}}}_A)f = 1^{\underline{\mathbf{C}}'}_{(A)f}$ $(A \in Ob(\underline{\mathbf{C}}))$.

Except for (a) these are the usual compatibility conditions for homomorphisms known from the homogeneous case.

3 Substructures, subalgebras, closed subsets, generation

One of the most important features of algebraic structures is the fact that by means of the process of generation usually very large subsets of a given structure can be described by specifying a comparatively small subset as a **generating subset**, whenever the subset under consideration can be described as a so-called closed subset. This fact may also serve as the most important motivation for dealing with partial algebras rather than describing them as relational systems, where generation is not so easily dealt with as in an algebraic context (cf. also Part II for further motivation and examples in this direction from computer science, where this is one of the most important reasons for the application of universal algebra at all).
As already mentioned at the end of the previous section, we tacitly assume some fixed type τ and a heterogeneous sort algebra $\underline{S}$ of type τ. Moreover, all partial algebras are of type τ, and each partial algebra $\underline{A}$ is provided with a fixed sort homomorphism $v_A:\underline{A} \to \underline{S}$ such that all homomorphisms between such partial $\underline{S}$-algebras are compatible with the corresponding sort homomorphisms. If we do not need it for some special reason we usually omit the "reminder" "`" referring to the existence of this sort homomorphism.

3.1 Substructures of partial algebras

3.1.1 Category theoretical concept of subobjects. In Category Theory a subobject of an object A in a category $\underline{C}$ is in most generality an ordered pair (B,f), where $f:B \to A$ is any monomorphism in C -- very often additional properties are requested. For our category **Alg**($\underline{S}$) of all partial $\underline{S}$-algebras of some similarity type τ together with their homomorphisms the simplest concept would mean that a **subobject** of a partial algebra $\underline{A}$ is a pair ($\underline{B}$,i), where $i:\underline{B} \to \underline{A}$ is an injective homomorphism (cf. 2.10.3). If ($\underline{B}$,i) is any subobject of $\underline{A}$ in this sense, then we may consider $A' := [B]i$ and $graph(\phi^{\underline{A}'}) :=$ $:= [graph(\phi^{\underline{B}})]i^{(K_\phi \cup \{K_\phi\})}$; therefore $i':\underline{B} \to \underline{A}'$ ($i = i' \cdot id_{A'}$) becomes an isomorphism, and ($\underline{A}'$,$id_{A'}$) is another subobject, which we shall call the weak (relative) subalgebra of $\underline{A}$ induced by ($\underline{B}$,i).

Two subobjects ($\underline{B}$,i) and ($\underline{C}$,j) of $\underline{A}$ are called equivalent, whenever there is an isomorphism $k:\underline{B} \to \underline{C}$ such that $kj = i$. Thus ($\underline{B}$,i) and ($\underline{A}'$,$id_{A'}$) above are equivalent, and ($\underline{A}'$,$id_{A'}$) is kind of a natural representative of the equivalence class of ($\underline{B}$,i). These observations motivate the definition below.

But it should also be observed that this concept of subobject of partial algebras is too weak to describe "generation". Therefore we strengthen the concept of subobject below by requiring additional properties for the monomorphisms under consideration, and the most appropriate one will be closedness. -- We shall almost always consider only the "natural representative" introduced above.

3.1.2 DEFINITION. Let $\underline{A}$ and $\underline{B}$ be any similar partial algebras such that $B \subseteq A$. Then $\underline{B}$ is said to be a

-- weak (relative) subalgebra of $\underline{A}$ iff $id_B:\underline{B} \to \underline{A}$ is an injective homomorphism,

-- relative subalgebra of $\underline{A}$ iff $id_B:\underline{B} \to \underline{A}$ is a **full** and injective homomorphism,

-- (closed) subalgebra of $\underline{A}$ iff $id_B:\underline{B} \to \underline{A}$ is a **closed** and injective homomorphism.

$C \subseteq A$ is said to be a closed subset of $\underline{A}$, iff for every fundamental operation $\phi^{\underline{A}}$ and for every sequence $\underline{a} \in dom\, \phi^{\underline{A}} \cap C^{K_\phi}$ one has $\underline{a}\phi^{\underline{A}} \in C$. We denote by

-- **Csubs**($\underline{A}$) the set of all closed subsets of $\underline{A}$,

-- **Wsuba**($\underline{A}$) the set of all weak subalgebras of $\underline{A}$,

-- **Rsuba**($\underline{A}$) the set of all relative subalgebras of $\underline{A}$,

-- **Suba**($\underline{A}$) the set of all (closed) subalgebras of $\underline{A}$, respectively.

3.1.3 PROPOSITION. Let $\underline{A}$, $\underline{B}$ be similar partial algebras, $B \subseteq A$. Then:

(i) $\underline{B}$ is a weak subalgebra of $\underline{A}$ iff for all $\phi \in \Omega$: $graph(\phi^{\underline{B}}) \subseteq graph(\phi^{\underline{A}})$.

(ii) $\underline{B}$ is a relative subalgebra of $\underline{A}$ iff for all $\phi \in \Omega$:

$graph(\phi^{\underline{B}}) = graph(\phi^{\underline{A}}) \cap (B^{K_\phi} \times B)$.

(iii) $\underline{B}$ is a subalgebra of $\underline{A}$ iff for all $\phi \in \Omega$: $graph(\phi^{\underline{B}}) = graph(\phi^{\underline{A}}) \cap (B^{K_\phi} \times A)$, i.e. iff $\underline{B}$ is a relative subalgebra of $\underline{A}$, and B is a closed subset of $\underline{A}$.

PROOF. (i) is obvious, (ii) follows from 3.1.2 and 2.4.4.(i), while (iii) can easily be derived from 3.1.2, 2.4.2.(ii) and 2.4.4.(ii). ■■■

3.1.4 REMARKS. **(i)**: It should be observed that each subalgebra is a relative subalgebra, and that each relative subalgebra is especially a weak subalgebra. Moreover, $\underline{A}$ is always a subalgebra of itself; and if $\underline{A}$ has no (non-empty) nullary constants, then the empty subset is always closed, too.

(ii): The importance of weak and relative subalgebras lies in their rôle as "small" building blocks, as we shall see in section 4 in connection with direct limits, and also in later sections; but they are of no grat importance in connection with the concept of generation, since there is a (weak) relative subalgebra of $\underline{A}$ for every subset of A. The situation is different for subalgebras, which only exist on closed subsets. Thus we will study in this section mainly the properties of subalgebras and closed subsets.

(iii): It should be noticed that -- as with homomorphisms -- the concept of a closed subset and hence of a subalgebra heavily depends on the similarity type chosen. E.g. if one defines groups -- as is usually done in group theory -- by only one binary operation, then the subalgebras in the sense of 3.1.2 are all subsemigroups of the given group, and one has to add at least the formation of the inverse as a fundamental operation in order to get all subgroups -- including the empty subalgebra -- as subalgebras. Only when we consider groups as universal algebras of type (2,1,0) (or in some equivalent way), we get coincidence for groups between our concept of subalgebras and the usual concept of subgroups. One reason for this conceptual difference is the fact that the universal algebraic concept of subalgebras uses as little of model theory as possible (i.e. tries to keep the axioms as simple as possible), while usually in group theory the axioms are logically relatively complex (containing existential quantifiers explicitly). The concept of a subalgebra, as usually used in universal algebra, fully describes substructures for classes which are described by equations but not by more complex axioms, and it is well known that one can describe groups by equations, when one chooses (2,1,0) as similarity type.

(iv): When considering any weak subalgebra $\underline{B}$ of some partial $\underline{S}$-algebra $\underline{A}$ (or just when writing $\underline{B} \subseteq \underline{A}$) the sort mapping v_B is always the restriction of v_A to B, and thus it is a homomorphism from $\underline{B}$ into $\underline{S}$. Thus it is usually not necessary to refer to $\underline{S}$ in connection with substructures of partial $\underline{S}$-algebras, anyway.

3.1.5 PROPOSITION. Let $\underline{A}$ be any partial algebra of similarity type τ, and let $\Omega' := \Omega \dot{\cup} \{\phi'_s \mid s \in S\} \dot{\cup} \{c_a \mid a \in A\}$ and $\tau' := \tau\hat{}(2|s \in S)\hat{}(0|a \in A)$ ("^" indicating concatenation of sequences), where each constant c_a has "output sort" $(a)v_A$. Then Cong($\underline{A}$) is exactly the set of all closed subsets of the following partial algebra $\underline{B}$ of similarity type τ', where:

$B := A \times A$ (i.e. $A^{\grave{}} \times A^{\grave{}}$ (!)), $c_a^{\underline{B}} := (a,a)$ for every $a \in A$, $\phi'^{\underline{B}}_s$ is a binary total heterogeneous operation defined on $\{((a,b),(c,d)) \mid a,b,c,d \in A_s = (s)v_A^{-1}\}$,

$$((a,b),(c,d))\phi'^{\underline{B}}_s := \begin{cases} (a,d), & \text{if } b = c, \\ (b,a), & \text{if } b \neq c; \end{cases}$$

and for all $\phi \in \Omega$ one has:

$\operatorname{dom} \phi^{\underline{B}} := \{((a_k,b_k) \mid k \in K_\phi) \mid (a_k \mid k \in K_\phi), (b_k \mid k \in K_\phi) \in \operatorname{dom} \phi^{\underline{A}}\}$, and if

$((a_k,b_k) \mid k \in K_\phi) \in \operatorname{dom} \phi^{\underline{B}}$, then $((a_k,b_k) \mid k \in K_\phi)\phi^{\underline{B}} := ((a_k \mid k \in K_\phi)\phi^{\underline{A}}, (b_k \mid k \in K_\phi)\phi^{\underline{A}})$.

v_B is the mapping induced by v_A.

PROOF. The definition of B guarantees that each closed subset of $\underline{B}$ is a binary relation on A, the constants $c_a^{\underline{B}}$ yield reflexivity, while the $\phi'^{\underline{B}}_s$ take care of symmetry and transitivity of the closed subsets of $\underline{B}$ on each phylum of A. Finally the definition of the partial operations $\phi^{\underline{B}}$ enforces compatibility of the equivalence relations with the fundamental operations of $\underline{A}$. ■■■

3.1.6. As a consequence of Proposition 3.1.5 everything which we shall say about generation of closed subsets or subalgebras of some partial algebra $\underline{A}$ will also apply to the generation of congruence relations in connection with the algebraic structure of $\underline{B}$ defined above on $A \times A$. Notice, however, that for the structure of $\underline{B}$ above the cardinal dimension is always $\mathbf{d^*} = \max(\mathbf{d}, \#\omega_0)$, where $\mathbf{d}$ is the cardinal dimension of τ, and its ordinal dimension is $\max(\delta, \omega_0)$, δ being the ordinal dimension of τ (cf. 1.C). Moreover, 3.1.5 yields examples of closed subsets, and as we have already observed in 3.1.4.(iii) our concept of subalgebra yields the correct subobjects of total algebras, whenever one considers a class of algebras defined by equations (identities). This can e.g. be done for semigroups, rings, groups, lattices, boolean algebras, vector spaces, hypercomplex systems, etc..

3.1.7 PROPOSITION. For every closed homomorphism into a total algebra the domain is a total algebra, too. ■■■

3.1.8 COROLLARY. Every subalgebra of a total algebra is again total. ■■■

3.1.9 PROPOSITION. Let $f:\underline{A} \to \underline{B}$ be any homomorphism, and let $\underline{C}$ and $\underline{D}$ be subalgebras of $\underline{A}$ and $\underline{B}$, respectively. Then:

(i) $[D]f^{-1}$ is a closed subset of $\underline{A}$, too: $[D]f^{-1} \in \mathbf{Csubs}(\underline{A})$.

(ii) If f is closed, then $[C]f$ is a closed subset of $\underline{B}$, but this need not be the case for arbitrary homomorphisms, even $[A]f$ need not be a closed subset of $\underline{B}$ in general.

PROOF. Ad (i): With the conditions of the Proposition assume $\underline{a} \in ([D]f^{-1})^{K_\phi} \cap \operatorname{dom} \phi^{\underline{A}}$; then $\underline{a}f \in D^{K_\phi} \cap \operatorname{dom} \phi^{\underline{B}}$, since f is a homomorphism, and $(\underline{a}f)\phi^{\underline{B}} = (\underline{a}\phi^{\underline{A}})f \in D$, since D is closed. Thus $\underline{a}\phi^{\underline{A}} \in [D]f^{-1}$, showing that $[D]f^{-1}$ is closed in $\underline{A}$.

Ad (ii): Let f be closed, and $\underline{b} \in ([C]f)^{K_\phi} \cap \operatorname{dom} \phi^{\underline{B}}$. Then, since f as well as C are closed, there exists $\underline{a} \in \operatorname{dom} \phi^{\underline{C}}$ such that $\underline{a}f = \underline{b}$ and $\underline{a}\phi^{\underline{A}} \in C$. Thus $(\underline{a}\phi^{\underline{A}})f = (\underline{a}f)\phi^{\underline{B}} = \underline{b}\phi^{\underline{B}} \in [C]f$. Hence $[C]f$ is closed in $\underline{B}$.
-- In order to show that arbitrary homomorphisms do not preserve closed subsets, we present Figure 3.1 with partial algebras of type (1). ■■■

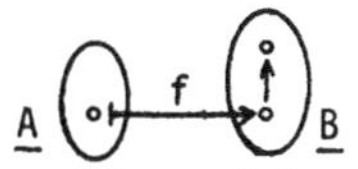

Figure 3.1.

3.1.10 COROLLARY. Let $f:\underline{A} \to \underline{B}$ be any homomorphism and $\underline{C}$ a total subalgebra of $\underline{A}$. Then $[C]f$ is the carrier set of a total subalgebra of $\underline{B}$.

PROOF. Examine the proof of 3.1.9.(ii) for the case when $\underline{C}$ is total. ■■■

3.1.11 PROPOSITION. Let B be a closed subset of $\underline{A}$, and C any subset of A.

(i) Then $C \in \mathbf{Csubs}(\underline{B})$ iff $C \subseteq B$ and $C \in \mathbf{Csubs}(\underline{A})$, in particular the concepts of "closed subset" and "subalgebra" are transitive.

(ii) If $f:\underline{A} \to \underline{D}$ is a closed homomorphism, then $f|B:\underline{B} \to \underline{D}$ is closed, too, where $\underline{B}$ is the subalgebra of $\underline{A}$ with carrier set B.

(iii) If $R \in Cong_C(\underline{A})$, then $R_B := R \cap (B \times B) \in Cong_C(\underline{B})$.

PROOF. All the three statements easily follow from the definitions of the related concepts. ■■■

3.2 Comparison of substructures

It is sometimes necessary to compare partial algebraic structures having the same carrier set. For that purpose we define:

3.2.1 DEFINITION. Let A be any set, $O' := (\phi')_{\phi \in \Omega}$ and $O'' := (\phi'')_{\phi \in \Omega}$ partial algebraic ($\underline{S}$-) structures of type τ on A. We say that O' (or (A,O')) is <u>weaker</u> or <u>poorer</u> than O" (or (A,O")), and that O" is <u>richer</u> than O', iff $id_A:(A,O') \to (A,O'')$ is an ($\underline{S}$-) homomorphism.

3.2.2 REMARKS. (i): Let $\underline{A} \in$ **Alg**$(\underline{S})$ be any partial algebra, $B \subseteq A$ any subset. Then the relative subalgebra $\underline{B}$ of $\underline{A}$ with carrier set B carries the richest ($\underline{S}$-) structure O' on B such that $id_B:(B,O') \to \underline{A}$ is a homomorphism; and it has the property that for every homomorphism $f:\underline{C} \to \underline{A}$, which satisfies $[C]f \subseteq B$, the mapping $f:C \to B$ is also a homomorphism $f:\underline{C} \to \underline{B}$ (cf. 2.7.3.(ii), moreover, recall, that a mapping consists of a domain, graph and codomain, and that in " $f:C \to B$ " f usually represents the graph, while at other occasions "f" stands for the whole triple). The relative substructure of $\underline{A}$ on B is the richest weak relative substructure of $\underline{A}$ on B, the discrete substructure on B is the weakest or poorest one. ■

(ii): Let $O^i := (\phi^i)_{\phi \in \Omega}$ be partial algebraic structures on a set A ($i \in I$). Then $O^* := \cap^*\{O^i | i \in I\} := (\cap\{\phi^i | i \in I\} | \phi \in \Omega)$ (see 23.1.2) is a partial algebraic structure on A defined by $\quad graph(\phi^*) := \cap\{graph(\phi^i) | i \in I\}$, $(\phi \in \Omega)$.
O^* is the richest structure on A such that, for each $i \in I$, $id_A:(A,O^*) \to (A,O^i)$ is a homomorphism; it is called the <u>intersection</u> or <u>infimum</u> of the given family of partial algebraic structures on A.
If, for each $\phi \in \Omega$, $graph(\phi^\circ) := \cup\{graph(\phi^i) | i \in I\}$ is still the graph of a partial operation, then $O^\circ := (\phi^\circ)_{\phi \in \Omega} = \cup^*(O^i | i \in I)$ (see 23.1.2) is the poorest partial algebraic structure on A such that, for each $i \in I$, $id_A:(A,O^i) \to (A,O^\circ)$ is a homomorphism. It is called -- whenever it exists -- the <u>supremum</u> of the given family of partial algebraic structures on A. -- Therefore the partial algebraic structures on any set A form a conditionally complete lower semilattice (see 24.1). ■

(iii): Let $\underline{B}_i := (B_i, (\phi^{\underline{B}_i})_{\phi \in \Omega})$ ($i \in I$) be any family of (weak) relative subalgebras of some partial algebra $\underline{A}$. Then
$\underline{B} := \cap^*(\underline{B}_i | i \in I) := ((\cap\{B_i | i \in I\}, (\phi^*)_{\phi \in \Omega}), \nu_B)$ with $graph(\phi^*) := \cap\{graph(\phi^{\underline{B}_i}) | i \in I\}$
is the largest and richest weak relative subalgebra $\underline{C}$ of $\underline{A}$ such that for each $i \in I$ $id_C:\underline{C} \to \underline{B}_i$ is a homomorphism. We call $\underline{B}$ the <u>intersection</u> of the (weak) relative subalgebras $\underline{B}_i$ ($i \in I$). If all $\underline{B}_i$ are relative subalgebras of $\underline{A}$, then so is $\underline{B}$. The sort homomorphism of $\underline{B}$ is the restriction to B of the given sort homomorphism for $\underline{A}$. ■■■

Because of its importance we formulate separately

3.2.3 PROPOSITION and NOTATION. Let $\underline{A}$ be any partial algebra, and let

$(\underline{B}_i \mid i \in I) \in (\mathbf{Suba}(\underline{A}))^I$ be any family of subalgebras of $\underline{A}$. Then

$$\underline{B} := \bigcap{}^*\{\underline{B}_i \mid i \in I\} \in \mathbf{Suba}(\underline{A}) \ .$$

In particular the intersection of an arbitrary family of closed subsets of $\underline{A}$ is again a closed subset of $\underline{A}$ (and if $I = \emptyset$, the empty set, then $B := A$).
Hence $\mathbf{Csubs}(\underline{A})$ is a closure system on A -- which does not depend on v_A --, and we denote by $\mathbf{C}_{\underline{A}}$ the corresponding closure operator (cf. the Appendix, section 24).

PROOF. Let $\underline{b} \in B^{K_\phi} \cap \operatorname{dom} \phi^{\underline{A}}$. Then $\underline{b} \in B_i^{K_\phi} \cap \operatorname{dom} \phi^{\underline{B}_i}$ for each $i \in I$, and $\underline{b}\phi^{\underline{A}} \in \bigcap\{B_i \mid i \in I\} = B$, i.e. B is closed. ■■■

3.3 The global process of generation

3.3.1. The statement of Proposition 3.2.3 and the results listed in the Appendix 24 now tell us that for every subset $M \subseteq A$ of any partial algebra $\underline{A}$ its closure with respect to C_A is given by $C_A\, M = \bigcap\{B \mid B \in \mathbf{Csubs}(\underline{A}),\ M \subseteq B\}$.
We call $C_A\, M$ the closed subset of A generated by M, and $\underline{C}_A\, M$ designates the subalgebra of $\underline{A}$ generated by M.
If $C_A\, M = A$, then M is called a generating subset or a set of generators of the partial ($\underline{S}$-) algebra $\underline{A}$, and $\underline{A}$ is said to be generated by M.
It should be observed that this general **global process of generation** does not depend on the fact that we might consider a partial $\underline{S}$-algebra; only the partial algebraic structure of $\underline{A}$ is responsible for the size and structure of $\underline{C}_A\, M$.

3.3.2 PROPOSITION. Let $\underline{A}$ be any partial algebra of similarity type τ with cardinal dimension $\mathbf{d}$, and let $\mathbf{c} := \max\{2, \mathbf{d}\}$. Then:

(i) $\mathbf{Csubs}(\underline{A})$ is a $\mathbf{c}$-inductive closure system on A.

(ii) $(\mathbf{Csubs}(\underline{A}), \subseteq)$ is a $\mathbf{c}$-compactly generated (and hence a $\mathbf{c}$-algebraic) lattice. Its $\mathbf{c}$-compact elements are exactly those closed subsets of $\underline{A}$ which have a generating subset of cardinality less than $\mathbf{c}$.
In particular, one has for every subset $M \subseteq A$:
(*) $$\underline{C}_A\, M = \bigcup\{\underline{C}_A\, N \mid N \subseteq M,\ \#N < \mathbf{c}\} \ .$$

PROOF. (ii) is an immediate consequence of (i) and the general results on closure systems indicated in the Appendix, section 24. In order to prove (i) let $\mathbf{B} \subseteq \mathbf{Csubs}(\underline{A})$ be a $\mathbf{c}$-directed system of closed subsets of $\underline{A}$, and let $B := \bigcup\mathbf{B}$. We have to show that B is closed in $\underline{A}$: Therefore let $\phi \in \Omega$ and $\underline{b} = (b_k \mid k \in K_\phi) \in B^{K_\phi} \cap \operatorname{dom} \phi^{\underline{A}}$. Since $B = \bigcup\mathbf{B}$ there exists for each $k \in K_\phi$ a set $B_k \in \mathbf{B}$ such that $b_k \in B_k$ (notice that $\mathbf{B} \neq \emptyset$, since it is $\mathbf{c}$-directed). $\mathbf{c}$-directedness of $\mathbf{B}$ now implies the existence of a set $B' \in \mathbf{B}$ such that $B_k \subseteq B'$ for each $k \in K_\phi$ (observe $\#K_\phi < \mathbf{c}$); since B' is closed, $\underline{b}\phi^{\underline{A}} \in B' \subseteq B$, showing that B is closed. ■■■

3.4 "Local" generation (generation "from below")

3.4.1. So far generation of a closed subset has been a "global process", i.e. we had to know the whole closure system of all closed subsets (or because of $\mathbf{c}$-inductivity -- $\mathbf{c} := \max(2, \#\omega_0)$ -- at least the $\mathbf{c}$-compact closed sets) in order to determine whether a given subset of a partial algebra $\underline{A}$ is closed or not, and to compute its closure in the latter case (cf. 3.3.2.(*)). In the case of partial algebras the decision about closedness or the construction of the closure can be accomplished by using the fundamental partial operations. And this can be done by an iterative "local process" without any knowledge about the closure system at all. This is achieved in analogy to the fact that the linear subspace generated by some set M in a vector space $\underline{V}$ over some field $\underline{K}$ is the set of all finite $\underline{K}$-linear combinations of elements of M. A similar process of "generation from below" is well known for semigroups, groups and other algebraic structures. One generalization of this procedure is described in the following definition. The constructions under consideration only use the partial algebraic structure and nothing in connection with a possibly existing sort homomorphism.

3.4.2 DEFINITION. Let $\underline{A}$ be any partial algebra of some similarity type $\tau = (K_\phi)_{\phi\in\Omega}$, and let $M \subseteq A$ be any subset. We define operators $\mathbf{D}_{\underline{A}}$ and $\mathbf{B}_{\underline{A}}$:

(i) $\mathbf{D}M := \mathbf{D}_{\underline{A}}M := M \cup \bigcup\{[M^{K_\phi}]\phi^{\underline{A}} \mid \phi \in \Omega\}$, and $\mathbf{B}M := \mathbf{B}_{\underline{A}}M := \mathbf{D}_{\underline{A}}M \setminus M$.

(ii) Moreover we define recursively for any ordinal α and any limit ordinal $\lambda \neq 0$:

$\mathbf{D}^0M := \mathbf{D}_{\underline{A}}{}^0M := M$; $\mathbf{B}^0M := \mathbf{B}_{\underline{A}}{}^0M := M$;

$\mathbf{D}^{\alpha+1}M := \mathbf{D}_{\underline{A}}{}^{\alpha+1}M := \mathbf{D}_{\underline{A}}(\mathbf{D}_{\underline{A}}{}^{\alpha}M)$; $\mathbf{B}^{\alpha+1}M := \mathbf{B}_{\underline{A}}{}^{\alpha+1}M := \mathbf{D}_{\underline{A}}{}^{\alpha+1}M \setminus \mathbf{D}_{\underline{A}}{}^{\alpha}M$;

$\mathbf{D}^{\lambda}M := \mathbf{D}_{\underline{A}}{}^{\lambda}M := \mathbf{D}_{\underline{A}}(\bigcup\{\mathbf{D}_{\underline{A}}{}^{\beta}M \mid \beta < \lambda\})$; $\mathbf{B}^{\lambda}M := \mathbf{B}_{\underline{A}}{}^{\lambda}M := \mathbf{D}_{\underline{A}}{}^{\lambda}M \setminus \bigcup\{\mathbf{D}_{\underline{A}}{}^{\beta}M \mid \beta < \lambda\} = \mathbf{D}_{\underline{A}}{}^{\lambda}M \setminus \bigcup\{\mathbf{B}_{\underline{A}}{}^{\beta}M \mid \beta < \lambda\}$.

In analogy to topology $\mathbf{B}_{\underline{A}}$ is called the <u>Baire-</u> or <u>Borel-operator</u> on $\underline{A}$, and $\mathbf{B}_{\underline{A}}{}^{\alpha}M$ is called the α-th <u>Baire-</u> or <u>Borel-class</u> of M. $\mathbf{D}_{\underline{A}}$ is called the <u>step-operator</u> for (or: on) $\underline{A}$. It is obvious that, for $\alpha \neq \beta$, $\mathbf{B}_{\underline{A}}{}^{\alpha}M \cap \mathbf{B}_{\underline{A}}{}^{\beta}M = \emptyset$; therefore we say for $a \in \mathbf{B}_{\underline{A}}{}^{\alpha}M$ that a <u>has step-number</u> or <u>rank</u> α <u>with respect to</u> M, in symbols: $\sigma_{\underline{A}}(a,M) := \alpha$.
In contrast to the "one-step" closure operator $\mathbf{C}_{\underline{A}}$ (i.e. $\mathbf{C}_{\underline{A}}\mathbf{C}_{\underline{A}}M = \mathbf{C}_{\underline{A}}M$), the operator $\mathbf{D}_{\underline{A}}$ is called a <u>many-step closure operator</u> on $\underline{A}$; the reason for this will become obvious from the following results.

3.4.3 LEMMA. Let $\underline{A}$ be any partial algebra, $M \subseteq A$. Then $\mathbf{C}_{\underline{A}}M = M$ iff $\mathbf{D}_{\underline{A}}M = M$; i.e. M is closed with respect to the operator $\mathbf{C}_{\underline{A}}$ iff M is closed with respect to the operator $\mathbf{D}_{\underline{A}}$. ■■■

This observation shows that $\mathbf{D}_{\underline{A}}$ might be the correct candidate for "local generation".

3.4.4 THEOREM. For every partial algebra $\underline{A}$ of similarity type τ, which has the ordinal dimension δ (cf. 1.3.1), and for every set $M \subseteq A$ one has:

(i) $\mathbf{C}_{\underline{A}}M = \mathbf{D}_{\underline{A}}{}^{\delta}M$;

(ii) If $\delta > 1$ then we already have $\mathbf{C}_{\underline{A}}M = \bigcup\{\mathbf{D}_{\underline{A}}{}^{\alpha}M \mid \alpha < \delta\}$.

3.4.5. We shall see below that Theorem 3.4.4 yields the optimal general result, but it should also be noticed that in concrete cases the iteration process may stop much earlier; e.g. if $\underline{V}$ is a linear space of finite dimension $m \neq 0$ over the field of real numbers, and if M is a basis of $\underline{V}$, then one has for $\alpha := \lceil \log_2 m \rceil + 1$ that $V = \mathbf{D}_{\underline{V}}{}^{\alpha}M$ ($\lceil r \rceil$ is here the least integer greater or equal to r); and this bound is sharp; therefore we cannot get below 3.4.4.(ii), if M is infinite.

PROOF of Theorem 3.4.4. If $\delta = 0$, then $\Omega = \emptyset$; thus $\mathbf{C}_{\underline{A}}M = M = \mathbf{D}_{\underline{A}}{}^0M$. If $\delta = 1$, then $C_{\underline{A}}M = M \cup C_{\underline{A}}\emptyset = D_{\underline{A}}{}^1M$, and obviously $C_{\underline{A}}M \neq M = D_{\underline{A}}{}^0M$, if $C_{\underline{A}}\emptyset \not\subseteq M$. ∎

Now assume $\delta > 1$. Because of Lemma 3.4.3 it suffices to show that $D_{\underline{A}}(\bigcup\{D_{\underline{A}}{}^{\alpha}M|\alpha < \delta\}) = \bigcup\{D_{\underline{A}}{}^{\alpha}M|\alpha < \delta\}$ (notice that just by the definition of $D_{\underline{A}}$ one has $N \subseteq D_{\underline{A}}N \subseteq C_{\underline{A}}N$ for every subset N of A, and inductively: $D_{\underline{A}}{}^{\alpha}N \subseteq C_{\underline{A}}N$ for every ordinal α and for every subset N of A). Therefore assume $\underline{b} := (b_k|k \in K_\phi) \in (\bigcup\{D_{\underline{A}}{}^{\alpha}M|\alpha < \delta\})^{K_\phi} \cap \operatorname{dom} \phi^{\underline{A}}$ for some $\phi \in \Omega$. Since δ is regular and $\#K_\phi < \#\delta$ there exists an $\alpha < \delta$ such that $\{b_k|k \in K_\phi\} \subseteq \bigcup\{D_{\underline{A}}{}^{\beta}M|\beta < \alpha\}$, and therefore $\underline{b}\phi^{\underline{A}} \in D_{\underline{A}}{}^{\alpha}M \subseteq \bigcup\{D_{\underline{A}}{}^{\beta}M|\beta < \delta\}$. This argument shows that $C_{\underline{A}}M \subseteq \bigcup\{D_{\underline{A}}{}^{\beta}M|\beta < \delta\}$. ∎∎∎

3.4.6. We would like to observe that obviously for each subset M of a partial algebra $\underline{A}$ one has $C_{\underline{A}}M = \bigcup\{B_{\underline{A}}{}^{\alpha}M|\alpha \leq \delta\}$, and that this is a disjoint union. This is the justification of the rank or step-number defined in 3.4.2. We illustrate the situation concerning the Baire-classes in Figure 3.2. In particular one has: If an element a of $C_{\underline{A}}M$ has step-number α, then there are $\phi \in \Omega$ and $\underline{a} \in \operatorname{dom} \phi^{\underline{A}}$ such that each component of $\underline{a}$ belongs to $C_{\underline{A}}M$, has a step-number less than α, and the supremum of all step-numbers of components of $\underline{a}$ is either α-1, if α is a successor ordinal, or it is α, if α is a limit ordinal. Moreover it should be noticed that $B_{\underline{A}}{}^{\alpha}M = \emptyset$ iff $\bigcup\{B_{\underline{A}}{}^{\beta}M|\beta < \alpha\}$ is a closed subset of $\underline{A}$.

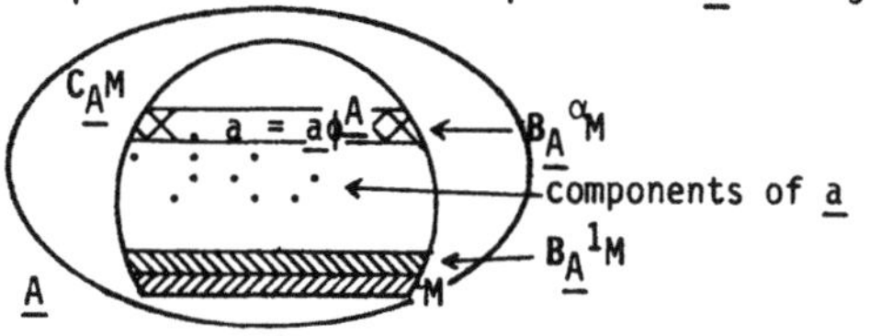

Figure 3.2.

3.5 The principle of algebraic induction

3.5.1 ALGEBRAIC INDUCTION. The process of generation from below along Baire-classes as described in the previous section may suggest to prove theorems for partial algebras by transfinite induction on step numbers, when a generating subset is given. In fact this is often done. But in most such cases it would be more elegant to apply the method of algebraic induction (in mathematical logic this method usually appears in the form of "induction on the structure of formulas"):

Let $\underline{A}$ be any partial algebra of some similarity type $\tau = (K_\phi)_{\phi \in \Omega}$, and let $M,N \subseteq A$ be any subsets (where N is usually defined to be the set of all elements of $\underline{A}$ having some given property -- usually described by a first order formula (cf. Chapter II)):

In order to prove " $C_{\underline{A}}M \subseteq N$ (or that a given property holds for all elements of $C_{\underline{A}}M$) it is sufficient to prove:

(i) $x \in N$ for every $x \in M$ (i.e. $M \subseteq N$) -- basis of the induction.

(ii) If for some $\phi \in \Omega$ and some $\underline{b} \in \operatorname{dom} \phi^{\underline{A}} \cap (C_{\underline{A}}M)^{K_\phi}$ one has $\underline{b} \in N^{K_\phi}$ (induction hypothesis), then $\underline{b}\phi^{\underline{A}} \in N$ (induction step).

Notice that one version of (ii) says that N is itself a closed subset of $\underline{A}$ containing M (cf. (i)); therefore it has to contain $C_{\underline{A}}M$, since $C_{\underline{A}}$ is a closure operator.

-- If $\tau = (1)$, $\underline{A} = (\mathbf{N};SUCC)$, $M = \{0\}$, where $nSUCC := n+1$ is the successor function for natural numbers, then the above principle of algebraic induction is nothing else than the principle of complete induction for the set of natural numbers, which thus turns out to be a special case of algebraic induction.

Applications of the principle of algebraic induction are -- as in the examples below -- sometimes not so evident; better examples can be found in section 5.

3.5.2 PROPOSITION. For each subset M of a partial algebra $\underline{A}$ one has

(*) $$C_{\underline{A}}M = M \cup \bigcup\{[(C_{\underline{A}}M)^{K_\phi}]\phi^{\underline{A}} \mid \phi \in \Omega\} ,$$

i.e. each element of the closure of a set M is either an element of M itself or it is the value under an application of a fundamental partial operation to an argument sequence within its closure.

PROOF. It is obvious that the right-hand side of (*) is contained in the left-hand side, and since it contains M and is easily seen to be closed, the other inclusion also holds. ■■■

3.5.3 PROPOSITION. Let $f,g:\underline{A} \to \underline{B}$ be two homomorphisms between similar partial algebras; then the "equalizer" $\{a \in A \mid af = ag\}$ is a closed subset of $\underline{A}$. ■■■

3.5.4 COROLLARY 1. Let $f,g:\underline{A} \to \underline{B}$ be two homomorphisms between similar partial algebras which coincide on some generating subset M of $\underline{A}$: $f|M = g|M$; then f and g are identical: $f = g$. ■■■

3.5.5 COROLLARY 2. Let $\underline{A}$ and $\underline{B}$ be similar partial algebras, M a generating subset of $\underline{A}$, $f_0:M \to B$ any mapping. Then there is at most one homomorphism $f:\underline{A} \to \underline{B}$ extending f_0, i.e. satisfying $f|M = f_0$. -- In particular, every homomorphism is uniquely determined by its values on (i.e. by its restriction to) a generating subset of its domain. ■■■

3.5.6 COROLLARY 3. Let $f:\underline{A} \to \underline{A}$ be any endomorphism; then the set $\{x \in A \mid xf = x\}$ of all fixed points of f is a closed subset of $\underline{A}$. ■■■

3.6 Some further results concerning generation, epimorphisms

We are now able to characterize epimorphisms (cf. 2.10.2):

3.6.1 PROPOSITION. Let $f:\underline{A} \to \underline{B}$ be any homomorphism; then the following statements are equivalent:

(i) f is an epimorphism.

(ii) $C_{\underline{B}}[A]f = B$, i.e. the image of f is "dense" in $\underline{B}$.

PROOF. "(ii) ⇒ (i)": Let $g,h:\underline{B} \to \underline{C}$ such that $fg = fh$. Then g and h coincide on the generating set [A]f of $\underline{B}$, and therefore they are equal according to 3.5.4. ■

"(i) ⇒ (ii)": Assume that [A]f does not generate $\underline{B}$; then we have to show that f is not an epimorphism. We achieve this by constructing a partial algebra $\underline{C}$ and two different homomorphisms g and h from $\underline{B}$ into $\underline{C}$ such that $fg = fh$:

Let $D := C_{\underline{B}}[A]f$, $E := B \setminus D$ and E' a copy of E disjoint from B, and $i:E' \to E$ a bijection, which we extend by the identity on D to a bijection $i':D \cup E' \to D \cup E$. Then we define:

$C := B \cup E'$;

$$cv_C := \begin{cases} cv_B & \text{, if } c \in B , \\ civ_B & \text{, if } c \in E' . \end{cases}$$

For each $\phi \in \Omega$ we set

$\mathrm{dom}\,\phi^{\underline{C}} := \{\underline{c} \in C^{K\phi} \mid \underline{c} \in \mathrm{dom}\,\phi^{\underline{B}}$ or $\underline{c}i' \in \mathrm{dom}\,\phi^{\underline{B}}\}$,

and if $\underline{c} \in \mathrm{dom}\,\phi^{\underline{C}}$, then either $\underline{c}\phi^{\underline{C}} := \underline{c}\phi^{\underline{B}}$ or $\underline{c}\phi^{\underline{C}} := ((\underline{c}i')\phi^{\underline{B}})i'^{-1}$.
Hence $h := i'^{-1}$ becomes a closed and injective homomorphism from $\underline{B}$ into $\underline{C}$; and setting $g := id_B$, we get indeed $fg = fh$, but $g \neq h$. ■■■

The following observation is closely connected to Proposition 3.1.9.

3.6.2 PROPOSITION. Let $f: \underline{A} \to \underline{B}$ be any homomorphism between similar partial algebras, and let $M \subseteq A$ be any subset. Then:

(i) $[C_{\underline{A}}M]f \subseteq C_{\underline{B}}[M]f$, and in general equality is false.

(ii) If f is a closed homomorphism (i.e. in particular, if $\underline{A}$ is total), then $[C_{\underline{A}}M]f = C_{\underline{B}}[M]f$.

(iii) If f is an epimorphism, and $g: \underline{B} \to \underline{C}$ is any homomorphism, then $[B]g \subseteq C_{\underline{C}}([A](fg))$, and if g is closed, then equality holds.

PROOF. Ad (i): By Proposition 3.1.9.(i) $[C_{\underline{B}}[M]f]f^{-1}$ is a closed subset of $\underline{A}$ which obviously contains M and therefore also $C_{\underline{A}}M$. Applying f again yields (i). ■
That usually equality does not hold in (i) is easily seen from the example in the proof of Proposition 3.1.9. ■
Ad (ii): If f is closed, then 3.1.9.(ii) implies that $[C_{\underline{A}}M]f$ is a closed subset of $\underline{B}$ which contains [M]f and therefore $C_{\underline{B}}[M]f$. The statement now follows from part (i). ■
Ad (iii): 3.6.1 and (i) imply $[B]g = [C_{\underline{B}}([A]f)]g \subseteq C_{\underline{C}}([[A]f]g) = C_{\underline{C}}([A](fg))$, and if g is closed, then (ii) implies equality. ■■■

3.6.3 REMARK. Closed homomorphisms have received their name mainly because they preserve closed subsets. But it should be noticed that contrary to closed continuous mappings in topology there are non-closed homomorphisms between partial algebras which also preserve closed subsets; thus there is no full analogy: Consider the type (1,1) for operation symbols ϕ and ϕ', and let $A = B = \{0,1,2\}$, $\mathrm{graph}(\phi^{\underline{A}}) := \mathrm{graph}(\phi^{\underline{B}}) :=$
$:= \{((0),1),((1),2),((2),0)\}$, $\mathrm{graph}(\phi'^{\underline{A}}) := \emptyset$ and $\mathrm{graph}(\phi'^{\underline{B}}) := \{((1),0)\}$ as shown in Figure 3.3 (f is the identity on $\{0,1,2\}$).

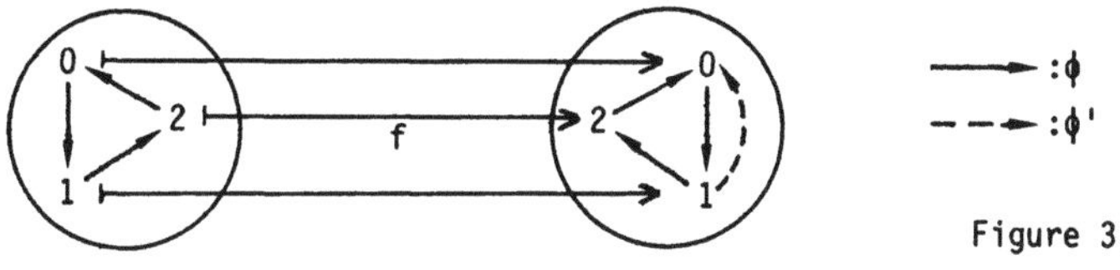

Figure 3.3.

But we do not know any syntactical description of homomorphisms preserving closed subsets, and what we call -- according to Jürgen Schmidt -- "closed homomorphisms" seems to be the smoothest approximation. -- The following concept of, say, "semiclosed homomorphisms" would also do in this connection, as can easily be seen:

For every $\phi \in \Omega$, $\underline{a} \in A^{K\phi}$ satisfying $\underline{a}f \in \mathrm{dom}\,\phi^{\underline{B}}$ there exists $\underline{a}' \in (C_{\underline{A}}([K_\phi]\underline{a}))^{K\phi} \cap \mathrm{dom}\,\phi^{\underline{A}}$ such that $\underline{a}f = \underline{a}'f$.
-- It should also be observed that especially the case of partial $\underline{S}$-algebras shows us that closed homomorphisms are exactly those homomorphisms modeling "totalness", and in

particular all homomorphisms between total $\underline{S}$-algebras are closed.

3.6.4 LEMMA. Let $\underline{A}$ and $\underline{B}$ be possibly different similar partial algebras on the same carrier set A, and let $\mathrm{graph}(\phi^{\underline{A}}) \subseteq \mathrm{graph}(\phi^{\underline{B}})$ for all $\phi \in \Omega$, i.e. $\underline{A}$ is a weak subalgebra of $\underline{B}$. Then:

(i) $\mathbf{Csubs}(\underline{B}) \subseteq \mathbf{Csubs}(\underline{A})$.

(ii) For each $M \subseteq A$ one has $\underline{C}_{\underline{A}}M \subseteq \underline{C}_{\underline{B}}M$.

PROOF. Each subset of A closed with respect to the structure of $\underline{B}$ is obviously also closed with respect to the structure of $\underline{A}$; this yields (i). Now, if $M \subseteq A$, then

$$\underline{C}_{\underline{A}}M = \cap\{C \in \mathbf{Csubs}(\underline{A}) \mid M \subseteq C\} \subseteq \cap\{C \in \mathbf{Csubs}(\underline{B}) \mid M \subseteq C\} = \underline{C}_{\underline{B}}M \text{ .} \quad ■■■$$

3.6.5 COROLLARY 1. If $\underline{A}$ is any weak subalgebra of $\underline{B}$, then one has for any $M \subseteq A$ that $\underline{C}_{\underline{A}}M \subseteq \underline{C}_{\underline{B}}M$.

PROOF. For $\underline{C} := (B, (\phi^{\underline{A}})_{\phi\in\Omega})$ the assumptions of 3.6.4 are satisfied for $\underline{C}$ and $\underline{B}$, and moreover one has $\underline{C}_{\underline{A}}M = \underline{C}_{\underline{C}}M$ for each $M \subseteq A$, hence the statement of 3.6.5. ■■■

3.6.6 COROLLARY 2. Let $\underline{A}$ be a weak subalgebra of $\underline{B}$ and $\underline{B}$ a weak subalgebra of $\underline{C}$, and assume $\underline{C}_{\underline{B}}A = B$, $\underline{C}_{\underline{C}}B = C$. Then A generates $\underline{C}$, i.e. $\underline{C}_{\underline{C}}A = C$.

PROOF. $\underline{C}_{\underline{C}}A = \underline{C}_{\underline{C}}\underline{C}_{\underline{C}}A \supseteq \underline{C}_{\underline{C}}\underline{C}_{\underline{B}}A = \underline{C}_{\underline{C}}B = C$. ■■■

3.6.7 COROLLARY 3. Let $f:\underline{A} \to \underline{B}$ be any **epimorphism** and M a generating subset of $\underline{A}$, then $[M]f$ is a generating subset of $\underline{B}$. ■■■

3.7 The analogue of E.Noether's First Isomomorphism Theorem

3.7.1 DEFINITION. Let $\underline{A}$ be any partial algebra, $R \in \mathrm{Cong}(\underline{A})$ any congruence relation on $\underline{A}$, and B any subset of A. Then the R-saturation $[B]R$ of B is the subset

$[B]R := \{a \in A \mid (a,b) \in R \text{ for some } b \in B\}$.

(Observe that the corresponding sort mapping $v_{[B]R}:[B]R \to S$ is the restriction to $[B]R$ of the sort homomorphism $v_{\underline{A}}:\underline{A} \to \underline{S}$.) By $[B]\underline{R}$ we denote the relative subalgebra of $\underline{A}$ with carrier set $[B]R$. Moreover, we let R_B designate the restriction of congruence R to the relative subalgebra $\underline{B}$ of $\underline{A}$, i.e. $R_B := R \cap (B \times B)$. Obviously $R_B \in \mathrm{Cong}(\underline{B})$.

3.7.2 PROPOSITION. Let B be a closed subset of a partial algebra $\underline{A}$ and R a closed congruence of $\underline{A}$. Then $[B]R$ is also a closed subset of $\underline{A}$.

PROOF. Let $\mathrm{nat}_R:\underline{A} \to \underline{A}/R$ be the natural projection. Then obviously $[B]R = [[B]\mathrm{nat}_R](\mathrm{nat}_R)^{-1}$. Since nat_R is closed (cf. 2.6.3), $[B]\mathrm{nat}_R$ is closed in $\underline{A}/R$ (cf. 3.1.9.(ii)); therefore $[B]R$ is closed in $\underline{A}$ according to 3.1.9.(i). ■■■

3.7.3 THEOREM. Let $\underline{A}$ be any partial algebra, $\underline{B}$ a relative subalgebra of $\underline{A}$, and $R \in \mathrm{Cong}(\underline{A})$ a congruence on $\underline{A}$. Then there exists a bijective homomorphism

$$f:\underline{B}/R_B \to [B]\underline{R}/R_{[B]R} \quad \text{defined by} \quad f:[b]R_B \mapsto [b]R \text{ for every } b \in B$$

(cf. Figure 3.4 and notice that $[b]R$ designates the congruence class of b with respect to R).

If B is a closed subset of $\underline{A}$, and if R is a closed congruence on $\underline{A}$, then f is an isomorphism.

PROOF. Consider the diagram in Figure 3.4. There exists an induced homomorphism $f:\underline{B}/R_B \to [B]\underline{R}/R_{[B]R}$ according to the Diagram Completion Lemma 2.7.1. f is also injective, since obviously $\ker nat_{R_B} = \ker(id_B \cdot nat_{R_{[B]R}})$. Since $id_B \cdot nat_{R_{[B]R}}$ is

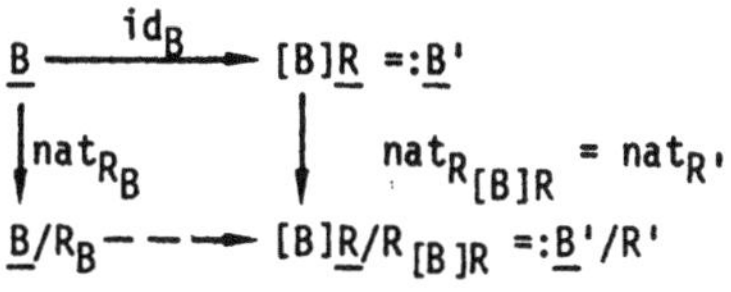

Figure 3.4.

also surjective, f is even bijective. Therefore there only remains to show that f is full, if B and R are closed (cf. 2.4.6.(iii)): For brevity set $B' := [B]R$, and $R' := R_{[B]R} = R_{B'}$, and assume $([b_k]R' \mid k \in K_\phi) \in \operatorname{dom} \phi^{\underline{B}'/R'}$, where each $b_k \in B$. Then there exists a sequence $\underline{a} := (a_k \mid k \in K_\phi) \in \operatorname{dom} \phi^{\underline{B}'}$ such that $\underline{a} \cdot nat_{R'} = ([b_k]R' \mid k \in K_\phi)$. By 3.7.2 B' is closed in $\underline{A}$, and by 3.1.11.(iii) R' is a closed congruence on $\underline{B}'$. Thus $\underline{b} := (b_k \mid k \in K_\phi) \in \operatorname{dom} \phi^{\underline{B}'}$. Since $id_B : \underline{B} \to \underline{B}'$ is closed, we get $\underline{b} \in \operatorname{dom} \phi^{\underline{B}}$. Hence $([b_k]R_B \mid k \in K_\phi) \in \operatorname{dom} \phi^{\underline{B}/R_B}$, showing that f is full (closed). ∎

3.7.4 COROLLARY. Let $\underline{B}$ be any subalgebra and R a closed congruence of a partial algebra $\underline{A}$ such that $[B]R = A$. Then $\underline{A}/R$ is isomorphic to $\underline{B}/R_B$: $\underline{A}/R \cong \underline{B}/R_B$. ∎

3.7.5 REMARK. We leave it as an exercise to show that both assumptions, namely the closedness of B as well as the closedness of R, are necessary assumptions in 3.7.3 in order to get an isomorphism; i.e. if we omit only one of them, then we get counterexamples. The partial algebra in Figure 3.5 may be a good candidate.

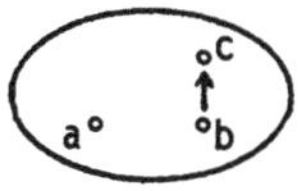

Figure 3.5.

4 Direct and reduced products, coproducts, direct limits and related constructions

In this section we discuss some basic constructions of new partial algebras from families of given ones. So far we started with one partial algebra and ended with a "smaller" one, whereas we shall now get large structures from a family of possibly relatively small ones. Products, coproducts and direct limits also play an important rôle in category theory, while reduced products will be needed in connection with model theoretic investigations.

We recall that we shall use the conventions introduced at the end of section 2 concerning partial $\underline{S}$-algebras encoding heterogeneous partial algebras.

4.1 (Direct) Products (homogeneous case)

4.1.1 DEFINITION. Let $\underline{\mathbf{C}}$ be any category and $(A_i \mid i \in I) \in (Ob(\underline{\mathbf{C}}))^I$ -- for some set I -- be any family of $\underline{\mathbf{C}}$-objects.

A <u>product</u> of the family $(A_i \mid i \in I)$ is a pair $(A, (p_i \mid i \in I))$ such that $A \in Ob(\underline{\mathbf{C}})$, and for each $i \in I$ p_i is a $\underline{\mathbf{C}}$-morphism $p_i : A \to A_i$, and whenever there is any $\underline{\mathbf{C}}$-object $B \in Ob(\underline{\mathbf{C}})$ with a "source" $(g_i : B \to A_i)_{i \in I}$ -- i.e. a family of $\underline{\mathbf{C}}$-morphisms starting from the same object -- then there exists a **unique** $\underline{C}$-morphism $g:B \to A$ -- the product morphism induced by the family $(g_i \mid i \in I)$ (see 4.1.4 below) -- satisfying $g_i = gp_i$ for each $i \in I$, i.e. making each diagram in Figure 4.1 commutative. -- If $I = \emptyset$, then the above definition means that for every $B \in Ob(\underline{C})$ there exists exactly one $\underline{C}$-morphism from B into A i.e. A is a so-called right zero or terminal object of the category $\underline{C}$ under consideration.

$(i \in I)$

Figure 4.1.

An immediate consequence of category theory is the following

4.1.2 PROPOSITION. If in a category $\underline{C}$ a product $(A,(p_i \mid i \in I))$ of some family $(A_i \mid i \in I)$ in $\underline{C}$ exists, then it is unique up to unique isomorphism.

PROOF. Let $(B,(q_i \mid i \in I))$ be another product of the given family, then we have two unique product morphisms: $p:A \to B$, induced by $(p_i \mid i \in I)$, and $q:B \to A$ induced by $(q_i \mid i \in I)$, which satisfy, for each $i \in I$, $pqp_i = pq_i = p_i = 1_A p_i$ and $qpq_i = qp_i = q_i = 1_B q_i$, and the uniqueness of the induced $\underline{C}$-morphisms implies $pq = 1_A$ and $qp = 1_B$, showing that p and q are isomorphisms, i.e. A and B are canonically isomorphic. ■

In what follows we at first restrict our considerations to the homogeneous case, and only later on we shall also get a description of products in the heterogeneous case (of partial $\underline{S}$-algebras).

4.1.3 CONSTRUCTION OF DIRECT PRODUCTS OF FAMILIES OF HOMOGENEOUS PARTIAL ALGEBRAS:
Let $(\underline{A}_i \mid i \in I) \in (\mathbf{Alg}(\tau))^I$ be any family of similar homogeneous partial algebras, where I is any set.

(a): Let $A := \{a:I \to \bigcup\{A_i \mid i \in I\} \mid (i)a \in A_i \text{ for each } i \in I\}$
be the set of all choice functions from I into the family $(A_i \mid i \in I)$ of all carrier sets, i.e. the so called cartesian (or direct) product of all carrier sets: $A =: \mathsf{X}(A_i \mid i \in I)$.

(b): For each $\phi \in \Omega$ and for every sequence $\underline{a} = (a_k \mid k \in K_\phi) \in A^{K_\phi}$ define $\underline{a} \in dom\, \phi^{\underline{A}}$ iff $(i)\underline{a} := ((i)a_k \mid k \in K_\phi) \in dom\, \phi^{\underline{A}_i}$ for every $i \in I$, and if $\underline{a} \in dom\, \phi^{\underline{A}}$, then set $\underline{a}\phi^{\underline{A}} := (((i)a_k \mid k \in K_\phi)\phi^{\underline{A}_i} \mid i \in I)$,
i.e. the partial operations $\phi^{\underline{A}}$ are defined **componentwise**.

(c): For each $i \in I$ the mapping $pr_i := pr_i^A : A \to A_i$, $a \mapsto (i)a$ $(a \in A)$ is called the i-th projection.

If $I = \emptyset$ is the empty set, then we assume $\underline{A}$ to be the one-element total algebra.

If I contains exactly one element, say, i, then we (can) identify $\underline{A}$ in a natural way with $\underline{A}_i$ (see the following Theorem).

$\underline{A}$, or more concisely $(\underline{A},(pr_i \mid i \in I))$, is called the direct product of the family $(\underline{A}_i \mid i \in I)$ of partial algebras, and we introduce the following notation:

$$\underline{A} =: \mathsf{X}(\underline{A}_i \mid i \in I),$$

and if I is finite and non-empty, say $I = \{1,2,\ldots,n\}$, then we write $\underline{A} =: \underline{A}_1 \times \underline{A}_2 \times \ldots \times \underline{A}_n$, and the elements of A are identified with the ordered sequences $(a_1,a_2,\ldots,a_n)$, where $a_i \in A_i$ for each $i \in I$.

If $\underline{A}_i = \underline{B}$ for each $i \in I$, then we write $\underline{A} =: \underline{B}^I$ and in the finite case, where

$I = \{1,\ldots,n\}$, we write $\underline{A} =: \underline{B}^n$.

(d): If $(f_i:\underline{A}_i \to \underline{B}_i \mid i \in I)$ is any family of homomorphisms, then the unique homomorphism (see below) $f:\underline{A} := \mathsf{X}(\underline{A}_i \mid i \in I) \to \mathsf{X}(\underline{B}_i \mid i \in I)$ induced by the family $(pr_i^{\underline{A}} \cdot f_i \mid i \in I)$ is denoted by $\mathsf{X}(f_i \mid i \in I)$.

4.1.4 THEOREM. Let $(\underline{A}_i \mid i \in I) \in (\mathbf{Alg}(\tau))^I$ be any family of similar partial algebras and $(\underline{A},(pr_i \mid i \in I))$ its direct product.

(i) For each $i \in I$ the i-th projection is a homomorphism $pr_i:\underline{A} \to \underline{A}_i$, and $(\phi^{\underline{A}})_{\phi\in\Omega}$ is in fact the richest structure ("initial structure". in the sense of BOURBAKI) on $\underline{A}$ such that all projections are homomorphisms.

(ii) Let $(g_i:\underline{B} \to \underline{A}_i \mid i \in I)$ be a source of homomorphisms. Then there is a unique homomorphism $g:\underline{B} \to \underline{A}$ -- defined by $bg := (bg_i \mid i \in I)$ for each $b \in B$ -- such that for each $i \in I$ one has $g_i = g \cdot pr_i$, i.e. for each $i \in I$ the diagram in Figure 4.1 -- with pr_i instead of p_i -- is commutative. We then also use the notation $g := \langle g_i \mid i \in I\rangle$.

(iii) $(\underline{A},(pr_i \mid i \in I))$ is a product in the category theoretical sense of the family $(\underline{A}_i \mid i \in I)$.

PROOF. That each projection is a homomorphism is obvious from the definition of the structure of $\underline{A}$ in 4.1.3, which also shows that the homomorphism condition is violated as soon as the structure of $\underline{A}$ is enriched at some place. Moreover, (iii) easily follows from 4.1.1, once we have shown (ii). ∎

Ad (ii): If $g:B \to A$ exists such that $g \cdot pr_i = g_i$ for each $i \in I$, then $bg = (bg_i \mid i \in I)$ for each $b \in B$ according to the definition of projections, hence there is at most one such g. And if we actually define g like that, it is a mapping having the required commutativity properties. But it is also a homomorphism; for let $\underline{b} \in \mathrm{dom}\, \phi^{\underline{B}}$, then $\underline{bg_i} \in \mathrm{dom}\, \phi^{\underline{A}_i}$ for each $i \in I$, therefore $\underline{bg} \in \mathrm{dom}\, \phi^{\underline{A}}$ by 4.1.3 and the definition of g. Moreover, $(\underline{b}\phi^{\underline{B}})g = ((\underline{bg_i})\phi^{\underline{A}_i} \mid i \in I) = (\underline{bg})\phi^{\underline{A}}$. ∎∎∎

4.1.5 COROLLARY. Let $f,g:\underline{B} \to \mathsf{X}(\underline{A}_i \mid i \in I)$ be any two homomorphisms such that $f \cdot pr_i = g \cdot pr_i$ for all $i \in I$, then $f = g$. ∎∎∎

4.1.6 REMARKS. (i): Notice that according to -- and with the notation of -- 4.1.4.(ii) g already exists for every family of **mappings**, and that it is a homomorphism just when all g_i are homomorphisms.

(ii): If $I \neq \emptyset$, then $\mathsf{X}(A_i \mid i \in I) \neq \emptyset$ iff, for each $i \in I$, $A_i \neq \emptyset$ because of the Axiom of Choice which actually is equivalent to this fact.

(iii): Notice that, for $\underline{A} := \mathsf{X}(\underline{A}_i \mid i \in I)$, $\mathrm{dom}\, \phi^{\underline{A}} = \emptyset$ iff $\mathrm{dom}\, \phi^{\underline{A}_i} = \emptyset$ for at least one $i \in I$.

(iv): Observe that in Proposition 3.1.5 we have actually enriched the structure of $\underline{A} \times \underline{A}$ by a sufficient amount of nullary constants and an additional binary operation in order to guarantee reflexivity, symmetry and transitivity of the closed subsets.

4.1.7 PROPOSITION. Let $(f_i:\underline{B} \to \underline{A}_i \mid i \in I)$ be a family of homomorphisms (I any set) and $f:\underline{B} \to \mathsf{X}(\underline{A}_i \mid i \in I)$ the induced product homomorphism. Then

$$\ker f = \bigcap(\ker f_i \mid i \in I) .$$

PROOF. One has $(b,b') \in \ker f$ iff $bf = b'f$, iff $(bf_i \mid i \in I) = (b'f_i \mid i \in I)$, iff $(b,b') \in \bigcap(\ker f_i \mid i \in I)$. ■■■

4.1.8 DEFINITION. Let $(f_i : B \to A_i \mid i \in I)$ be a source of mappings. It is said to <u>separate</u> B (to be a <u>separating family</u> of B), iff for any two different elements b and b' of B there exists an $i \in I$ such that $bf_i \neq b'f_i$.

4.1.9 PROPOSITION. Let $(f_i : \underline{B} \to \underline{A}_i \mid i \in I)$ be a source of homomorphisms. Then the following statements are equivalent:

(i) $(f_i : B \to A_i \mid i \in I)$ separates B.

(ii) The induced homomorphism $f : \underline{B} \to \mathsf{X}(\underline{A}_i \mid i \in I)$ is injective.

(iii) $\bigcap(\ker f_i \mid i \in I) = \Delta_B$. ■■■

We add some further results concerning direct products and homomorphisms, which will be of great importance in section 5.

4.1.10 PROPOSITION. Let $\underline{A},\underline{B} \in \mathbf{Alg}(\tau)$, and $f : A \to B$; then:

(i) If f is a homomorphism $(f : \underline{A} \to \underline{B})$, then graph(f) is a closed subset of $\underline{A} \times \underline{B}$.

(ii) If $\underline{B}$ is total, and if graph(f) is a closed subset of $\underline{A} \times \underline{B}$, then f is a homomorphism $f : \underline{A} \to \underline{B}$.

(iii) In general, if graph(f) is a closed subset of $\underline{A} \times \underline{B}$, then f need not be a homomorphism from $\underline{A}$ into $\underline{B}$.

PROOF. (i) is obvious; and if $\underline{B}$ is total, then the implication concerning the existence of the values of operations becomes trivial, while closedness of graph(f) guarantees the compatibility condition for homomorphisms. Finally consider Figure 4.2 in connection with (iii). ■■■

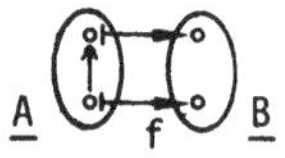

Figure 4.2.

4.1.11 LEMMA. Let $f : \underline{A} \to \underline{B}$ be a homomorphism, g any mapping out of A into B such that $\mathrm{graph}(g) \subseteq \mathrm{graph}(f)$, and let <u>dom</u> g be the relative subalgebra of $\underline{A}$ on the domain of g.

(i) $g : \underline{\mathrm{dom}}\ g \to \underline{B}$ is a homomorphism.

(ii) graph g is a closed subset of $\underline{A} \times \underline{B}$ iff dom g is a closed subset of $\underline{A}$.

PROOF. (i) is obvious. Thus assume graph(g) to be closed in $\underline{A} \times \underline{B}$, and let $\underline{a} \in \mathrm{dom}\, \phi^{\underline{A}} \cap (\mathrm{dom}\, g)^{K_\phi}$. Then $\underline{ag} = \underline{af} \in \mathrm{dom}\, \phi^{\underline{B}}$, and therefore -- graph(g) being closed -- $(\underline{a}\phi^{\underline{A}}, (\underline{ag})\phi^{\underline{B}}) \in \mathrm{graph}(g)$, i.e. $\underline{a}\phi^{\underline{A}} \in \mathrm{dom}\, g$, showing that dom g is closed in $\underline{A}$. -- The converse is proved in a similar way. ■■■

4.1.12 COROLLARY. Let $f : \underline{A} \to \underline{B}$ be a homomorphism and $M \subseteq A$ any subset of A. Then the following statements are equivalent:

(i) $C_{\underline{A}}M = A$.

(ii) $C_{\underline{A}\times\underline{B}}\mathrm{graph}(f|M) = \mathrm{graph}(f)$. ■■■

The following technical lemma will become quite useful later on (see 5.4.1 and 21.1):

4.1.13 LEMMA. Let $\underline{A},\underline{B},\underline{C},\underline{D} \in \mathbf{Alg}(\tau)$, $\underline{C}$ a weak subalgebra of $\underline{A}$ such that C generates $\underline{A}$, $\underline{D}$ a relative subalgebra of $\underline{B}$. Moreover, let $f : \underline{C} \to \underline{D}$ be a homomorphism which allows a homomorphic extension $\overline{f} : \underline{A} \to \underline{B}$, and let $\mathrm{graph}(f^{\#}) := C_{\underline{A}\times\underline{D}}\mathrm{graph}(f)$. Then one has:

(i) graph($f^{\#}$) is the graph of a homomorphism $f^{\#}:\underline{\mathrm{dom}}\ f^{\#} \to \underline{D}$ ($\underline{\mathrm{dom}}\ f^{\#}$ being the relative subalgebra of $\underline{A}$ with carrier set dom $f^{\#}$), and $\underline{\mathrm{dom}}\ f^{\#}$ is generated by C.

(ii) If $\underline{E}$ is a C-generated relative subalgebra of $\underline{A}$, and if $g:\underline{E} \to \underline{D}$ is a homomorphic extension of f (i.e. $g|C = f$), then $\mathrm{graph}(g) \subseteq \mathrm{graph}(f^{\#})$.

(iii) If $\underline{A}$ is total, then $f^{\#}$ is a closed homomorphism from $\underline{\mathrm{dom}}\ f^{\#}$ into $\underline{D}$; and if $f^{\#}$ is closed and $g:\underline{E} \to \underline{D}$ is any closed homomorphic extension of f such that $\underline{E}$ is a C-generated relative subalgebra of $\underline{A}$, then $g = f^{\#}$.

PROOF. Ad (i): Since $\overline{f}:\underline{A} \to \underline{B}$ is a homomorphism, graph($\overline{f}$) is a closed subset of $\underline{A} \times \underline{B}$ such that $\mathrm{graph}(f) \subseteq \mathrm{graph}(\overline{f})$. Therefore

$$\mathrm{graph}(f^{\#}) := C_{\underline{A}\times\underline{D}}\mathrm{graph}(f) \subseteq C_{\underline{A}\times\underline{B}}\mathrm{graph}(f) \subseteq \mathrm{graph}(\overline{f}) .$$

Hence $f^{\#}:\underline{\mathrm{dom}}\ f^{\#} \to \underline{B}$ is a homomorphism according to Lemma 4.1.11; but since $[\mathrm{dom}\ f^{\#}]f^{\#} \subseteq D$, we may consider $f^{\#}$ as a homomorphism into $\underline{D}$. Since $\mathrm{graph}(f^{\#}) = C_{\underline{A}\times\underline{D}}\mathrm{graph}(f^{\#}|C)$, Corollary 4.1.12 implies that C generates $\underline{\mathrm{dom}}\ f^{\#}$. ∎

Ad (ii): Now, let $g:\underline{E} \to \underline{D}$ as in (ii) of the Lemma, then Corollary 4.1.12 implies $\mathrm{graph}(g) = C_{\underline{E}\times\underline{D}}\mathrm{graph}(f) \subseteq C_{\underline{A}\times\underline{D}}\mathrm{graph}(f) = \mathrm{graph}(f^{\#})$. ∎

Ad (iii): Finally, let $\underline{A}$ be total, and $\underline{a} \in (\mathrm{dom}\ f^{\#})^{K_\phi}$ such that $\underline{a}f^{\#} \in \mathrm{dom}\ \phi^{\underline{D}}$. Obviously, $\underline{a} \in \mathrm{dom}\ \phi^{\underline{A}}$ -- $\underline{A}$ being total --, therefore $(\underline{a}\phi^{\underline{A}},(\underline{a}f^{\#})\phi^{\underline{D}}) \in \mathrm{graph}(f^{\#}) = C_{\underline{A}\times\underline{D}}\mathrm{graph}(f)$. Thus $\underline{a} \in \mathrm{dom}\ \phi^{\underline{\mathrm{dom}}\ f^{\#}}$, i.e. $f^{\#}$ is a closed homomorphism. -- Now, let $f^{\#}$ be closed, and let $g:\underline{E} \to \underline{D}$ be another closed homomorphic extension of f such that $C_{\underline{E}}C = E$. Then $E \subseteq \mathrm{dom}\ f^{\#}$, and $\mathrm{graph}(g) \subseteq \mathrm{graph}(f^{\#})$ because of (ii). That $\mathrm{dom}\ f^{\#} \subseteq E$ can be proved by algebraic induction: $C \subseteq E$, and if $b \in \mathrm{dom}\ f^{\#}$, $b = \underline{b}\phi^{\underline{\mathrm{dom}}\ f^{\#}}$ such that $\underline{b} \in E^{K_\phi}$, then one has $\underline{b}f^{\#} = \underline{b}g \in \mathrm{dom}\ \phi^{\underline{D}}$. Since g is closed, $\underline{b} \in \mathrm{dom}\ \phi^{\underline{E}}$, and $b = \underline{b}\phi^{\underline{E}} = \underline{b}\phi^{\underline{A}} \in E$. Thus $E = \mathrm{dom}\ f^{\#}$ and $g = f^{\#}$. ∎∎∎

4.2 Pullbacks and products of partial $\underline{S}$-algebras

4.2.1. Before we can characterize the products in the category **Alg**($\underline{S}$) of all partial $\underline{S}$-algebras, we have to describe so-called pullbacks of homogeneous partial algebras. In this connection we also need the concept of a sink of homomorphisms or mappings, i.e. of such a family, say, $(f_i:\underline{A}_i \to \underline{B}|i \in I)$ of homomorphisms which all have the same target or codomain (here $\underline{B}$).

4.2.2 DEFINITION. Let $\underline{\mathfrak{C}}$ be any category, I be any non-empty set, and let $\underline{f} := (f_i:A_i \to B|i \in I)$ be any sink of $\underline{\mathfrak{C}}$-morphisms. A multiple pullback of $\underline{f}$ is a source $\underline{g} := (g_i:P \to A_i|i \in I)$ such that

(i) $g_if_i = g_jf_j$ for all $i,j \in I$, and

(ii) for every source $\underline{h} := (h_i:Q \to A_i|i \in I)$, which satisfies $h_if_i = h_jf_j$ for all $i,j \in I$ there exists a unique $\underline{\mathfrak{C}}$-morphism $h:Q \to P$ such that $h_i = hg_i$ for all $i \in I$, i.e. the diagram in Figure 4.3 is commutative.

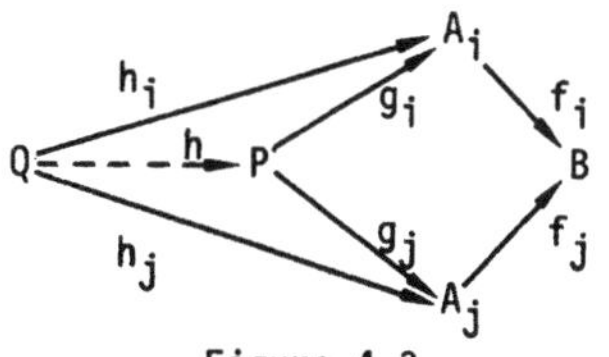

Figure 4.3.

If I has exactly two elements, then one is speaking of a pullback rather than of a multiple pullback.

4.2.3 COROLLARY. Let I be any non-empty set, $\underline{f} = (f_i:A_i \to B|i \in I)$ any sink of $\underline{C}$-morphisms, and $\underline{g} = (g_i:P \to A_i|i \in I)$ as well as $\underline{g}' = (g_i':P' \to A_i|i \in I)$ be any two pullbacks of $\underline{f}$. Then there exists a unique isomorphism $p:P' \to P$ such that $g_i' = pg_i$ for all $i \in I$, i.e. a pullback -- whenever it exists -- is uniquely determined up to unique isomorphism.

The proof is similar to the one of the uniqueness of products (cf. 4.1.2), threrefore we omit the details. ∎∎∎

4.2.4 THEOREM. Let I be any non-empty set, and let $\underline{f} = (f_i:\underline{A}_i \to \underline{B}|i \in I)$ be any sink of homomorphisms. Then there exists a multiple pullback of $\underline{f}$. Moreover:

(i) If $(\underline{A},(pr_i:\underline{A} \to \underline{A}_i|i \in I) = (X(\underline{A}_i|i \in I),(pr_i|i \in I))$ is the direct product of the family $(\underline{A}_i|i \in I)$ of corresponding homogeneous partial algebras, and if $\underline{P}$ is the relative subalgebra of $\underline{A}$ with carrier set

$$P:= \{a \in A|(apr_i)f_i = (apr_j)f_j \text{ for all } i,j \in I\},$$

then $\underline{P}$ is a subalgebra of $\underline{A}$, and $(\underline{P},(p_i:= (pr_i|P):\underline{P} \to \underline{A}_i|i \in I)$ is a pullback of $\underline{f}$ (in the heterogeneous case we define $v_P:= p_i f_i v_B$ for any $i \in I$.)

(ii) If $i' \in I$, and f_i is surjective, full and surjective, injective or closed for all $i \in I \setminus \{i'\}$, then $p_{i'}$ has the same property.

PROOF. Ad (i): Let $\underline{h}:= (h_i:\underline{Q} \to \underline{A}_i|i \in I)$ be any source such that $h_i f_i = h_j f_j$ for all $i,j \in I$. Then there exists a homomorphism $h:\underline{Q} \to \underline{A}$, where $\underline{A}:= X(\underline{A}_i|i \in I)$, such that $h \cdot pr_i = h_i$ for all $i \in I$. First we show that $[Q]h \subseteq P$: Thus consider $q \in Q$ and $\mathfrak{q}:= qh \in A$; for all $i,j \in I$ we have $qh_i f_i = qh_j f_j$, i.e. $(\mathfrak{q}pr_i)f_i = (qh)pr_i \cdot f_i = qh_i f_i = qh_j f_j = (qh)pr_j \cdot f_j = (\mathfrak{q}pr_j)f_j$, i.e. $\mathfrak{q} \in P$. Moreover, the definition of v_P does not depend -- by definition of $\underline{P}$ -- on the special i chosen from I and is a homomorphism into $\underline{S}$ such that e.g. $hv_P = h \cdot pr_i \cdot f_i \cdot v_B = (h_i f_i)v_B = v_Q$, since we ended with $\underline{S}$-homomorphisms. From the definition of $\underline{P}$ it now easily follows that $(\underline{P},(p_i|i \in I))$ is indeed a pullback of $\underline{f}$.

We still have to show that P is a closed subset of $\underline{A}$: In order to realize this let us observe that each $P_{ij}:= \{a \in A|(apr_i)f_i = (apr_j)f_j\}$ $(i,j \in I)$ is a closed subset of $\underline{A}$ according to 3.5.3, and therefore $P = \bigcap\{P_{ij}|i,j \in I\}$ is also a closed subset of $\underline{A}$. ∎

Ad (ii): Now assume $i' \in I$ (observe that $I \neq \emptyset$), and let all f_i be surjective (injective, full or closed) for all $i \in I \setminus \{i'\} =: I'$.

If $A_{i'} = \emptyset$, then also A and P are empty, and $p_{i'}$ is an isomorphism having all these properties, and also for $I = \{i'\}$ $p_{i'}$ is an isomorphism. Thus we may assume $\#I \geq 2$ and $A_{i'} \neq \emptyset$. If, nevertheless, $A = \emptyset$, then $A_i = \emptyset$, for some $i \in I'$; but in the case of surjectivity this would imply $B = \emptyset$ contradicting $A_{i'} \neq \emptyset$ and the fact that $f_{i'}$ is a mapping from $A_{i'}$ into B -- the cases of injectivity or of being full and surjective are no problem, either. And if all f_j $(j \in I')$ are closed, then $\underline{B}$ and hence $\underline{A}_{i'}$ cannot have any existing nullary constants, and the empty mapping $p_{i'}$ is closed, too. Thus we may additionally assume that A and hence all A_i are non-empty:

(a): Let all f_i $(i \in I')$ be surjective, and consider $a' \in A_{i'}$, $a'f_{i'} =: b' \in B$. Then we can find (by the Axiom of Choice) a sequence $a \in A$ such that $(i')a = a'$ and $(i)af_i = b'$ for each $i \in I$, i.e. $a \in P$, and $ap_{i'} = a'$, showing that $p_{i'}$ is surjective. ∎

(b): Let all f_i $(i \in I')$ be full and surjective, and let $\underline{a} \in P^{K_\phi}$ such that

$\underline{a}p_{i'} \in \mathrm{dom}\ \phi^{\underline{A}_{i'}}$. Then $\underline{a}p_{i'}f_{i'} \in \mathrm{dom}\ \phi^{\underline{B}}$, and because of the surjectivity and fullness of all f_i $(i \in I')$ we can find sequences $\underline{b}^{(i)} \in \mathrm{dom}\ \phi^{\underline{A}_i}$ for each $i \in I'$ such that $\underline{b}^{(i)}f_i = \underline{a}p_{i'}f_{i'}$, and all these sequences compose together with $\underline{a}p_{i'}$ to a sequence $\underline{h} \in \mathrm{dom}\ \phi^{\underline{A}}$ satisfying $\underline{h}p_{i'} = \underline{a}p_{i'}$, and which actually belongs to $\underline{P}$, showing that $p_{i'}$ is full -- surjectivity already follows from (a). ∎

The cases of injectivity or closedness can be treated in a similar way, and since they will follow from more general results in section 10.1 we leave the details as an exercise at this place. ∎∎∎

The preceding result also contains the construction of products within the category **Alg**($\underline{S}$), since the product of $(\underline{A}_i|i \in I)$ in **Alg**($\underline{S}$) is by the definition of this category nothing else than the pullback of the family $(v_{A_i}:\underline{A}_i \to \underline{S}|i \in I)$ within the category **Alg**(τ) of homogeneous partial algebras. Hence we have the

4.2.5 COROLLARY. In the category **Alg**($\underline{S}$) products exist for arbitrary families of partial $\underline{S}$-algebras indexed by a set, and for non-empty index sets they can be computed within **Alg**(τ) as the multiple pullback of the corresponding family of sort homomorphisms. For the empty index set the corresponding product in **Alg**($\underline{S}$) is given by $(\underline{S},\emptyset)$, where $v_S := id_S$. The corresponding product object will also be denoted by $\underline{A} := \mathbf{X}(\underline{A}_i|i \in I)$, and we hope that this will not cause any confusion.
Also arbitrary multiple pullbacks exist in **Alg**($\underline{S}$), and they can be constructed by just computing the corresponding multiple pullback within **Alg**(τ) according to 4.2.4. ∎∎∎

4.2.6 REMARKS. (i): Theorem 4.2.4 and its proof stand as an example how one can compute arbitrary so-called "limits" in the categories **Alg**(τ) and **Alg**($\underline{S}$) as substructures of the corresponding direct product of the objects involved. For the general concept of "limits" in category theory we refer to books on category theory like [HS 73] or [Ml 71].

(ii): All the results in 4.1 on products and the relationship between homomorphisms and their graphs considered as substructures of the direct product between domain and codomain now carry over to partial $\underline{S}$-algebras without any significant changes -- one only has to realize that the corresponding subobjects are always already contained in the multiple pullback object with respect to the sink of the sort homomorphisms, i.e. in the product object of the partial $\underline{S}$-algebras under consideration.

4.3 ("Global") Coproducts and multiple pushouts

In 4.1 we have seen that the concept of direct products is actually a category theoretical one. Now, in category theory one gets for every notion a dual one by reversing the arrows. This leads to the following

4.3.1 DEFINITION. Let $\underline{\mathbf{C}}$ be any category, and $(A_i|i \in I)$ any family of $\underline{\mathbf{C}}$-objects -- I any set. A pair $((j_i:A_i \to A|i \in I),A)$ is called a <u>coproduct</u> (or <u>free sum</u>) of this family, if, for each $i \in I$, j_i is a $\underline{\mathbf{C}}$-morphism such that for any sink $\underline{g} := (g_iA_i \to B|i \in I)$ there exists a unique $\underline{\mathbf{C}}$-morphism $g:A \to B$ satisfying, for each $i \in I$, $g_i = j_ig$, i.e. all the diagrams like Figure 4.4 commute.

If $I = \emptyset$, then this definition means that for every $B \in Ob(\underline{\mathfrak{C}})$ there exists exactly one $\underline{\mathfrak{C}}$-morphism from A into B, i.e. A is a so-called <u>left-zero</u> or <u>initial object</u> of the category under consideration.

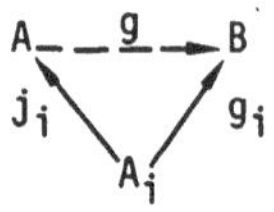

Figure 4.4.

In the same way as for products one proves

4.3.2 PROPOSITION. If in a category $\underline{\mathfrak{C}}$ a coproduct $((j_i|i \in I),A)$ of some family $(A_i|i \in I)$ in $\underline{\mathfrak{C}}$ exists, then it is unique up to unique isomorphism. ∎∎∎

4.3.3 REMARKS. (i): The expression "free sum" is used in algebra especially, when all j_i are injective, but we do not require this in general. Despite this fact these morphisms are often called the "injections" or "embedding morphisms" corresponding to a coproduct construction. The investigation whether these "embeddings" are really injective is then usually rather involved and its answer for a given class of total or partial algebras is an important structural insight.

(ii): While the construction of products is usually the same in the subcategories "of interest" of partial algebras as in the global category, this is not the case for coproducts, which -- for a given subclass in which they exist at all -- are usually only special epimorphic images of the global ones. This , too, makes the theory of coproducts usually much more complicated than the one of products, and within this book we shall not need coproducts very often.

(iii): Coproducts are special cases of "colimits", and in this book we shall rather use a more special sort of colimits, namely **directed colimits,** also called **direct limits** (this notation has historical reasons), which we shall introduce in the next subsection.

4.3.4 CONSTRUCTION of COPRODUCTS of PARTIAL ALGEBRAS:

Let $(\underline{A}_i|i \in I)$ be any set-indexed family of partial algebras of similarity type $\tau = (K_\varphi)_{\varphi\in\Omega}$. Let $\Omega' := \{\varphi \in \Omega | K_\varphi \neq \emptyset\}$, $\Omega_0 := \Omega \smallsetminus \Omega'$, and $\tau' := (K_\varphi)_{\varphi\in\Omega'}$.

For each $i \in I$ we call $\underline{A}_i' := ((A_i,(\varphi^{\underline{A}_i})_{\varphi\in\Omega'}),v_{A_i})$ the Ω'-<u>reduct</u> of $\underline{A}_i$.

(a): We first define the coproduct of the Ω'-reducts of the $\underline{A}_i$:

$$\underline{A}' := ((A' := \bigcup\{\{i\} \times A_i | i \in I\},(\varphi^{\underline{A}'})_{\varphi\in\Omega'}),v_{A'}),$$

where $(i,a)v_{A'} := av_{A_i}$ for all $(i,a) \in A'$, and for all $\varphi \in \Omega'$:

$\underline{a} = ((i_k,a_k)|k \in K_\varphi) \in \mathrm{dom}\,\varphi^{\underline{A}'}$ iff $i_k = i_{k'} =: i$ for all $k,k' \in K_\varphi$, and $(a_k|k \in K_\varphi) \in \mathrm{dom}\,\varphi^{\underline{A}_i}$, and if $\underline{a} \in \mathrm{dom}\,\varphi^{\underline{A}'}$ corresponding to the index $i \in I$, then $\underline{a}\varphi^{\underline{A}'} := (i,(a_k|k \in K_\varphi)\varphi^{\underline{A}_i})$ -- we shall indicate this by sloppily writing $\mathrm{graph}(\varphi^{\underline{A}'}) := \bigcup(\{i\} \times \mathrm{graph}(\varphi^{\underline{A}_i})|i \in I)$ for every $\varphi \in \Omega'$.

Finally we define, for each $i \in I$, $j_i' : A_i \to A'$, $a \mapsto (i,a)$ $(a \in A_i)$.

(b): We now take care of the appropriate identification of the existing nullary constants and set

$$R' := \{((i,\varphi^{\underline{A}_i}),(i',\varphi^{\underline{A}_{i'}}))|i,i' \in I,\ \varphi \in \Omega_0;\ \varphi^{\underline{A}_i},\varphi^{\underline{A}_{i'}} \text{ exist}\};$$

Moreover, let $R := \mathrm{Cong}_{\underline{A}'}(R')$ be the congruence relation on $\underline{A}'$ generated by R'. Finally, we set $A := A'/R$, $v_A := v_{A'/R}$, $j_i := j_i' \cdot nat_R : A_i \to A$, $\varphi^{\underline{A}} := \varphi^{\underline{A}'/R}$ for all $\varphi \in \Omega'$ and $\varphi^{\underline{A}} := \varphi^{\underline{A}_i} j_i$ for all $\varphi \in \Omega_0$ for which $\varphi^{\underline{A}_i}$ exists -- for any $i \in I$.

If $I = \emptyset$, then we define $\underline{A} := \underline{\emptyset}$, the empty partial algebra.

4.3.5 PROPOSITION. With the assumptions and notation of 4.3.4 $((j_i|i \in I),\underline{A})$ is a coproduct in all of **Alg**($\underline{S}$) of the family $(\underline{A}_i|i \in I)$.
$\underline{A}$ is usually denoted by $\Sigma(\underline{A}_i|i \in I)$.
Notice that there is no difference between the coproduct of $(\underline{A}_i|i \in I)$ within **Alg**(τ) and of the one of $(\underline{A}_i|i \in I)$ in **Alg**($\underline{S}$) except for the sort homomorphism.

PROOF. If $I = \emptyset$, then the statements are obvious. Further, we can observe that $\underline{A}$ and all the mappings j_i $(i \in I)$ are well defined, that each j_i is a homomorphism with respect to the similarity type τ' preserving constants whenever they exist according to the definition of R' and R; therefore they are homomorphisms $j_i:\underline{A}_i \to \underline{A}$. Now, let $(g_i:\underline{A}_i \to \underline{B}|i \in I)$ be any family of homomorphisms. Then there is obviously a unique τ'-homomorphism $g':\underline{A}' \to \underline{B}'$ (where $\underline{B}'$ is the τ'-reduct of $\underline{B}$) mapping (i,a) onto ag_i, such that $g_i = j_i'g'$ for every $i \in I$. But whenever $\phi^{\underline{A}_i}$ and $\phi^{\underline{A}_{i'}}$ exist for some $\phi \in \Omega_0$, then $\phi^{\underline{A}_i}g_i = \phi^{\underline{A}_{i'}}g_{i'}$, and $((i,\phi^{\underline{A}_i}),(i',\phi^{\underline{A}_{i'}})) \in \ker g'$, showing that $R' \subseteq \ker g'$, i.e. $R \subseteq \ker g'$. Therefore there exists a unique homomorphism $g:\underline{A} \to \underline{B}$ such that $g' = nat_R \cdot g$ (according to the Diagram Completion Lemma 2.7.1). Finally, for every $i \in I$ one has $g_i = j_i'g' = j_i' \cdot nat_R \cdot g = j_i g$. The uniqueness of g is obvious from the above construction, from which it also follows, that there is no difference in the objects and mappings, when we consider partial algebras in the given family with or without their sort homomorphisms. ■■■

4.3.6 NOTATION. For finite index sets we usually write $\underline{A}_1 + \underline{A}_2$ and $\underline{A}_1 + \underline{A}_2 + \ldots + \underline{A}_n$ instead of $\Sigma(\underline{A}_i|i \in \{1,2\})$ respectively $\Sigma(\underline{A}_i|i \in \{1,2,\ldots,n\})$.

4.3.7 EXAMPLES. **(i)**: Let $\tau := (1,0,0)$, $\Omega := \{\uparrow, o, x\}$, and consider Figure 4.5, where the dotted lines indicate the non-trivial congruence classes of R on the disjoint union of carrier sets (cf. R and R' in 4.3.4). The canonical injections should be obvious, but notice that the one starting from $\underline{A}_1$ is not injective.

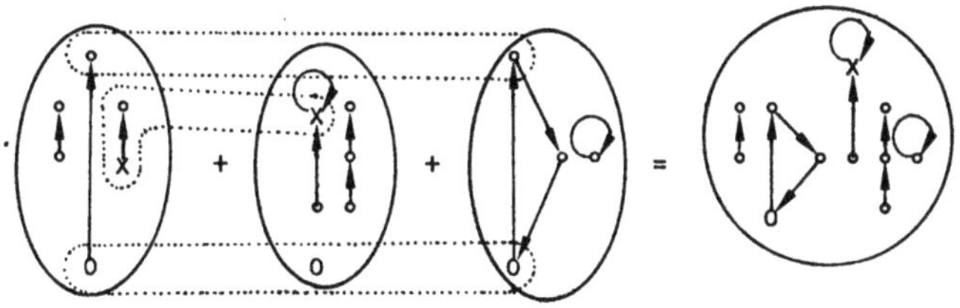

Figure 4.5.

(ii): In the category of all abelian groups the carrier set of $\Sigma(\underline{A}_i|i \in I)$ is given by $\{a \in \mathbf{X}(\underline{A}_i|i \in I)|(i)\, a \neq 0_i$ for only finitely many $i \in I\}$, $(a)j_i := a_{(a,i)}$ such that $(i)a_{(a,i)} = a$, and $(k)a_{(a,i)} = 0_k$, otherwise. Notice that $\{a_{(a,i)}|a \in A_i, i \in I\}$ is up to a natural bijection equal to $\mathbf{U}(A_i \times \{i\})/Cong\{((0_i,i),(0_j,j))|i,j \in I\}$. In particular one has $\underline{A} + \underline{B} = \underline{A} \times \underline{B}$.
(iii): But notice, that the above construction in (ii) does not yield the coproduct in the case of groups in general, where the carrier set of $\underline{G} + \underline{H}$ is usually described as the set of all "words" in elements of G and H -- which are assumed to be disjoint -- such that no two elements of the same group are allowed to stand together (the neutral

elements are not allowed at all, and the new neutral element is represented by the empty sequence). Multiplication of such sequences is the usual concatenation followed by a reduction process, reducing the new sequence recursively by replacing two neighbouring elements from the same group by their product in this group, or by deleting both of them, if this product becomes a neutral element.

Before we conclude this subsection we briefly discuss multiple pushouts, the concept dual to multiple pullbacks:

4.3.8 DEFINITION. Let $\underline{\mathbf{C}}$ be any category, $(A,(f_i:A \to A_i|i \in I)$ a source of $\underline{\mathbf{C}}$-morphisms. A sink $((g_i:A_i \to P|i \in I),P)$ of $\underline{\mathbf{C}}$-morphisms -- sometimes only the target object P of this family -- is called a <u>multiple pushout</u> of the given sink iff

(a) $f_i g_i = f_j g_j$, and

(b) if $((h_i:A_i \to B|i \in I),B)$ is another sink of $\underline{\mathbf{C}}$-morphisms satisfying $f_i h_i = f_k h_k$ for all $i,k \in I$, then there is a unique $\underline{\mathbf{C}}$-morphism $h:P \to B$ such that $g_i h = h_i$ for all $i \in I$. -- One gets for all $i,j \in I$ a commutative diagram dual to the one in Figure 4.3.

As for products (cf. 4.1.2) and pullbacks one gets the

4.3.9 COROLLARY. If a multiple pushout exists, then it is unique up to unique isomorphism. ■■■

The concrete construction of multiple pushouts is sketched in the following

4.3.10 PROPOSITION. Let $(\underline{C},(f_i:\underline{C} \to \underline{A}_i|i \in I))$ be a source of homomorphisms between similar partial algebras, $((j_i:\underline{A}_i \to \underline{A}|i \in I),\underline{A}:= \Sigma(\underline{A}_i|i \in I))$ its coproduct, $R':= \{(cf_i,cf_k)|c \in C , i,k \in I\}$, and $R:= \mathrm{Cong}_{\underline{A}}(R')$ the congruence generated by R'. Then a multiple pushout of the given family is obtained in $((h_i:= j_i \cdot nat_R:\underline{A}_i \to \underline{A}/R|i \in I),\underline{A}/R)$. ■■■

4.3.11. Observe that we have constructed a pushout of $(\underline{B},(g,h))$ in the proof of 3.6.1.

4.4 Direct limits (= directed colimits)

4.4.1. In the last subsection we have briefly discussed coproducts and multiple pushouts, which form special instances of the category theoretical concept of **colimits.** The most important kind of colimit for our purpose is the one of a (**k**-) directed colimit (for historical reasons also called (**k**-) direct limit) (for some cardinal **k**.) In this subsection we only discuss the fundamentals of this concept and return to it in Chapter III, when we have additional tools around, since **k**-direct limits are a very important concept in connection with model theoretic results. Nevertheless we think it useful to have this concept already in this introductory chapter. For the concept of **k**-directed partially ordered sets see section 24.

4.4.2 DEFINITION. Let $\underline{\mathbf{C}}$ be any category and $\mathbf{k} \geq \#\omega_0$ any infinite cardinal.

(i) A <u>directed system</u> is any family $\mathbf{A}_{\underline{I}}:= (A_i,f_{il}:A_i \to A_l|i,l \in I , i \leq l)$, where $\underline{I} = (I;\leq)$ is a directed poset, $A_i \in \mathrm{Ob}(\underline{\mathbf{C}})$ for each $i \in I$, and each f_{il} is a $\underline{\mathbf{C}}$-

morphism ($i \leq l$), such that

(DS1) $f_{ii} = 1_{A_i}$ for every $i \in I$.

(DS2) If $i \leq l \leq m$ in $\underline{I}$, then $f_{il}f_{lm} = f_{im}$.

If $(I;\leq)$ is in addition **k**-directed, then $\mathbb{A}_{\underline{I}}$ will be called a <u>**k**-directed system</u>.

(ii) Let $\mathbb{A}_{\underline{I}}$ be a directed system in $\underline{\mathbb{C}}$; then $\underline{f} := ((f_i : A_i \to A \mid i \in I), A)$ is called its <u>directed colimit</u> or its <u>direct limit</u> (other notations are: <u>colimiting cocone</u> or <u>inductive limit</u>) -- and this is indicated by writing $\underline{f} =: \varinjlim \mathbb{A}_{\underline{I}}$ (often only $A = \varinjlim \mathbb{A}_{\underline{I}}$) -- iff (DL1) through (DL3) below hold:

(DL1) $A \in \mathrm{Ob}(\underline{\mathbb{C}})$, $f_i \in \mathrm{Mor}(\underline{\mathbb{C}})$ for each $i \in I$.

(DL2) If $i \leq l$ in $\underline{I}$, then $f_{il}f_l = f_i$ (i.e. $\underline{f}$ is <u>compatible</u> with $\mathbb{A}_{\underline{I}}$).

(DL3) If $\underline{g} := ((g_i : A_i \to B \mid i \in I), B)$ is any system **compatible** with $\mathbb{A}_{\underline{I}}$, (i.e. $f_{il}g_l = g_i$ for all $i \leq l$ in I), then there exists exactly one $\underline{\mathbb{C}}$-morphism $g : A \to B$ such that $f_i g = g_i$ for every i in I (i.e. one has commutativity of the diagram in Figure 4.6); g is then called the <u>direct limit</u> of <u>g</u> or the <u>colimit morphism induced</u> by <u>g</u>.

If $\underline{I}$ is **k**-directed, then $\underline{f}$ above is also called a <u>**k**-direct limit</u> (<u>**k**-directed colimit</u>). Often only the object A above is denoted as $\varinjlim \mathbb{A}_{\underline{I}}$.

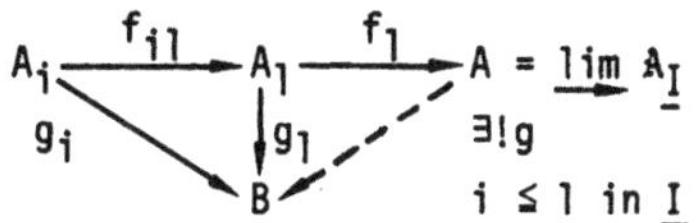

Figure 4.6.

As before one easily proves:

4.4.3 PROPOSITION. Every (**k**-) direct limit is unique up to unique isomorphism. ■■■

4.4.4 PROPOSITION. (Existence of direct limits for partial algebras)

Let τ be any similarity type of cardinal dimension **d** and $\mathbf{d}^* := \max\{\mathbf{d}, \#\omega_0\}$, let $\underline{I} := (I;\leq)$ be any directed poset, and let $\mathbb{A}_{\underline{I}} := (\underline{A}_i, f_{il} \mid i \leq l \text{ in } I)$ be a directed system of partial algebras in $\mathbf{Alg}(\underline{S})$. Then one has:

(i) $\varinjlim \mathbb{A}_{\underline{I}} =: ((f_i \mid i \in I), \underline{A})$ exists and is given by $((j_i \cdot \mathrm{nat}_R \mid i \in I), \Sigma(\underline{A}_i \mid i \in I))$, where $R := \mathrm{Cong}_{\Sigma(\underline{A}_i \mid i \in I)}(R')$ (see 4.3.4f), and

$R' := \{(aj_i, bj_l) \mid i, l \in I,\ a \in A_i,\ b \in A_l, \text{ and there is } m \in I\colon i, l \leq m \text{ and } af_{im} = bf_{lm}\}$.

(ii) If $\underline{I}$ is $\mathbf{d}^*$-directed, then in (i) $R = R'$, and we may even define

$A := (\bigcup\{\{i\} \times A_i \mid i \in I\})/R''$, where

$R'' := \{((i,a),(l,b)) \mid i, l \in I,\ a \in A_i,\ b \in A_l, \text{ and there is } m \in I\colon i, l \leq m \text{ and } af_{im} = bf_{lm}\}$,

$f_i : a \mapsto [(i,a)]R''$ $(i \in I$, $a \in A_i)$, $v_A : [(i,a)]R'' \mapsto av_{A_i}$, and for $\phi \in \Omega$ we have

$\mathrm{dom}\,\phi^{\underline{A}} := \{([(i_k,a_k)]R'' \mid k \in K_\phi) \mid \text{there is } l \in I \text{ with } (a_k f_{i_k l} \mid k \in K_\phi) \in \mathrm{dom}\,\phi^{\underline{A}_l}\}$,

and if $([(i_k,a_k)]R'' \mid k \in K_\phi) \in \mathrm{dom}\,\phi^{\underline{A}}$ with respect to some $l \in I$, then

$([(i_k,a_k)]R'' \mid k \in K_\phi)\phi^{\underline{A}} := [(l, (a_k f_{i_k l} \mid k \in K_\phi)\phi^{\underline{A}_l})]R''$.

PROOF. Since in what follows we shall always assume $\mathbb{A}_{\underline{I}}$ to be $\mathbf{d}^*$-directed, we shall only prove (ii) and leave the details of a proof of (i) as an exercise to the reader -- as well as the proof of $R' = R$ stated in (ii).

Ad (ii): Thus we assume that $\underline{I}$ is $\mathbf{d}^*$-directed. We use the notation of 4.3.4, especially the one concerning Ω', τ', Ω_0 and $\underline{A}'$, and we claim:

(a): R'' is a congruence on $\underline{A}'$ containing the relation R' from 4.3.4 identifying all pairs arising from corresponding existing fundamental constants (therefore $\underline{A}$ is a well-

defined partial $\underline{S}$-algebra):
(DS1) implies reflexivity of R", symmetry is already contained in the definition of R", while transitivity of R" follows from directedness of $\underline{I}$: If $((i,a),(i',b)),((i',b),(i'',c)) \in R''$, then there are $l' \geq i,i'$ and $l'' \geq i',i''$ such that $af_{il'} = bf_{i'l'}$ and $bf_{i'l''} = cf_{i''l''}$. Since $\underline{I}$ is directed, there is $m \geq l',l''$ in I, thus $af_{im} = af_{il'}f_{l'm} = bf_{i'l'}f_{l'm} = bf_{i'm} = bf_{i'l''}f_{l''m} = cf_{i''l''}f_{l''m} = cf_{i''m}$.
Therefore R" is an equivalence relation. If, for $\varphi \in \Omega_0$ and for some i,i' in I, $\varphi^{\underline{A}_i}$ and $\varphi^{\underline{A}_{i'}}$ exist, then there is $l \geq i,i'$ in I, such that $\varphi^{\underline{A}_i}f_{il} = \varphi^{\underline{A}_{i'}}f_{i'l}$, i.e. $((i,\varphi^{\underline{A}_i}),(i',\varphi^{\underline{A}_{i'}})) \in R''$. It remains to show that R" is compatible with all (non-nullary!) operations of $\underline{A}$:
Let $\varphi \in \Omega'$, and $((i,a_k),(l,b_k)) \in R''$ $(k \in K_\varphi)$, and $(a_k|k \in K_\varphi) \in \text{dom}\, \varphi^{\underline{A}_i}$, $(b_k|k \in K_\varphi) \in \text{dom}\, \varphi^{\underline{A}_l}$. By assumption there is for each $k \in K_\varphi$ some $m_k \geq i,l$ in I such that $a_kf_{im_k} = b_kf_{lm_k}$. Since $\underline{I}$ is **d***-directed and $\#K_\varphi <$ **d*** , there is some $m \geq m_k$ $(k \in K_\varphi)$ in I. Thus we get
$(a_k|k \in K_\varphi)\varphi^{\underline{A}_i}f_{im} = (a_kf_{im}|k \in K_\varphi)\varphi^{\underline{A}_m} = (b_kf_{lm}|k \in K_\varphi)\varphi^{\underline{A}_m} = (b_k|k \in K_\varphi)\varphi^{\underline{A}_l}f_{lm}$,
i.e. $((i,(a_k|k \in K_\varphi)\varphi^{\underline{A}_i}),(l,(b_k|k \in K_\varphi)\varphi^{\underline{A}_l})) \in R''$. ∎
(b): The structure defined in 4.4.4 on $\underline{A}$ is indeed the quotient structure of $\underline{A}'$ with respect to R" as far as non-nullary operations are concerned: This is an immediate consequence of the definition of $\underline{A}$ and of what we have already shown for (a).
(c): $((f_i|i \in I),\underline{A})$ is a **d***-direct limit of $\mathbb{A}_{\underline{I}}$:
That $\underline{A}$ is indeed a partial $\underline{S}$-algebra follows immediately from (a), and that all f_i are homomorphisms satisfying (DL2) is a consequence of the definition of R" and the proof of (a). In order to show (DL3) assume that $((g_i|i \in I),\underline{B})$ is a sink compatible with $\mathbb{A}_{\underline{I}}$. Then there is $g':\underline{A}' \to \underline{B}'$ (in the notation of 4.3.4 and of 4.3.5) such that $j_i'g' = g_i$ $(i \in I)$. Because of the Diagram Completion Lemma 2.7.1 it remains to show that $R'' \subseteq \ker g'$:
Thus assume $((i,a),(l,b)) \in R''$. Then there is $m \geq i,l$ in I such that $af_{im} = bf_{lm}$, therefore $(i,a)g' = aj_i'g' = ag_i = af_{im}g_m = bf_{lm}g_m = bg_l = bj_l'g' = (l,b)g'$, i.e. $((i,a),(l,b)) \in \ker g'$.
Notice that everything is compatible with the sort homomorphisms. ∎∎∎

4.4.5 REMARK. The following example shows that in 4.4.4.(ii) **d***-directedness is really needed:
Let $A_i := (\{0\} \times \mathbf{N}_0) \cup (\{1\} \times \mathbf{N}_i) \cup \{(0,\infty),(1,\infty)\}$, $i \in \mathbf{N}$, $\infty > i$ for all $i \in \mathbf{N}$, where $\mathbf{N}_i := \{l \in \mathbf{N}_0 | l \geq i\} = \{i,i+1,i+2,...\}$, and

$$f_{ij}: \begin{cases} (0,n) \mapsto (0,n) , & n \geq 0 , \\ (1,n) \mapsto (0,n) , & \text{if } i \leq n < j , \\ (1,n) \mapsto (1,n) , & \text{if } n \geq j \end{cases}$$ i.e. "successive identification";

$\tau := (\mathbf{N})$, $\text{dom}\, \varphi^{\underline{A}_i} := \{\underline{a}_0{}^{(i)},\underline{a}_1{}^{(i)}\}$, where
$(n)\underline{a}_0{}^{(i)} := (0,n)$ for all $n \in \mathbf{N}$,
$(n)\underline{a}_1{}^{(i)} := (0,n)$ for $0 \leq n < i$,
$(n)\underline{a}_1{}^{(i)} := (1,n)$ for $n \geq i$;
$\underline{a}_0{}^{(i)}\varphi^{\underline{A}_i} := (0,\infty)$, $\underline{a}_1{}^{(i)}\varphi^{\underline{A}_i} := (1,\infty)$.

$A'/R'' = \{[(0,(0,n))]R'',[(0,(0,\infty))]R'',[(0,(1,\infty))]R'' \mid n \in \mathbf{N}_0\}$,
$A = \{[(0,(0,n))]R'',[(0,(0,\infty))]R'' \mid n \in \mathbf{N}_0\}$, where the elements $[(0,(0,\infty))]R''$ and $[(0,(1,\infty))]R''$ are different in A'/R''. Thus R'' is no congruence relation on $\underline{A}'$, and the index set $(\mathbf{N};\le)$ is not $\#\omega_1$-directed.

The following observations are sometimes quite useful:

4.4.6 PROPOSITION. Let $\mathbb{A}_I := (\underline{A}_i, f_{il}:\underline{A}_i \to \underline{A}_l \mid i \le l \text{ in } I)$ with direct limit $\underline{f} := ((f_i \mid i \in I),\underline{A})$ for a $\mathbf{d^*}$-directed set $\underline{I}$, let $\underline{g} := ((g_i \mid i \in I),\underline{B})$ be any sink compatible with $\mathbb{A}_I$, and let $g:\underline{A} \to \underline{B}$ be the direct limit of $\underline{g}$. Then we have:

(i) If all f_{il} are isomorphisms, injective, closed, surjective, epimorphisms or full, then the f_i have the same property $(i \in I)$.

(ii) If all g_i are isomorphisms, injective, closed, surjective, epimorphisms or full, then g has the same property.

PROOF. Since the other statements will be proved in a more general context -- using concepts from model theory -- we only prove here the statements about fullness, leaving the other proofs as exercises at this point.

Ad (i): Let all f_{il} be full, let $\underline{a} := (a_k \mid k \in K_\phi) \in A_i^{K_\phi}$ such that $\underline{a}f_i \in \operatorname{dom} \phi^{\underline{A}}$. Then there are $l \in I$ and $\underline{b} := (b_k \mid k \in K_\phi) \in \operatorname{dom} \phi^{\underline{A}_l}$ such that $\underline{b}f_l = \underline{a}f_i$. Then directedness and $\mathbf{d^*}$-directedness of $\underline{I}$ imply -- as in the proof of (a) for 4.4.4 -- the existence of $m \ge m_k \ge i,l$ $(k \in K_\phi)$ such that $a_k f_{im} = b_k f_{lm}$. But $(b_k f_{lm} \mid k \in K_\phi) \in \operatorname{dom} \phi^{\underline{A}_m}$ and fullness of f_{im} implies the existence of $\underline{a}' \in \operatorname{dom} \phi^{\underline{A}_i}$ such that $\underline{a}'f_{im} = \underline{a}f_{im}$, and also $\underline{a}'f_i = \underline{a}'f_{im}f_m = \underline{a}f_{im}f_m = \underline{a}f_i$, i.e. f_i is full. ∎

Ad (ii): Let all g_i be full, and consider $\underline{c} \in A^{K_\phi}$ such that $\underline{c}g \in \operatorname{dom} \phi^{\underline{B}}$. Then there exist $i \in I$ and $\underline{a} \in A_i^{K_\phi}$ such that $\underline{c} = \underline{a}f_i$, i.e. $\underline{c}g = \underline{a}f_i g = \underline{a}g_i$. Since g_i is full, there is $\underline{a}' \in \operatorname{dom} \phi^{\underline{A}_i}$ such that $\underline{a}g_i = \underline{a}'g_i$. Hence $\underline{c}g = \underline{a}f_i g = \underline{a}g_i = \underline{a}'g_i = (\underline{a}'f_i)g$, where $\underline{a}'f_i \in \operatorname{dom} \phi^{\underline{A}}$ showing that g is full. ∎∎∎

We now present some important examples of the representation of partial algebras by "small" (weak) substructures:

4.4.7 PROPOSITION. Let τ be a similarity type of cardinal dimension $\mathbf{d}$, let $\mathbf{k} \ge \mathbf{d^*}$ be any regular infinite cardinal, and let M be some generating subset of some partial algebra $\underline{A}$. Then consider the following sets:

$I_0 := I_0(\underline{A},M) := \{(\mathbf{C}_{\underline{A}}N) \mid N \subseteq M ,\ \#N < \mathbf{k}\}$;

$I_1 := I_1(\underline{A},M) := \{\underline{B} \mid \underline{B}$ is a relative subalgebra of $\underline{A}$, $\#B < \mathbf{k}$ and $B = \mathbf{C}_{\underline{B}}(B \cap M)\}$;

$I_2 := I_2(\underline{A},M) := \{\underline{B} \mid \underline{B}$ is a weak subalgebra of $\underline{A}$, $\#B < \mathbf{k}$ and $B = \mathbf{C}_{\underline{B}}(B \cap M)\}$;

$I_3 := I_3(\underline{A},M) := \{\underline{B} \mid \underline{B} \in \mathbf{Wsuba}(\underline{A})$, $\#B < \mathbf{k}$, $\#(\dot{\bigcup}\{\operatorname{dom} \phi^{\underline{B}} \mid \phi \in \Omega\}) < \mathbf{k}$ and $B = \mathbf{C}_{\underline{B}}(B \cap M)\}$.

In each case define $\underline{B} \le \underline{B}'$ iff $B \subseteq B'$ and $\operatorname{graph}(\phi^{\underline{B}}) \subseteq \operatorname{graph}(\phi^{\underline{B}'})$ for all $\phi \in \Omega$; and for $\underline{B} \le \underline{B}'$ let $f_{BB'} := \mathrm{id}_{BB'}:\underline{B} \to \underline{B}'$ be the identity mapping on B.

Then $(I_n;\le)$ is a $\mathbf{k}$-directed set, and $\mathbb{A}_n(\underline{A},M) := (\underline{B}, f_{BB'} \mid \underline{B} \le \underline{B}' \text{ in } I_n)$ is a $\mathbf{k}$-directed system with $\mathbf{k}$-direct limit $((\mathrm{id}_{BA} \mid \underline{B} \in I_n),\underline{A})$ for $0 \le n \le 3$.

PROOF. Since $\mathbf{k}$ is assumed to be a regular infinite cardinal ($\ge \mathbf{d^*}$), $\mathbf{k}$-directedness of each poset $(I_n;\le)$ $(0 \le n \le 3)$ is easily realized. Notice, too, that -- in a cofinal way (see 23.3.14) -- each object in I_{n+1} is less than or equal to (i.e. contained in) some object of I_n ($0 \leq n \leq 2$). We therefore only prove the statement for $\underline{I}_3$, the other proofs run similarly (exercise!) or will follow from later more general observations:

(DL1) and (DL2) are obvious. In order to prove (DL3) let $\underline{g} := (g_B : \underline{B} \to \underline{C} \mid \underline{B} \in I_3)$ be a sink compatible with $A_3(\underline{A}, M)$. If we can show that $\text{graph}(g) := \bigcup\{\text{graph}(g_{\underline{B}}) \mid \underline{B} \in I_3\}$ is the graph of a homomorphism from $\underline{A}$ into $\underline{C}$, then the proof is finished, since the required commutativities are then obvious. -- Obviously, g is a function, namely if $a \in \text{dom } g$, $a \in B$, $a \in B'$ for some $\underline{B}, \underline{B}' \in I_3$, then $ag := ag_{\underline{B}} = ag_{\underline{B}'}$ because of the compatibility condition. Moreover, $M \subseteq \text{dom } g$ is obvious, since $\overline{k} \geq 2$. We prove $A \subseteq \text{dom } g$ and the homomorphism condition for g at the same time by algebraic induction: If $a \in A \setminus \bigcup\{[A^{K_\phi}]\phi^{\underline{A}} \mid \phi \in \Omega\}$, then $a \in M \subseteq \text{dom } g$, and no homomorphism condition has to be proved. Thus assume $a = \underline{b}\phi'^{\underline{A}}$, $\underline{b} \in \text{dom } \phi'^{\underline{A}}$, and that for each $k \in K_{\phi'}$ there is $\underline{B}_k \in I_3$ such that $b_k \in B_k$. Then define $\underline{B} := ((\{a\} \cup \bigcup\{B_k \mid k \in K_{\phi'}\}, (\phi^{\underline{B}})_{\phi \in \Omega}), v_A \mid B)$, where

$$\text{graph}(\phi^{\underline{B}}) := \begin{cases} \{(\underline{b}, a)\} \cup \bigcup\{\text{graph}(\phi'^{\underline{B}_k}) \mid k \in K_{\phi'}\}, & \text{if } \phi = \phi', \\ \bigcup\{\text{graph}(\phi^{\underline{B}_k}) \mid k \in K_\phi\}, & \text{if } \phi \in \Omega \setminus \{\phi'\}. \end{cases}$$

Then it is obvious -- $\mathbf{k} \geq \mathbf{d}^*$ being regular -- that $\#B < \mathbf{k}$, $\#(\text{dom } \phi^{\underline{B}}) < \mathbf{k}$, and $B = C_{\underline{B}}(B \cap M)$ (cf. 3.6.6), i.e. $\underline{B} \in I_3$. Since $g_{\underline{B}} : \underline{B} \to \underline{C}$ is a homomorphism, we get $a \in \text{dom } g$ as well as $ag = (\underline{b}\phi^{\underline{A}})g = (\underline{b}\phi^{\underline{B}})g_{\underline{B}} = (\underline{b}g_{\underline{B}})\phi^{\underline{C}} = (\underline{b}g)\phi^{\underline{C}}$. ■■■

4.4.8 COROLLARY 1. If $\underline{A} = C_{\underline{A}}M \in \mathbf{Alg}(\underline{S})$, then $\underline{A} = \bigcup^*\{\underline{B} \mid \underline{B} \in I_n\}$, $0 \leq n \leq 3$ (the union being taken for each component of the structure, cf. 3.2.2). ■■■

A useful tool is the following result, which follows from the proof of 4.4.7 for $\underline{I}_0$:

4.4.9 COROLLARY 2 (TACKING LEMMA). Let $\underline{A}$ and $\underline{B}$ be similar partial algebras -- cardinal dimension **d**, M a generating subset of $\underline{A}$. Then the following statements are equivalent for every mapping $f : M \to B$:

(i) For every subset N of M of cardinality less than $\mathbf{d}^*$ the restriction $f|N$ of f to N has a homomorphic extension $f_N : \underline{C_A N} \to \underline{B}$.

(ii) f has a homomorphic extension $f^* : \underline{A} \to \underline{B}$. ■■■

4.4.10 REMARKS. (i): Observe that in 4.4.7 and 4.4.8 the assumption $\mathbf{k} \geq \mathbf{d}^*$ is necessary for the general statement, since for a regular cardinal $\mathbf{k} < \mathbf{d}^*$ it is possible -- e.g. in a Peano algebra, which we shall discuss in section 5 -- that we need $\geq \mathbf{k}$ elements of $M \subseteq A$ in order to generate some element of $\underline{A}$.

(ii): The examples in 4.4.7 as well as the following one in 4.4.11 concerning congruence relations and quotients are of special model theoretic interest, as we shall see in Chapter III. Especially for partial algebras, a combination of these two principles: stepwise generation of the elements and the structure on one side and stepwise enlarging of the congruence relations, form an important tool in connection with the generation of algebraic structures. In the theory of total algebras one usually already starts with a total "free" (cf. section 5) algebra and one only has to generate an appropriate congruence, while for partial algebras one additionally has to construct the elements and the admissible structure.

4.4.11 PROPOSITION. Let $\underline{A}$ be any partial algebra, R a congruence on $\underline{A}$ generated by some (admissible) relation R', $\mathbf{k} \geq \mathbf{d^*}$ a regular cardinal, and
$I_4 := \{Cong_{\underline{A}}(R^\circ) | R^\circ \subseteq R', \#R^\circ < \mathbf{k}\}$, ordered by set theoretical inclusion.
If $R'' \subseteq \overline{R}^*$ in I_4, then define $p_{R''R^*}: \underline{A}/R'' \to \underline{A}/R^*$ to be the homomorphism induced according to the Diagram Completion Lemma 2.7.1.
Then $\mathbb{A}_4(\underline{A},R') := (\underline{A}/R'', p_{R''R^*} | R'' \subseteq R^* \text{ in } I_4)$ is a **k**-directed system with direct limit $((nat_{R/R''}: \underline{A}/R'' \to \underline{A}/R | R'' \in I_4), \underline{A}/R)$.
If $R' = R$ then we may choose an arbitrary $\mathbf{k} \geq \#\omega_0$.

PROOF. (DL1) and (DL2) are obvious. Let $(f_{R''}: \underline{A}/R'' \to \underline{C} | R'' \in I_4)$ be a sink of homomorphisms compatible with $\mathbb{A}_4(\underline{A},R')$. It is then easily seen (cf. 2.5.4.(ii)) that $R \subseteq \bigcup\{ker(nat_{R''} \cdot f_{R''}) | R'' \in I_4\}$, and 2.7.1 yields a unique homomorphism $f: \underline{A}/R \to \underline{C}$ satisfying $nat_{R''} \cdot f = f_{R''}$ for each $R'' \in I_4$. ■

4.5 Reduced products and ultraproducts

4.5.1. Closely connected with the construction of direct products (and direct limits) is the one of reduced products. These play an important rôle in model theory, especially in the form of ultraproducts, but we shall need them in their general form in Chapter III.

4.5.2 FILTERS and ULTRAFILTERS. Let I be any set. We recall that a <u>filter</u> on I is any system $F \subseteq \mathbf{Subs}(I)$ of subsets of I such that the following axioms hold:
(F1) $F \neq \emptyset$ (or, equivalently in view of (F3): $I \in F$).
(F2) $F, F' \in F$ implies $F \cap F' \in F$.
(F3) $F \in F$ and $F \subseteq F' \subseteq I$ imply $F' \in F$.
If, in addition, one has
(F4) $\emptyset \notin F$ (or equivalently in view of (F3): $F \neq \mathbf{Subs}(I)$),
then we call F a <u>proper filter</u> on I. And if, moreover, F satisfies
(F5U) For all $M \subseteq I$ either $M \in F$ or $I \setminus M \in F$,
then F is called an <u>ultrafilter</u> on I; i.e. an ultrafilter is always a proper filter which is obviously maximal with respect to this property, and it is well known that every maximal proper filter on I is indeed an ultrafilter on I.
Obviously, for every $M \subseteq I$, $\uparrow M := \{N \subseteq I | M \subseteq N\}$ is a filter on I which is proper if $M \neq \emptyset$, and which is an ultrafilter if $M = \{i\}$ is a one-element set. $\uparrow M$ is called the <u>principal filter</u> of I generated by M.
Other important examples of filters are the so-called <u>generalized Fréchet filters</u> on I: Let $\#\omega_0 \leq \mathbf{c} \leq \#I$ be any cardinal number, then $F_{\mathbf{c}} := \{F \subseteq I | \#(I \setminus F) < \mathbf{c}\}$ is a non-principal proper filter for every infinite set I -- finite sets possess only principal filters. Although no non-principal ultrafilters are explicitly known, it follows from Zorn's Lemma that every proper filter is contained in some ultrafilter.
Let **k** be any cardinal and F a filter on I. F is called **k**-<u>complete</u>, if in addition to (F2)
(F2,**k**) For every subset $\mathbf{M} \subseteq F$, where $\#\mathbf{M} < \mathbf{k}$, one has $\bigcap\mathbf{M} \in F$, i.e. the intersection of less than **k** elements of F also belongs to F.
Obviously every filter is **k**-complete for $3 \leq \mathbf{k} \leq \#\omega_0$. If $\#\omega_0 \leq \mathbf{c} \leq \#I$, and if **c** is

additively inaccessible, then F_c is c-complete. The intersection of **k**-complete filters on I is also a **k**-complete filter on I.

With respect to the fact that direct limits are available, the easiest way to define reduced products is the following one, which we shall usually use later on:

4.5.3 DEFINITION. Let I be any set, $(\underline{A}_i|i \in I)$ any family of similar partial algebras, and let **F** be any **k**-complete filter on I. We first introduce on **F** the dual ordering with respect to inclusion, i.e. for $F, F' \in \mathbf{F}$ set:
$F \le F'$ iff $F \supseteq F'$.
Thus we set $\underline{\mathbf{F}} := (\mathbf{F};\le) = (\mathbf{F};\supseteq)$.
For each $F \in \mathbf{F}$ set $\underline{A}_F := \mathbf{X}(\underline{A}_i|i \in F)$; and for $F \supseteq G \in \mathbf{F}$ define
$p_{FG}: A_F \to A_G$ by $A_F \ni a \mapsto a|G$, i.e. $(a_i|i \in F)p_{FG} := (a_i|i \in G)$.
We call p_{FG} the <u>restriction mapping</u> with respect to F and G. Obviously, p_{FG} is a homomorphism from $\underline{A}_F$ into $\underline{A}_G$ induced by the family $(pr_i:\underline{A}_F \to \underline{A}_i|i \in G)$ of projections. Hence it is easy to realize that $\mathbb{A}_F := (\underline{A}_F, p_{FG}|F \supseteq G \text{ in } \mathbf{F})$ is a **k**-directed system, and its **k**-direct limit $((p_F|F \in \mathbf{F}), \varinjlim \mathbb{A}_F)$ is called a (<u>category theoretical</u>) **k**-<u>reduced product</u> of $(\underline{A}_i|i \in I)$ with respect to the filter **F**.
In particular, we introduce for the resulting object the notation
$\mathbf{X}^{\mathbf{k}}(\underline{A}_i|i \in I)/\mathbf{F} := \varinjlim \mathbb{A}_F$ $(=: \mathbf{X}(\underline{A}_i|i \in I)/\mathbf{F}$ if $\mathbf{k} \ge \mathbf{d^*}$, since then the structure is independent of how much **k** is larger than **d***, as the construction of direct limits in 4.4.4 shows).
If **F** is an ultrafilter, then $\mathbf{X}^{\mathbf{k}}(\underline{A}_i|i \in I)/\mathbf{F}$ is called an <u>ultraproduct</u>.

4.5.4 REMARKS. (i): In model theory one usually defines reduced products in another way, which we want to sketch briefly:
Let $(\underline{A}_i|i \in I)$ be any family of partial algebras which have non-empty phyla for all $s \in S$ -- if any phylum is empty for some $i \in I$, then the corresponding phylum of $\mathbf{X}(A_i|i \in I)$ will be empty --, **F** a **d***-complete filter on I. On the cartesian product $A := \mathbf{X}(A_i|i \in I)$ we define, for $a, b \in A$, $I_{ab} := \{i \in I|(i)a = (i)b\}$ (the <u>equalizer</u> of a and b), and moreover:
(*) $R_F := \{(a,b) \in A \times A | I_{ab} \in \mathbf{F}\}$.
The assumptions on **F** easily allow us to show that R_F is a congruence on $\underline{A}$.
But the structure of the reduced product $\underline{A}/\mathbf{F}$, whose carrier set is A/R_F, is for partial algebras usually richer than the quotient structure (for **total algebras** we may define $\underline{A}/\mathbf{F} := \underline{A}/R_F$): Let us denote the elements of $A/\mathbf{F} := A/R_F$ by $a/\mathbf{F}$, and sequences of them by $\underline{a}/\mathbf{F}$.
One defines for $\phi \in \Omega$ and for any sequence $\underline{a}/\mathbf{F} \in (A/\mathbf{F})^{K_\phi}$:
$\underline{a}/\mathbf{F} \in \text{dom } \phi^{\underline{A}/\mathbf{F}}$ iff there is some "corresponding" sequence $\underline{a}' \in A^{K_\phi}$ such that $\underline{a}/\mathbf{F} = \underline{a}'/\mathbf{F}$ and $\{i \in I|(i)\underline{a}' \in \text{dom } \phi^{\underline{A}_i}\} \in \mathbf{F}$; and if $\underline{a}/\mathbf{F} \in \text{dom } \phi^{\underline{A}/\mathbf{F}}$ with "corresponding" sequence $\underline{a}'$, then $(\underline{a}/\mathbf{F})\phi^{\underline{A}/\mathbf{F}} := (b_i|i \in I)/\mathbf{F}$, where $b_i := ((i)\underline{a}')\phi^{\underline{A}_i}$, if $(i)\underline{a}' \in \text{dom } \phi^{\underline{A}_i}$, and any element of appropriate sort from A_i is taken for b_i, otherwise.
(ii): With our assumptions in (i) of non-empty phyla the two definitions of reduced products are equivalent, while with empty phyla allowed in (i), the definitions may be inequivalent, since A_I may have an empty phylum, while this is non-empty in A_F for some

$F \in \mathbf{F}$ (see [ABN 80], p. 345ff). ∎

Because of such features we prefer the category theoretical definition.

(iii): With the notation from (i) and 4.5.3 one has:

If $F := \uparrow M$ is a principal filter, then $\underline{A}/F \cong \mathbf{X}(\underline{A}_i \mid i \in M)$. ∎

If $\{i \in I \mid \underline{A}_i$ is total$\} \in F$, then $\underline{A}/F$ is total, even when some $\underline{A}_i$ are discrete. ∎

Another example, where $\underline{A}$ is **discrete,** but $\underline{A}/F$ is **total,** is given as follows:

Let $\Omega := \mathbf{N}$, the set of natural numbers, $\tau := (1)_{n \in \mathbf{N}}$, $I := \mathbf{N}$, $A_n := \mathbf{Q}$, the set of rational numbers ($n \in \mathbf{N}$), and $\phi_m^{\underline{A}_n}$ is nowhere defined, if $m = n$, and it is multiplication by m, otherwise. F will be the Frêchet filter of all subsets of I with finite complement.

(iv): The relationship between $\underline{B} := \mathbf{X}(\underline{A}_i \mid i \in I)/F$ (heterogeneous case) and $\underline{C} := \mathbf{X}(\underline{A}_i \mid i \in I)/F$ (homogeneous case) is described as follows:

$\underline{B}$ is always a subalgebra of $\underline{C}$.

Namely because of 4.2.4.(i) for each $\mathbf{X}(\underline{A}_i \mid i \in F)$ the carrier set is a closed subset of $\mathbf{X}(\underline{A}_i \mid i \in F)$ ($F \in \mathbf{F}$), and in a generalization of 4.4.6.(ii) we shall see in section 11 that this implies that B is a closed subset of $\underline{C}$. ∎

4.6 Some properties of the operators H,..., S,.. and P,..

4.6.1. In the previous subsections we have introduced several constructions of new partial algebraic structures from (families of) given ones. These constructions lead to algebraic closure operators on the class of all partial ($\underline{S}$-) algebras of some given type, which especially gain importance in connection with implicational model theory in Chapters II and III, where we shall also need their behaviour with respect to other operators. Some of their properties will be investigated in this subsection.

4.6.2 DEFINITION. Let $\mathbf{K} \subseteq \mathbf{Alg}(\underline{S})$ be any subclass. We define

$H(\mathbf{K}) := H_w(\mathbf{K}) := \{\underline{B} \in \mathbf{Alg}(\underline{S}) \mid$ there are $\underline{A} \in \mathbf{K}$ and a surjective homomorphism $f: \underline{A} \to \underline{B}\} =$
  = **class of all weak homomorphic images** of **K**-algebras;

$H_f(\mathbf{K}) := \{\underline{B} \in \mathbf{Alg}(\underline{S}) \mid$ there are $\underline{A} \in \mathbf{K}$ and a full and surjective homomorphism $f: \underline{A} \to \underline{B}\} =$
  = **class of all full homomorphic images** of **K**-algebras;

$H_c(\mathbf{K}) := \{\underline{B} \in \mathbf{Alg}(\underline{S}) \mid$ there are $\underline{A} \in \mathbf{K}$ and a closed and surjective homomorphism $f: \underline{A} \to \underline{B}\} =$
  = **class of all closed homomorphic images** of K-algebras;

$I(\mathbf{K}) := \{\underline{B} \in \mathbf{Alg}(\underline{S}) \mid$ there are $\underline{A} \in \mathbf{K}$ and an isomorphism $f: \underline{A} \to \underline{B}\} =$
  = **class of all isomorphic copies** of **K**-algebras;

$S(\mathbf{K}) := S_c(\mathbf{K}) := \{\underline{B} \in \mathbf{Alg}(\underline{S}) \mid$ there are $\underline{A} \in \mathbf{K}$ and a closed monomorphism $f: \underline{B} \to \underline{A}\} =$
  = **class of all closed subalgebras** (up to isomorphism) of **K**-algebras;

$S_f(\mathbf{K}) := S_r(\mathbf{K}) := \{\underline{B} \in \mathbf{Alg}(\underline{S}) \mid$ there are $\underline{A} \in \mathbf{K}$ and a full monomorphism $f: \underline{B} \to \underline{A}\} =$
  = **class of all relative subalgebras** (up to isomorphism) of **K**-algebras;

$S_w(\mathbf{K}) := \{\underline{B} \in \mathbf{Alg}(\underline{S}) \mid$ there are $\underline{A} \in \mathbf{K}$ and a monomorphism $f: \underline{B} \to \underline{A}\} =$
  = **class of all weak subalgebras** (up to isomorphism) of **K**-algebras;

$P(\mathbf{K}) := \{\underline{B} \in \mathbf{Alg}(\underline{S}) \mid$ there are a set I and a family $(\underline{A}_i \mid i \in I) \in \mathbf{K}^I$ such that $\underline{B} \cong \mathbf{X}(\underline{A}_i \mid i \in I)\} =$
  = **class of all products** of (set indexed) families of **K**-algebras;

$P^r(K) := \{\underline{B} \in Alg(\underline{S}) \mid$ there are a set I, a d^*-complete filter F on I and a family $(\underline{A}_i \mid i \in I) \in K^I$ such that $\underline{B} \cong X(\underline{A}_i \mid i \in I)/F\}$ =
= **class of all d*-reduced products** (up to isomorphism) of (set indexed) families of K-algebras;

$P_u(K) := \{\underline{B} \in Alg(\underline{S}) \mid$ there are a set I, a d^*-complete ultrafilter F on I and a family $(\underline{A}_i \mid i \in I) \in K^I$ such that $\underline{B} \cong X(\underline{A}_i \mid i \in I)/F\}$ =
= **class of all ultraproducts** (up to isomorphism and for d*-complete ultrafilters) of (set indexed) families of K-algebras;

$o(K) := K \cup \{\emptyset\}$ = K and the empty algebra;

$e(K) := K \cup I(\underline{S})$ = K and all products for the empty index set.

Operators P_+ and $P_+{}^r$ are defined like the corresponding operators P and P^r, except that only non-empty index sets and proper filters are allowed.

Let $\underline{O}$ be the semigroup generated by the above operators -- **application from the left !** --, i.e. of all finite sequences $Y_1Y_2\ldots Y_n$ for some natural number $n \geq 1$, where each Y_i $(1 \leq i \leq n)$ is one of the above operators.

For $Y, Y' \in O$ we define

$Y \subseteq Y'$ iff for every $K \subseteq Alg(\underline{S})$ one has $Y(K) \subseteq Y'(K)$;

and we set $Y = Y'$, if $Y \subseteq Y'$ and $Y' \subseteq Y$.

For convenience we add to the generating set of operators of $\underline{O}$ the operator i: $i(K) := K$ for all $K \subseteq Alg(\underline{S})$, i.e. the identity operator.

For later use we define

4.6.3 DEFINITION. Let $K \subseteq Alg(\underline{S})$. K is called <u>primitive</u>, iff $HSP(K) = K$, and <u>quasi-primitive</u>, iff $SP(K) = K$.

4.6.4 LEMMA. Let $Y \in O$, then one has on $Alg(\underline{S})$ (if P_u is concerned, τ has to be finitary):

(i) $iY = Yi = Y$, and $K \subseteq K' \subseteq Alg(\underline{S})$ implies $Y(K) \subseteq Y(K')$.

(ii) $i \subseteq Y$; and
$V, W \in O$ and $V \subseteq W$ imply $YV \subseteq YW$ and $VY \subseteq WY$.

(iii) $IY = YI$.

(iv) $IY = Y$, if $Y \in \{H, H_f, H_c, S, S_r, S_w, I, P, P^r, P_u, P_+, P_+{}^r\}$ or if $Y \in O$ is a sequence which somewhere contains one of these operators.

(v) $YY = Y$ for all $Y \in \{i, H, H_f, H_c, S, S_r, S_w, I, P, P^r, P_u, P_+, P_+{}^r, o, e\}$.

(vi) $P = P_+e = eP_+$ and $P^r = P_+{}^re = eP_+{}^r$.

(vii) $I \subseteq H_c \subseteq H_f \subseteq H$, $I \subseteq S \subseteq S_r \subseteq S_w$, $I \subseteq P_+ \subseteq P \subseteq P^r$, $I \subseteq P_+ \subseteq P_+{}^r \subseteq P^r$, $I \subseteq P_u \subseteq P^r$.

(viii) Let $Y \in \{H, H_f, H_c\}$, $V \in \{S, S_r, S_w\}$, $W \in \{P, P^r, P_u, P_+, P_+{}^r\}$, then $VY \subset YV$, $WY \subseteq YW$, and $WV \subseteq VW$.

(ix) With the notation from (viii) we have that Y, V, W, YV, YW and YVW are all closure operators on $Alg(\underline{S})$.

(x) When we only allow nonempty phyla or only sort algebras whose carrier sets have less than d* elements, then $P_+{}^r \subseteq HP_+$ and $P^r \subseteq HP$. And when we consider only finitary partial algebras, then $P^r \subseteq SPP_u$ and $P_+{}^r \subseteq SP_+P_u$.

SKETCH of PROOF. The statements (i), (ii), (iii), (iv), (vi) and (vii) are easily realized from the definitions. ∎

In what follows **K** is any subclass of **Alg**($\underline{S}$).

Ad (v): $\mathbf{Y} \subseteq \mathbf{YY}$ is obvious. $\mathbf{YY} \subseteq \mathbf{Y}$ is easily seen, if $\mathbf{Y} \in \{\mathbf{i},\mathbf{H},\mathbf{H_f},\mathbf{H_c},\mathbf{S},\mathbf{S_r},\mathbf{S_w},\mathbf{I},\mathbf{o},\mathbf{e}\}$. Let us argue for products first: Let for each $i \in I$ $(\underline{A}_i := \mathbf{X}(\underline{B}_{ij} | j \in J_i),(p_{ij} | j \in J_i))$ be products, and let $(\underline{A} := \mathbf{X}(\underline{A}_i | i \in I),(p_i | i \in I))$ be their product. We claim that $(\underline{A},(p_i p_{ij} | i \in I , j \in J_i))$ is a product of $(\underline{B}_{ij} | i \in I , j \in J_i)$: -- Let $(f_{ij}:\underline{C} \to \underline{B}_{ij} | i \in I , j \in J_i)$ be a source of homomorphisms. Then we have for each $i \in I$ a unique $f_i:\underline{C} \to \underline{A}_i$ satisfying $f_i p_{ij} = f_{ij}$ $(j \in J_i)$. These f_i induce a unique homomorphism $f:\underline{C} \to \underline{A}$ satisfying $fp_i = f_i$ for each $i \in I$; thus $fp_i p_{ij} = f_i p_{ij} = f_{ij}$ for each $i \in I$ and $j \in J_i$. If $I \neq \emptyset$ and some $J_i \neq \emptyset$, then also $\mathbf{U}(\{i\} \times J_i | i \in I) \neq \emptyset$.

In the case of reduced products let -- with the above notation -- $\mathbf{F}_i$ be a $\mathbf{d^*}$-complete filter on J_i $(i \in I)$, and $\mathbf{F}$ be a $\mathbf{d^*}$-complete filter on I. Then it is not too difficult, but combined with some computation, to see that $\mathbf{G} := \{\mathbf{U}(\{i\} \times G_i | i \in I) | G_i \subseteq J_i \ (i \in I)$ and $\{i \in I | G_i \in \mathbf{F}_i\} \in \mathbf{F}\}$ is a $\mathbf{d^*}$-complete filter on $J := \mathbf{U}(\{i\} \times J_i | i \in I)$, and that $\mathbf{X}(\underline{B}_{ij} | (i,j) \in J)/\mathbf{G} \cong \mathbf{X}(\mathbf{X}(\underline{B}_{ij} | j \in J_i)/\mathbf{F}_i | i \in I)/\mathbf{F}$, where one can use in the case of non-empty phyla that $(b_{ij} | (i,j) \in J)/\mathbf{G} \mapsto ((b_{ij} | j \in J_i)/\mathbf{F}_i | i \in I)/\mathbf{F}$, and $(b_{ij} | (i,j) \in J) R_{\mathbf{G}} (b_{ij}' | (i,j) \in J)$ iff $\{i | I_{(b_{ij} | j \in J_i)(b_{ij}' | j \in J_i)} \in \mathbf{F}_i\} \in \mathbf{F}$, iff $((b_{ij} | j \in J_i)/\mathbf{F}_i | i \in I)/\mathbf{F} = ((b_{ij}' | j \in J_i)/\mathbf{F}_i | i \in I)/\mathbf{F}$ (for notation see 4.5.4); for the structure the argumentation is similar. -- In general one has to use the argumentation for direct limits. Then one should, for each $G \in \mathbf{G}$, define $F_G := \{i \in I | \{j \in J_i | (i,j) \in G\} \in \mathbf{F}_i\}$ (therefore $G \in \mathbf{G}$ iff $F_G \in \mathbf{F}$), and observe -- using results from section 11 -- that one has a canonical homomorphism $q_G:\underline{B}_G \to \underline{A}_{F_G}$ -- where $\underline{A}_i := \mathbf{X}(\underline{B}_{ij} | j \in J_i)/\mathbf{F}_i$ for each $i \in I$ --, which is a morphism between directed systems and therefore induces a homomorphism between their direct limits, for which it has to be proved that it is an isomorphism.

Assume now that the similarity type τ is finitary, that $\mathbf{F}$ and all $\mathbf{F}_i$ $(i \in I)$ are ultrafilters, and let $\mathbf{G}$ be the filter defined above. We claim that $\mathbf{G}$ is also an ultrafilter: Thus assume $M \subseteq J$, $M \notin \mathbf{G}$; then $F := \{i \in I | M_i := \{j \in J_i | (i,j) \in M\} \in \mathbf{F}_i\} \notin \mathbf{F}$. Let $\overline{M} := J \setminus M$, $\overline{F} := I \setminus F$. Then $\overline{F} \in \mathbf{F}$, since $\mathbf{F}$ is an ultrafilter, and for each $i \in \overline{F}$ $M_i \notin \mathbf{F}_i$. Therefore $\overline{M}_i := J_i \setminus M_i = \{j \in J_i | (i,j) \in \overline{M}\} \in \mathbf{F}_i$, since $\mathbf{F}_i$ is an ultrafilter. Thus $\overline{M} \in \mathbf{G}$, showing that $\mathbf{G}$ is an ultrafilter (observe that $\mathbf{G}$ cannot be equal to **Subs**(J)).

In the remaining cases the argumentation is similar. ∎

Ad (viii): $\mathbf{VY} \subseteq \mathbf{YV}$: Let $\underline{A}$ be a **V**-substructure of a **Y**-homomorphic image $\underline{B}$ of $\underline{C} \in \mathbf{K}$, i.e. we have $f:\underline{C} \to \underline{B}$ in **Y** (and w.l.o.g. $\underline{A} \subseteq \underline{B}$). Supply $D := [A]f^{-1}$ with the richest weak substructure (getting $\underline{D}$) of $\underline{C}$ which allows $f' := f|D:D \to A$ to be still a homomorphism from $\underline{D}$ into $\underline{A}$. Then it is not too difficult to prove that $\underline{D}$ is a **V**-substructure of $\underline{C}$ and $\underline{A}$ a **Y**-homomorphic image of $\underline{D}$; therefore $\mathbf{VY} \subseteq \mathbf{YV}$. -- In order to show $\mathbf{VY} \neq \mathbf{YV}$ choose τ to be the type of commutative rings with unit: $\tau := (2,1,0,2,0)$ ($\underline{S}$ the one-element total algebra, i.e. homogeneous case), let $\underline{D} := \underline{\mathbf{Z}}$ be the ring of integers and $\underline{A} := \underline{\mathbf{Z}}/3\mathbf{Z}$ its quotient ($f' := nat_{R_{3Z}}$), and consider $\underline{D}$ as a subring of the ring $\underline{\mathbf{Q}}$ of rational numbers. Obviously $\underline{A} \in \mathbf{YV}(\{\underline{\mathbf{Q}}\})$, but $\underline{A} \notin \mathbf{VY}(\{\underline{\mathbf{Q}}\})$, since $\underline{\mathbf{Q}}$ only allows homomorphisms onto itself or into the one-element ring, and $\underline{A}$ is not a substructure of any of them.

$\mathbf{WY} \subseteq \mathbf{YW}$: Let $\underline{A}$ be a **W**-product of **Y**-homomorphic images of **K**-algebras, say, for $(\underline{C}_i|i \in I) \in \mathbf{K}^I$ we have $f_i:\underline{C}_i \to \underline{B}_i$ $(i \in I)$ and a **d***-complete filter **F** on I (which is just I in the case of products and an ultrafilter in the (finitary) case of ultraproducts) such that $\underline{A} \cong \mathbf{X}(\underline{B}_i|i \in I)/F =: \underline{B}$ with an isomorphism $g:\underline{A} \to \underline{B}$. If $\underline{C} := \mathbf{X}(\underline{C}_i|i \in I)/F$, then let, for each $F \in \mathbf{F}$, $f_F:\underline{C}_F \to \underline{B}_F$ be the direct product of the family $(f_i|i \in F)$. Then $(id_{\mathbf{F}},(f_F|F \in \mathbf{F}))$ is a morphism between directed systems in the sense of section 11 which induces g, and in subsections 11.1 and 11.3 it will be shown that g inherits injectivity, closedness, fullness and surjectivity from the f_F, which -- in turn -- inherit these properties from the f_i (i.e. if all of the f_i have one of these properties, then so do all f_F and therefore also g). Hence this argument is also suitable to prove $\mathbf{WV} \subseteq \mathbf{VW}$.

In general inequality also holds in the two statements under consideration; for more information see H.HÖFT [Hö 72], D.PIGOZZI [Pig 66] or Zs.MARKUSZ [Mr 83], [Mr 83a]. ∎

Ad (ix): Remembering (i) and (ii) one easily sees that each $\mathbf{Y} \in \mathbf{O}$ is extensive and monotone. Idempotency of **I**, **Y**, **V** and **W** follows from (v). Using all the other properties "proved" so far we get:

$\mathbf{YVW}(\mathbf{K}) \subseteq \mathbf{YV}(\mathbf{WY})\mathbf{VW}(\mathbf{K}) \subseteq \mathbf{Y}(\mathbf{VY})(\mathbf{WV})\mathbf{W}(\mathbf{K}) \subseteq \mathbf{YYVVWW}(\mathbf{K}) \subseteq \mathbf{YVW}(\mathbf{K})$, showing equality. ∎

Ad (x): Let $(\underline{A}_i|i \in I) \in \mathbf{K}^I$ be any family of **K**-algebras and F an at least **d***-complete filter on I. When all phyla are non-empty, then $\underline{C} := \mathbf{X}(\underline{A}_i|i \in I)/F$ is obviously a (weak) homomorphic image of $\underline{A} := \mathbf{X}(\underline{A}_i|i \in I)$ (via p_I, see 4.5.3). If empty phyla are allowed, but $\#S < \mathbf{d^*}$, then for every $s \in S$ for which $C_s \neq \emptyset$, there exists an $F_s \in \mathbf{F}$ such that $A_{is} \neq \emptyset$ for all $i \in F_s$. Set $F_s := I$ for all other $s \in S$. Then **d***-directedness of F implies $F := \cap(F_s|s \in S) \in \mathbf{F}$, and $\underline{C}$ is then a homomorphic image of $\underline{A}_F := \mathbf{X}(\underline{A}_i|i \in F)$. -- That the assumptions are necessary is shown by the following example, where we consider only sets and let $I =: S$ be any infinite set, $A := ((\{0\} \times S) \cup (\{1\} \times S), v_A)$, where $(j,s)v_A := s$ $(j \in \{0,1\}$, and $s \in S)$, and $A_i := A \setminus \{(0,i),(1,i)\}$ with the restriction of v_A as sort mapping v_i $(i \in I)$, i.e. all phyla contain zero or two elements and exactly one phylum is empty. Let **F** be the Fréchet filter on I of all subsets of I, whose complement has cardinality less than I. Then all A_F $(F \in \mathbf{F})$ -- in fact all products with non-empty index set of any family $(B_j|j \in J) \in \{A_i|i \in I\}^J$ -- have at least one empty phylum, while in $\mathbf{X}(A_i|i \in I)/F$ each phylum is infinite. -- Now let τ be finitary, and consider $\underline{A} \in \mathbf{P}^r(\mathbf{K})$, i.e. we have a set I, a filter **F** on I and a family $(\underline{A}_i|i \in I) \in \mathbf{K}^I$ such that $\underline{A} \cong \mathbf{X}(\underline{A}_i|i \in I)/F$. If $I = \emptyset$, then $\underline{A}$ is isomorphic to the so-called empty product and hence belongs to $\mathbf{SPP}_u(\mathbf{K})$. Therefore let I be non-empty. If **F** is not proper, then the result is the same as if I were empty; therefore **F** can also be assumed to be proper. Now let $u(\mathbf{F}) := \{\mathbf{G}|\mathbf{G} \supseteq \mathbf{F}$ is ultrafilter on I$\}$. Then the Axiom of Choice tells us that $u(\mathbf{F}) \neq \emptyset$ and $\mathbf{F} = \cap u(\mathbf{F})$. We claim that $\underline{A}$ is isomorphic to a subalgebra of $\mathbf{X}(\mathbf{X}(\underline{A}_i|i \in I)/\mathbf{G}|\mathbf{G} \in u(\mathbf{F}))$: Let $\mathbb{A}_{\mathbf{F}} := (\underline{A}_F, p_{FF'}|F \supseteq F' \in \mathbf{F})$ be the direct system belonging to $\underline{A}$, and let $\mathbb{A}_{\mathbf{G}} := (\underline{A}_G, q_{GG'}|G \supseteq G' \in \mathbf{G})$ be the direct system for any $\mathbf{G} \in u(\mathbf{F})$ with direct limit $((q_G|G \in \mathbf{G}),\underline{A}(\mathbf{G}))$. Since $\mathbb{A}_{\mathbf{F}}$ is a subsystem of $\mathbb{A}_{\mathbf{G}}$ the subfamily $(q_F|F \in \mathbf{F})$ induces a homomorphism $q_{\mathbf{G}}:\underline{A} \to \underline{A}(\mathbf{G})$. Hence we get a product homomorphism $q:\underline{A} \to \mathbf{X}(\underline{A}(\mathbf{G})|\mathbf{G} \in u(\mathbf{F}))$. With arguments similar to the ones in subsection 4.5 -- and using $\mathbf{F} = \cap u(\mathbf{F})$ -- one can show that q is injective and closed. ∎

This ends the sketch of the proof of Lemma 4.6.4. ∎∎∎

The concepts of freeness, and more generally of **K**-universal solutions for some class **K** of (partial) algebras, play a fundamental rôle in universal algebra and also in the theory of partial algebras. Starting from these general concepts we mainly discuss in this section the case when **K** = **TAlg**($\underline{S}$) is the class of all total algebras. The results presented here then form the basis of the treatment of more general situations and of a model theoretic approach to partial algebras in later chapters. As before we usually consider partial $\underline{S}$-algebras without specially mentioning it, except when we explicitly restrict considerations to the homogeneous case.

5.1 The general concepts of universal solution and freeness

5.1.1 DEFINITION. Let $\underline{\mathbf{C}}$ be any category, $\mathbf{K} \subseteq Ob(\mathbf{C})$ any class of $\underline{\mathbf{C}}$-objects and $f:A \to B$ a $\underline{\mathbf{C}}$-morphism.

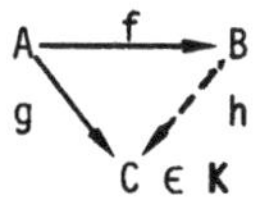

Figure 5.1.

(a) f is called **K**-extendable (cf. HERRLICH, STRECKER [HS 73], Definition 37.8; in [NSa 77], [AN 83] and other papers of ANDRÉKA, NÉMETI and SAIN f is said to be valid in **K** and **K** is said to be injective w.r.t. (with respect to) f), iff for every $C \in \mathbf{K}$ and for every $\underline{\mathbf{C}}$-morphism $g:A \to C$ there exists a $\underline{\mathbf{C}}$-morphism $h:B \to C$ such that $fh = g$ (i.e. the diagram in Figure 5.1 is commutative).

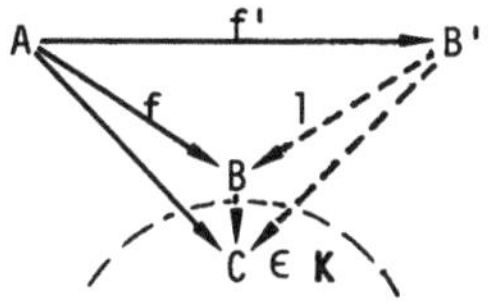

Figure 5.2

(b) f is called **K**-universal, if it is a **K**-extendable **epimorphism**, and if for every **K**-extendable epimorphism $f':A \to B'$ there exists a $\underline{\mathbf{C}}$-morphism $l:B' \to B$ such that $f'l = f$ (cf. Figure 5.2).

(c) If $f:A \to B$ is a **K**-universal epimorphism, then (f,B) is called a **K**-universal solution of A, and if, in addition, $B \in \mathbf{K}$, then (f,B) is called a **K**-universal **K**-solution of $\underline{A}$.

5.1.2 EXAMPLES. Let us present already at this point some simple examples: For every $\underline{A} \in \mathbf{Alg}(\underline{S})$ and for every $\mathbf{K} \subseteq \mathbf{Alg}(\underline{S})$ the identity mapping $id_A:\underline{A} \to \underline{A}$ is **K**-extendable. -- Let $\underline{F}(M,\mathbf{Gr})$ be the free group with free generating set M, then $id_M:M \to \underline{F}(M,\mathbf{Gr})$ is a **Gr**-universal homomorphism (where M is considered as a discrete partial algebra), and $(id_M,\underline{F}(M,\mathbf{Gr}))$ is a **Gr**-universal **Gr**-solution of M. -- We leave it as an exercise for the reader to realize that for the class **R** of all rings and for any ring $\underline{R} \in \mathbf{R}$ and any element $x \notin R$ the usual polynomial ring $\underline{R}[x]$ together with the epimorphism $id_{R\cup\{x\}}:\underline{R} \cup \{x\} \to \underline{R}[x]$ is the **R**-universal **R**-solution of $\underline{R} \cup \{x\} = \underline{R} + \underline{\{x\}_d}$.

5.1.3 PROPOSITION. Let $\underline{\mathbf{C}}$ be any category, **K** any class of $\underline{\mathbf{C}}$-objects and $A \in Ob(\underline{\mathbf{C}})$.

(i) Any **K**-universal $\underline{\mathbf{C}}$-morphism for A is unique up to unique isomorphism, i.e. if $f:A \to B$ and $f':A \to B'$ both are **K**-universal epimorphisms, then there exists a unique isomorphism $i:B \to B'$ such that $fi = f'$.

(ii) If $f:A \to B$ is a **K**-extendable $\underline{\mathbf{C}}$-epimorphism and $B \in \mathbf{K}$, then f is **K**-universal, and (f,B) is a **K**-universal **K**-solution of A.

(iii) The composition of **K**-extendable $\underline{\mathbf{C}}$-morphisms is **K**-extendable, and each isomorphism is **K**-extendable.

PROOF. Ad (i): Let $f:A \to B$ and $f':A \to B'$ be K-universal epimorphisms. Then there exist according to 5.1.1.(b) morphisms $g:B \to B'$ and $h:B' \to B$ such that $fg = f'$, $f'h = f$, $f1_B = f = f'h = fgh$. Since f is an epimorphism, this implies $1_B = gh$. Similarly we can show $hg = 1_{B'}$; therefore g and h are isomorphisms inverse to each other. Uniqueness follows, since both f and f' are epimorphisms. ∎

Ad (ii): Since $B \in K$, 5.1.1.(b) and (c) are easily verified. ∎

Ad (iii): These statements are immediately realized. ∎∎∎

5.1.4 DEFINITION and NOTATION. (i): The result of 5.1.3.(i) allows to introduce a special notation for K-universal epimorphisms, whenever they exist: We write $r_{\underline{A},\mathbf{K}}:\underline{A} \to \underline{F}(\underline{A},\mathbf{K})$, if this is a K-universal solution of $\underline{A} \in \mathbf{Alg}(\underline{S})$ for some $\mathbf{K} \subseteq \mathbf{Alg}(\underline{S})$. If $\underline{A}$ is discrete, we usually write $\underline{F}(A,\mathbf{K})$ instead of $\underline{F}(\underline{A},\mathbf{K})$, and if $\underline{F}(A,\mathbf{K}) \in \mathbf{K}$, then we call this partial algebra the K-free K-algebra, K-freely generated by A (even when $r_{A,\mathbf{K}}$ is not injective), and this expression is often also used, when $\underline{A}$ is not discrete (instead of "K-universal K-solution of $\underline{A}$").

(ii): Let $M \subseteq \underline{A} \in \mathbf{Alg}(\underline{S})$: If $id_M:M \to \underline{C}_{\underline{A}}M$ is K-extendable, i.e. if every mapping $f:M \to \underline{B} \in \mathbf{K}$ extends to a homomorphism $\overline{f}:\underline{C}_{\underline{A}}M \to \underline{B}$, then M is called a K-independent (or K-free) subset of $\underline{A}$; if $\mathbf{K} = \{\underline{B}\}$, then we write "$\underline{B}$-independent" rather than "$\{\underline{B}\}$-independent". -- If M generates $\underline{A}$ and is $\underline{A}$-independent, then M is called a basis of $\underline{A}$, and $\underline{A}$ itself is called a (relatively) free partial algebra. Moreover we introduce the independence class of M with respect to $\underline{A}$ as

$$ind_{\underline{A}}(M) := \{\underline{B} \in \mathbf{Alg}(\underline{S}) \mid M \text{ is } \underline{B}\text{-independent in } \underline{A}\} .$$

Similarly, if E is any class of homomorphisms, we introduce the injectivity class

$$Inj(\mathbf{E}) := \{\underline{A} \in \mathbf{Alg}(\underline{S}) \mid \text{each } e \in \mathbf{E} \text{ is } \{\underline{A}\}\text{-extendable}\} .$$

5.1.5 REMARKS. (i): Although the notions which we have introduced above belong more or less to the same concept, they have developed somewhat independently and do not show this interconnection directly. Notice, for instance, that $ind_{\underline{A}}(M) = Inj(\{id_M:M \to \underline{C}_{\underline{A}}M\})$.

(ii): Notice that all direct products on empty index sets (i.e. isomorphic to $(\underline{S},id_S)$) belong to Inj(E) and therefore also to $ind_{\underline{A}}(M)$ for every class E of homomorphisms. More generally we have the result in 5.1.6 below.

5.1.6 PROPOSITION. Let $\mathbf{K} \subseteq \mathbf{Alg}(\underline{S})$ be any subclass.

(i) Every K-extendable homomorphism is also **P(K)**-extendable, and if it is an epimorphism, then it is even **SP(K)**-extendable.

(ii) Every K-independent subset M of some partial algebra $\underline{A}$ is also **HSP(K)**-independent, and if $M \neq \emptyset$, then it is even **HSPo(K)**-independent. M is $\{\underline{B}\}$-independent for all $\underline{B} \in \mathbf{Alg}(\underline{S})$ satisfying $[M]v_M \not\subseteq [B]v_B$.

(iii) Every injectivity class for epimorphisms is quasi-primitive, and every independence class is primitive.

PROOF. (iii) is an immediate consequence of (i) and (ii), respectively. ∎

Ad (i): Let $e:\underline{B} \to \underline{C}$ be K-extendable, let I be any set, $(\underline{A}_i \mid i \in I) \in \mathbf{K}^I$, $(\underline{A},(pr_i \mid i \in I))$ a product of this family and $f:\underline{B} \to \underline{A}$ any homomorphism. Since $\underline{A}_i \in \mathbf{K} \subseteq Inj(\{e\})$, there exist $g_i:\underline{C} \to \underline{A}_i$ such that $eg_i = f \cdot pr_i$ $(i \in I)$. Then $(g_i \mid i \in I)$ induces a homomorphism $g:\underline{C} \to \underline{A}$ satisfying $g_i = g \cdot pr_i$ $(i \in I)$. Hence

$f\cdot pr_i = e\cdot g_i = e\cdot g\cdot pr_i$ $(i \in I)$, i.e. $f = eg$, showing that $\underline{A} \in Inj(\{e\})$. -- If e is an epimorphism, $\underline{D} \in \mathbf{K}$, $\underline{A}$ a subalgebra of $\underline{D}$, and $f:\underline{B} \to \underline{A}$ any homomorphism, then $f\cdot id_A:\underline{B} \to \underline{D}$ implies that there exists $g:\underline{C} \to \underline{D}$ such that $f\cdot id_A = e\cdot g$. By 3.6.2.(iii) we have $[C]g \subseteq A$, and therefore already $g:\underline{C} \to \underline{A}$, satisfying $eg = f$. ■

Ad (ii): Because of (i) and 5.1.5.(i) we only have to show that a **K**-independent set is also **H(K)**-independent: Let $\underline{D} \in \mathbf{K}$, $h:\underline{D} \to \underline{E}$ any surjective homomorphism, and $f:M \to E$ any mapping (M a **K**-independent subset of $\underline{A}$). Since h is surjective, the Axiom of Choice yields a mapping $g':M \to D$ such that $g'h = f$. Since M is $\underline{D}$-independent in $\underline{A}$, we get a homomorphism $g'':\underline{C}_A M \to \underline{D}$ such that $id_M\cdot g'' = g'$: Then $g''h:\underline{C}_A M \to \underline{E}$ is a homomorphism satisfying $id_M\cdot g''\cdot h = g'\cdot h = f$, i.e. extending f. -- If $M \neq \emptyset$, then there does not exist a mapping into the empty partial algebra $\underline{\emptyset}$; therefore M is trivially $\underline{\emptyset}$-independent in any partial algebra containing M, and a similar argument yields the more general statement. ■■■

The last argument even yields 5.1.7, while 5.1.8 follows from 5.1.6.(i):

5.1.7 COROLLARY 1. Let $e:\underline{B} \to \underline{C}$ be any homomorphism and $\underline{D} \in \mathbf{Alg}(\underline{S})$ such that $[B]v_B \setminus [D]v_D \neq \emptyset$, then $\underline{D} \in Inj(\{e\})$. ■■■

5.1.8 COROLLARY 2. Every **K**-universal epimorphism is also **SP(K)**-universal. And if $(r_{\underline{A},\mathbf{K}}, \underline{F}(\underline{A},\mathbf{K}))$ is a **K**-universal solution of $\underline{A}$, then it is also an **SP(K)**-universal solution of $\underline{A}$. ■■■

5.2 Partial Peano algebras

5.2.1. So far we have only dealt with very abstract concepts, and it is not yet quite clear, in what instances the morphisms and objects which we have discussed in 5.1 will exist. Of special interest is the case where the subclass **K** of $\mathbf{Alg}(\underline{S})$ equals $\mathbf{TAlg}(\underline{S})$, the subclass of all total ($\underline{S}$-) algebras of type τ. Especially the $\mathbf{TAlg}(\underline{S})$-free $\mathbf{TAlg}(\underline{S})$-algebras $\underline{F}(M,\mathbf{TAlg}(\underline{S}))$ play a very important rôle e.g. as **term algebras**, i.e. as starting points for model theoretic investigations. -- In order to show their existence and to get a better knowledge of them, we introduce the concept of partial Peano algebras.

5.2.2 DEFINITION. Let $\underline{B} \in \mathbf{Alg}(\underline{S})$ be any partial algebra of similarity type $\tau = (K_\phi)_{\phi\in\Omega}$, and let $X \subseteq B$ be any subset. Then $\underline{B}$ is called a <u>partial Peano algebra on X</u> (<u>with Peano basis</u> X), iff the following **generalized Peano axioms** are valid in $\underline{B}$:

(P1) For every $\phi \in \Omega$ and for every $\underline{a} \in \mathrm{dom}\,\phi^{\underline{B}}$, $\underline{a}\phi^{\underline{B}} \notin X$ (i.e. $X \cap \mathbf{U}([\mathrm{dom}\,\phi^{\underline{B}}]\phi^{\underline{B}} \mid \phi \in \Omega) = \emptyset$).

(P2) For any $\phi,\phi' \in \Omega$ and for any $\underline{a} \in \mathrm{dom}\,\phi^{\underline{B}}$, $\underline{a}' \in \mathrm{dom}\,\phi'^{\underline{B}}$ one has:
$$\underline{a}\phi^{\underline{B}} = \underline{a}'\phi'^{\underline{B}} \text{ implies } \phi = \phi' \text{ and } \underline{a} = \underline{a}'.$$

(P3) (**Axiom of Induction**) $C_{\underline{B}}X = B$.

5.2.3. Notice that (P2) says that each partial operation is injective and that the images of different partial operations are disjoint, while (P3) means that X generates $\underline{B}$; and because of (P1) X is obviously the smallest possible generating subset of $\underline{B}$.
If a partial Peano algebra is total (as an $\underline{S}$-algebra), then it is called a <u>Peano algebra</u> (or for reasons to be discussed later a <u>word algebra</u>, or an ($\underline{S}$-) <u>term algebra</u> on X).

5.2.4 EXAMPLES. (i): A discrete partial algebra $\underline{A}_d$ is always a partial Peano algebra on A. ∎

(ii): Let $':\mathbf{N} \to \mathbf{N}$, $n' := n+1$ be the successor function on the set $\mathbf{N}$ of natural numbers, then $(\mathbf{N};')$ is a Peano algebra of type (1) on {0}, and in this case (P1), (P2) and (P3) are nothing else than the usual Peano axioms for the natural numbers. ∎

(iii): The intuitive construction of partial Peano algebras of a given type runs as follows (the heterogeneous case is similar, but observe the correct phyla): Let X be any set; for each $\phi \in \Omega$ and for (some) $\underline{b} \in X^{K_\phi}$ choose a new element $\underline{b}\phi$ not contained in X and different from $\underline{b}'\phi'$ if $\phi \neq \phi'$ or $\underline{b} \neq \underline{b}'$. Then "do the same thing" with the collection of new and old elements as long as you wish -- and can --, observing that to sequences (and corresponding operation symbols) which have already been considered in some step there will not be assigned a new element in a later step. Then it is easy to realize that at each step this procedure yields a partial Peano algebra on X. ∎

More examples are obtained with the following Lemma, whose proof is a nice example for an algebraic induction.

5.2.5 LEMMA. Let $\underline{B}$ be any partial Peano algebra on $X \subseteq B$, $\underline{A}$ any weak subalgebra of $\underline{B}$, and let $Y := A \setminus \mathbf{U}([\mathrm{dom}\ \phi^{\underline{A}}]\phi^{\underline{A}} | \phi \in \Omega)$. Then $\underline{A}$ is a partial Peano algebra on Y.

PROOF. (P1) is satisfied because of the definition of Y; and (P2) holds in $\underline{A}$, because it is true in $\underline{B}$. Thus it remains to prove (P3), and this is done by proving by algebraic induction for $x \in B$ the statement

$H(x) :=$ " $x \in A$ implies $x \in \mathbf{C}_{\underline{A}}Y$ ":

Assume $x \in X$; if $x \in A$, then $x \in Y$ by (P1) for $\underline{B}$ and the definition of Y. -- Thus, assume $x \in B$, $x = \underline{b}\phi^{\underline{B}}$ for some $\phi \in \Omega$ and $\underline{b} \in \mathrm{dom}\ \phi^{\underline{B}}$, and let the statement $H((k)\underline{b})$ be true for every $k \in K_\phi$. Moreover, let us assume that $x \in A$ (otherwise H(x) is already true). Then either $x \in Y$ -- yielding $x \in \mathbf{C}_{\underline{A}}Y$ --, or there exist $\phi' \in \Omega$ and $\underline{b}' \in \mathrm{dom}\ \phi'^{\underline{A}}$ such that $x = \underline{b}'\phi'^{\underline{A}}$ (by the definition of Y). Since $\mathrm{graph}(\phi'^{\underline{A}}) \subseteq \mathrm{graph}(\phi'^{\underline{B}})$, and since $\underline{B}$ satisfies (P2), we get $\phi' = \phi$ and $\underline{b}' = \underline{b}$. Thus $(k)\underline{b} \in A$ for each $k \in K_\phi$. Therefore, by the induction hypothesis one has $(k)\underline{b} \in \mathbf{C}_{\underline{A}}Y$ $(k \in K_\phi)$. But this implies $x = \underline{b}\phi^{\underline{A}} \in \mathbf{C}_{\underline{A}}Y$, since $\mathbf{C}_{\underline{A}}Y$ is closed in $\underline{A}$. This proves $A \subseteq \mathbf{C}_{\underline{A}}Y$. ∎∎∎

The following results will positively solve the question of existence of total ($\underline{S}$-) Peano algebras in a constructive way.

5.2.6 THEOREM. For every type τ, every sort-algebra $\underline{S}$ of type τ and every $\underline{S}$-set X there exists a total ($\underline{S}$-) Peano algebra $\underline{P}(X,\underline{S})$ with Peano basis X. (The proof contains a construction).

PROOF. (a): We first consider the homogeneous case, using the construction of R.KERKHOFF in [Ke 65]: If $X = \emptyset$ and if τ contains no nullary constants, then set $\underline{P}(\emptyset,\underline{S}) := \underline{\emptyset}$; this is obviously a Peano algebra over X. Else define

(*) $Z := (\{0\} \times X) \cup (\{1\} \times \Omega) \cup \mathbf{U}(K_\phi | \phi \in \Omega)$,

-- the disjointness of the images of X and Ω will later become important -- and let

(**) $Z^* := \{(z_1,z_2,\ldots,z_n) | n \in \omega_0,\ n \geq 1,\ z_1,z_2,\ldots,z_n \in Z\}$,

be the set of all non-empty finite sequences of elements of Z. Moreover, set $G := \mathbf{Subs}(Z^*)$, the set of all subsets of Z*, and introduce on G a total algebraic structure as follows (for $\varphi \in \Omega$ and $(M_k|k \in K_\varphi) \in G^{K_\varphi}$):

(***) $(M_k|k \in K_\varphi)\varphi^{\underline{G}} := \{((1,\varphi))\} \cup \{(k,z_1,\dots,z_n)|k \in K_\varphi,\ (z_1,\dots,z_n) \in M_k\}$

-- notice that $\varphi^{\underline{G}} = \{((1,\varphi))\}$, if $K_\varphi = \emptyset$. We have a natural injection $j:X \to G$ by $x \mapsto \{((0,x))\}$, and define $\underline{W}$ as the subalgebra of $\underline{G}$ generated by $[X]j$. Observe that there is an algebra $\underline{W}'$ containing X and isomorphic to $\underline{W}$ over a homomorphic extension of j (see 23.2.8 in the Appendix), and it is easy to realize, that $\underline{W}'$ is a Peano algebra over X once we have shown that $\underline{W}$ is a Peano algebra over $[X]j$:

(P3) is satisfied by the definition of $\underline{W}$. ■

Ad (P1): If $(M_k|k \in K_\varphi)\varphi^{\underline{W}} \in [X]j$, then $(M_k|k \in K_\varphi)\varphi^{\underline{W}} = \{((0,x))\}$ for some $x \in X$. By (***) $(M_k|k \in K_\varphi)\varphi^{\underline{W}}$ contains exactly one sequence of length 1, namely $((1,\varphi))$, which is different from $((0,x))$. Thus $(M_k|k \in K_\varphi)\varphi^{\underline{W}} \notin [X]j$, a contradiction. ■

Ad (P2): Assume that $(M_k|k \in K_\varphi)\varphi^{\underline{W}} = (N_l|l \in K_{\varphi'})\varphi'^{\underline{W}}$. Since each element of W contains only one sequence of length 1, and since this is $((1,\varphi))$ respectively $((1,\varphi'))$ above, we may conclude $\varphi = \varphi'$ and $K_\varphi = K_{\varphi'}$. Now, let $k \in K_\varphi$ and $(z_1,\dots,z_n) \in M_k$, then $(k,z_1,\dots,z_n) \in (N_l|l \in K_{\varphi'})\varphi'^{\underline{W}}$, i.e. $(z_1,\dots,z_n) \in N_k$ showing that $M_k \subseteq N_k$ $(k \in K_\varphi)$. Symmetry of the situation now yields equality everywhere, which was to be shown. ■

(b): We now consider the heterogeneous situation and present a more intuitive description of a total ($\underline{S}$-) Peano algebra connected with the above construction: Since each fundamental operation is defined in $\underline{S}$ on at most -- and usually exactly -- one sequence, it is associated with at most (exactly) one output sort. For given $X = (X,v_X:X \to S)$ we have already associated an element of S with each element of X. If $w \in W' \setminus X$, then $w = (w_k|k \in K_\varphi)\varphi^{\underline{W}'}$ and we associate with w the value of $\varphi^{\underline{S}}$ in $\underline{S}$, if it exists, and no element of S, otherwise. Thus we get a mapping $v_{W'}$ out of W' into S. Redefine the structure on W' to $(\varphi^{\underline{Y}}|\varphi \in \Omega)$ such that $\underline{Y}$ becomes a weak subalgebra of $\underline{W}'$ and $\underline{a} \in \mathrm{dom}\,\varphi^{\underline{Y}}$ iff $\underline{a} \in \mathrm{dom}\,\varphi^{\underline{W}'}$ and $\underline{a}v_{W'} \in \mathrm{dom}\,\varphi^{\underline{S}}$. By 5.2.5 $\underline{Y}$ is a partial Peano algebra, and it is not too difficult to realize that $\underline{C}_{\underline{Y}}X$ is a total $\underline{S}$-algebra satisfying the axioms (P1), (P2) and (P3). ■■■

5.2.7 REMARKS. (i): In the finitary case the proof is usually given by considering the set W of all empty or non-empty "words", i.e. finite sequences from elements in $(\{0\} \times X) \cup (\{1\} \times \Omega)$, such that the structure is defined by $(z_1,\dots,z_n)\varphi^{\underline{W}} := z_1\hat{}\dots\hat{}z_n\hat{}((1,\varphi))$ ($\varphi \in \Omega$ and $(z_1,\dots,z_n) \in W^{K_\varphi}$, " ^ " here meaning concatenation of sequences). An isomorphic copy of the desired total Peano algebra on X is then obtained in the subalgebra of $\underline{W}$ generated by $\{0\} \times X$. ■

(ii): Another proof of the existence of total ($\underline{S}$-) Peano algebras will follow from the Generalized Recursion Theorem in subsection 5.4.

One of the most important features of Peano algebras is the following

5.2.8 RECURSION THEOREM. Let $\underline{P}$ be any partial Peano algebra on X, $\underline{C} \in$ **TAlg**($\underline{S}$) any total algebra and $f:X \to C$ any mapping. Then there exists a unique homomorphic extension $f':\underline{P} \to \underline{C}$ of f, i.e. $f'|X = f$.

PROOF. Let $f^* := C_{\underline{P}\times\underline{C}}\mathrm{graph}(f)$. When we have shown that $f^* = \mathrm{graph}(f')$ is the graph of a mapping f', then 4.1.10.(ii) (which is easily extended to partial $\underline{S}$-algebras)

implies that f' will be a homomorphism (between partial $\underline{S}$-algebras), and from 4.1.12 and (P3) we may infer that dom f' = P . Therefore we prove by algebraic induction on $p \in P$ that $H(p) :=$ " There exists exactly one $c \in C$ such that $(p,c) \in f^*$ " is true: We first assume $p \in X$: Then $(p,pf) \in f^*$; if also $(p,c) \in f^*$ and $c \neq pf$, then this can only happen, when there are $\varphi \in \Omega$ and $(p_k,c_k) \in f^*$ $(k \in K_\varphi)$ such that $(p_k | k \in K_\varphi)\varphi^{\underline{P}} = p \in X$, and this is impossible by (P1). Hence we have proved the basis of induction. -- Assume now $p \in P \setminus X$, $p = (p_k | k \in K_\varphi)\varphi^{\underline{P}}$, and let $H(p_k)$ be true for each $k \in K_\varphi$. Moreover, let, for each $k \in K_\varphi$, c_k be the unique element from C such that $(p_k,c_k) \in f^*$. Since -- by (P2) -- p can be represented as a value of a fundamental operation in exactly one way, and since each p_k satisfies H, p satisfies H, too, i.e. $c := (c_k | k \in K_\varphi)\varphi^{\underline{C}}$ is the unique element of C satisfying $(p,c) \in f^*$. Hence the principle of algebraic induction yields that f* is the graph of a mapping. ■■■

Consequences of this result, e.g. the fact that $\underline{P}(X,\underline{S})$ is canonically isomorphic to $\underline{F}(X,\mathbf{TAlg}(\underline{S}))$, will be discussed later in a more general context.

5.3 Free completions

In this subsection we describe and discuss **TAlg**($\underline{S}$)-universal **TAlg**($\underline{S}$)-solutions of partial algebras $\underline{A}$. The corresponding object $\underline{F}(\underline{A},\mathbf{TAlg}(\underline{S}))$ for $\underline{A}$ will be shown later to be canonically isomorphic to the so-called **(absolutely) free completions** of $\underline{A}$.

5.3.1 DEFINITION. Let $\underline{A} \in \mathbf{Alg}(\underline{S})$ be any partial algebra. A total algebra $\underline{B} \in \mathbf{TAlg}(\underline{S})$ is called an <u>(absolutely) free completion</u> of $\underline{A}$ iff the following axioms hold:

(FC0) $A \subseteq B$ and for each $\varphi \in \Omega$: $\mathrm{graph}(\varphi^{\underline{A}}) \subseteq \mathrm{graph}(\varphi^{\underline{B}})$.

(FC1) For each $\varphi \in \Omega$ and for each $\underline{b} \in B^{K_\varphi}$ one has: $\underline{b}\varphi^{\underline{B}} \in A$ implies $\underline{b}\varphi^{\underline{B}} = \underline{b}\varphi^{\underline{A}}$ (thus in particular $\underline{b} \in \mathrm{dom}\ \varphi^{\underline{A}}$ $(\subseteq A^{K_\varphi})$).

(FC2) For all $\varphi,\varphi' \in \Omega$ and for all $\underline{b} \in B^{K_\varphi}$, $\underline{b}' \in B^{K_{\varphi'}}$ one has: $\underline{b}\varphi^{\underline{B}} = \underline{b}'\varphi'^{\underline{B}} \notin A$ implies $\varphi = \varphi'$ and $\underline{b} = \underline{b}'$.

(FC3) $\mathbf{C}_{\underline{B}}A = B$ (i.e. the completion $\underline{B}$ of $\underline{A}$ is <u>minimal</u>).

5.3.2 REMARK. The axioms (FC0) through (FC3) are closely related to the generalized Peano axioms in 5.2.2: (FC0) says that $\underline{A}$ is a weak subalgebra of $\underline{B}$, while (FC1) implies that it is even a relative subalgebra of $\underline{B}$ -- relative subalgebras satisfying (FC1) will be called <u>normal</u> according to J.SCHMIDT [Sch 70]. (FC1) means that no value of a fundamental operation which lies in A "can come from the outside" (cf. (P1)). (FC2) says that "outside of $\underline{A}$" (P2) is satisfied, while (FC3) -- the axiom of induction -- corresponds to (P3). Altogether we get that "outside of $\underline{A}$" a free completion like $\underline{B}$ behaves like a partial Peano algebra over A.

5.3.3 COROLLARY. $\underline{B} \in \mathbf{TAlg}(\underline{S})$ is a total ($\underline{S}$-) Peano algebra, iff $\underline{B}$ is the free completion of a discrete partial algebra. ■■■

The existence of free completions also follows from the one of Peano algebras:

5.3.4 THEOREM. For every partial algebra $\underline{A}$ there exists a free completion $\underline{B}$.

PROOF. We shall actually indicate two methods of proof, both providing effective constructions relative to the one of Peano algebras.

(i): Let $\underline{P} := \underline{P}(A,\underline{S})$ be a total Peano algebra on A. Define a relation R' on P as follows: $(p,q) \in R'$, iff there is $\underline{b} \in \mathrm{dom}\,\phi^{\underline{A}}$ such that $p = \underline{b}\phi^{\underline{A}}$ and $q = \underline{b}\phi^{\underline{P}}$. Let $R := \mathrm{Cong}_P(R')$ be the congruence of $\underline{P}$ generated by R'. Then $\underline{P}/R$ satisfies (FC0) through (FC3) with respect to its relative subalgebra on $([A]\mathrm{nat}_R)$, which is isomorphic to $\underline{A}$ over $\mathrm{nat}_R|A$. More details and an intrinsic description of R can be found in [G 68a].

(ii): Let $\underline{W}$ be a total Peano algebra on A, whose existence follows from 5.2.6, and define for $\phi \in \Omega$ and $\underline{w} \in W^{K\phi}$:

$$\underline{w}\phi^{\#} := \begin{cases} \underline{w}\phi^{\underline{A}}, & \text{if } \underline{w} \in \mathrm{dom}\,\phi^{\underline{A}}, \\ \underline{w}\phi^{\underline{W}}, & \text{else.} \end{cases}$$

Let $\underline{W}^{\#} := ((W,(\phi^{\#})_{\phi\in\Omega}),v_W)$ and $\underline{B} := \underline{C}_{\underline{W}^{\#}}A$. Then (FC0), (FC1), (FC2) and (FC3) follow directly from the definition of $\underline{B}$ and the fact that $\underline{W}$ satisfies the Peano axioms (P1), (P2) and (P3). ■■■

We are now able to present a complete characterization of partial Peano algebras:

5.3.5 COROLLARY. $\underline{A}$ is a partial Peano algebra, iff each (and it suffices to say: some) free completion of $\underline{A}$ is a total Peano algebra.

PROOF. It has been proved in 5.2.5 that weak relative subalgebras of Peano algebras are always partial Peano algebras. On the other hand it is easy to check that any free completion of a partial Peano algebra satisfies the axioms (P1), (P2) and (P3). ■■■

5.3.6 REMARK. The first version of the proof of 5.3.4 yields the more familiar construction, while the second one can be interpreted in the way that from each congruence class in the first construction a canonical representative is chosen. Thus we get by (ii) sort of a "representation by canonical terms" in the sense of [GoTWa 78] (cf. also section 17.6), which sometimes gives easier and better insight into the structure of the free completion than the factor algebra.

5.3.7 RECURSION THEOREM for FREE COMPLETIONS. Let $\underline{A}$ be any partial algebra, $\underline{B}$ a free completion of $\underline{A}$, and $\underline{C}$ any total algebra similar to $\underline{A}$. Then there exists for every homomorphism $f:\underline{A} \to \underline{C}$ a unique homomorphism $g:\underline{B} \to \underline{C}$ extending f (i.e. $g|A = f$).

PROOF. Since $\underline{B}$ is a free completion of $\underline{A}$, axioms (FC2) and (FC3) tell us that $\underline{B}^* := ((B,(\phi^*)_{\phi\in\Omega}),v_B)$ is a partial Peano algebra on A, when we define ϕ^* to be the restriction of $\phi^{\underline{B}}$ to $B^{K\phi} \setminus \mathrm{dom}\,\phi^{\underline{A}}$. Hence the Recursion Theorem 5.2.8 yields a homomorphic extension $f':\underline{B}^* \to \underline{C}$ extending f. Since $f'|A = f$ is a homomorphism from $\underline{A}$ into $\underline{C}$ by assumption, f' is compatible with all of the structure of $\underline{B}$, i.e. $f':\underline{B} \to \underline{C}$ is a homomorphism. ■■■

5.3.8 COROLLARY. Let $\underline{A} \in$ **Alg**$(\underline{S})$ be any partial algebra. Then $\underline{A}$ possesses up to unique isomorphism exactly one free completion, say $\underline{B}$, and $(\mathrm{id}_A,\underline{B})$ is "the" **TAlg**$(\underline{S})$-universal **TAlg**$(\underline{S})$-solution of $\underline{A}$, i.e. $\underline{B} \cong \underline{F}(\underline{A},\mathbf{TAlg}(\underline{S}))$ over id_A.

PROOF. From 5.3.7 we infer that $\mathrm{id}_A:\underline{A} \to \underline{B}$ is a **TAlg**$(\underline{S})$-extendable epimorphism for every free completion $\underline{B}$ of $\underline{A}$; therefore it is a **TAlg**($\underline{S}$)-universal **TAlg**($\underline{S}$)-solution according to 5.1.3(ii), while 5.1.3.(i) yields the uniqueness of the free completion up to unique isomorphism. ■■■

5.4 The Generalized Recursion Theorem

In connection with the recursion theorems proved so far we get a very useful tool, which will become the basis of a model theoretic approach to the theory of partial algebras.

5.4.1 GENERALIZED RECURSION THEOREM. Let $f:\underline{A} \to \underline{B}$ be any homomorphism between similar partial algebras $\underline{A}$ and $\underline{B}$. Then there exists a closed homomorphic extension $(\tilde{f})_{\underline{B}} := \tilde{f}:\underline{\mathrm{dom}}\ \tilde{f} \to \underline{B}$ of f such that $\underline{\mathrm{dom}}\ \tilde{f}$ is an A-generated relative subalgebra of the free completion $\underline{F}(\underline{A},\mathbf{TAlg}(\underline{S}))$ of $\underline{A}$; moreover, we have:

(i) $\mathrm{graph}(\tilde{f}) = \mathbf{C}_{\underline{F}(\underline{A},\mathbf{TAlg}(\underline{S}))\times\underline{B}}\ \mathrm{graph}(f)$.

(ii) $\tilde{f}$ is the largest homomorphic extension of f to an A-generated relative subalgebra (what we shall call an A-initial extension) of $\underline{F}(\underline{A},\mathbf{TAlg}(\underline{S}))$.

(iii) Let $\overline{f}:\underline{F}(\underline{A},\mathbf{TAlg}(\underline{S})) \to \underline{F}(\underline{B},\mathbf{TAlg}(\underline{S}))$ be the homomorphic extension of f which exists according to 5.3.7. Then $\mathrm{dom}\ \tilde{f} = [B]\overline{f}^{-1}$ and $\tilde{f} = \overline{f}|\mathrm{dom}\ \tilde{f}$.

PROOF. According to the Recursion Theorem 5.3.7 there exists a homomorphic extension $\overline{f}:\underline{F}(\underline{A},\mathbf{TAlg}(\underline{S})) \to \underline{F}(\underline{B},\mathbf{TAlg}(\underline{S}))$. Moreover, $\underline{A}$ and $\underline{B}$ are relative subalgebras of $\underline{F}(\underline{A},\mathbf{TAlg}(\underline{S}))$ and $\underline{F}(\underline{B},\mathbf{TAlg}(\underline{S}))$, respectively. Thus -- according to Lemma 4.1.13 -- $\mathrm{graph}(\tilde{f}) := \mathbf{C}_{\underline{F}(\underline{A},\mathbf{TAlg}(\underline{S}))\times\underline{B}}\ \mathrm{graph}(f)$ is the graph of a homomorphic extension of f to an A-generated relative subalgebra $\underline{\mathrm{dom}}\ \tilde{f}$ of $\underline{F}(\underline{A},\mathbf{TAlg}(\underline{S}))$; and since $\underline{F}(\underline{A},\mathbf{TAlg}(\underline{S}))$ is total, $\tilde{f}:\underline{\mathrm{dom}}\ \tilde{f} \to \underline{B}$ is closed. And 4.1.13.(iii) implies that $\tilde{f}$ is the largest homomorphic extension of f to an A-generated relative subalgebra of $\underline{F}(\underline{A},\mathbf{TAlg}(\underline{S}))$ with respect to $\underline{B}$. Thus it remains only to show (iii): Obviously $\mathrm{dom}\ \tilde{f} \subseteq [B]\overline{f}^{-1}$. Therefore -- because of (ii) -- we only have to show that A generates $[B]\overline{f}^{-1}$. This is done by algebraic induction on $q \in F(\underline{A},\mathbf{TAlg}(\underline{S}))$ with respect to the property $H(q)$:= " $q \in [B]\overline{f}^{-1}$ implies $q \in \mathbf{C}_{[\underline{B}]\overline{f}^{-1}} A$ " :

$H(q)$ is true for all $q \in A$. Let $q \in [B]\overline{f}^{-1}$, $q \notin A$. Then $q = (c_k|k \in K_\phi)\phi^{\underline{F}(\underline{A},\mathbf{TAlg}(\underline{S}))}$. Let $H(c_k)$ be true for all $k \in K_\phi$. Since $q\overline{f} \in B$, we have $q\overline{f} = (c_k\overline{f}|k \in K_\phi)\phi^{\underline{F}(\underline{B},\mathbf{TAlg}(\underline{S}))} = (c_k\overline{f}|k \in K_\phi)\phi^{\underline{B}}$, as $\underline{F}(\underline{B},\mathbf{TAlg}(\underline{S}))$ satisfies (FC1). Therefore $c_k\overline{f} \in B$, i.e. $c_k \in [B]\overline{f}^{-1}$, for each $k \in K_\phi$. Because of $H(c_k)$ we have $c_k \in \mathbf{C}_{[\underline{B}]\overline{f}^{-1}} A$, and therefore also $q = (c_k|k \in K_\phi)\phi^{\underline{F}(\underline{A},\mathbf{TAlg}(\underline{S}))} \in \mathbf{C}_{[\underline{B}]\overline{f}^{-1}} A$. ■■■

5.4.2 REMARK and NOTATION. The Generalized Recursion Theorem yields the foundation for a model theory of partial algebras, as we shall see especially in Chapter II. Therefore the closed initial homomorphic extension of 5.4.1 will occur quite often. Hence it seems natural that it gets a short notation. We shall preserve the notation " $\tilde{f}$ " in connection with homomorphisms $f:\underline{A} \to \underline{B}$ only for this closed A-initial homomorphic extension of f described in Theorem 5.4.1 and if it might not be clear what target algebra is used, we shall write $(\tilde{f})_{\underline{B}}$ to indicate that $\underline{B}$ is the target -- the starting algebra should always be clear, else we write $(\tilde{f})_{\underline{A},\underline{B}}$. If f is composed, f = gh , then we shall write $\tilde{f} = (gh)^{\sim}$ for the largest closed initial extension.

5.4.3 REMARK. Theorem 5.4.1 provides us with another possibility for describing $\underline{F}(M,\mathbf{TAlg}(\underline{S}))$ with respect to $\underline{F}(M,\mathbf{TAlg}(\tau))$ and the set $M = (M,v_M)$ of variables, namely:

$$\underline{F}(M,\mathbf{TAlg}(\underline{S})) = \underline{\mathrm{dom}}\ \tilde{v}_M .$$

The straightforward proof is left to the reader. ∎

5.5 The algebraic quasi-ordering on partial algebras

5.5.1 DEFINITION. Let $\underline{A} \in \mathbf{Alg}(\underline{S})$ be any partial algebra. Then one can define on A relations $\lhd_{\underline{A}}$ and $\preceq_{\underline{A}}$ as follows:

$\lhd_{\underline{A}} := \{(a,a') \in A \times A \mid$ there are $\phi \in \Omega$, $k_0 \in K_\phi$ and $\underline{b} \in \mathrm{dom}\ \phi^{\underline{A}}$ such that $a' = \underline{b}\phi^{\underline{A}}$ and $a = (k_0)\underline{b}\}$.

If $a \lhd_{\underline{A}} a'$, then we say that a is an immediate predecessor of a'. $\preceq_{\underline{A}}$ is the reflexive and transitive closure of $\lhd_{\underline{A}}$. Thus, by definition, $\preceq_{\underline{A}}$ is a quasi-order on A which is called the algebraic quasi-order on A -- and if it is a partial order, then it is called the algebraic partial order on A -- induced by $\underline{A}$ (" on $\underline{A}$ "). (Note that the relationship of $\underline{A}$ to $\underline{S}$ does not have any direct influence on this quasi-order, it only restricts the structure of $\underline{A}$.)

If $M \subseteq A$, we denote by $\downarrow M$ the initial segment in $\underline{A}$ with respect to the above quasi-order, i.e.

$\downarrow M := \{x \in A \mid x \preceq_{\underline{A}} m$ for some $m \in M\}$;

in particular we denote $\downarrow\{m\}$ also by $\downarrow m$.

$\underline{\downarrow M}$ designates the relative subalgebra of $\underline{A}$ with carrier set $\downarrow M$. $\downarrow M$ is often called the set of all predecessors of elements of M in $\underline{A}$.

If $\underline{A} = \underline{F}(M,\mathbf{TAlg}(\underline{S}))$ is a Peano algebra, then the elements of A are also called terms, and for any term $t \in A$ the elements of $\downarrow t$ are called subterms of t.

5.5.2 PROPOSITION. Let $\underline{A}$ be a partial Peano algebra on some set X. Then one has in connection with the rank function $\sigma_{\underline{A}}(.,X)$:

$$a <_{\underline{A}} b \text{ implies } \sigma_{\underline{A}}(a,X) < \sigma_{\underline{A}}(b,X) \text{ for all } a,b \in A$$

(notice that $a <_{\underline{A}} b$ iff $a \preceq_{\underline{A}} b$ and $a \neq b$, as usual). Therefore $\preceq_{\underline{A}}$ is then a partial order on A.

PROOF. Let $\underline{A}$ be a partial Peano algebra on some set X, and let $a \lhd_{\underline{A}} b$ for some $a,b \in A$. Then $b = \underline{a}\phi^{\underline{A}}$ and $a = (k)\underline{a}$ for some $\phi \in \Omega$, $\underline{a} \in \mathrm{dom}\ \phi^{\underline{A}}$ and $k \in K_\phi$. Hence, by (P1) $b \notin X$, i.e. $\sigma_{\underline{A}}(b,X) > 0$. Then (P3), Theorem 3.4.4 and observation 3.4.6 imply that there exist $\phi' \in \Omega$ and $\underline{c} \in \mathrm{dom}\ \phi'^{\underline{A}}$ such that $\sigma_{\underline{A}}((l)\underline{c},X) < \sigma_{\underline{A}}(b,X)$ for every $l \in K_{\phi'}$, and that $b = \underline{c}\phi'^{\underline{A}}$. (P2) then immediately implies $\phi = \phi'$ and $\underline{a} = \underline{c}$, i.e. $a = (k)\underline{c}$ satisfies $\sigma_{\underline{A}}(a,X) < \sigma_{\underline{A}}(b,X)$. A transitivity argument shows the general implication, which immediately implies that $a <_{\underline{A}} b$ and $b <_{\underline{A}} a$ cannot hold at the same time. Hence $\preceq_{\underline{A}}$ is anti-symmetric. ∎∎∎

5.5.3 COROLLARY 1. In a partial Peano algebra $\underline{B}$ $\prec_{\underline{B}}$ only allows **finite** strictly descending chains.

PROOF. Notice that the step numbers are ordinals, which have this property. ∎∎∎

5.5.4 COROLLARY 2. Let $\underline{A}$ be any partial algebra, $\underline{B}$ a relative subalgebra of $F(\underline{A},\mathbf{TAlg}(\underline{S}))$, and $A \subseteq B$. Then A generates $\underline{B}$, iff $\downarrow B = B$ (w.r.t. $\prec_{\underline{F}(\underline{A},\mathbf{TAlg}(\underline{S}))}$).

A similar statement holds, when $\underline{B}$ is any relative subalgebra of a partial Peano algebra $\underline{C}$ on the set A, and B contains A.

PROOF. Since $\underline{F}(\underline{A},\mathbf{TAlg}(\underline{S}))$ becomes a partial Peano algebra, when we forget about the structure of $\underline{A}$ in it, we have only to prove the second part: If there is an element b in $\downarrow B \setminus B$, then there is $c \in B$ which has predecessors in $\underline{C}$ which do not belong to B. Thus c would belong to every generating set of $\underline{B}$, but it cannot belong to A (whose elements do not have predecessors in $\underline{C}$), i.e. then A cannot generate $\underline{B}$. Conversely, if $\downarrow B = B$, then one can prove by algebraic induction that A generates $\underline{B}$ (cf. the proof of 5.2.5). ∎∎∎

5.5.5 LEMMA.

(i) Let $\underline{B}$ be any partial Peano algebra on M, $t \in B$. $\underline{\downarrow t}$ is a relative subalgebra of $\underline{B}$ generated by $M_t := \downarrow t \cap M$; and for any $a \in \downarrow t$ $t \notin C_{\underline{D}}(M \cap D)$, where $D := \downarrow t \setminus \{a\}$.

(ii) Let $w: M \to \underline{A}$ be any mapping, and $t \in F(M,\mathbf{TAlg}(\underline{S}))$. Then $t \in \mathrm{dom}\,\tilde{w}$ iff $\downarrow t \subseteq \mathrm{dom}\,\tilde{w}$; moreover, $t\tilde{w}$ only depends on $w|M_t$.

PROOF. Ad (i): The first part easily follows from 5.5.4 and the fact that already $\downarrow t \subseteq F(M_t,\mathbf{TAlg}(\underline{S}))$, while the second one can be proved by algebraic induction on t.

Ad (ii): From 5.5.4 we infer that $\underline{\downarrow t}$ is a relative subalgebra of $\underline{F}(M,\mathbf{TAlg}(\underline{S}))$ which is generated by M_t. Since $\tilde{w}$ is the largest M-initial homomorphic extension of w (cf. 5.4.1.(ii)), the first statement follows, while the second one follows from the fact that M_t generates $\underline{\downarrow t}$. ∎∎∎

5.5.6 REMARK. It should be observed that, for $M \subseteq N$ and $t \in F(M,\mathbf{TAlg}(\underline{S}))$, t can also be considered as an element of $F(N,\mathbf{TAlg}(\underline{S}))$, but while there may be many mappings $w: M \to \underline{A}$ for which $t \in \mathrm{dom}\,\tilde{w}$, there may be no mapping from N into $\underline{A}$ with this property, namely when there is no mapping at all from N into $\underline{A}$; and this happens iff $[A]v_A \not\supseteq [N]v_N$. (In the homogeneous case it can only happen that M but not N allows a mapping into $\underline{A}$ when M and A both are empty, but then there is no element from $\underline{F}(M,\mathbf{TAlg}(\tau))$, which can be mapped into A.) Thus it will be necessary -- at least for the heterogeneous case -- to fix to each term $t \in F(M,\mathbf{TAlg}(\underline{S}))$ the Peano basis with respect to which it is considered, i.e. (M:t) will be the correct notation for a **heterogeneous term** (notice that in Part II (by H.REICHEL) terms are written as (v:t) rather than (M:t), where v means our v_M).

5.6 Tree representation of terms

5.6.1. In this subsection we assume that the reader is familiar with the basic terminology of graph theory (cf. e.g. F.HARARY [Hy 72], Chapters 4 and 16). Else this subsection may be skipped, since we only want to represent terms (the elements of a Peano algebra) as trees, which sometimes may provide for a better understanding of this important concept.

5.6.2. We shall consider rooted (vertex-) labeled directed trees $\underline{G} := ((V,E,r),l)$, where V is the set of **points** or **vertices**, E the set of **directed lines** or **directed edges**, which are ordered pairs of elements of V, $r \in V$ is the **root** and $l: V \to X$ (X any set, the **set of labels**) is a **labeling** of the vertices with elements of X such that (V,E) is a directed (out-) tree with root r (i.e. all edges are directed away from r, and -- even when we forget about directions -- each vertex can be reached from r in exactly one (finite) path). Two labeled rooted trees $((V,E,r),l)$ and $((V',E',r'),l')$ are isomorphic, iff there is a bijection $f: V \to V'$ such that $rf = r'$, $(v,v^*) \in E$ iff $(vf,v^*f) \in E'$ and $fl' = l$. If X is any set and **c** any cardinal, then we denote by $Tr(X,\mathbf{c})$ the set of all isomorphism types (e.g. a set containing exactly one element from each isomorphism class) of rooted trees labeled with elements from X, and where the vertex set has cardinality less than **c**.

5.6.3. Let $\tau = (K_\phi)_{\phi \in \Omega}$ be any similarity type of infinite cardinal dimension $\mathbf{d^*}$, and let M be any set, which we assume w.l.o.g. to be disjoint from Ω. Then we supply the set $Tr(M \cup \Omega, \mathbf{d^*})$ with an algebraic structure as follows (we only treat the homogeneous case, in the heterogeneous case one has to observe the sort mapping of M and the output sorts of the fundamental operations -- as in part (b) of the proof of Theorem 5.2.6):

Let $\phi \in \Omega$, $\underline{t} := (((V_k,E_k,r_k),l_k: V_k \to M \cup \Omega) \mid k \in K_\phi) \in (Tr(M \cup \Omega, \mathbf{d^*}))^{K_\phi}$ be any K_ϕ-sequence of rooted trees. Then $\underline{t}\phi^{Tr(M\cup\Omega,\mathbf{d^*})} := ((V,E,r),l)$, where

$V := \{(\phi,\phi)\} \cup \mathbf{U}(V_k \times \{k\} \mid k \in K_\phi)$, $r := (\phi,\phi)$,

$E := \{((\phi,\phi),(r_k,k)) \mid k \in K_\phi\} \cup \mathbf{U}(\{((a,k),(b,k)) \mid (a,b) \in E_k\} \mid k \in K_\phi)$,

$l: V \to X \cup \Omega: \quad (\phi,\phi) \mapsto \phi, \quad (a,k) \mapsto al_k, \ (a \in V_k, \ k \in K_\phi)$.

An example for $\tau := (2,1,0)$, $\Omega := \{\phi,\phi',\phi''\}$, $M := \{m,m',m''\}$ can be seen in Figure 5.3, where we have drawn the trees as usual in graph theory and labeled the vertices directly with their labels.

Notice that we have already chosen such trees as arguments which shall later on represent terms (i.e. elements of $\underline{F}(M,\mathbf{TAlg}(\tau))$; moreover, it should be observed that in the diagrams we have attached only the labels to the vertices, not the names of the vertices, and that by "reading such a tree" one starts at the left hand **leaf** (end vertex, vertex with out-degree 0) which also is the left hand leaf of the first argument, and then proceedes successively to the right.

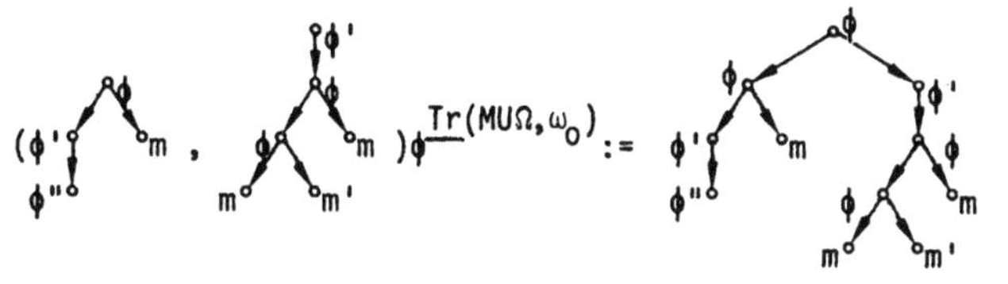

Figure 5.3

5.6.4. Let $\overline{M} := \{\underline{G}_m \in Tr(M \cup \Omega, d^*) | m \in M\}$, where $\underline{G}_m := ((\{m\}, \emptyset, m), l_m := id_{\{m\}})$. Then it is not too difficult to realize that $j: M \to \overline{M}$, $m \mapsto \underline{G}_m$ $(m \in M)$ induces an isomorphism $\overline{j}: \underline{F}(M, \mathbf{TAlg}(\tau)) \to \underline{C}_{Tr(M \cup \Omega, d^*)}{}^{\overline{M}} =: \underline{Trees}(M,\tau)$ (exercise!).
For instance the trees in Figure 5.3 have the following preimages (represented as "words" in $(X \cup \Omega)^*$, i.e. by "inverse Polish parenthesis-free notation"): $\phi''\phi'm\phi$, $mm'\phi m\phi\phi'$, and $\phi''\phi'm\phi mm'\phi m\phi\phi'\phi$. This might give an idea of how "to read the trees" as elements of $\underline{Trees}(M,\tau)$ $(\cong \underline{F}(M, \mathbf{TAlg}(\tau)))$. -- Via the isomorphism above one may now also get the trees corresponding to terms in $\underline{F}(M, \mathbf{TAlg}(\underline{S}))$, since this is actually an M-generated relative subalgebra of $\underline{F}(M, \mathbf{TAlg}(\tau))$.

5.6.5. (i): Notice that one can easily prove, say by algebraic induction, that for every tree $\underline{G} \in Trees(M,\tau)$ every path is finite, i.e. every maximal directed path starting from the root ends with a leaf, and the leaves are labeled with elements from $M \cup \Omega_0$ (Ω_0 again being the set of all fundamental nullary operation symbols), while all the other vertices are labeled by elements from $\Omega \setminus \Omega_0$. ∎
(ii): It is also easily realized that for a finitary similarity type τ the value $\sigma_{\underline{Trees}(M,\tau)}(\underline{G}, \overline{M})$ of the rank function on $\underline{Trees}(M,\tau)$ for $\underline{G} \in Trees(M,\tau)$ computes either as the length of the longest path in $\underline{G}$ from the root to a leaf, if for no such longest path the leaf is labeled by a fundamental constant, or one has to add 1. ∎
(iii): Let $\underline{G} = ((V,E,r),l) \in Trees(M,\tau)$ be any rooted tree (identified with the corresponding isomorphism class), $v \in V$ any vertex of $\underline{G}$. Let $\underline{G}_v := ((V_v, E_v, v), l|V_v)$ be given by
$V_v := \{w \in V |$ there is a directed path in $\underline{G}$ from v to w$\}$,
and $E_v := \{(a,b) \in E | a,b \in V_v\}$. Then $\underline{G}_v$ is also a rooted tree belonging to $Trees(M,\tau)$, and with the isomorphism $\overline{j}$ from 5.6.4 one obviously has for any $t \in F(M, \mathbf{TAlg}(\underline{S}))$ that $[\downarrow t]\overline{j} = \{\underline{G}_v | v \in V(\underline{G})\}$, where $\underline{G} = t\overline{j}$. ∎

5.7 (Global) Term operations and polynomials

One very important consequence of the Generalized Recursion Theorem 5.4.1 is the fact that it allows us to define term operations and polynomials.

5.7.1 DEFINITION. Let M be any set, $\underline{A} \in \mathbf{Alg}(\underline{S})$ any partial algebra, and $t \in F(M, \mathbf{TAlg}(\underline{S}))$ any term. Then t defines on $\underline{A}$ a partial M-ary term operation $(M:t)^{\underline{A}}$ as follows:
dom $(M:t)^{\underline{A}} := \{w \in A^M | t \in \text{dom } \tilde{w}\}$, and $w \in \text{dom}(M:t)^{\underline{A}}$ implies $(w)(M:t)^{\underline{A}} := (t)\tilde{w}$.
$(M:t)^{\underline{A}}$ is called the (global) term operation on $\underline{A}$ corresponding to the term t -- or more precisely to (M:t). In this connection M is called the set of variables under consideration, while the mappings $w: M \to \underline{A}$ are sometimes called valuations of M in $\underline{A}$ or assignments of values in A to the variables in M. Moreover, t (or (M:t)) is said to be evaluable in $\underline{A}$ with respect to the assignment $w: M \to A$ iff $t \in \text{dom } \tilde{w}$ (iff $w \in \text{dom } (M:t)^{\underline{A}}$).
For every $t' \in \downarrow t$, (M:t') is called an M-subterm of t (or (M:t)).

5.7.2 REMARKS. (i): Notice that in the homogeneous case it is not necessary to mention the set of variables in connection with a term, while we have already mentioned in

5.5.6, why this is necessary in the heterogeneous case.

(ii): The tree model is probably the most intuitive representation of a term. In this connection a valuation in $\underline{A}$ corresponds to the process that each label of a leaf of such a tree is replaced by an admissible element of A, and each other label, which then is an operation symbol, say ϕ, is replaced by the corresponding fundamental operation $\phi^{\underline{A}}$ of $\underline{A}$. The operations are successively evaluated as long as possible, proceeding from the leaves to the root. Whenever we reach a vertex such that the corresponding operation is not evaluable on the values of its successors -- or when it is a leaf, if there is no admissible element for it in A --, then the whole term is not evaluable. Hence a term (M:t) is evaluable, iff each M-subterm of (M:t) is evaluable. Notice that the term operations are obtained by repeated "superposition" of fundamental operations, starting with the so-called projection operations (cf. section 15 for more details), since for each variable $m \in M$ the mapping $w \mapsto mw = m\tilde{w} = w(M:m)^{\underline{A}}$ $(w \in A^M)$ is a total M-ary operation on $\underline{A}$, called the m-th M-ary <u>projection</u> and denoted by $(e_m{}^M)^{\underline{A}}$, i.e. $(e_m{}^M)^{\underline{A}}: A^M \to A$, $(w)(e_m{}^M)^{\underline{A}} := (m)\tilde{w}$ -- in most cases one writes $e_m{}^M$ instead of $(e_m{}^M)^{\underline{A}}$, when the context yields the corresponding partial algebra $\underline{A}$. Notice that $(e_m{}^M)^{\underline{A}}$ is independent from any structure on A, but it depends on v_A and v_M, since for $[M]v_M \not\subseteq [A]v_A$ it is empty.

5.7.3 EXAMPLE. In order to get a better feeling for the evaluation of terms we consider a type $\tau = (2,1,0)$ for $\Omega = (\phi, \phi', \phi'')$ and $M := \{m, m', m''\}$ as in 5.6.3, and we write terms as words, as in 5.6.4, and rewrite it with brackets for better structuring; moreover we represent it as a tree (cf. Figure 5.4): Let the term t (rather (M:t), but we are in a homogeneous case) be given as follows:
$t := mm'm''\phi'\phi\phi\phi'\phi''\phi \triangleq (((m,(m',(m'')\phi')\phi)\phi)\phi',\phi'')\phi$.

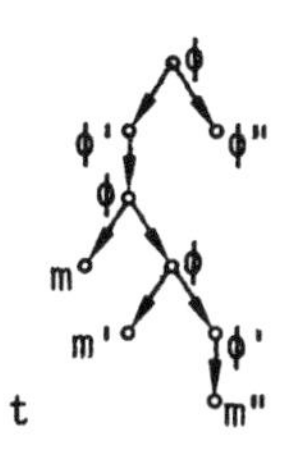

Figure 5.4

For the evaluations of t we consider the partial algebra $\underline{A}$ sketched in Figure 5.5, where we assume $\phi^{\underline{A}}$ to be commutative. We evaluate t e.g. for w and w', where $graph(w) := \{(m,b),(m',a),(m'',c)\}$ and $graph(w') := \{(m,a),(m',b),(m'',c)\}$ by replacing the labels of the tree successively by its evaluations, starting with the leaves (cf. Figures 5.6 and 5.7).

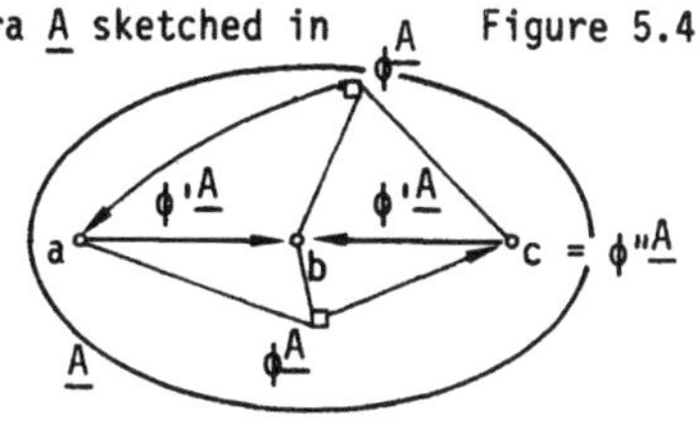

Figure 5.5

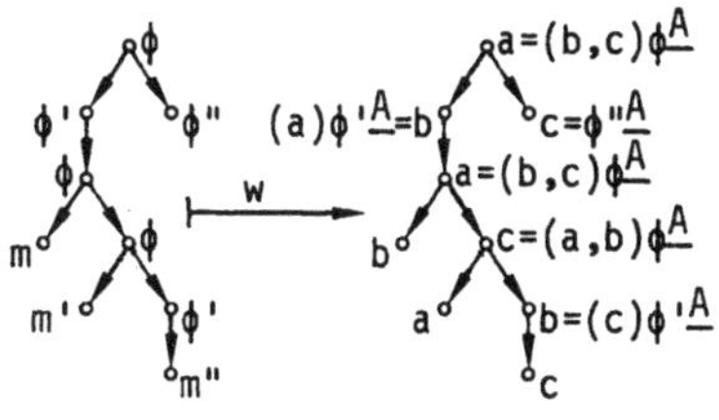

Figure 5.6 $wt^{\underline{A}} = a$

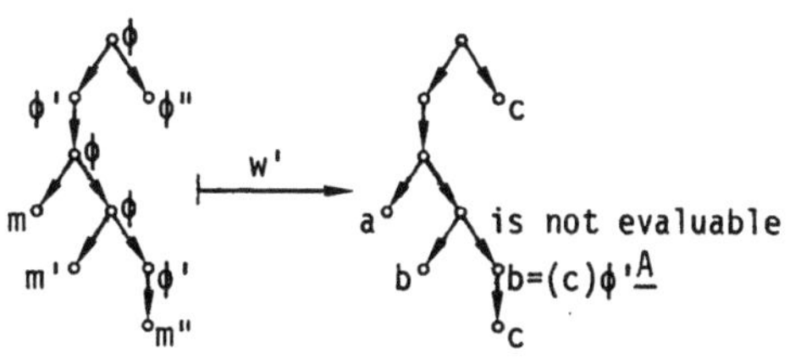

Figure 5.7 $w't^{\underline{A}}$ does not exist.

In connection with this example it is relatively easy to determine $\mathrm{dom}\ t^{\underline{A}}$, and we suggest it as an exercise, giving only some hints: Starting with the root one realizes that here the value exists iff the output of its successor labeled with ϕ' is b (since always $\phi''^{\underline{A}} = c$). Observing the structure of $\underline{A}$ one thus realizes that $\mathrm{dom}\, t^{\underline{A}} = \{(b,a,a),(b,a,c),(b,c,a),(b,c,c)\}$, and $wt^{\underline{A}} = a$ for every $w \in \mathrm{dom}\, t^{\underline{A}}$. ∎

The following results are of great importance in the sequel:

5.7.4 LEMMA. Let M be any set, $f:\underline{A} \to \underline{B}$ any homomorphism, and $w:M \to A$ any valuation. Then $\mathrm{dom}\, \tilde{w} \subseteq \mathrm{dom}(wf)^{\sim}$ -- equality holding when f is closed --, and $\ker \tilde{w} \subseteq \ker(wf)^{\sim}$.

PROOF. Obviously $\tilde{w}f$ is a homomorphic extension of wf to an M-generated relative subalgebra of $\underline{F}(M,\mathbf{TAlg}(\underline{S}))$. Since $(wf)^{\sim}$ is the largest such extension, we get $\mathrm{dom}\, \tilde{w} \subseteq \mathrm{dom}\,(wf)^{\sim}$, and if f is closed, then $\tilde{w}f$ is already a closed such extension, and equality has to hold (cf. 5.4.1.(ii)). Since $\tilde{w}f = (wf)^{\sim}|\mathrm{dom}\, \tilde{w}$, $\ker \tilde{w} \subseteq \ker \tilde{w}f \subseteq \ker(wf)^{\sim}$. ∎∎∎

5.7.5 COROLLARY. Each homomorphism is compatible with each induced term operation, i.e.: If $f:\underline{A} \to \underline{B}$, $t \in F(M,\mathbf{TAlg}(\underline{S}))$, $w:M \to A$ and $a \in A$, then $w(M:t)^{\underline{A}} = a$ implies $(wf)(M:t)^{\underline{B}} = af$ -- i.e. if $w(M:t)^{\underline{A}}$ exists, then so does $(wf)(M:t)^{\underline{B}}$ and one has $(w(M:t)^{\underline{A}})f = (wf)(M:t)^{\underline{B}}$.

PROOF. With the notation of the Corollary and the assumption that $w(M:t)^{\underline{A}}$ exists one has $t \in \mathrm{dom}\, \tilde{w} \subseteq \mathrm{dom}(wf)^{\sim}$; therefore $(t)(wf)^{\sim}$ exists and satisfies because of $t \in \mathrm{dom}\, \tilde{w} = \mathrm{dom}\, \tilde{w}f$: $(wf)(M:t)^{\underline{B}} = (t)(wf)^{\sim} = (t)\tilde{w}f = (w(M:t)^{\underline{A}})f$ $(= af)$. ∎∎∎

5.7.6 REMARK. 5.7.5 may form the basis for the possibility of extending rational equivalence to partial algebras, but we do not go into details in this book, since -- as far as we know -- this concept has not yet been investigated for partial algebras.

5.7.7 PROPOSITION. Let $\underline{A}$ be any partial algebra of cardinal dimension **d** and let M be a generating subset of $\underline{A}$. Then:

(i) For every $a \in A$ there exists a term $t \in F(M,\mathbf{TAlg}(\underline{S}))$ such that $a = (\mathrm{id}_M)(M:t)^{\underline{A}}$.

(ii) For every $a \in A$ there exists a subset N of M satisfying $\#N < \max\{2,\mathbf{d}\}$ with a term $t \in F(N,\mathbf{TAlg}(\underline{S}))$ such that $a = (\mathrm{id}_N)(N:t)^{\underline{A}}$.

PROOF. Let $i := \mathrm{id}_M : M \to A$, then $\tilde{i}$ is surjective, since M generates $\underline{A}$. Thus $a\tilde{i}^{-1} \neq \emptyset$; let $t \in a\tilde{i}^{-1}$, then $t\tilde{i} = a = (i)(M:t)^{\underline{A}}$. -- Since $A = C_{\underline{A}}M = \bigcup(C_{\underline{A}}N \mid N \subseteq M,\ \#N < \max\{2,\mathbf{d}\})$ (cf. 3.3.2.(*)), the second statement follows in a similar way. ∎∎∎

Closely connected with the concept of global term operations is the one of global polynomials with coefficients in some partial algebra $\underline{A}$ and some set X of variables:

5.7.8 DEFINITION and REMARK. Let $\underline{A} \in \mathbf{Alg}(\underline{S})$ be any partial algebra and X any set (of variables), and we may assume w.l.o.g. that $A \cap X = \emptyset$. Then $\underline{A}_X := \underline{A} + \underline{X}_d$ is a partial algebra, which is the coproduct in $\mathbf{Alg}(\underline{S})$ of $\underline{A}$ and the discrete partial algebra on X. $\underline{F}(\underline{A}_X,\mathbf{TAlg}(\underline{S})) =: \underline{F}(\underline{A} \cup X,\mathbf{TAlg}(\underline{S}))$ is called the <u>algebra of global partial polynomials with coefficients in</u> $\underline{A}$, and each $p \in F(\underline{A}_X,\mathbf{TAlg}(\underline{S}))$ is called a (<u>global partial</u>) <u>polynomial</u> on $\underline{A}$, since by the Generalized Recursion Theorem 5.4.1 we have:

Let $f:\underline{A} \to \underline{B}$ be any given homomorphism (e.g. $\underline{B} = \underline{A}$ and $f = \mathrm{id}_A$), then we may define for $p \in F(\underline{A}_X, \mathbf{TAlg}(\underline{S}))$ and for $w: X \to B$: $\operatorname{dom} p^{\underline{B},f} := \{w \in B^X \mid p \in \operatorname{dom}(fUw)^{\sim}\}$ -- " fUw " meaning the mapping having as graph the disjoint union of the graphs of f and w --, and if $w \in \operatorname{dom} p^{\underline{B},f}$, then $(w)p^{\underline{B},f} := (p)(fUw)^{\sim}$.

If $\underline{B}$ is a total homogeneous algebra, then this concept generalizes the one of polynomials in rings.

Because of the close relationship between global polynomials and global term operations -- as far as the technicalities are concerned --, we do not present here more details on polynomials (but cf. section 15). They have been discussed more thoroughly in the thesis of M.SIEGMUND-SCHULTZE [Si 84]. When dealing with them one always has to recall the Generalized Recursion Theorem 5.4.1 and its main consequences.

5.8 The Epimorphism Theorem

Another very important tool provided by the Generalized Recursion Theorem will be discussed in this subsection. It is an analogue to the Homomorphism Theorem which deals with quotient mappings, while this new one concerns epimorphisms. As a preparation we present the following characterization of epimorphisms:

5.8.1 LEMMA. The homomorphism $f: \underline{A} \to \underline{B}$ is an epimorphism iff its greatest A-initial extension $\tilde{f}$ in $\underline{F}(\underline{A}, \mathbf{TAlg}(\underline{S}))$ is surjective.

PROOF. Let $f: \underline{A} \to \underline{B}$ be an epimorphism, i.e. $\mathbf{C}_{\underline{B}}[A]f = B$. Then $\tilde{f}: \underline{\operatorname{dom}}\,\tilde{f} =: \underline{C} \to \underline{B}$ -- being closed -- satisfies $[C]\tilde{f} = [\mathbf{C}_{\underline{C}}A]\tilde{f} = \mathbf{C}_{\underline{B}}[A]\tilde{f} = \mathbf{C}_{\underline{B}}[A]f = B$ (cf. 3.6.2.(ii)), i.e. $\tilde{f}$ is surjective. If, conversely, $\tilde{f}: \underline{C} \to \underline{B}$ is surjective, then the same calculation as above yields that f is dense, hence an epimorphism. ∎

The following result is due to J.SCHMIDT [Sch 70].

5.8.2 DIAGRAM COMPLETION LEMMA for EPIMORPHISMS. Let $f: \underline{A} \to \underline{B}$ be an epimorphism, $g: \underline{A} \to \underline{C}$ any homomorphism.

(a): Then the following statements are equivalent:

- (i) There exists a unique homomorphism $h: \underline{B} \to \underline{C}$ such that $fh = g$.
- (ii) $\ker \tilde{f} \subseteq \ker \tilde{g}$ (within $\underline{F}(\underline{A}, \mathbf{TAlg}(\underline{S}))$).

(b): If $h: \underline{B} \to \underline{C}$ satisfying $fh = g$ exists, then:

- (iii) h is an epimorphism, iff g is an epimorphism.
- (iv) h is surjective, iff $\tilde{g}|\operatorname{dom} \tilde{f}$ is surjective (hence especially, when g is surjective).
- (v) h is injective (i.e. a monomorphism), iff $\ker \tilde{f} = \ker \tilde{g} \cap (\operatorname{dom} \tilde{f})^2$.
- (vi) h is closed, iff $\operatorname{dom} \tilde{f} = \operatorname{dom} \tilde{g}$.
- (vii) h is closed and injective, iff $\ker \tilde{f} = \ker \tilde{g}$.
- (viii) h is an isomorphism, iff g is an epimorphism and $\ker \tilde{f} = \ker \tilde{g}$.
- (ix) h is full, iff $\tilde{g}|\operatorname{dom} \tilde{f}$ is full.

PROOF. Ad (a): Let us consider the diagram in Figure 5.8, and let us assume that $h: \underline{B} \to \underline{C}$ exists such that $fh = g$. Then the diagram is commutative in all fully drawn parts (i_1, i_2 being identity injections).

Since h exists, we have $g = fh = i_1\tilde{f}h$, i.e. $\tilde{f}h$ is an A-initial extension of g. Since $\tilde{g}$ is the largest such extension, we have $\underline{D} := \mathrm{dom}\,\tilde{f} \subseteq \underline{E} := \mathrm{dom}\,\tilde{g}$, therefore the embedding $i := id_D : \underline{D} \to \underline{E}$ exists and satisfies $i_2 = i_1 i$. Hence $i\tilde{g} = \tilde{g}|D = \tilde{f}h$, and therefore $\ker \tilde{g} \supseteq \ker(\tilde{g}|D) = \ker(\tilde{f}h) \supseteq \ker \tilde{f}$. -- Conversely, if $\ker \tilde{f} \subseteq \ker \tilde{g}$, then $\mathrm{dom}\,\tilde{f} \subseteq \mathrm{dom}\,\tilde{g}$, i.e. i above

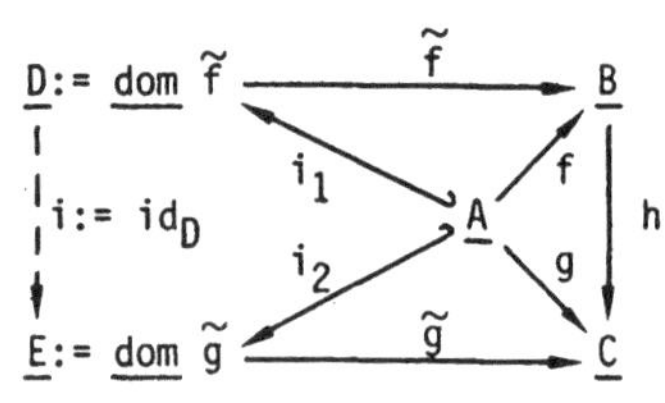

Figure 5.8.

exists, $i:\underline{D} \to \underline{E}$. $\tilde{f}$ is full (even closed) and surjective (cf. 5.8.1), and we can apply 2.7.2 getting a unique homomorphism $h:\underline{B} \to \underline{C}$ such that $\tilde{f}h = i\tilde{g}$. Then $fh = i_1\tilde{f}h = i_1 i\tilde{g} = i_2\tilde{g} = g$. Thus the equivalence in (a) is proved. ∎

Ad (b): Let $h:\underline{B} \to \underline{C}$ exist such that $fh = g$. (iii) follows, since f is an epimorphism: $fhv = fhw$ implies $hv = hw$, and hence $v = w$, if h is an epimorphism; if $hv = hw$, then $fhv = fhw$; thus $v = w$, if $g = fh$ is an epimorphism. ∎

It is not too difficult to derive facts (iv) through (viii) from the Diagram Completion Lemma 2.7.1 and its Corollary 2.7.2, since $\tilde{f}$ is closed (hence full) and surjective, occasionally using 5.8.1; therefore the proofs are left as exercises. -- Finally, (ix) is a consequence of 2.4.5.(iii).(a) and (c), since $\tilde{f}$ is full and surjective. ∎∎∎

As a corollary we get what J.SCHMIDT has called the "General Homomorphism Theorem", and what we call the

5.8.3 EPIMORPHISM THEOREM. Let $f:\underline{A} \to \underline{B}$ and $g:\underline{A} \to \underline{C}$ be any two epimorphisms. Then the following statements are equivalent:

(i) $\ker \tilde{f} = \ker \tilde{g}$.

(ii) There exists an isomorphism $h:\underline{B} \to \underline{C}$ such that $fh = g$. ∎∎∎

The following result (cf. [BJPa 78]) will become important e.g. in section 10:

5.8.4 COROLLARY. Let $f:\underline{A} \to \underline{B}$ be any homomorphism, $\underline{E} := \mathrm{dom}\,\tilde{f}$, and $e:\underline{A} \to \underline{E}$ the natural injection.

(i) Then f is the composition of the **TAlg**($\underline{S}$)-extendable epimorphism e followed by the closed homomorphism $\tilde{f}$.

(ii) Assume that $f = e'g$, where $e':\underline{A} \to \underline{C}$ is any **TAlg**($\underline{S}$)-extendable epimorphism and $g:\underline{C} \to \underline{B}$ a closed homomorphism. Then there exists an isomorphism $j:\underline{C} \to \underline{E}$ such that $e'j = e$ and $j\tilde{f} = g$.

PROOF. Since every homomorphism $f':\underline{A} \to \underline{D}$ into some total algebra $\underline{D}$ allows a homomorphic extension $\overline{f'}:\underline{F}(\underline{A},\mathbf{TAlg}(\underline{S})) \to \underline{D}$ such that $f'' := \overline{f'}|E$ satisfies $ef'' = f'$, e is **TAlg**($\underline{S}$)-extendable, and since A generates $\underline{E}$, e is an epimorphism. Hence (i) is proved, since $\tilde{f}$ is indeed closed. ∎

Ad (ii): Assume $f = e'g$ as described in the assumption of (ii). If we can show that $\ker \tilde{e} = \ker (e')^\sim$, then 5.8.3 implies the existence of an isomorphism $j:\underline{C} \to \underline{E}$ such that $e'j = e$, and then $e'g = e\tilde{f} = e'j\tilde{f}$; therefore $g = j\tilde{f}$, since e' is an epimorphism. Now, $\ker \tilde{e} = \Delta_E$, and $\tilde{e}\tilde{f}$ as well as $(e')^\sim g$ are closed A-initial extensions of f; thus $\mathrm{dom}(e')^\sim = \mathrm{dom}\,\tilde{e} = E$. Since e' is **TAlg**($\underline{S}$)-extendable, and since e can be considered as a homomorphism from $\underline{A}$ into $\underline{F}(\underline{E},\mathbf{TAlg}(\underline{S}))$, we get a homomorphism

$j':\underline{C} \to \underline{F}(\underline{E},\mathbf{TAlg}(\underline{S}))$ such that $e'j' = e$, and therefore $(e')^{\sim}j' = \tilde{e}$, showing that $(e')^{\sim}$ has to be injective, too, i.e. indeed $\ker \tilde{e} = \ker (e')^{\sim}$. ∎∎∎

5.8.5 REMARKS. **(i):** In [Sch 70] " ker f " is used for what we denote by " $\ker \tilde{f}$ ". In [BWo 82] we have called " $\ker \tilde{f}$ " in an even more general context the J.SCHMIDT-kernel (briefly: S-kernel) of the homomorphism f.
(ii): Theorem 5.8.3 characterizes the epimorphisms starting from a given partial algebra $\underline{A}$ up to unique isomorphism by closed congruences on A-initial relative subalgebras of $\underline{F}(\underline{A},\mathbf{TAlg}(\underline{S}))$; thus we have up to isomorphism only a **set** of epimorphisms starting from $\underline{A}$ and therefore only a set of epimorphic images (up to isomorphism). In the language of category theory this means that the category (**Alg**($\underline{S}$), class of all homomorphisms) is <u>co-(well-powered)</u>.
(iii): In section 10 we will realize that 5.8.4 just describes a factorization system on the category of all partial $\underline{S}$-algebras (see also [BJPa 78]).

5.9 The initial closed congruence relations of a free completion

For a still better understanding of Theorem 5.8.3 let us realize that 5.8.1 implies the

5.9.1 PROPOSITION. Let $\underline{A}$ be any partial algebra, $\underline{C}$ an A-generated relative subalgebra of $\underline{F}(\underline{A},\mathbf{TAlg}(\underline{S}))$, R a closed congruence on $\underline{C}$, $nat_R:\underline{C} \to \underline{C}/R$ the natural projection, and $nat_R|A:\underline{A} \to \underline{C}/R$ its restriction to $\underline{A}$. Then $nat_R|A$ is an epimorphism. ∎∎∎

5.9.2 DEFINITION. Let $\underline{B}$ be any partial algebra, $A \subseteq B$ any subset. Then we denote by **Rsuba**$(A,\underline{B})$ the set of all A-generated relative subalgebras of $\underline{B}$ (together with the restriction of the sort homomorphism v_B); and by $Cong_c(A,\underline{B})$ we denote the set of all closed congruence relations on A-generated relative subalgebras of $\underline{B}$:
$Cong_c(A,\underline{B}) := \bigcup(Cong_c(\underline{C}) \mid \underline{C} \in \mathbf{Rsuba}(A,\underline{B}))$, $\underline{Cong_c}(A,\underline{B}) := (Cong_c(A,\underline{B});\subseteq)$.
Notice that, for $R,R' \in Cong_c(A,\underline{B})$, $R \subseteq R'$ always implies $\operatorname{dom} R \subseteq \operatorname{dom} R'$, where $\operatorname{dom} R := \{x \in B \mid (x,y) \in R \text{ for some } y \in B\}$.
The elements of **Rsuba**$(A,\underline{B})$ are sometimes also called A-<u>initial segments</u> of $\underline{B}$, while the elements of $Cong_c(A,\underline{B})$ are called A-<u>initial closed congruences</u> of $\underline{B}$.
The motivation for this denotation is taken from 5.5.4.

5.9.3 PROPOSITION. Let $\underline{A} \in \mathbf{Alg}(\underline{S})$, then:
(i) **Rsuba**$(A,\underline{F}(\underline{A},\mathbf{TAlg}(\underline{S})))$ is a closure system on $\underline{F}(\underline{A},\mathbf{TAlg}(\underline{S}))$ with smallest element $\underline{A}$ (cf. 3.2.2.(iii)).
(ii) $Cong_c(A,\underline{F}(\underline{A},\mathbf{TAlg}(\underline{S})))$ is a closure system on $F(\underline{A},\mathbf{TAlg}(\underline{S})) \times F(\underline{A},\mathbf{TAlg}(\underline{S}))$ with smallest element Δ_A.
(iii) For a family $\mathbf{C} \subseteq Cong_c(A,\underline{F}(\underline{A},\mathbf{TAlg}(\underline{S})))$, $\operatorname{dom} \bigcap\mathbf{C} = \bigcap(\operatorname{dom} R \mid R \in \mathbf{C})$.
(iv) $\underline{Cong_c}(A,\underline{F}(\underline{A},\mathbf{TAlg}(\underline{S})))$ is **d***-inductive.

PROOF. Because of 5.5.4, and since the intersection of initial segments of a quasi-ordered set is again an initial segment, (i) is easily realized (cf. also 3.2.2.(iii)). (iii) is obvious, too. In order to prove (ii) we have to show that, for $\mathbf{C} \subseteq Cong_c(A,\underline{F}(\underline{A},\mathbf{TAlg}(\underline{S})))$, $R := \bigcap\mathbf{C}$ is a closed congruence on $\underline{\operatorname{dom}} R =: \underline{B}$: Thus assume

$\underline{a},\underline{b} \in B^{K_\phi}$, $\underline{a} \in \operatorname{dom} \phi^{\underline{B}}$, and $((k)\underline{a},(k)\underline{b}) \in R$ for each $k \in K_\phi$. Since each $R' \in \mathbf{C}$ is closed, we get $\underline{b} \in \operatorname{dom} \phi^{\underline{\operatorname{dom} R'}}$ $(R' \in \mathbf{C})$, therefore $\underline{b} \in \operatorname{dom} \phi^{\underline{B}}$, and $(\underline{a}\phi^{\underline{B}},\underline{b}\phi^{\underline{B}}) \in \mathbf{C} = R$. (iv) is proved in a similar way observing the proof of 2.5.4.(ii). ■■■

5.9.4 COROLLARY. Let $(f_i:\underline{A} \to \underline{B}_i | i \in I)$ be a family of homomorphisms such that $\tilde{f}_i:\underline{\operatorname{dom}}\, \tilde{f}_i \to \underline{B}_i$ are their closed initial extensions, and let $f:\underline{A} \to \mathbf{X}(\underline{B}_i | i \in I)$ be the induced homomorphism. Then $\underline{\operatorname{dom}}\, \tilde{f} = \bigcap^*(\underline{\operatorname{dom}}\, \tilde{f}_i | i \in I)$ and $\ker \tilde{f} = \bigcap(\ker \tilde{f}_i | i \in I)$.

PROOF. Let $\underline{C} := \bigcap(\underline{\operatorname{dom}}\, \tilde{f}_i | i \in I)$ and $g_i := \tilde{f}_i | C:\underline{C} \to \underline{B}_i$. Then there exists the induced homomorphism $g:\underline{C} \to \mathbf{X}(\underline{B}_i | i \in I) =: \underline{B}$. We claim that $g = \tilde{f}$: $\underline{C}$ is obviously A-generated, and by 4.1.7 $\ker g = \bigcap(\ker \tilde{f}_i \cap C^2 | i \in I) = (\bigcap(\ker \tilde{f}_i | i \in I)) \cap C^2 = \bigcap(\ker \tilde{f}_i | i \in I)$ (cf. the definition of $\underline{C}$) is a closed congruence. In order to show $g = \tilde{f}$ we still have to show that g is closed: Let $\underline{c} \in C^{K_\phi}$ such that $\underline{c}g \in \operatorname{dom} \phi^{\underline{B}}$. Then $\underline{c}g_i \in \operatorname{dom} \phi^{\underline{B}_i}$ $(i \in I)$, i.e. $\underline{c} \in \bigcap(\operatorname{dom} \phi^{\underline{\operatorname{dom}}\, \tilde{f}_i} | i \in I) = \operatorname{dom} \phi^{\underline{C}}$. ■■■

5.10 On the cardinalities of M-generated partial algebras

From the results obtained so far we can also infer some observations concerning the cardinalities of partial algebras.

5.10.1 LEMMA. Let M be any set. Then

$$\#F(M,\mathbf{TAlg}(\underline{S})) \leq 2^{\#M + \#\Omega + \#\omega_0 + \Sigma(\#K_\phi | \phi \in \Omega)} = 2^{\sup(\#M,\#\Omega,\#\omega_0,\#K_\phi | \phi \in \Omega)} .$$

PROOF. In the proof of Theorem 5.2.6 we have realized $F(M,\mathbf{TAlg}(\underline{S}))$ as a subset of the power set $\mathbf{Subs}(Z^*)$ of the set Z^* of all finite sequences with elements in $M \,\dot{\cup}\, \Omega \cup \mathbf{U}\{K_\phi | \phi \in \Omega\} =: Z$. Now $\#Z \leq \#\Omega + \#M + \Sigma(\#K_\phi | \phi \in \Omega) =: \mathbf{s}'$. Let $\mathbf{s} := \mathbf{s}' + \#\omega_0$. Then it is well known from cardinal arithmetic that $\#(Z^n) \leq \mathbf{s}^n = \mathbf{s}$ for every natural number n, since $\mathbf{s}$ is infinite. Hence $\#(Z^*) = \Sigma(\#(Z^n | n \in \mathbf{N}) \leq (\#\omega_0) \cdot \mathbf{s} = \mathbf{s}$. Thus $\#F(M,\mathbf{TAlg}(\underline{S})) \leq \#\mathbf{Subs}(Z^*) = 2^{\#(Z^*)} \leq 2^{\mathbf{s}}$. There also follows from cardinal arithmetic that $\#\Omega + \#M + \#\omega_0 + \Sigma(\#K_\phi | \phi \in \Omega) = \sup(\#M,\#\Omega,\#\omega_0,\#K_\phi | \phi \in \Omega)$, since both cardinals are infinite and $\Sigma(\#K_\phi | \phi \in \Omega) \leq \#\omega_0 + \#\Omega \cdot \sup(\#K_\phi | \phi \in \Omega) = \#\omega_0 + \#\Omega + \sup(\#K_\phi | \phi \in \Omega)$. ■■■

5.10.2 PROPOSITION. Let $\underline{A}$ be any partial algebra of similarity type τ, and let M be a generating subset of $\underline{A}$. Then $\#A \leq \#F(M,\mathbf{TAlg}(\underline{S}))$.

PROOF. Because of the Generalized Recursion Theorem 5.4.1 $i := \mathrm{id}_M:M \to A$ extends to a closed homomorphism $\tilde{i}:\underline{\operatorname{dom}}\, \tilde{i} \to \underline{A}$. Since M generates $\underline{A}$, $\tilde{i}$ is surjective, and therefore $\#A \leq \#(\operatorname{dom} \tilde{i}) \leq \#F(M,\mathbf{TAlg}(\underline{S}))$. ■■■

5.10.3 COROLLARY. If $\underline{A} \in \mathbf{Alg}(\underline{S})$ and M generates $\underline{A}$, then we have for the rank $\mathbf{r}$ of τ (cf. 1.11):

$$\#A \leq 2^{\max\{\#M,\#\Omega,\#\omega_0,\mathbf{r}\}} . \quad ■■■$$

These results show that an M-generated partial algebra cannot become too large with respect to #M and the data of its similarity type.

5.11 On the existence of K-universal (K-) solutions

We now have enough tools at hand to investigate the existence of **K**-universal solutions for a given partial algebra $\underline{A}$ with respect to a class **K** of partial algebras of the same type, and to construct them when they exist. The results of the previous subsections and the Diagram Completion Lemma will immediately motivate the following

5.11.1 DEFINITION. Let $\underline{A} \in \mathbf{Alg}(\underline{S})$, $\mathbf{K} \subseteq \mathbf{Alg}(\underline{S})$ be given.
(i): $\mathbf{Rsuba}(\underline{A};;\mathbf{K}) := \{\underline{\mathrm{dom}}\ \tilde{f} \mid f:\underline{A} \to \underline{B}$ for some $\underline{B} \in \mathbf{K}\}$ designates the set of all A-initial segments of $\underline{F}(\underline{A},\mathbf{TAlg}(\underline{S}))$ induced by homomorphisms from $\underline{A}$ into some **K**-algebra.
$\mathrm{Cong}_c(\underline{A};;\mathbf{K}) := \{\ker \tilde{f} \mid f:\underline{A} \to \underline{B}$ for some $\underline{B} \in \mathbf{K}\}$ designates the set of all A-initial congruences induced by some homomorphism from $\underline{A}$ into some **K**-algebra.
(ii): $\mathrm{char}(\underline{A};;\mathbf{K}) := \bigcap \mathrm{Cong}_c(\underline{A};;\mathbf{K})$ is called the $\underline{A}$-<u>characteristic</u> of **K**.

As an immediate consequence of 5.9.3 we get

5.11.2 LEMMA. Let $\underline{A} \in \mathbf{Alg}(\underline{S})$, $\mathbf{K} \subseteq \mathbf{Alg}(\underline{S})$. Then $\bigcap{}^*\mathbf{Rsuba}(\underline{A};;\mathbf{K}) = \underline{\mathrm{dom}}(\mathrm{char}(\underline{A};;\mathbf{K}))$, and $\mathrm{char}(\underline{A};;\mathbf{K})$ is a closed congruence relation on $\bigcap{}^*\mathbf{Rsuba}(\underline{A};;\mathbf{K})$. ■■■

5.11.3 REMARKS. (i): The concept of an $\underline{A}$-characteristic of some class **K** is a generalization of the concept of an M-characteristic of a partial algebra or a class of partial algebras -- M any set --, as it was introduced by J.SCHMIDT in [Sch 62] and [Sch 64] as a generalization of the characteristic of a ring.
(ii): The two successive semicolons in " $\mathrm{char}(\underline{A};;\mathbf{K})$ " shall remind the reader of the fact that we deal not only with $\underline{A}$, but we should have written more precisely " $\mathrm{char}(\underline{A},\underline{F}(\underline{A},\mathbf{TAlg}(\underline{S}));\mathbf{K})$ ".
(iii): $\mathrm{dom}(\mathrm{char}(\underline{A};;\mathbf{K}))$ is the set of all those elements of $F(\underline{A},\mathbf{TAlg}(\underline{S}))$ which are "interpreted" by every homomorphism from $\underline{A}$ into some **K**-algebra. Hence, if $\underline{A}$ is discrete, $\mathrm{dom}(\mathrm{char}(A;;\mathbf{K})$ is the set of exactly those terms $t \in F(A,\mathbf{TAlg}(\underline{S}))$ for which $(A:t)^{\underline{B}}$ is a **total** term operation in **every** **K**-algebra $\underline{B}$. Therefore it could be considered as a "<u>completeness degree</u>" for **K**. The meaning of $\mathrm{char}(\underline{A};;\mathbf{K})$ will become more obvious in the next section about the equational theory of partial algebras.
(iv): Notice that in the case, when $\mathbf{Rsuba}(\underline{A};;\mathbf{K})$ -- and hence $\mathrm{Cong}_c(\underline{A};;\mathbf{K})$ -- is empty, i.e. when there is no homomorphism from $\underline{A}$ into any **K**-algebra, then $\bigcap{}^*\mathbf{Rsuba}(\underline{A};;\mathbf{K}) = \underline{F}(\underline{A},\mathbf{TAlg}(\underline{S}))$ and $\mathrm{char}(\underline{A};;\mathbf{K}) = \nabla_{F(\underline{A},\mathbf{TAlg}(\underline{S}))}$, since these are the largest elements of the corresponding complete lattices (ordered by inclusion).

5.11.4 THEOREM. Let $\underline{A} \in \mathbf{Alg}(\underline{S})$, $\mathbf{K} \subseteq \mathbf{Alg}(\underline{S})$. Then the restriction r to A of the natural homomorphism $\mathrm{nat}_{\mathrm{char}(\underline{A};;\mathbf{K})}: \bigcap{}^*\mathbf{Rsuba}(\underline{A};;\mathbf{K}) \to \bigcap{}^*\mathbf{Rsuba}(\underline{A};;\mathbf{K})/\mathrm{char}(\underline{A};;\mathbf{K}) =: \underline{F}$ (i.e. $r:\underline{A} \to \underline{F}$) is a **K**-universal epimorphism.

PROOF. 5.9.1 implies that r is an epimorphism. From the definition of r and the Diagram Completion Lemma 5.8.2 for epimorphisms it immediately follows that $r:\underline{A} \to \underline{F}$ is **K**-extendable. There remains the extremality property to be shown: Let $r':\underline{A} \to \underline{F}'$ be any other **K**-extendable epimorphism, and let $\underline{C} := \underline{\mathrm{dom}}(r')^{\sim}$, $R := \ker(r')^{\sim}$. Then 5.8.2 implies that $R \subseteq \ker \tilde{f}$ for every $f:\underline{A} \to \underline{B}$ ($\underline{B} \in \mathbf{K}$), i.e. $R \subseteq \mathrm{char}(\underline{A};;\mathbf{K})$, therefore another application of 5.8.2 yields the existence of a homomorphism $h:\underline{F}' \to \underline{F}$ such that $r'h = r$. This shows that r is **K**-universal. ■■■

For the following we need a technical lemma.

5.11.5 LEMMA. Let $\underline{A} \in \mathbf{Alg}(\underline{S})$, $\mathbf{K} \subseteq \mathbf{Alg}(\underline{S})$. Then there exists a subset $\mathbf{K}' \subseteq \mathbf{K}$ such that for every $\underline{B} \in \mathbf{K}$ and for every homomorphism $f:\underline{A} \to \underline{B}$ there exist $\underline{B}' \in \mathbf{K}'$ and $f':\underline{A} \to \underline{B}'$ such that $\operatorname{dom} \tilde{f} = \operatorname{dom}(f')^{\sim}$ and $\ker \tilde{f} = \ker(f')^{\sim}$, i.e. such that
(*) $\{(\underline{\operatorname{dom}}\, \tilde{f})/\ker \tilde{f} | f:\underline{A} \to \underline{B} , \underline{B} \in \mathbf{K}\} = \{(\underline{\operatorname{dom}}\, \tilde{f})/\ker \tilde{f} | f:\underline{A} \to \underline{B}' , \underline{B}' \in \mathbf{K}'\}$.

PROOF. Let $\mathbf{M} := \{(\underline{C},R) | \underline{C} \in \mathbf{Rsuba}(A,\underline{F}(\underline{A},\mathbf{TAlg}(\underline{S})) , R \in \operatorname{Cong}_C(\underline{C})\}$, and let $Q \subseteq \mathbf{M} \times \mathbf{K}$ be the following relation:
$((\underline{C},R),\underline{B}) \in Q$, iff $(\underline{C},R) \in \mathbf{M}$, $\underline{B} \in \mathbf{K}$, and there exists a homomorphism $f:\underline{A} \to \underline{B}$ such that $\underline{\operatorname{dom}}\, \tilde{f} = \underline{C}$ and $\ker \tilde{f} = R$.
By Proposition II.7.8 in A.LEVY [Lev 79] (cf. also section 23) there exists a subrelation $Q' \subseteq Q$ such that, for each $(\underline{C},R) \in \mathbf{M}$, $[(\underline{C},R)]Q' := \{\underline{B} \in \mathbf{K} | ((\underline{C},R),\underline{B}) \in Q'\}$ is a **set** and $[(\underline{C},R)]Q' \neq \emptyset$ iff $[(\underline{C},R)]Q \neq \emptyset$ for each element of $\mathbf{M}$.
Since $\mathbf{M}$ is a set, we get that
$\mathbf{K}' := \{\underline{B} \in \mathbf{K} | \underline{B} \in [(\underline{C},R)]Q'$ for some $(\underline{C},R) \in \mathbf{M}\} = \bigcup\{[(\underline{C},R)]Q' | (\underline{C},R) \in \mathbf{M}\}$
is a **set**. Because of the definitions of $\mathbf{M}$, Q and Q' the set $\mathbf{K}'$ has the required properties. ■■■

From 5.11.5.(*) we immediately get the

5.11.6 COROLLARY. Let $\underline{A} \in \mathbf{Alg}(\underline{S})$, $\mathbf{K} \subseteq \mathbf{Alg}(\underline{S})$, and $\mathbf{K}' \subseteq \mathbf{K}$ any set such that 5.11.5.(*) is satisfied. Then one has
$\operatorname{char}(\underline{A};;\mathbf{K}) = \bigcap\{\ker \tilde{f} | f:\underline{A} \to \underline{B} , \underline{B} \in \mathbf{K}\} = \bigcap\{\ker \tilde{f} | f:\underline{A} \to \underline{B} , \underline{B} \in \mathbf{K}'\} = \operatorname{char}(\underline{A};;\mathbf{K}')$ and
$\underline{\operatorname{dom}}(\operatorname{char}(\underline{A};;\mathbf{K})) = \bigcap^*\mathbf{Rsuba}(\underline{A};;\mathbf{K}) = \underline{\operatorname{dom}}(\operatorname{char}(\underline{A};;\mathbf{K}'))$. ■■■

5.11.7 PROPOSITION. With the assumptions and the notation of Theorem 5.11.4 we have $\underline{F} \in \mathbf{SP}(\mathbf{K})$.

PROOF. From Lemma 5.11.5 we know that there is a subset $\mathbf{K}'$ of $\mathbf{K}$ such that (*) of 5.11.5 is satisfied. Let $\underline{P} := \underline{\operatorname{dom}}(\operatorname{char}(\underline{A};;\mathbf{K}))$ and $R := \operatorname{char}(\underline{A};;\mathbf{K})$. Then 5.11.6 yields $\underline{P} = \bigcap^*\mathbf{Rsuba}(\underline{A};;\mathbf{K}) = \bigcap^*\mathbf{Rsuba}(\underline{A};;\mathbf{K}')$ and $R = \operatorname{char}(\underline{A};;\mathbf{K}')$; therefore $\underline{F} = \underline{P}/R$. Consider the index set (!) $I := \{(f,\underline{B}) | \underline{B} \in \mathbf{K}' , f:\underline{A} \to \underline{B}\}$, and define $\underline{D} := \mathsf{X}(\underline{B} | (f,\underline{B}) \in I)$. By the definition of $\operatorname{char}(\underline{A};\mathbf{K}')$ and $\underline{P}$ each $f:\underline{A} \to \underline{B}$ $((f,\underline{B}) \in I)$ extends to a homomorphism $\bar{f}_{(f,\underline{B})}:\underline{P} \to \underline{B}$ (since $\operatorname{char}(\underline{A};;\mathbf{K}') \subseteq \ker \tilde{f}$). Thus we get an induced homomorphism $u:\underline{P} \to \underline{D}$ such that $u \cdot pr_{(f,\underline{B})} = \bar{f}_{(f,\underline{B})}$ for each $(f,\underline{B}) \in I$. Obviously $\ker u = \bigcap\{\ker \bar{f}_{(f,\underline{B})} | (f,\underline{B}) \in I\} = \operatorname{char}(\underline{A};;\mathbf{K}') = R$, i.e. -- by 5.9.4 and its proof -- u is closed. The First Homomorphism Theorem 2.7.4 implies that u induces an isomorphism $i:\underline{P}/R = \underline{F} \to \underline{C}_D[P]u$. Hence $\underline{F}$ is isomorphic to a subalgebra of a product of $\mathbf{K}$-objects, i.e. $\underline{F} \in \mathbf{SP}(\bar{\mathbf{K}})$. ■■■

Observing the definitions of **K**-universal solutions and quasi-primitive classes we immediately get

5.11.8 EXISTENCE THEOREM for K-UNIVERSAL K-SOLUTIONS.
Let $\mathbf{K} \subseteq \mathbf{Alg}(\underline{S})$ be any **quasi-primitive** class of partial algebras. Then there exists for every partial algebra $\underline{A} \in \mathbf{Alg}(\underline{S})$ its **K**-universal **K**-solution, and one realization of it is $(r_{\underline{A},\mathbf{K}},\underline{F}(\underline{A},\mathbf{K})) := (nat_{\operatorname{char}(\underline{A};;\mathbf{K})}|A, \bigcap^*\mathbf{Rsuba}(\underline{A};;\mathbf{K})/\operatorname{char}(\underline{A};;\mathbf{K}))$. ■■■

Reviewing the preceding arguments one easily realizes that for 5.11.8 the closedness of **K** with respect to products on the empty index set is only needed in the case, when there does not exist a homomorphism from $\underline{A}$ into any **K**-algebra. Thus we have the

5.11.9 COROLLARY 1. Let $\mathbf{K} \subseteq \mathbf{Alg}(\underline{S})$ such that $\mathbf{SP}_+(\mathbf{K}) = \mathbf{K}$. Then a **K**-universal **K**-solution exists for every partial algebra $\underline{A} \in \mathbf{Alg}(\underline{S})$ which allows at least one homomorphism into some **K**-algebra. ■■■

5.11.10. The observation 5.11.9 will become important in connection with classes defined by universal Horn formulas, which we shall consider in Chapters II and III.

5.11.11 COROLLARY 2. Let $\mathbf{K} = \mathbf{SP}_+(\mathbf{K}) \subseteq \mathbf{Alg}(\underline{S})$, and let M be any set. If $\{\underline{A} \in \mathbf{K} \mid [M]v_M \subseteq [A]v_A\} \neq \emptyset$, then the **K**-universal solution of M exists. And if for each $s \in S$ satisfying $\#M_s \geq 2$ there is $\underline{A} \in \mathbf{K}$ such that $[M]v_M \subseteq [A]v_A$ and $\#A_s \geq 2$, then the **K**-free **K**-algebra **K**-freely generated by M exists. ■■■

5.11.12 REMARKS. (i): Notice that the last statement of 5.11.11 just means that $r_{M,K}$ is injective. In the case of **homogeneous** partial algebras the assumptions in 5.11.11 just mean " $\mathbf{K} \neq \emptyset$ " respectively " if $\#M \geq 2$, then there is $\underline{A} \in \mathbf{K}$ with $\#A \geq 2$ ".
(ii): If $\underline{F}(\emptyset,\mathbf{K})$ exists, then it is called the <u>initial</u> **K**-<u>algebra</u>. Obviously $\underline{F}(\emptyset,\mathbf{K})$ exists, whenever $\mathbf{SP}_+(\mathbf{K}) = \mathbf{K} \neq \emptyset$. Initial **K**-algebras play an important rôle in connection with the algebraic specification of abstract data types in computer science (see 17.6, and for more details and examples e.g. Part II by H.REICHEL, or [GoWTa 78] for the total case).

6 The equational theory of partial algebras

Before we introduce in Chapter II a general model theory for partial algebras, we develop here their equational theory, for which the foundations have been laid in the preceding sections. In Chapter III some of the following results will also be obtained in a more general context. The motivation and background of existence-equations -- the basis of our model theoretic approach -- has been intensively discussed in [ABN 80] and [B 82].

6.1 Existence-equations

6.1.1 DEFINITION. (i): Let $\tau = (K_\phi)_{\phi\in\Omega}$ be any fixed similarity type of cardinal dimension **d**, ($\mathbf{d}^* = \max\{\mathbf{d},\#\omega_0\}$), let $\underline{S}$ be any sort algebra (specifying a heterogeneous similarity type as indicated in 2.11), and let X be any fixed set such that each phylum of X has (at least) cardinality $\mathbf{d}^*$ -- for most applications $\max\{2,\mathbf{d}\}$ would suffice. Moreover, let $\mathbf{d}^\&:= \max\{\mathbf{d}^*,(\#S)^+\}$, where $(\#S)^+$ is the least cardinal greater than #S.
X will be called the (<u>global</u>) <u>set of variables</u> under consideration. Moreover, let $\underline{T} := \underline{F}(X,\mathbf{TAlg}(\underline{S}))$ be the Peano algebra on X, and in this context we shall call $\underline{T}$ the <u>global term algebra</u>, its elements are called (<u>global</u>) <u>terms</u> (with variables in X). In the homogeneous case, $\underline{T}$ would be sufficient, but since we also want to include the

heterogeneous case in the form of partial $\underline{S}$-algebras with possibly empty phyla, we also have to consider the set $T_{loc} := \bigcup(\{M\} \times F(M,\mathbf{TAlg}(\underline{S})) \mid M \subseteq X\ ,\ \#M < d^{\&})$ of all local terms with variables in X, which are written as (M:t) (for $t \in F(M,\mathbf{TAlg}(\underline{S}))$). If $\underline{A} \in \mathbf{Alg}(\underline{S})$ is any partial algebra and $M \subseteq X$ any subset of variables, then every mapping $w: M \to \underline{A}$ will be called a (local) valuation or assignment for M in $\underline{A}$, and $\tilde{w}: \mathrm{dom}\ \tilde{w} \to \underline{A}$ -- its closed M-initial extension in $\underline{F}(M,\mathbf{TAlg}(\underline{S}))$ (or $\underline{F}(X,\mathbf{TAlg}(\underline{S}))$) -- will be called the corresponding evaluation (or corresponding interpretation) of terms.

(ii): The elements $(M,t,t') \in \{M\} \times F(M,\mathbf{TAlg}(\underline{S}))^2$ which satisfy $tv_{F(M,\mathbf{TAlg}(\underline{S}))} = t'v_{F(M,\mathbf{TAlg}(\underline{S}))}$ will be called existence-equations (briefly: E-equations), and will formally be written as $(M:t \stackrel{e}{=} t')$ -- here we do not impose any cardinality restriction on M!

For each $M \subseteq X$, $(\#M < d^{\&})$ we introduce the sets

$Eeq_M := \{(M:t \stackrel{e}{=} t') \mid t,t' \in F(M,\mathbf{TAlg}(\underline{S}))$ and $tv_{F(M,\mathbf{TAlg}(\underline{S}))} = t'v_{F(M,\mathbf{TAlg}(\underline{S}))}\}$

of all E-equations on M, and

$Eeq^*_X := \bigcup(Eeq_M \mid M \subseteq X\ ,\ \#M < d^{\&})$

of all small E-equations with variables in X (i.e. connected with a small set $(\#M < d^{\&})$ of variables).

We say that an E-equation $(M:t \stackrel{e}{=} t')$ is satisfied in a partial algebra $\underline{A} \in \mathbf{Alg}(\underline{S})$ under a valuation $v: M \to A$ -- and we denote this by " $\underline{A} \models (M:t \stackrel{e}{=} t')[v]$ " --, iff $(t,t') \in \ker \tilde{v}$ (i.e. iff **both** evaluations $v(M:t)^{\underline{A}}$ and $v(M:t')^{\underline{A}}$ **exist** in $\underline{A}$ **and are equal**: $v(M:t)^{\underline{A}} = v(M:t')^{\underline{A}}$ (or, equivalently, $t\tilde{v} = t'\tilde{v}$)).

If $\underline{A} \models (M:t \stackrel{e}{=} t')[v]$, then v is called a solution of the E-equation $(M:t \stackrel{e}{=} t')$.

We say that an E-equation $(M:t \stackrel{e}{=} t')$ is valid in $\underline{A}$ -- and denote this by " $\underline{A} \models (M:t \stackrel{e}{=} t')$ " --, iff $\underline{A}$ satisfies $(M:t \stackrel{e}{=} t')$ for **every** assignment $v: M \to A$, i.e. $\underline{A} \models (M:t \stackrel{e}{=} t')[v]$ holds for every $v \in A^M$.

(iii): Let $\mathbf{K} \subseteq \mathbf{Alg}(\underline{S})$ be any class of partial algebras, let $G \subseteq Eeq^*_X$ be any set of E-equations, and let $M \subseteq X$, $(\#M < d^{\&})$; then we define:

$\mathbf{Eeq}_M(\mathbf{K}) := \{(M:t \stackrel{e}{=} t') \in Eeq_M \mid \underline{A} \models (M:t \stackrel{e}{=} t')$ for every $\underline{A} \in \mathbf{K}\}$, the set of all E-equations on M, which are valid in **K**, i.e. in all **K**-algebras;

$\mathbf{Eeq}(\mathbf{K}) := \{(M:t \stackrel{e}{=} t') \in Eeq^*_X \mid \underline{A} = (M:t \stackrel{e}{=} t')$ for every $\underline{A} \in \mathbf{K}\}$, the set of all E-equations with (a small set of) variables in X, which are valid in **K**;

$\mathbf{Mod}(G) := \{\underline{A} \in \mathbf{Alg}(\underline{S}) \mid \underline{A} \models (M:t \stackrel{e}{=} t')$ for every $(M:t \stackrel{e}{=} t') \in G\}$.

For " $\underline{A} \in \mathbf{Mod}(G)$ " we also write " $\underline{A} \models G$ " (" G is valid in $\underline{A}$ ").

Mod(G) is called the class of all models of the system G of E-equations, and **Eeq(K)** is called the (E-) equational theory of the class **K**.

6.1.2 REMARKS and further concepts. **(i):** Notice that " $\underline{A} \models (M:t \stackrel{e}{=} t')[v]$ " means that $(t,t') \in \ker \tilde{v}$, i.e. the values of $(M:t)^{\underline{A}}$ and $(M:t')^{\underline{A}}$ exist with respect to v and are equal, i.e. " $\underline{A} \models (M:t \stackrel{e}{=} t')[v]$ " includes **three statements**:

(a): $t \in \mathrm{dom}\ \tilde{v}$, (b): $t' \in \mathrm{dom}\ \tilde{v}$, (c): $t\tilde{v} = t'\tilde{v}$.

Thus the statement " $\underline{A} \models (M:t \stackrel{e}{=} t')[v]$ is false " means that at least one of the statements (a), (b) or (c) is false.

(ii): If t and t' are identical terms, then " $\underline{A} \models (M:t \stackrel{e}{=} t)[v]$ " still says that $t \in \mathrm{dom}\ \tilde{v}$, i.e. that $v(M:t)^{\underline{A}}$ **exists.** This is the reason, why we speak of "**existence-**equations" rather than of "equations": Namely the "diagonal-expressions" $(M:t \stackrel{e}{=} t)$, which are trivially satisfied and valid in all total algebras, get a real meaning for partial algebras. Because of this semantic property we call E-equations of the form " $(M:t \overset{e}{=} t)$ " term-existence-expressions (briefly: TE-expressions).

From results which we shall derive in later chapters we shall realize that exactly the fact that TE-expressions get a real (non-trivial) meaning for partial algebras -- while being of no real importance for total algebras -- causes the wealth of concepts, e.g. of significant properties of substructures and homomorphism, for partial algebras in comparison to total algebras. With a more extended first order language and the corresponding model theory at hand we shall see in Chapters II and III that these concepts and properties can be organized quite well by model theoretic concepts especially using TE-expressions (and their negations).

(iii): It should also be observed that **validity** of an E-equation $(M:t \overset{e}{=} t')$ in a partial algebra $\underline{A}$ does not only mean that $(M:t)^{\underline{A}}$ and $(M:t')^{\underline{A}}$ take the same values, whenever they are defined, but also that they both are **total** M-ary term operations on $\underline{A}$, i.e. $\underline{A} \models (M:t \overset{e}{=} t')$ iff $\mathrm{dom}(M:t)^{\underline{A}} = \mathrm{dom}(M:t')^{\underline{A}} = A^M$ **and** $(M:t)^{\underline{A}} = (M:t')^{\underline{A}}$.

When we have to describe really partial term operations, we shall have to use **elementary implications** between conjunctions of E-equations, as we shall introduce them in the next chapter.

(iv): It will become clear in the next subsections (cf. 6.4.9.(iii)), why we consider $\mathbf{d}^{\&}$ rather than $\mathbf{d}^*$ as the cardinality restriction for the subsets of variables. It would suffice to consider sets M in X such that $\#M_s < \mathbf{d}^*$ for each $s \in S$.

6.1.3. On the Galois-connection related to validity of E-equations:

Validity of E-equations establishes a Galois-connection between Eeq^*_X (respectively Eeq_M for any $M \subseteq X$) and $\mathbf{Alg}(\underline{S})$ (cf. section 24.4). Thus the operators **Eeq, Mod, ModEeq** and **EeqMod** (or **Eeq** always replaced by $\mathbf{Eeq}_M$ and Eeq^*_X by Eeq_M -- notice that in the combinations one first has to apply the right-hand operator and only afterwards the left-hand operator) have the following properties for $G \subseteq G' \subseteq \mathrm{Eeq}^*_X$ (Eeq_M) and $\mathbf{K} \subseteq \mathbf{K'} \subseteq \mathbf{Alg}(\underline{S})$:

(1): $\mathbf{Eeq}(\mathbf{K'}) \subseteq \mathbf{Eeq}(\mathbf{K})$ and $\mathbf{Mod}(G') \subseteq \mathbf{Mod}(G)$;

(2): $\mathbf{K} \subseteq \mathbf{ModEeq}(\mathbf{K})$ and $G \subseteq \mathbf{EeqMod}(G)$;

(3): $\mathbf{ModEeq}(\mathbf{K}) \subseteq \mathbf{ModEeq}(\mathbf{K'})$ and $\mathbf{EeqMod}(G) \subseteq \mathbf{EeqMod}(G')$;

(4): $\mathbf{EeqModEeq}(\mathbf{K}) = \mathbf{Eeq}(\mathbf{K})$ and $\mathbf{ModEeqMod}(G) = \mathbf{Mod}(G)$,
$\mathbf{ModEeqModEeq}(\mathbf{K}) = \mathbf{ModEeq}(\mathbf{K})$ and $\mathbf{EeqModEeqMod}(G) = \mathbf{EeqMod}(G)$.

(5): $\mathbf{Eeq}_M\mathbf{ModEeq}(\mathbf{K}) = \mathbf{Eeq}_M\mathbf{ModEeq}_M(\mathbf{K}) = \mathbf{Eeq}_M(\mathbf{K})$.

The easy straightforward proofs, which also follow from the general principles of Galois-connections (cf. section 24.4) are left as exercises to the reader. In connection with (5) it has to be observed that because of the fact that we also treat heterogeneous partial algebras with possibly empty phyla an E-equation cannot be separated from the set of variables with which it has been connected. Notice that in the homogeneous case this would not have been necessary. Namely, every mapping from some non-empty subset of X into some homogeneous partial algebra can be extended to all of X. One only has to treat the empty subset separately, since we allow empty models. ∎

Thus **EeqMod** is a closure operator on Eeq^*_X , and **ModEeq** behaves like a closure operator on $\mathbf{Alg}(\underline{S})$ -- observe that $\mathbf{Alg}(\underline{S})$ is not a set. It is the main aim of this section to characterize the closed subsets of Eeq^*_X, which are called (E-) equational theories, and the closed subclasses of **Alg**($\underline{S}$), which are called (E-) equationally definable classes or E-varieties (here we cannot drop the prefix "E-", since we shall consider later also other variety concepts).

The next lemma is an immediate consequence of the above definitions and section 5.

6.1.4 LEMMA. Let $\underline{A} \in \mathbf{Alg}(\underline{S})$, $\mathbf{K} \subseteq \mathbf{Alg}(\underline{S})$, $M \subseteq X$ ($\#M < \mathbf{d}^{\&}$) and $G \subseteq \mathrm{Eeq}^*_X$. Then

(i) $\mathbf{Eeq}_M(\{\underline{A}\}) = \cap\{\ker \tilde{v} | v \in A^M\} = \mathrm{char}(M;;\{\underline{A}\})$;

$\mathbf{Eeq}(\{\underline{A}\}) = \cup\{\mathbf{Eeq}_M(\{\underline{A}\}) | M \subseteq X\} = \cup\{\mathrm{char}(M;;\{\underline{A}\}) | M \subseteq X , \#M < \mathbf{d}^{\&}\}$.

(ii) $\mathbf{Eeq}_M(\mathbf{K}) = \cap\{\mathbf{Eeq}_M(\{\underline{A}\}) | \underline{A} \in \mathbf{K}\} = \mathrm{char}(M;;\mathbf{K})$;

$\mathbf{Eeq}(\mathbf{K}) = \cup\{\mathbf{Eeq}_M(\mathbf{K}) | M \subseteq X\} = \cup\{\mathrm{char}(M;;\mathbf{K}) | M \subseteq X , \#M < \mathbf{d}^{\&}\}$.

In the **homogeneous case** it is sufficient to consider $\emptyset$ and X instead of all (small) subsets of X.

(iii) $\underline{A} \in \mathbf{Mod}(G)$, iff $G \subseteq \mathbf{Eeq}(\{\underline{A}\})$,

i.e. iff $G \subseteq \cup\{\cap\{\ker \tilde{v} | v \in A^M\} | M \subseteq X , \#M < \mathbf{d}^{\&}\}$, or -- in other words -- iff, for each $M \subseteq X$ with $\#M < \mathbf{d}^{\&}$, $G \cap \mathrm{Eeq}_M \subseteq \mathrm{char}(M;;\{\underline{A}\})$. ■■■

6.2 Description of the model classes (E-varieties)

6.2.1 PROPOSITION. Let $G \subseteq \mathrm{Eeq}^*_X$, $\mathbf{K} := \mathbf{Mod}(G)$. Then:

(i) $\mathbf{H}(\mathbf{K}) \cup \mathbf{S}(\mathbf{K}) \cup \mathbf{P}^{r,\mathbf{d}^{\&}}(\mathbf{K}) \subseteq \mathbf{K}$;

(ii) $(\underline{\mathrm{dom}}\ \mathbf{Eeq}_M(\mathbf{K}))/\mathbf{Eeq}_M(\mathbf{K}) \cong \underline{F}(M,\mathbf{K}) \in \mathbf{K}$.

PROOF. Ad (i): (a): Let $\underline{A} \in \mathbf{K}$, $f:\underline{A} \to \underline{B}$ a surjective homomorphism and $v:M \to B$ an assignment for some $M \subseteq X$, ($\#M < \mathbf{d}^{\&}$). Since f is surjective, there exists -- using the Axiom of Choice -- an assignment $w:M \to A$ such that $v = wf$. Now, by 6.1.4.(iii) and 6.1.3.(1) $\underline{A} \in \mathbf{Mod}(G)$ implies $G \cap \mathrm{Eeq}_M \subseteq \mathrm{char}(M;;\mathbf{Mod}(G)) \subseteq \mathrm{char}(M;;\{\underline{A}\}) \subseteq \ker \widetilde{wf} \subseteq \subseteq \ker \tilde{v}$, and therefore $G \subseteq \mathbf{Eeq}(\{\underline{B}\})$ easily follows. Thus $\mathbf{H}(\mathbf{K}) \subseteq \mathbf{K}$. ■

(b): Let $\underline{A} \in \mathbf{K}$, $\underline{B}$ a subalgebra of $\underline{A}$, $M \subseteq X$ ($\#M < \mathbf{d}^{\&}$) and $v:M \to \underline{B}$. Then $\ker \tilde{v}$ does not depend on the fact that we have chosen $\underline{B}$ instead of $\underline{A}$ as the target. Thus $G \cap \mathrm{Eeq}_M \subseteq \mathbf{Eeq}_M(\{\underline{A}\}) \subseteq \mathbf{Eeq}_M(\{\underline{B}\})$, i.e. $\underline{B} \in \mathbf{K}$. ■

(c): Let $(\underline{A}_i | i \in I)$ be a family of **K**-algebras, F a $\mathbf{d}^{\&}$-complete filter on I and $((p_F | F \in \mathbf{F}), \underline{A})$ the direct limit of the $\mathbf{d}^{\&}$-directed system $(\underline{A}_F := \mathsf{X}(\underline{A}_i | i \in F), p_{FF'}:\underline{A}_F \to \underline{A}_{F'} | F \supseteq F' \in \mathbf{F})$ (i.e. a reduced product), and let $v:M \to A$ be any assignment for $M \subseteq X$, $\#M < \mathbf{d}^{\&}$. Since each element of A has a preimage in some A_F, since $\#M < \mathbf{d}^{\&}$, and since **F** is $\mathbf{d}^{\&}$-complete, there are $F \in \mathbf{F}$ and -- using the Axiom of Choice -- a mapping $w:M \to A_F$ such that $wp_F = v$. Since $G \cap \mathrm{Eeq}_M \subseteq \cap\{\mathrm{char}(M;;\{\underline{A}_i\}) | i \in F\} \subseteq \cap\{\ker(w \cdot pr_i)^\sim | i \in F\} = \ker \tilde{w}$, we also get $G \cap \mathrm{Eeq}_M \subseteq \ker \widetilde{wp_F} \subseteq \ker \tilde{v}$, showing that $\underline{A} \in \mathbf{Mod}(G)$. ■

(ii) is an immediate consequence of (i) and the results in 5.11. ■■■

Because of $\mathbf{H}(\mathbf{Mod}(G)) \subseteq \mathbf{Mod}(G)$ for any set G of E-equations we have in particular

6.2.2 COROLLARY. Let $G \subseteq \mathrm{Eeq}^*_X$. Then $\mathbf{Mod}(G)$ contains $(\underline{S}, id_S)$ -- the empty product --, and for every $\underline{A}$ also each "inner extension" of $\underline{A}$, i.e. each $\underline{B} \in \mathbf{Alg}(\underline{S})$ for which $B = A$ and $id_A:\underline{A} \to \underline{B}$ is a homomorphism. ■■■

Many of the technical results needed in the proof of the following main results are collected in

6.2.3 LEMMA. Let N and M be any sets and $K \subseteq \mathbf{Alg}(\underline{S})$. Moreover, let
$p := nat_{char(M;;K)} : \underline{dom}\ char(M;;K) \to (\underline{dom}\ char(M;;K))/char(M;;K) = \underline{F}(M,K)$
(cf. 5.11.8). Then:

(i) Let $\underline{P}$ and $\underline{Q}$ be partial Peano algebras on M and N, respectively, and let $j:N \to M$ be an injective mapping, which extends to a homomorphism $\overline{j}:\underline{Q} \to \underline{P}$. Then $\overline{j}$ is injective.

(ii) If $N \subseteq M$, then $char(N;;K) \subseteq char(M;;K) \cap (C_{\underline{F}(M,\mathbf{TAlg}(\underline{S}))}N)^2$, and if
(*) $[M]v_M \subseteq [dom\ char(N;;K)]v_{dom\ char(N;;K)}$,
then we have equality.

(iii) If $j:N \to M$ is an injective mapping, then it induces a homomorphism $\overline{j}:\underline{F}(N,K) \to \underline{F}(M,K)$; and if (ii).(*) is satisfied, then $\overline{j}$ is closed and injective.

(iv) For every set Y and terms $t,t' \in F(Y,\mathbf{TAlg}(\underline{S}))$ with $t\tilde{v}_Y = t'\tilde{v}_Y$ there are a set $M \subseteq X$, an injective mapping $j:M \to Y$ and terms $q,q' \in F(M,\mathbf{TAlg}(\underline{S}))$ such that j has a homomorphic extension $\overline{j}:\underline{F}(M,\mathbf{TAlg}(\underline{S})) \to \underline{F}(Y,\mathbf{TAlg}(\underline{S}))$ satisfying $\#M < \mathbf{d}^{\&}$, $q\overline{j} = t$, $q'\overline{j} = t'$, and such that for every $\underline{A} \in \mathbf{Alg}(\underline{S})$ and every $v:Y \to A$ we have $\underline{A} \models (Y:t \overset{e}{=} t')[v]$ iff $\underline{A} \models (M:q \overset{e}{=} q')[jv]$. Moreover
(**) $\underline{A} \models (Y:t \overset{e}{=} t')$ iff $\underline{A} \models (M:q \overset{e}{=} q')$.

(v) If for some $s \in S$ there is $\underline{A} \in \mathbf{K}$ such that the phylum A_s of $\underline{A}$ has at least two elements, then the restriction $p|M_s$ is injective.

(vi) Let X and Y both be sets (of variables) satisfying that each phylum has cardinality at least $\mathbf{d}^*$, and let -- for the moment --, for each $K \subseteq \mathbf{Alg}(\underline{S})$, $\mathbf{Eeq}^*_X(K)$ and $\mathbf{Eeq}^*_Y(K)$ designate the sets of all E-equations valid in K with variables in X, respectively Y. Then the operators $\mathbf{ModEeq}^*_X$ and $\mathbf{ModEeq}^*_Y$ are identical.

PROOF. If $K = \emptyset$, then $char(M;;K) = \underline{F}(M,\mathbf{TAlg}(\underline{S}))^2$, and all statements become trivial. Hence we can assume in what follows that $K \neq \emptyset$.
Ad (i): This can either be proved by algebraic induction, or one may argue as follows: One can find a set $N' \supseteq N$ such that j extends to a bijection $j':N' \to M$, which then extends to an isomorphism $\overline{j'}:\underline{F}(Q\dot{\cup}(N' \setminus N),\mathbf{TAlg}(\underline{S})) \to \underline{F}(\underline{P},\mathbf{TAlg}(\underline{S}))$ between total Peano algebras and is an extension of $\overline{j}$, which therefore has to be injective. ∎
Ad (ii): Since every mapping from M into $\underline{A} \in K$ can be restricted to N, but not necessarily every mapping from N into A extended to one from M into $\underline{A}$, we easily get the inclusion stated in (ii). Assume (*) and $(t,t') \in char(M;;K) \cap (C_{\underline{F}(M,\mathbf{TAlg}(\underline{S}))}N)^2$, and let $v:N \to \underline{A}$ $(\underline{A} \in K)$ be any assignment for N. Condition (*) now says -- together with the Axiom of Choice -- that we can get an extension $w:M \to \underline{A}$ of v, whose closed homomorphic extension $\tilde{w}:\underline{dom}\ \tilde{w} \to \underline{A}$ obviously coincides on $(dom\ \tilde{w}) \cap F(N,\mathbf{TAlg}(\underline{S})) = dom\ \tilde{v}$ with $\tilde{v}$ -- concerning the equality with respect to the domains observe that graph(w) U graph($\tilde{v}$) is the graph of an M-initial homomorphic extension of w, and $\underline{F}(M,\mathbf{TAlg}(\underline{S}))$ is a Peano algebra. Hence $\ker \tilde{v} = \ker \tilde{w} \cap (C_{\underline{F}(M,\mathbf{TAlg}(\underline{S}))}N)^2$. The definition of char(N;;K) then provides the other desired inclusion. ∎
Ad(iii): This is an immediate consequence of (ii), (i) and 5.8.2. ∎
Ad (iv): Let $(Y:t \overset{e}{=} t')$ be any E-equation with respect to an arbitrary set Y. Let $var_Y(t,t') := Y \cap (\downarrow t \cup \downarrow t')$ be the set of all elements of Y really needed (cf. 5.5.5)

to generate t and t'. Then $\#var(t,t') < \mathbf{d^*}$. From each nonempty phylum we pick an element $y_s \in Y_s$ $(s \in S)$ and define $Y(t,t') := var_Y(t,t') \cup \{y_s | s \in S , Y_s \neq \emptyset\}$. Obviously $\#Y(t,t') < \mathbf{d^{\&}}$ and we have an injective mapping $j':Y(t,t') \to X$. Define $M := [Y(t,t')]j'$ and $j := j'^{-1}:M \to Y$. Obviously j' has a homomorphic extension $\overline{j'}:C_{\underline{F}(Y,\mathbf{TAlg}(\underline{S}))}{}^{Y(t,t')} \to C_{\underline{F}(X,\mathbf{TAlg}(\underline{S}))}{}^{M}$, which is an isomorphism. Define $q := t\overline{j'}$ and $q' := t'\overline{j'}$. Then the rest of the statement is easily realized using that $\overline{j'}$ is an isomorphism. -- (**) then easily follows. ■

Ad (v): With the assumptions of (v) we have for any two different elements m and m' of the phylum M_s a mapping $w_{m,m'}:M \to \underline{A}_{m,m'} \in \mathbf{K}$ such that $mw_{m,m'} \neq m'w_{m,m'}$. Hence the induced mapping, say w, from M into the product of all these K-algebras separates all these pairs, i.e. it is injective on M_s. Since there is a homomorphism g from $\underline{F}(M,\mathbf{K})$ into this product such that $w = r_{M,\mathbf{K}}g$, $r_{M,\mathbf{K}}|M_s$ has to be injective, and it is easy to realize that p is its closed M-initial extension. ■

Ad (vi): Assume $\underline{A} \in \mathbf{ModEeq^*}_X(\mathbf{K}) \setminus \mathbf{ModEeq^*}_Y(\mathbf{K})$. Then we have $(M:t \stackrel{e}{=} t') \in \mathbf{Eeq^*}_Y(\mathbf{K})$, which is not valid in $\underline{A}$. But this is easily seen to contradict (iv) and the fact that $\underline{A} \in \mathbf{ModEeq^*}_X(\mathbf{K})$. ■■■

6.2.4 REMARK. 6.2.3.(vi) justifies that we may fix for a set of variables for E-equations any set X which only satisfies that each phylum has at least cardinality $\mathbf{d^*}$, as we have done at the beginning.

We are now able to prove the generalization to partial ($\underline{S}$-) algebras of G.Birkhoff's first main theorem on equational theories (cf. [Bi 35]).

6.2.5 FIRST BIRKHOFF THEOREM. Let $\mathbf{K} \subseteq \mathbf{Alg}(\underline{S})$ be any class of partial algebras and let X be any set of variables such that each phylum of X has cardinality at least $\mathbf{d^*}$. Then:

(a) $\mathbf{ModEeq}(\mathbf{K}) = \mathbf{HSP}(\mathbf{K})$.

(b) The following statements are equivalent:

  (i) $\mathbf{K} = \mathbf{ModEeq}(\mathbf{K})$.

  (ii) $\mathbf{K} = \mathbf{HSP}(\mathbf{K})$.

PROOF. (b) is an immediate consequence of (a), and $\mathbf{HSP}(\mathbf{K}) \subseteq \mathbf{ModEeq}(\mathbf{K})$ follows from 6.2.1. and 4.6.4.(viii). -- Thus assume $\underline{A} \in \mathbf{ModEeq}(\mathbf{K})$. From 5.11.7 and 5.11.8 we know that $\underline{F}(A,\mathbf{K})$ -- the K-free K-algebra on the carrier set $[A]r_{A,\mathbf{K}}$ -- exists and belongs to $\mathbf{SP}(\mathbf{K})$. Hence there is to show that $i := id_A:A \to \underline{A}$ induces a surjective homomorphism $\overline{i}:\underline{F}(A,\mathbf{K}) \to \underline{A}$. And with respect to 5.11.8 and 5.8.2.(a) we only have to show that $char(\underline{A};;\mathbf{K}) \subseteq \ker \overline{i}$. Thus assume $(A:t \stackrel{e}{=} t') \in char(\underline{A};;\mathbf{K})$. Then we have, according to 6.2.3.(iv), $M \subseteq X$, $q,q' \in C_{\underline{F}(M,\mathbf{TAlg}(\underline{S}))}{}^{M}$ and an injective mapping $j:M \to A$ such that $\#M < \mathbf{d^{\&}}$ and $\underline{C} \models (M:q \stackrel{e}{=} q')$ iff $\underline{C} \models (A:t \stackrel{e}{=} t')$ for all $\underline{C} \in \mathbf{Alg}(\underline{S})$. Since $(A:t \stackrel{e}{=} t') \in char(\underline{A};;\mathbf{K})$ iff $\underline{D} \models (A:t \stackrel{e}{=} t')$ for all $\underline{D} \in \mathbf{K}$, this implies $(M:q \stackrel{e}{=} q') \in \mathbf{Eeq}(\mathbf{K})$, and therefore $\underline{A} \models (M:q \stackrel{e}{=} q')$, respectively $\underline{A} \models (A:t \stackrel{e}{=} t')$; thus $(t,t') \in \ker \overline{i}$, which was to be shown. ■■■

6.2.6 REMARKS. (i): The preceding result shows that the E-varieties are exactly the primitive classes. -- In the homogeneous case one often takes assignments only for the whole set X. Then every E-variety always contains the empty partial algebra since this then allows no assignment and hence each E-equation is trivially valid in it -- or one also has to forbid the empty algebra. If we proceed in the same way for many-sorted (partial) algebras, we would get a whole class of non-isomorphic "trivial models" contained in each E-variety, namely all those with at least one empty phylum -- or one has to allow only partial $\underline{S}$-algebras with surjective sort homomorphism, i.e. forbid empty phyla, which causes trouble in connection with substructures and especially with the concept of generation. This shows that with each concept one runs into trouble, and one just has to make a decision, as we have done in this book.

(ii): It should be noticed that 6.2.3.(ii).(*) is in general really necessary for the statements in 6.2.3.(ii) and (iii), i.e. the homomorphism $\overline{j}$ from 6.2.3.(iii) need in general be neither injective nor closed, nor full. In order to see this, let us consider the following example:

$\tau := (0,0,0,1,1,2)$, $\Omega := (\phi,\phi',\phi'',\nu,\rho,\sigma)$, $\mathbf{K} := \{\underline{A},\underline{B}\}$, where $\underline{S}$, $\underline{A}$ and $\underline{B}$ are sketched in Figure 6.1. It is easy to realize that (assuming $b' \in X$) $\underline{A} = \underline{F}(\emptyset,\mathbf{K})$ and $\underline{B} = \underline{F}(\{b'\},\mathbf{K})$, since there is no assignment from $\{b'\}$ into $\underline{A}$, and the only possible homomorphism $f:\underline{A} \to \underline{B}$ is neither injective nor full nor is its image a subalgebra of $\underline{B}$. In the homogeneous case such an effect can only be obtained, when $\mathbf{K}$ contains the empty algebra and there is a nullary fundamental constant which is defined in each non-empty $\mathbf{K}$-algebra. -- Better features can be obtained, when one considers hierarchical equoids, as this is done in Part II, but they do not form an E-variety.

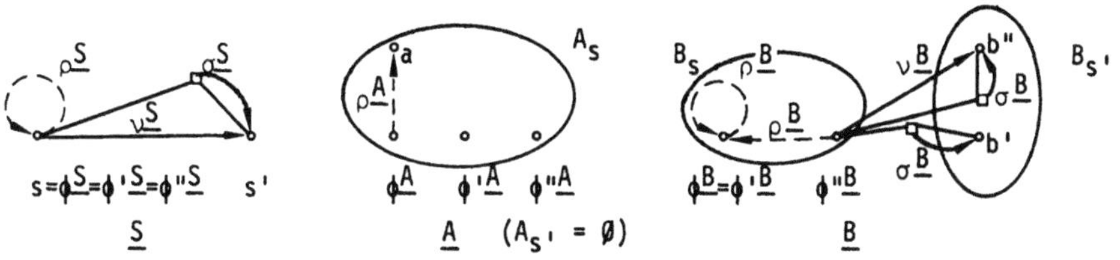

Figure 6.1.

(iii): Notice that stack automata (which we have considered in the introduction) with natural numbers as input form an E-variety, where we have the following specification (in the notation of Part II):

```
STACK(NAT) is definition
           sorts  States, Nat
           oprn   0 ↦ Nat
                  SUCC(Nat) ↦ Nat
                  PRED(Nat) ↦ Nat
                  Λ ↦ States
                  TOP(States) ↦ Nat
                  POP(States) ↦ States
                  PUSH(States,Nat) ↦ States
           axioms x:States , y:Nat
                  (Ø:0 ≗ 0)
                  ({y}:PRED(SUCC(y)) ≗ y)
                  (Ø:Λ ≗ Λ)
                  ({x,y}:TOP(PUSH(x,y)) ≗ y)
                  ({x,y}:POP(PUSH(x,y)) ≗ x)
end STACK(NAT) .
```

Let **K** be the class of all stack automata with natural numbers as input, i.e. all partial algebras of the above type satisfying the above axioms. As the stack automaton with natural numbers as input elements one then declares $\underline{F}(\emptyset,\mathbf{K})$, the so-called initial algebra of **K**. It is really partial, since POP(Λ), TOP(Λ) and PRED(0) are not defined.

(iv): Notice that the class of all small categories is not an E-variety, since it is not closed with respect to homomorphic images. -- We shall see later that they form an ECE-variety, where an ECE-equation is an elementary implication with a conjunction of TE-expressions in the premise.

6.3 More facts about freeness, fully invariant congruences

6.3.1. In 5.1.4 and 5.11.11 we have already mentioned **K**-free algebras, and in 6.2.5 they played an important rôle in the proof, showing that in an E-variety **K** each **K**-algebra $\underline{A}$ is a homomorphic image of the **K**-free **K**-algebra $\underline{F}(A,\mathbf{K})$ -- which then exists, as one can derive from the proof of 6.2.5 -- and it is then characterized in 5.11.8. From 5.1.6 and 6.2.5 we may derive that every independence class is an E-variety. If we consider only homogeneous partial algebras and do not allow the empty partial algebra, then Remark 6.2.6 easily yields that also the converse is true -- this is also the case, when E-equations and assignments are always defined with respect to the whole set X of variables.

In this subsection we want to learn more on free partial algebras, since the results e.g. in 6.1.4 and 5.11.4 show that the **K**-free **K**-algebras encode the E-equations valid in **K**. Hence we shall use them for the characterization of closed sets of E-equations.

6.3.2 PROPOSITION. Let $\mathbf{K} \subseteq \mathbf{Alg}(\underline{S})$ and let M be any set. Then $\underline{\mathrm{dom}}\ \mathrm{char}(M;;\mathbf{K})$ is freely and **K**-freely generated by M.

PROOF. Because of 5.9.3 M generates $\underline{P} := \underline{\mathrm{dom}}\ \mathrm{char}(M;;\mathbf{K})$. If $v{:}M \to \underline{A} \in \mathbf{K}$ is any assignment, then there is $\bar{v}{:}\underline{F}(M,\mathbf{K}) \to \underline{A}$ such that, for $p := \mathrm{nat}_{\mathrm{char}(M;;\mathbf{K})}$, $p\bar{v}$ extends v, i.e. M is **K**-independent in $\underline{P}$. -- Now assume $v{:}M \to \underline{P}$ to be any mapping, and let $\tilde{v}{:}\underline{\mathrm{dom}}\ \tilde{v} \to \underline{P}$ be its closed M-initial homomorphic extension. Then $\tilde{v}p{:}\underline{\mathrm{dom}}\ \tilde{v} \to \underline{F}(M,\mathbf{K}) \in \mathbf{K}$ is a closed extension of vp, and -- by the definition of P and the closedness of p: $P \subseteq \mathrm{dom}(vp)^{\sim} = \mathrm{dom}\ \tilde{v}p = \mathrm{dom}\ \tilde{v}$; therefore $\tilde{v}|P{:}\underline{P} \to \underline{P}$ is a homomorphic extension of v to $\underline{P}$, showing that M is independent in $\underline{P}$. ■■■

A closely related result is the following one:

6.3.3 PROPOSITION. Let $\underline{F} \in \mathbf{Alg}(\underline{S})$ be a free partial algebra with basis M, and let $\underline{P} := \underline{\mathrm{dom}}(\mathrm{id}_M)^{\sim}$ be the corresponding relative subalgebra of $\underline{F}(M,\mathbf{TAlg}(\underline{S}))$ with respect to $\mathrm{id}_M{:}M \to \underline{F}$.

Then M freely generates $\underline{P}$, and for every $t \in P$, $(M{:}t)^{\underline{F}}$ is a total M-ary term operation on $\underline{F}$, and only elements of P induce total M-ary term operations on $\underline{F}$.

PROOF. For the proof that M is a basis of $\underline{P}$ let $f{:}M \to P$ be any mapping, and let $\tilde{f}{:}\underline{\mathrm{dom}}\ \tilde{f} \to \underline{P}$ be its closed M-initial extension. Moreover, let $p := (\mathrm{id}_M)^{\sim}{:}\underline{P} \to \underline{F}$. Since M freely generates $\underline{F}$, fp has a homomorphic extension $g{:}\underline{F} \to \underline{F}$ $(g|M = fp)$. Now $\tilde{f}p$ is a closed M-initial homomorphic extension of fp, and pg is just an M-initial one; therefore

$P \subseteq \text{dom}\,\tilde{f}$, i.e. $\tilde{f}|P:\underline{P} \to \underline{P}$ is a homomorphic extension of f with respect to $\underline{P}$. -- Now, let $t \in P$ be arbitrary, and $v:M \to F$ any assignment. Then there exists a homomorphic extension $\bar{v}:\underline{F} \to \underline{F}$ of v within $\underline{F}$, and $p\bar{v}:\underline{P} \to \underline{F}$ is an M-initial homomorphic extension of v as a mapping out of P into F. Thus $t \in \text{dom}\,\tilde{v}$, i.e. $v \in \text{dom}\,(M:t)^{\underline{F}}$. If, on the other hand, $\text{dom}(M:t)^{\underline{F}} = F^M$ for some $t \in \underline{F}(M,\mathbf{TAlg}(\underline{S}))$, then $id_M \in \text{dom}\,(M:t)^{\underline{F}}$, i.e. $t \in \text{dom}\,(id_M)^{\sim} = \text{dom}\,p = P$. ■■■

6.3.4 COROLLARY. With the notation and the assumptions of 6.3.3 we have $\ker(id_M)^{\sim} = \text{char}(M;;\{\underline{F}\}) = \text{char}(M;;\text{ind}_{\underline{F}}(M))$. ■■■

6.3.5 DEFINITION. A congruence R on a partial algebra $\underline{A}$ is called <u>fully invariant</u> iff, for every endomorphism $f:\underline{A} \to \underline{A}$, $[R]f^{(2)} \subseteq R$.
By $\text{Cong}_{fi}(\underline{A})$ (respectively $\text{Cong}_{cfi}(\underline{A})$) we denote the set of all fully invariant (closed) congruence relations on $\underline{A}$.

6.3.6 PROPOSITION. For every $\underline{A} \in \mathbf{Alg}(\underline{S})$ $\text{Cong}_{fi}(\underline{A})$ is a closure system on $A \times A$.

PROOF. Obviously $A \times A = \nabla_A$ is fully invariant. Closedness with respect to arbitrary meets can be shown using that the image of a meet is always contained in the meet of the images. ■■■

6.3.7 LEMMA. Let $\underline{A} \in \mathbf{Alg}(\underline{S})$ and $R \in \text{Cong}_{fi}(\underline{A})$, $R \cap (A_s \times A_s) \neq A_s \times A_s$ for some $s \in S$. If $M \subseteq A$ freely generates $\underline{A}$, then $R \cap (M_s \times M_s) = \Delta_{M_s}$.

PROOF. If $(m,n) \in R \cap (M_s \times M_s)$, $m \neq n$, then for every $a,b \in A_s$ the mapping $f:M \to A$ defined by $mf:= a$, $lf:= b$ for all $l \in M_s \setminus \{m\}$, and $lf:= l$ otherwise, has a homomorphic extension $\bar{f}:\underline{A} \to \underline{A}$ satisfying $(m\bar{f},n\bar{f}) = (a,b) \in R$, contradicting the assumptions on R. ■■■

6.3.8 PROPOSITION. Let $f:\underline{A} \to \underline{B}$ be a homomorphism.

(i) If f is full and surjective, if $\ker f$ is fully invariant, and if M is a basis of $\underline{A}$, then $[M]f$ is a basis of $\underline{B}$. And if, for some $s \in S$, $\#B_s > 1$, then $f|M_s$ is injective. Moreover, M is $\underline{B}$-free in $\underline{A}$, and for each homomorphism $g:\underline{A} \to \underline{B}$ we have $\ker f \subseteq \ker g$.

(ii) If $\underline{B}$ has a basis N, and if f maps some generating subset M of $\underline{A}$ into N such that $f|M_s$ is injective for every $s \in S$ for which $\#B_s > 1$, then $\ker f$ is fully invariant.

PROOF. Ad (i): Let us first observe that $[M]f$ generates $\underline{B}$, since $B = [A]f = [C_{\underline{A}}M]f \subseteq C_{\underline{B}}([M]f)$. If all phyla of B have at most one element, then (i) is trivial. Hence assume that $\#B_s > 1$ for some $s \in S$; thus 6.3.7 implies that $f|M_s$ is injective. Let $g:[M]f \to B$ be any mapping. Then -- using the Axiom of Choice and the surjectivity of f -- we can find a mapping $h:M \to A$ such that $hf = (f|M)g$. Since M is a basis of $\underline{A}$, h has a homomorphic extension $\bar{h}:\underline{A} \to \underline{A}$. Since $\ker f$ is fully invariant, we have $[\ker f]\bar{h}^{(2)} \subseteq \ker f$, and $\ker f \subseteq \ker \bar{h}f$ easily follows. Since f is full and surjective, the Diagram Completion Lemma 2.7.1 implies the existence of a homomorphism $\bar{g}:\underline{B} \to \underline{B}$ such that $f\bar{g} = \bar{h}f$. Thus $(f|M)\bar{g} = (\bar{h}|M)f = hf = (f|M)g$, i.e. $\bar{g}|[M]f = g$, showing that $\bar{g}$ extends g. This shows that $[M]f$ freely generates $\underline{B}$. -- Now, assume that

$g:M \to B$ is a mapping from M into B. Since $f|M_s$ is injective for every $s \in S$ for which $\underline{B}$ has a non-trivial phylum, there is $h:[M]f \to B$ such that $g = (f|M)h$. Since $[M]f$ is a basis of $\underline{B}$, h extends to an endomorphism $\overline{h}$ of $\underline{B}$, and $f\overline{h}$ is the homomorphic extension of g we have looked for. Moreover $\ker f \subseteq \ker f\overline{h}$, implying the last statement. ■

Ad (ii): Let the assumptions of (ii) be satisfied, let h be any endomorphism of $\underline{A}$ and assume $(a,b) \in \ker f$. We have to show $ahf = bhf$. Since $f|M_s$ is injective, whenever $\#B_s > 1$, we can find a mapping $g:[M]f \to B$ such that $(f|M)g = (h|M)f$ ($g: mf \mapsto mhf$, $m \in M$). Next we extend g to a mapping $g':N \to B$, which then can be extended to a homomorphism $\overline{g}:\underline{B} \to \underline{B}$, i.e. $\overline{g}|[M]f = g$. Hence $(f|M)\overline{g} = (f|M)g = (h|M)f$, implying $f\overline{g} = hf$, and $ahf = af\overline{g} = bf\overline{g} = bhf$, what we wanted to show. ■■■

6.3.9 COROLLARY 1. For any subclass $\mathbf{K} \subseteq \mathbf{Alg}(\underline{S})$ and for any set M $\mathrm{char}(M;;\mathbf{K}) = \ker(r_{M,\mathbf{K}})^{\sim}$, and this is a closed and fully invariant congruence relation on $\underline{\mathrm{dom}}\ \mathrm{char}(M;;\mathbf{K})$.

PROOF. Closedness of $\mathrm{char}(M;;\mathbf{K})$ and $\mathrm{char}(M;;\mathbf{K}) = \ker(r_{M,\mathbf{K}})$ are already known (cf. 5.11.2 and 6.2.3.(v)). Since $\underline{F}(M,\mathbf{K})$ is $\mathbf{ModEeq}(\mathbf{K})$-freely generated by $[M]r_{M,\mathbf{K}}$, it is also freely generated by $[M]r_{M,\mathbf{K}}$. If, for some $s \in S$, $\#F(M,\mathbf{K})_s > 1$, then 6.2.3.(v) implies that $r_{M,\mathbf{K}}|M_s$ is injective. Thus 6.3.8.(ii) implies full invariance of $\mathrm{char}(M;;\mathbf{K})$. ■■■

6.3.2 and 6.3.4.(ii) together with 6.3.9 imply

6.3.10 COROLLARY 2. For each subclass $\mathbf{K} \subseteq \mathbf{Alg}(\underline{S})$ and for each set M, $\mathbf{Eeq}_M(\mathbf{K})$ is a closed and fully invariant congruence relation on a relative subalgebra, say $\underline{P}$, of $\underline{F}(M,\mathbf{TAlg}(\underline{S}))$ such that M freely generates $\underline{P}$. ■■■

6.4 Characterization of E-equational theories

6.4.1. We now have the tools to describe the closed sets of E-equations. The description is a little bit more involved than it would be in the homogeneous case.

In the following results we compare sets G_M and G_N of E-equations on different sets M and N such that N is a subset of M; " $G_N \subseteq G_M$ " will mean that $\{(t,t')|(N:t \stackrel{e}{=} t') \in G_N\} \subseteq \{(t,t')|(M:t \stackrel{e}{=} t') \in G_M\}$; $\mathrm{dom}\ G_M := \{t,t'|(M:t \stackrel{e}{=} t') \in G\}$.

6.4.2 SECOND BIRKHOFF THEOREM. Let X be any set of variables such that each phylum has cardinality not less than $\mathbf{d^*}$. Let $G \subseteq \mathrm{Eeq}^*_X$ and for each $M \subseteq X$ ($\#M < \mathbf{d}^{\&}$) let $G_M := G \cap F(M,\mathbf{TAlg}(\underline{S}))^2$. Then the following statements are equivalent:

(i) $\mathbf{EeqMod}(G) = G$.

(ii) For each $M \subseteq X$ with $\#M < \mathbf{d}^{\&}$, G_M is a closed and fully invariant M-initial congruence on a relative subalgebra $\underline{\mathrm{dom}}\ G_M$ of $\underline{F}(M,\mathbf{TAlg}(\underline{S}))$, which is freely generated by M; moreover, for each $N \subseteq M$ and for the homomorphisms $i_{NM}:\underline{\mathrm{dom}}\ G_N \to \underline{\mathrm{dom}}\ G_M$ induced by the inclusion mapping from N into M:

(a) $[G_N]i_{NM}^{(2)} \subseteq G_M$ (cf. 6.4.1);

(b) if $[M]v_M \subseteq [\mathrm{dom}\ G_N]v_{F(N,\mathbf{TAlg}(\underline{S}))}$, then

$$[G_N]i_{NM}^{(2)} = G_M \cap (\mathbf{C}_{\underline{F}(M,\mathbf{TAlg}(\underline{S}))}N)^2 .$$

(iii) $G = \mathbf{Eeq}(\{(\underline{\mathrm{dom}}\ G_M)/G_M \mid M \subseteq X\ ,\ \#M < \mathbf{d}^*\})$.

PROOF. Since $\mathbf{Eeq}_M(\mathbf{K}) = \mathrm{char}(M;;\mathbf{K})$ (in the sense of 6.4.1), (ii) easily follows from (i) by 6.2.3.(ii) and (iii) in connection with 6.3.9. Moreover, it is easily seen in connection with 6.1.3 that (iii) implies (i); thus it remains only to show that (ii) implies (iii):

Let, for each $M \subseteq X$, $\underline{F}_M := (\underline{\mathrm{dom}}\ G_M)/G_M$, $r_M: M \to \underline{F}_M$ be the restriction to M of $p_M := \mathrm{nat}_{G_M}: \underline{\mathrm{dom}}\ G_M \to \underline{F}_M$, and let $\mathbf{K} := \{\underline{F}_M \mid M \subseteq X\ ,\ \#M < \mathbf{d}^{\&}\}$. Then the assumptions in (ii) imply that, for each $M \subseteq X$ with $\#M < \mathbf{d}^{\&}$, $G_M = \ker p_M$; thus $\mathbf{Eeq}(\mathbf{K}) \subseteq G$. Hence it remains to show that for each $Y,M \subseteq X$, $\#Y,\#M < \mathbf{d}^{\&}$, and each assignment $w: Y \to \underline{F}_M$ we have $G_Y \subseteq \ker \widetilde{w}$ (the reader should draw a diagram):

In order to use assumptions (a) and (b) we define $Z := M \cup Y$. Obviously $\#Z < \mathbf{d}^{\&}$. Then Z allows a mapping into F_M -- e.g. $w': Z \to F_M$ such that $w'|Y := w$ and $w'|(M \setminus Y) := id_{M \setminus Y}$ --, and we have $[Z]v_Z \subseteq [\mathrm{dom}\ G_M]v_{\mathrm{dom}\ G_M}$. Thus, according to (b) and 6.2.3.(iii) i_{MZ} induces an injective and closed homomorphism $i_{MZ}^*: \underline{F}_M \to \underline{F}_Z$ and $\ker \widetilde{w} = \ker \widetilde{w} i_{MZ}^* = \ker(w^*)^\sim$, where $w^* := w i_{MZ}^*$. Moreover $\widetilde{w}' i_{MZ}^*$ obviously extends w^* and, as before, $\ker(w')^\sim i_{MZ}^* = \ker(w')^\sim \supseteq \ker(r_Z)^\sim = G_Z$ -- the inclusion following from the assumptions and 6.3.8.(i). Hence, if $(Y: t \overset{e}{=} t') \in G_Y$, then -- because of (a) -- $t\widetilde{w} = t i_{YZ}(w')^\sim = t' i_{YZ}(w')^\sim = t'\widetilde{w}$, i.e. $(t,t') \in \ker \widetilde{w}$. ■■■

6.4.3 REMARKS. (i): In the homogeneous case, and when we forget about the empty partial algebra, then we can always only consider E-equations of the form $(X: t \overset{e}{=} t')$ -- X the global set of variables --, and we get $\mathbf{Mod}(G) = \mathbf{ModEeq}(\{\underline{F}(X,\mathbf{Mod}(G))\})$, moreover, for every class $\mathbf{K} \subseteq \mathbf{Alg}(\tau)$, $\mathbf{ModEeq}(\mathbf{K}) = \mathbf{ModEeq}_X(\mathbf{K}) = \mathbf{ModEeq}(\{\underline{F}(X,\mathbf{K})\})$, and $\mathbf{Eeq}_X(\mathbf{K})$ is a closed and fully invariant congruence relation on the relative subalgebra $\underline{\mathrm{dom}}(\mathbf{Eeq}_X(\mathbf{K}))$ which is freely generated by X. Hence we have a strong analogy to the original Second Birkhoff Theorem for total algebras, where $\mathbf{Eeq}(\mathbf{K})$ is always a fully invariant congruence relation on $\underline{F}(X,\mathbf{TAlg}(\tau))$, and every fully invariant congruence R on $\underline{F}(X,\mathbf{TAlg}(\tau))$ is just $\mathbf{Eeq}(\{\underline{F}(X,\mathbf{TAlg}(\tau))/R\})$. -- In the heterogeneous case the situation becomes a little bit more complicated. Namely we can restrict our considerations to the "global" E-equations with respect to X only when we forget about all models which have at least one empty phylum (or we have to consider these partial algebras as "trivial models" contained in each model class of an E-equational theory in this restricted sense). Then again the E-equational theories would be in one-to-one correspondence to the closed and fully invariant congruences on relative subalgebras of $\underline{F}(X,\mathbf{TAlg}(\underline{S}))$ freely generated by X; while in the general situation which we have considered above, we need a set **K** with usually more than one partial algebra in order to get for a given set G of E-equations that $\mathbf{EeqMod}(G) = \mathbf{Eeq}(\mathbf{K})$.

(ii): Observe that the fully invariant congruence relations of a partial algebra $\underline{A}$ are exactly the congruences of the partial algebra $((A,(\phi^{\underline{A}})_{\phi\in\Omega}{}^\frown(e|e \in \mathrm{End}(\underline{A}))),v_A)$, i.e. where we enrich the structure by all endomorphisms of $\underline{A}$. -- Since $\underline{P}_M := \underline{\mathrm{dom}}\ \mathbf{Eeq}_M(\mathbf{K})$ is always freely generated by M, we get for any $t \in P$ and for any endomorphism e of $\underline{P}$ that $te = (e|M)(M:t)^{\underline{P}}$, and the condition of full invariance for $R \in \mathrm{Cong}_{fi}(\underline{P})$ is equivalent with the fact that for any $(t,t') \in R$ and for every $v: M \to P$ (which by assumption induces an endomorphism $\overline{v}: \underline{P} \to \underline{P}$) we have $(v(M:t)^{\underline{P}}, v(M:t')^{\underline{P}}) \in R$, i.e. R is "closed with respect to substitution" as this fact is usually described, since $v(M:t)^{\underline{P}}$

is obtained from the term t by replacing each variable x occurring in t by xv.
We realize that this fact of "closedness with respect to substitution", which is well known for equational theories of total algebras, also extends to E-equational theories of partial algebras. We shall see below, how one can exploit this in order to "construct" E-equational theories from a given set of E-equations by a set of derivation rules. In comparison with the total case the situation for homogeneous partial algebras is more complicated, as one has to construct dom(**EeqMod**(G)) besides **EeqMod**(G) (with respect to a set X of variables). In the heterogeneous case one then also has to observe the "interaction" of E-equations on different sets of variables.

The following lemma will also be useful in later sections.

6.4.4 LEMMA. Let $G \subseteq Eeq^*_X$ be any set of E-equations, and let
$\downarrow G := \{(M:t'') \mid \text{there is } (M:t \overset{e}{=} t') \in G \text{ such that } t'' \in (\downarrow t \cup \downarrow t')\}$
be the set of all M-subterms of M-terms in G.
Let Y be any set, let $\underline{Q} := ((F(Y,\mathbf{TAlg}(\underline{S})),((M:t)^{\underline{F}(Y,\mathbf{TAlg}(\underline{S}))})_{(M:t)\in\downarrow G}, \nu_{F(Y,\mathbf{TAlg}(\underline{S}))})$,
$P := \mathbf{C}_{\underline{Q}}Y$, and let $\underline{P}$ be the corresponding relative subalgebra of $\underline{F}(Y,\mathbf{TAlg}(\underline{S}))$.
Then $\underline{P}$ is freely generated by Y.

PROOF. Let $v: Y \to P$ be any mapping; we have to show that $P \subseteq \mathrm{dom}\,\tilde{v}$, (where $\tilde{v} = (\tilde{v})_{\underline{P}}$), and we do this by algebraic induction on $x \in P$: If $x \in Y$, then $x \in \mathrm{dom}\,\tilde{v}$. Assume $x \in P \setminus Y$; then there are $(M:t) \in \downarrow G$ and $\underline{a} \in P^M$ such that $x = \underline{a}(M:t)^{\underline{F}(Y,\mathbf{TAlg}(\underline{S}))}$. Assuming as induction hypothesis that $\underline{a} \in (\mathrm{dom}\,\tilde{v})^M$, we get for the global homomorphic extension $\bar{v}:\underline{F}(Y,\mathbf{TAlg}(\underline{S})) \to \underline{F}(Y,\mathbf{TAlg}(\underline{S}))$ that $(\underline{a}\bar{v})(M:t)^{\underline{F}(Y,\mathbf{TAlg}(\underline{S}))} = (\underline{a}(M:t)^{\underline{F}(Y,\mathbf{TAlg}(\underline{S}))})\bar{v}$ (cf. 5.7.5), but by assumption and the definition of $\underline{P}$, $(\underline{a}\bar{v})(M:t)^{\underline{F}(Y,\mathbf{TAlg}(\underline{S}))}$ as well as $\underline{a}(M:t)^{\underline{F}(Y,\mathbf{TAlg}(\underline{S}))}$ both belong to $\underline{P}$; therefore $x = \underline{a}(M:t)^{\underline{F}(Y,\mathbf{TAlg}(\underline{S}))} \in P \cap \mathrm{dom}\,\tilde{v}$. ■■■

6.4.5 REMARK. Thus the above procedure may be chosen to generate the elements of $T_Y(G) :=$ dom $\mathbf{Eeq_YMod}(G)$, but unfortunately the smallest fully invariant congruence relation $\bar{G}$ generated by G in $\underline{P}$ from Lemma 6.4.4 in general need not be closed, as the reader may find out by considering the following (homogeneous) example:
$\tau := (2,2)$, $\Omega := (\phi,\phi')$, $G := \{(xx\phi,x),(xx\phi xx\phi\phi',xx\phi xx\phi\phi')\}$.
Nevertheless we have the following

6.4.6 PROPOSITION. With the same assumptions and notation as in Lemma 6.4.4 we get: Let $\bar{G}$ be the smallest fully invariant congruence relation on $\underline{P}$ -- $\underline{P}$ being defined as in 6.4.4 -- generated by G, i.e. containing
$R_{Y,G} := \{(v(M:t)^{\underline{P}},v(M:t')^{\underline{P}}) \mid (M:t \overset{e}{=} t') \in G \text{ and } v:M \to P\}$.
Moreover, let $G^* := \mathbf{Eeq_YMod}(G)$ and $\underline{P}^* := \underline{\mathrm{dom}}\ G^*$.
Then there is a natural isomorphism $j:\underline{P}/\bar{G} \to \underline{P}^*/G^* = \underline{F}(Y,\mathbf{Mod}(G))$ induced by $[y]\bar{G} \mapsto [y]G^*$ $(y \in Y)$.

PROOF. Obviously $\bar{G} \subseteq G^*$; therefore the homomorphism j exists according to the Diagram Completion Lemma. Combining 6.3.2 with 6.3.9 we realize that $G' := \ker(\mathrm{nat}_{\bar{G}})^{\sim}$ is a closed and fully invariant congruence on a relative subalgebra -- freely generated by Y -- of $\underline{F}(Y,\mathbf{TAlg}(\underline{S}))$, which obviously satisfies $R_{Y,G} \subseteq \bar{G} \subseteq G' \subseteq G^*$. Since 6.4.2 implies that G^* is the smallest such congruence with these properties, we get $G' = G^*$, and 5.8.2 implies that j is an isomorphism. ■■■

6.4.7. Thus, if one only wants to construct $\underline{F}(M,\mathbf{Mod}(G))$ and is not interested in $\mathbf{Eeq_M Mod}(G)$, then 6.4.6 suffices. If one really has to construct $\mathbf{EeqMod}(G)$ from G, one may apply to G the following complete system of **rules of derivation of E-equations.** We only formulate them for the case of E-equations with $\mathbf{d}^{\&}$-small variable sets, but it is easy to see that one can also get the following results allowing also larger subsets of X. One should also observe, that the rules below usually are of only theoretical interest, if the similarity type allows infinitary operations, or if we have infinitely many sorts. Nevertheless the following results hold even for the general case.

6.4.8 RULES of DERIVATION for E-EQUATIONS:
Let X be any set of variables such that each phylum of X has cardinality at least $\mathbf{d}^*$. Let $G \subseteq Eeq^*_X$ be any set of E-equations. An E-equation $(M:t \stackrel{e}{=} t') \in Eeq^*_X$ is said to be <u>derivable</u> <u>from</u> G -- in symbols: $G \vdash (M:t \stackrel{e}{=} t')$ --, iff there is a $\mathbf{d}^{\&}$-small subset G' of G (i.e. $\#G' < \mathbf{d}^{\&}$) such that $(M:t \stackrel{e}{=} t')$ is derivable from G' in a sequence of length less than $\mathbf{d}^{\&}$ of applications of the following <u>rules</u> <u>of</u> <u>derivation</u> <u>of</u> E-equations. Notice that letters x stand for elements of X, t, t', t'', t_i, t_k, etc. stand for terms (i.e. roughly speaking, for elements of $\underline{F}(X,\mathbf{TAlg}(\underline{S}))$), M, N, M_i, etc. stand for $\mathbf{d}^{\&}$-small subsets of X, while H, H_i, etc. stand for $\mathbf{d}^{\&}$-small subsets of G, and for any index set I we assume that $\#I < \mathbf{d}^{\&}$. Above a line we write the list of premises of a rule, while below a line there stands the conclusion -- if a rule does not have any line, then we have an empty set of premises. The rules should be read in the following way: If one knows about the derivability of all the sequences above a line, then one may conclude that the expression to the right of " $\vdash$ " below the line is derivable from the left hand side.

(E0)
$$\frac{\begin{array}{c} H_i \vdash (M_i:t_i \stackrel{e}{=} t_i') \quad (i \in I) \\ H \cup \{(M_i:t_i \stackrel{e}{=} t_i') \mid i \in I\} \vdash (M:t \stackrel{e}{=} t') \end{array}}{H \cup \bigcup\{H_i \mid i \in I\} \vdash (M:t \stackrel{e}{=} t')}$$
Transitivity of " $\vdash$ ".

(E1)
$$\vdash (M:x \stackrel{e}{=} x) \qquad (x \in N \subseteq X).$$

(E2)
$$H \vdash (M:t \stackrel{e}{=} t'), \text{ whenever } (M:t \stackrel{e}{=} t') \in H .$$

(E3)
$$\frac{H \vdash (M:t \stackrel{e}{=} t')}{H \vdash (M:t' \stackrel{e}{=} t)}$$

(E4)
$$\frac{\begin{array}{c} H_1 \vdash (M:t \stackrel{e}{=} t') \\ H_2 \vdash (M:t' \stackrel{e}{=} t'') \end{array}}{H_1 \cup H_2 \vdash (M:t \stackrel{e}{=} t'')}$$

(E5)
$$\frac{H \vdash (M:t \stackrel{e}{=} t')}{H \vdash (M:t \stackrel{e}{=} t)}$$

(E6)
$$\frac{H \vdash (M:(t_k \mid k \in K_\phi)\phi \stackrel{e}{=} (t_k \mid k \in K_\phi)\phi)}{H \vdash (M:t_k \stackrel{e}{=} t_k)} \qquad (\phi \in \Omega,\ k \in K_\phi).$$

(E7)
$$\frac{H_k \vdash (M:t_k \overset{e}{=} t'_k) \quad (k \in K_\phi) \qquad H \vdash (M:(t_k|k \in K_\phi)\phi \overset{e}{=} (t_k|k \in K_\phi)\phi)}{H \cup \mathbf{U}\{H_k|k \in K_\phi\} \vdash (M:(t_k|k \in K_\phi)\phi \overset{e}{=} (t'_k|k \in K_\phi)\phi)}$$

(E8) Let $f:M \to \underline{F}(M,\mathbf{TAlg}(\underline{S}))$, $x \mapsto t_x$ $(x \in M)$ be any assignment -- observe $xv_X = t_x\tilde{v}_X$ $(x \in M)$ --, and let $\bar{f}:\underline{F}(M,\mathbf{TAlg}(\underline{S})) \to \underline{F}(M,\mathbf{TAlg}(\underline{S}))$ be its homomorphic extension -- i.e. in each $t \in F(M,\mathbf{TAlg}(\underline{S}))$ each $x \in M$ is substituted by t_x --; then

$$\frac{H \vdash (M:t \overset{e}{=} t') \qquad H_x \vdash (M:t_x \overset{e}{=} t_x) \quad (x \in M)}{H \cup \mathbf{U}\{H_x|x \in M\} \vdash (M:t\bar{f} \overset{e}{=} t'\bar{f})}$$

(E9)
$$\frac{H \vdash (M:t \overset{e}{=} t')}{H \vdash (M \cup N:t \overset{e}{=} t')}$$

(E10) Let $M \subseteq N \subseteq X$ such that for $t_i \in F(M,\mathbf{TAlg}(\underline{S}))$ $(i \in I)$ and $t,t' \in F(N,\mathbf{TAlg}(\underline{S}))$: $\downarrow t \cup \downarrow t' \subseteq M$ and $[N]v_X \subseteq [M \cup \mathbf{U}\{t_i|i \in I\}]\tilde{v}_X$; then

$$\frac{H \vdash (N:t \overset{e}{=} t') \qquad H_i \vdash (M:t_i \overset{e}{=} t_i) \quad (i \in I)}{H \cup \mathbf{U}\{H_i|i \in I\} \vdash (M:t \overset{e}{=} t')}$$

6.4.9 REMARKS. (i): Rules (E8), (E9) and (E10) together can be replaced by the following rule

(E8*) Let $f:M \to \underline{F}(N,\mathbf{TAlg}(\underline{S}))$, $x \mapsto t_x$ $(x \in M)$ be any assignment -- observe $xv_X = t_x\tilde{v}_X$ $(x \in M)$ --, and let $\bar{f}:\underline{F}(M,\mathbf{TAlg}(\underline{S})) \to \underline{F}(N,\mathbf{TAlg}(\underline{S}))$ be its homomorphic extension -- i.e. in each $t \in F(M,\mathbf{TAlg}(\underline{S}))$ each $x \in M$ is substituted by t_x --; then

$$\frac{H \vdash (M:t \overset{e}{=} t') \qquad H_x \vdash (N:t_x \overset{e}{=} t_x) \quad (x \in M)}{H \cup \mathbf{U}\{H_x|x \in M\} \vdash (N:t\bar{f} \overset{e}{=} t'\bar{f})}$$

The argumentation in the derivation of (E8*) from the other rules and from our knowledge about homomorphisms between term algebras follows the idea of the proof of the Second Birkhoff Theorem, the other direction is even simpler; we leave the details as an exercise. ∎

(ii): In the case of total ($\underline{S}$-) algebras the above system of rules could be simplified: (E0), (E2), (E3), (E4) and (E9) would remain unchanged, (E5) and (E6) would become superfluous, and we could simplify the other rules to:

(E1t)
$$\vdash (M:t \overset{e}{=} t) \qquad t \in F(M,\mathbf{TAlg}(\underline{S})) \text{ , } M \subseteq X$$

(E7t)
$$\frac{H_k \vdash (M:t_k \overset{e}{=} t'_k) \quad (k \in K_\phi)}{\mathbf{U}\{H_k|k \in K_\phi\} \vdash (M:(t_k|k \in K_\phi)\phi \overset{e}{=} (t'_k|k \in K_\phi)\phi)}$$

(E8t) Let $f:M \to \underline{F}(M,\mathbf{TAlg}(\underline{S}))$ be any assignment, and let $\overline{f}:\underline{F}(M,\mathbf{TAlg}(\underline{S})) \to \underline{F}(M,\mathbf{TAlg}(\underline{S}))$ be its homomorphic extension -- i.e. in each $t \in F(M,\mathbf{TAlg}(\underline{S}))$ each $x \in M$ is substituted by xf --; then

$$\frac{H \vdash (M:t \overset{e}{=} t')}{H \vdash (M:t\overline{f} \overset{e}{=} t'\overline{f})}$$

(E10t) Let $M \subseteq N \subseteq X$ and $t,t' \in F(M,\mathbf{TAlg}(\underline{S}))$ such that $[N]v_X \subseteq [F(M,\mathbf{TAlg}(\underline{S}))]v_X$; then

$$\frac{H \vdash (N:t \overset{e}{=} t')}{H \vdash (M:t \overset{e}{=} t')}$$

These are -- except for some differences in the formulation -- exactly the rules presented in J.A.GOGUEN & J.MESEGUER [GoMe 81].

(iii): Let us present a very simple example showing that we really have to take care of the cardinality #S of the set S of sorts in connection with the cardinalities of our sets of variables:

Let $S := \mathbf{N}$ be the set of natural numbers, $\Omega := \{\phi_n, \phi_n' \mid n \in \mathbf{N}\}$, and $\tau := (0)_{\phi \in \Omega}$ a type only containing nullary constants such that ϕ_n and ϕ_n' are of sort n, and let $\mathbf{K} := \{\underline{A}_n \mid n \in \mathbf{N}\}$, where $A_n := \{\phi_m, \phi_m' \mid m \in \mathbf{N} \setminus \{n\}\}$, $v_n := v_{A_n}$ being the restriction of $v_{\underline{F}(\emptyset,\mathbf{TAlg}(\underline{S}))}$ to A_n, and all possible constants are defined in $\underline{A}_n$: $\phi_m^{\underline{A}_n} := \phi_m$ and $\phi_m'^{\underline{A}_n} := \phi_m'$. If we consider only E-equations with a finite set of variables, then **K** would satisfy no non-trivial E-equation, and we would have $\mathbf{ModEeq}_{finite}(\mathbf{K}) = \mathbf{TAlg}(\underline{S})$, while in $\mathbf{H\dot{S}P}(\mathbf{K})$ those partial algebras, all of whose phyla are non-empty, are isomorphic to $\underline{S}$, contradicting the First Birkhoff Theorem (and we actually have the non-trivial E-equations $(\mathbf{N}:\phi_m \overset{e}{=} \phi_m')$ $(m \in \mathbf{N})$).

But then it is easy to realize in connection with rules (E8) and (E10) that we sometimes also need infinite premises of the rules in order to be able to apply the rules for substitution and restriction of variables properly.

6.4.10 THEOREM of SOUNDNESS and COMPLETENESS. The set of rules presented in 6.4.8 for the derivation of E-equations is **correct (sound)** and **complete**, i.e. if $G \subseteq \mathrm{Eq}^*_X$ is any set of E-equations, and if G' is the set of all E-equations derivable from G according to 6.4.8, then $G' = \mathbf{EeqMod}(G)$.

PROOF. Soundness just means (a): $G' \subseteq \mathbf{EeqMod}(G) =: \overline{G}$,
while completeness means (b): $\mathbf{EeqMod}(G) \subseteq G'$.
We sketch the proof by briefly discussing the meaning of the rules:
(E0) means that we can use E-equations already derived to derive new ones, while (E2) just means that $G \subseteq G'$ -- not yielding any new E-equations. -- For the further proof let us define
$T(G) := \{(M:t) \mid t \in \mathrm{dom}(\tilde{v}_M)$, $M \subseteq X$, $\#M < \mathbf{d}^{\&}$, $G \vdash (M:t \overset{e}{=} t)\}$.
Then (E1) tells us that $(M:x) \in T(G)$ for each $\mathbf{d}^{\&}$-small subset M of X and for each **variable** $x \in M$, and that G' is reflexive on the set of variables. -- (E5) says that $T(G) = \mathrm{dom}\, G'$, and together with (E3) and (E4) it means that G' is reflexive, symmetric

and transitive on its domain. Hence $G'_M := G' \cap Eeq_M$ is an equivalence relation on $T(G)_M := \{(M:t) \mid (M:t) \in T(G)\}$. -- (E6) says that for each $\mathbf{d}^{\&}$-small subset M of X $T(G)_M$ is an initial segment of $\underline{F}(M,\mathbf{TAlg}(\underline{S}))$, while (E7) tells us that G'_M is a closed congruence on $\underline{\text{dom}}\ G'_M$. -- (E8) yields full invariance of G'_M and freeness of M in $\underline{T}(G)_M$ at the same time, since in connection with (E5) it tells us that the "substituted terms" are also in $T(G)_M$. -- (E9) just guarantees that property 6.4.2.(ii).(a) is satisfied by G'. -- Finally, (E10) means that property 6.4.2.(ii).(b) is true for G'. -- Since we have seen that the rules do nothing else but to enforce properties on G', which have to hold for $\overline{G}$, they are sound and (a) is proved. ∎

On the other hand, the fact that G' has all the properties required for $\overline{G}$ in 6.4.2 also proves (b). ∎∎∎

6.4.11 REMARKS. (i): In section 14.3 we shall return to this topic in a more general context.

(ii): The above rules now allow a stepwise procedure to construct $\underline{F}(M,\mathbf{Mod}(G))$ for some given set G of E-equations, but for lack of space we do not go into more details here.

(iii): It is often quite useful to try to look in connection with the description of free partial algebras for term operations which might be used to generate the carrier set as a partial Peano algebra, while the fundamental partial operations are then defined using the given E-equations (see also section 17.6, and see [B 68], Satz 2.2, for an example of such a construction).

(iv): For the example in 6.4.5 we derive some E-equations: Let us recall: $\tau := (2,2)$, $\Omega := (\phi,\phi')$, $G := \{(xx\phi,x),(xx\phi xx\phi\phi',xx\phi xx\phi\phi')\}$.

Behind the derivation we always mention the line and the rules which we have used:

1. $G \vdash xx\phi \overset{e}{=} x$ (E2)
2. $G \vdash xx\phi xx\phi\phi' \overset{e}{=} xx\phi xx\phi\phi'$ (E2)
3. $G \vdash xx\phi xx\phi\phi' \overset{e}{=} xx\phi'$ (E7) for 1., 1. and 2.
4. $G \vdash xx\phi \overset{e}{=} xx\phi$ (E5) for 1.
5. $G \vdash xx\phi xx\phi\phi \overset{e}{=} xx\phi$ (E8) for 1., 4. ($xx\phi$ substituted for x in 1.)
6. $G \vdash xx\phi xx\phi\phi \overset{e}{=} x$ (E4) for 1., 5.
7. $G \vdash xx\phi' \overset{e}{=} xx\phi xx\phi\phi'$ (E3) for 3.
8. $G \vdash xx\phi' \overset{e}{=} xx\phi'$ (E5) for 7.
9. $G \vdash xx\phi' xx\phi'\phi \overset{e}{=} xx\phi'$ (E8) for 1. and 8.

etc..

CHAPTER II

A BASIS FOR A TWO-VALUED MODEL THEORY FOR PARTIAL ALGEBRAS

In Chapter I we have dealt with the equational part of model theory, which is usually considered to be the core of universal algebra. But we have already stressed that the equational theory is in general not strong enough to formulate relevant axioms for partial algebras -- one also needs at least elementary implications and negations of TE-expressions, i.e. in general universal Horn formulas. This chapter and the following one will be dedicated to develop the fundaments of the theory of universal Horn formulas, based to a large extent on category theoretical methods, as they have been developed independently e.g. by B.BANASCHEWSKI and H.HERRLICH (cf. [BaH 76]) for the total case and by R.JOHN (see [J 75a]) as well as by H.ANDRÉKA and I.NÉMETI (partly supported by I.SAIN; see e.g. [AN 76], [AN 77], and [NSa 77], which will be especially important in Chapter III).

In this chapter we mainly present the basic concepts which will allow a better understanding of many features and concepts occurring in connection with partial algebras:

In **section 7** we present a comprehensive language for a two-valued model theory for (possibly infinitary) partial algebras, which is a "dialect" of the usual first order language when all arities and #S are finite. We do not treat relations separately, since we do not want to burden the presentation too much, and relations can also be considered as partial projections.

In **section 8** we collect some basic observations concerning elementary implications and universal Horn formulas, encode them algebraically and category theoretically, and stress reflection (resp. preservation) of universal Horn formulas as a basic tool for the description of concepts for partial algebras; e.g. preservation of formulas by algebraic operators is studied.

In **section 9** preservation and reflection of formulas by mappings is studied in more detail, while in **section 10** the category theoretical concept of factorization systems is studied because of the close relationship of the diagonal-fill-in property to the reflection of implications. It is shown that any category of all partial algebras of some given type allows quite a lot of relevant factorization systems.

7 A "first order" language for (infinitary) partial (S-) algebras

As in Chapter I we always tacitly assume that every partial algebra (or set), say $\underline{A}$, is endowed with a sort homomorphism $v_A:\underline{A} \to \underline{S}$ into a heterogeneous sort algebra $\underline{S}$ (see 2.11.2), and that all mappings are compatible with these sort mappings. We present here a tentative formulation of a two-valued logic for (possibly infinitary) partial algebras, whose presentation is strongly oriented to compatibility with the category theoretical formulation in [AN 79a]. The finitary homogeneous case has intensively been discussed in [B 82] (the basic ideas for such a logic are already contained in H.THIELE [Th 66]), while the finitary heterogeneous case is investigated in a recent paper of H.KAPHENGST [Ka 82]. We have been informed that "our" logic here has been intensively studied by Zs.MARKUSZ in [Mr 83] and [Mr 83a].

7.1 The syntax

7.1.1 DEFINITION. Let $\tau = (K_\phi)_{\phi\in\Omega}$ be any similarity type, **d** its cardinal dimension, let $\underline{S}$ be any heterogeneous sort algebra of type τ, let γ be any ordinal, $\#\gamma = \mathbf{c}$, and **q** and **v** be any cardinals, **v** regular, such that, for $\mathbf{d}^{\&} = \max(\mathbf{d}, \#\omega_0, (\#S)^+)$, $\mathbf{d}^{\&}, \mathbf{q}, \mathbf{c} \leq \mathbf{v}$; and let X be any set such that for each sort $s \in S$ each phylum X_s has cardinality **v**. We are going to define a language $L := L_{\mathbf{c},\mathbf{q},\mathbf{d},\mathbf{v}}(\tau, \underline{S}, (X, v_X)) =: L_{\mathbf{c},\mathbf{q}}(\underline{S}, X^`)$ as follows:

(i) The alphabet of L is given by

(a) The logical symbols, i.e.

- the elements x of the set X of <u>variables</u>; each variable x has a sort xv_X;
- all subsets of X of cardinality less than **v**;
- <u>sentential connectives</u> $\neg$, $\wedge$, Λ^* (for $\vee$, $\Rightarrow$, $\Leftrightarrow$ and V^* see 7.1.2.(v));
- <u>quantifiers</u> $\forall$, $\exists$, $\forall^*$, $\exists^*$;
- the identity symbol $\overset{e}{=}$;
- brackets (,) , colon : (vertical line | and curly brackets { , });

(b) non-logical constants

- constants ϕ for partial operations ($\phi \in \Omega$), which shall also include individual constants as nullary operation symbols (see 7.1.2.(iii));

(- $\in$, the "element symbol").

(ii) <u>Terms</u> of L are of the form (M:t) for elements $t \in F(M, \mathbf{TAlg}(\underline{S}))$, where $M \subseteq X$ with $\#M < \mathbf{v}$.

(iii) <u>Atomic formulas</u> of L are all E-equations $(M{:}t \overset{e}{=} t')$, where $t,t' \in F(M, \mathbf{TAlg}(\underline{S}))$ with $t\tilde{v}_M = t'\tilde{v}_M$ and $M \subseteq X$, $\#M < \mathbf{v}$ (more correctly one should write " $(M{:}t) \overset{e}{=} (M{:}t')$ ", but we shall always use an abbreviation like the above one).

(iv) <u>Formulas</u> of L are defined recursively as follows:

- each atomic formula of L is a formula of L;
- if $(M{:}\Phi)$ is a formula of L, then $(M{:}\neg\Phi)$ is a formula of L;
- if $(M{:}\Phi)$ and $(M'{:}\Phi')$ are formulas of L, then $(M \cup M'{:}(\Phi \wedge \Phi'))$ is a formula of L;
- if F is a non-empty set of formulas of L, $\#F < \mathbf{c}$, then $(M{:}\Psi)$ is a formula of L, where $M := \bigcup\{N \mid (N{:}\Phi) \in F\}$, and $\Psi := \Lambda^*\{\Phi \mid (N{:}\Phi) \in F\}$ (note that $\#M < \mathbf{v}$);
- if $M \subseteq X$, $\#M < \mathbf{v}$, $x \in M$ and $Y \subseteq M$ with $\#Y < \mathbf{q}$, and if $(M{:}\Phi)$ is a formula of L, then $(M \setminus \{x\}{:}\forall x\Phi)$ and $(M \setminus Y{:}\forall^* Y\Phi)$ are formulas of L;
- if $(M{:}\Phi)$ is a formula of L, and if $N \subseteq X$ such that $\#N < \mathbf{v}$, then $(M \cup N{:}\Phi)$ is a formula of L.

7.1.2 REMARKS. **(i)**: Actually a formula as defined above consists of two parts: the set of variables it is related to, and the second component, which we shall call the <u>body of the formula</u> (or <u>basic formula</u>). We have to refer explicitly to M because of the heterogeneous situation allowing empty phyla (see MARKUSZ [Mr 83],[Mr 83a] for more details).

(ii): The last possibility of the formation of formulas has been included in order to show that variables which have been taken away in connection with quantification can nevertheless occur in the set of variables of a formula, where the body has been quantified.

(iii): We could treat individual constants separately, as it is often done, but we try to get along without separate constants, to avoid further details; therefore we shall have to include axioms of the form $(\emptyset:c_a \stackrel{e}{=} c_a)$, when c_a is a constant which has to be interpreted -- see the definition of an interpretation below.

(iv): Notice that the logical symbols $\neg$, $\wedge$, $\vee$, $\bigwedge^*$, $\forall x$ and $\forall^* Y$ can be considered as operations on L, and except for $\bigwedge^*$ they have a fixed arity and satisfy the Peano axioms, while we have simplified the use of $\bigwedge^*$, which is often (usually in formal presentations) replaced by a family $(\bigwedge_\xi | 0 < \xi < \gamma)$, where γ is the smallest ordinal of cardinality **c**, and $\bigwedge_\xi$ has arity ξ. For we wanted to avoid the inclusion of too much arithmetic of ordinals (a more rigorous presentation of such languages can be found e.g. in [K 64]; for a treatment of infinite arities of non-logical constants see [HnMoT 85]). We do not want to deal here with more rigorous logical problems (cf. [ANSa 82]), but we just want to have a language at hand, which is able to express some features encountered in connection with partial algebras. It should be observed that, already with our approach, one can make proofs or definitions by algebraic induction or recursion.

(v): In order to have not too many basic logical operators we shall take in what follows as basic operators only $\neg$ (negation), $\bigwedge^*$ (**c**-restricted conjunction) and $\forall^*$ (**q**-restricted generalization) and define the others used above and further usual operators as follows, whenever the right hand sides are defined:

$(\Phi \wedge \Phi') := \bigwedge^*\{\Phi, \Phi'\}$	(**binary conjunction**);
$(\Phi \vee \Phi') := \neg\bigwedge^*\{\neg\Phi, \neg\Phi'\} = \bigvee^*\{\Phi, \Phi'\}$	(**binary disjunction**);
$\bigvee^* F := \neg\bigwedge^*\{\neg\Phi \mid \Phi \in F\}$	(**c-restricted disjunction**);
$(\Phi \Rightarrow \Phi') := \neg\bigwedge^*\{\Phi, \neg\Phi'\} = (\neg\Phi \vee \Phi')$	(**implication**);
$(\Phi \Leftrightarrow \Phi') := ((\Phi \Rightarrow \Phi') \wedge (\Phi' \Rightarrow \Phi))$	(**equivalence**);
$\forall x \Phi := \forall^*\{x\}\Phi$	(**generalization of one variable**);
$\exists x \Phi := \neg\forall^*\{x\}\neg\Phi$	(**existential quantification with one variable**);
$\exists^* Y \Phi := \neg\forall^* Y \neg\Phi$	(**q-restricted existential quantification**).

Often we shall include " $\Rightarrow$ " and " $\exists^*$ " into our basic considerations, especially when we start considering semantics. We denote the extended language in the same way as the original one.

(vi): Actually L is a disjoint family $L = (L_M | M \subseteq X,\ \#M < \mathbf{v})$, where each L_M is the set of all L-formulas whose set of variables is M. Only quantification (and adding new variables) change the set of variables.

(vii): In later applications the above parameters will often be chosen as follows: $\mathbf{c} = \mathbf{q} = \mathbf{v} = \mathbf{d}^{\&}$.

7.1.3 DEFINITION. For every formula Σ of $L := L_{\mathbf{c},\mathbf{q},\mathbf{d},\mathbf{v}}(\tau, \underline{S}, (X, v_X))$ we define the sets $var(\Sigma)$, $fvar(\Sigma)$, $\downarrow\Sigma$ and $subf(\Sigma)$ of **variables**, **free variables**, **subterms** and **subformulas** of Σ, respectively, recursively on the structure of Σ (we omit one pair of brackets, if there would be two of them):

(i) If $\Sigma = (M:t \stackrel{e}{=} t')$, i.e. Σ is atomic, then $\downarrow\Sigma := \{(M:t'') | t'' \in \downarrow t \cup \downarrow t'\}$, $fvar(\Sigma) := var(\Sigma) := X \cap \{t'' | (M:t'') \in \downarrow\Sigma\}$, $subf(\Sigma) := \{\Sigma\}$.

(ii) If $\Sigma = (M{:}\neg\Phi)$, then $\downarrow\Sigma := \downarrow(M{:}\Phi)$, $\mathrm{fvar}(\Sigma) := \mathrm{fvar}(M{:}\Phi)$, $\mathrm{var}(\Sigma) := \mathrm{var}(M{:}\Phi)$, $\mathrm{subf}(\Sigma) := \{\Sigma\} \cup \mathrm{subf}(M{:}\Phi)$.

(iii) If $\Sigma = (M{:}\Lambda^* F)$, $\#F < \mathbf{c}$, F a set of basic L-formulas, then $\downarrow\Sigma := \mathbf{U}\{\downarrow(M{:}\Phi) \mid \Phi \in F\}$, $\mathrm{var}(\Sigma) := \mathbf{U}\{\mathrm{var}(M{:}\Phi) \mid \Phi \in F\}$, $\mathrm{fvar}(\Sigma) := \mathbf{U}\{\mathrm{fvar}(M{:}\Phi) \mid \Phi \in F\}$, $\mathrm{subf}(\Sigma) := \{\Sigma\} \cup \mathbf{U}\{\mathrm{subf}(M{:}\Phi) \mid \Phi \in F\}$.

(iv) If $\Sigma = (M{:}\forall^* Y\Phi)$, $Y \subseteq X$, $\#Y < \mathbf{q}$, then $\downarrow\Sigma := \downarrow(M \cup Y{:}\Phi)$, $\mathrm{var}(\Sigma) := \mathrm{var}(M \cup Y{:}\Phi)$, $\mathrm{fvar}(\Sigma) := \mathrm{fvar}(M \cup Y{:}\Phi) \setminus Y$, $\mathrm{subf}(\Sigma) := \{\Sigma\} \cup \mathrm{subf}(M \cup Y{:}\Phi)$.

If $\Sigma = (M{:}\Phi)$, then we define $\downarrow\Phi := \{t \mid (M{:}t) \in \downarrow(M{:}\Phi)\}$, $\mathrm{var}(\Phi) := \mathrm{var}(M{:}\Phi)$, $\mathrm{fvar}(\Phi) := \mathrm{fvar}(M{:}\Phi)$, $\mathrm{subf}(\Phi) := \{\Psi \mid (N{:}\Psi) \in \mathrm{subf}(M{:}\Phi)\}$.

The following results are easily proved either by observing that $t \in \underline{F}(M,\mathbf{TAlg}(\underline{S}))$ implies $\mathrm{fvar}(M{:}t) \subseteq M$, or by induction on the structure of a formula:

7.1.4 LEMMA. For every formula $\Sigma = (M{:}\Phi)$ one has:

(i) $\mathrm{fvar}(\Sigma) \subseteq \mathrm{var}(\Sigma) = X \cap \downarrow\Phi$; and $\mathrm{fvar}(\Sigma) = \mathrm{var}(\Sigma)$ holds iff for every subformula $\Xi \in \mathrm{subf}(\Sigma)$, which is of the form $\Xi = (N{:}\forall^* Y\Phi')$, there holds $Y \cap \mathrm{var}(N \cup Y{:}\Phi') = \emptyset$.

(ii) $\mathrm{fvar}(M{:}\Phi) \subseteq M$, but **not necessarily** $\mathrm{var}(M{:}\Phi) \subseteq M$.

(iii) $\#M, \#\mathrm{var}(\Sigma) < \mathbf{v}$ and $\#X = \mathbf{v}$. ■■■

7.1.5 DEFINITION. (i): Let $(N{:}\Phi) \in L$, let $M \subseteq X$, $\#M < \mathbf{v}$, and let $f{:}N \to \underline{F}(M,\mathbf{TAlg}(\underline{S}))$ be any ($\underline{S}$-) mapping (!) and $\overline{f}{:}\underline{F}(N,\mathbf{TAlg}(\underline{S})) \to \underline{F}(M,\mathbf{TAlg}(\underline{S}))$ its homomorphic extension. We say that f is <u>admissible</u> for $(N{:}\Phi)$, if for every subformula of $(N{:}\Phi)$ of the form $(N'{:}\forall^* Y\Psi)$ we have $Y \cap \mathbf{U}\{\downarrow(xf) \mid x \in \mathrm{fvar}(N'{:}\forall^* Y\Psi)\} = \emptyset$.

(ii): Let $f{:}N \to \underline{F}(M,\mathbf{TAlg}(\underline{S}))$ be an admissible mapping for $(N{:}\Phi)$, then we say that we get $(M{:}(\Phi)f)$ by (<u>simultaneous</u> <u>uncritical</u>) <u>substitution</u> from $(N{:}\Phi)$ with respect to f, if $(\Phi)f$ is the following formula:

(a) If $\Phi = t \overset{e}{=} t'$, then $(\Phi)f := t\overline{f} \overset{e}{=} t'\overline{f}$;

(b) if $\Phi = \neg\Psi$, then $(\Phi)f := \neg(\Psi)f$;

(c) if $\Phi = \Lambda^* F$, then $(\Phi)f := \Lambda^*\{(\Psi)f \mid \Psi \in F\}$;

(d) if $\Phi = \forall^* Y\Psi$, then $(\Phi)f := \forall^* Y((\Psi)f_Y)$, where $f_Y{:}N \cup Y \to \underline{F}(M \cup Y,\mathbf{TAlg}(\underline{S}))$, defined by $xf_Y := xf$, if $x \in N \setminus Y$, and $xf_Y := x$, if $x \in Y$.

We write $\mathrm{Subst}((N{:}\Phi),f,(M{:}(\Phi)f))$ in order to indicate this situation.

7.1.6 REMARKS. (i): Notice that f in 7.1.5 is always admissible for $(N{:}\Phi)$, when Φ does not contain any quantifier -- i.e. no basic subformula of the form $\forall^* Y\Psi$; such a formula will be called <u>quantifier free</u>.

(ii): Moreover observe that f_Y in 7.1.5.(d) is admissible for $(N \cup Y{:}\Psi)$, if f is admissible for $(N{:}\Phi)$, since the concept of "subformula" is transitive. Admissibility just means that the corresponding substitution does not cause any collision between free and "bound" variables occurring in the new formula. A variable $x \in X$ <u>is bound in a formula</u> $(N{:}\Phi)$, if Φ has a subformula $\forall^* Y\Psi$ such that $x \in Y \cap \mathrm{fvar}(\Psi)$.

(iii): One could define a <u>generalized substitution</u> for $(N{:}\Phi)$, which is always applicable, by first "changing bound variables" to avoid collisions in connection with f; but we do not go into details here, since later on we shall mainly deal with formulas without quantifiers, or only with such of the form $\forall^* Y\Psi$, where Ψ is quantifier free.

(iv): It should be observed that the relation "to be a subformula of some formula" is a quasi-order and indeed even a partial order on L (the proof is similar to the one for terms; but observe that " Λ^* " only depends on the set of its arguments and not on a special sequence formed by them; therefore the original Peano algebra has already been factored with respect to a -- fully invariant and closed -- congruence.)

(v): If we only deal with homogeneous partial algebras, then we can simplify considerations by only considering formulas $(X:\Phi)$ for the whole set of variables and allowing the empty partial algebra as a trivial model for every theory, i.e. we can then restrict considerations to the basic formulas.

7.2 The semantics

Based on the semantics already introduced in section 6 for E-equations we now briefly sketch the fundaments of semantics for all of $L = L_{\mathbf{c},\mathbf{q}}(\underline{S},X^{`})$.

7.2.1 DEFINITION. Let $\underline{A} \in \mathbf{Alg}(\underline{S})$, and let $(M:\Phi)$ be an L-formula.

(i): Any mapping $w:M \to A$ (compatible with the sort mappings!) is called an assignment or valuation for $(M:\Phi)$ with values in $\underline{A}$.

(ii): Let $w:M \to A$ be an assignment with values in $\underline{A}$. The corresponding interpretation I^w (or mostly only $\tilde{w}$) is then defined as follows:

- $xI^w := x\tilde{w} = xw$ for every variable x in M, and xI^w is undefined for $x \in X \setminus M$).
- $\phi I^w := \phi^{\underline{A}}$ for every fundamental operation symbol $\phi \in \Omega$.
- If $(N:t)$ is any term, then $(N:t)$ is interpreted by I^w iff $M = N$ and $t \in \operatorname{dom} \tilde{w}$; and if this is the case, then $(M:t)I^w := t\tilde{w} = w(M:t)^{\underline{A}}$ according to section 5.

We could extend the interpretation in an obvious way to formulas, but the following definition of satisfaction of a formula in $\underline{A}$ with respect to an interpretation is understandable without this.

7.2.2 DEFINITION. Let $(M:\Phi)$ be an L-formula, and let $\underline{A} \in \mathbf{Alg}(\underline{S})$.

(i): Let $w:M \to A$ be any valuation with values in $\underline{A}$; then we define recursively, what it means that $\underline{A}$ satisfies $(M:\Phi)$ under the valuation w (i.e. that $(M:\Phi)$ is satisfied in $\underline{A}$ for w) -- in symbols: $\underline{A} \models (M:\Phi)[w]$:

(a) If $\Phi = t \overset{e}{=} t'$, then $\underline{A} \models (M:t \overset{e}{=} t')[w]$ iff $(t,t') \in \ker \tilde{w}$ (cf. 6.1.1.(ii)).

(b) If $\Phi = \neg\Psi$, then $\underline{A} \models (M:\neg\Psi)[w]$ iff it is not true that $\underline{A} \models (M:\Psi)[w]$.

(c) If $\Phi = \Lambda^*F$, $\#F < \mathbf{c}$, F a set of basic L-formulas, then
$\underline{A} \models (M:\Lambda^*F)[w]$ iff **for every** $\Psi \in F$ there holds $\underline{A} \models (M:\Psi)[w]$.

(d) If $\Phi = \forall^*Y\Psi$, $\#Y < \mathbf{q}$, then $\underline{A} \models (M:\forall^*Y\Psi)[w]$ iff for every mapping $w':M \cup Y \to A$ satisfying $w'|(M \setminus Y) = w|(M \setminus Y)$ one has $\underline{A} \models (M \cup Y:\Psi)[w']$.

(iii): $(M:\Phi)$ is valid in $\underline{A}$ -- in symbols: $\underline{A} \models (M:\Phi)$ --, iff one has $\underline{A} \models (M:\Phi)[w]$ **for every assignment** $w:M \to A$.

7.2.3. Thus we have $\underline{A} \models (M:\Phi)$, whenever there is no assignment from M into $\underline{A}$, i.e when A has empty phyla, and some of the corresponding ones of M are non-empty.

7.2.4 LEMMA. Let $(M:\Phi) \in L$, $\underline{A} \in \mathbf{Alg}(\underline{S})$, $Y \subseteq M$, $\#Y < \mathbf{q}$.

(i) $\underline{A} \models (M:\Phi)$ iff $\underline{A} \models (M \setminus Y:\forall^*Y\Phi)$; hence, if $\#M < \mathbf{q}$, then $\underline{A} \models (M:\Phi)$ iff $\underline{A} \models (\emptyset:\forall^*M\Phi)$.

(ii) If $w:M \to A$ is any assignment for $(M:\exists^*Y\Phi)$, and if $N := M \cup Y$, then $\underline{A} \models (M:\exists^*Y\Phi)[w]$ iff **there exists** an assignment $w':N \to A$ such that $w'|(N \setminus Y) = w|(N \setminus Y)$ and $\underline{A} \models (N:\Phi)[w']$.

(iii) If $\Phi = \neg(t \overset{e}{=} t')$ for some $t,t' \in F(M,\mathbf{TAlg}(\underline{S}))$, then we have for any $w:M \to A$: $\underline{A} \models (M:\neg(t \overset{e}{=} t'))[w]$ iff $t \notin \mathrm{dom}\,\tilde{w}$ **or** $t' \notin \mathrm{dom}\,\tilde{w}$ **or** $t\tilde{w} \neq t'\tilde{w}$; in particular, $\underline{A} \models (M:\neg(t \overset{e}{=} t))[w]$ iff $t \notin \mathrm{dom}\,\tilde{w}$ (i.e. iff t is not interpreted by I^w).

PROOF. (i), (ii) and (iii) are straightforward consequences of the preceding definitions, and we only prove (ii): We have for any assignment $w:M \to A$ that $\underline{A} \models (M:\exists^*Y\Phi)[w]$ iff it is not true that $\underline{A} \models (M:\forall^*Y\neg\Phi)[w]$, iff it is not true that for all assignments $w':N = M \cup Y \to A$ satisfying $w'|(N \setminus Y) = w|(N \setminus Y)$ it is the case that $\underline{A} \models (N:\neg\Phi)[w']$, iff there exists an assignment $w':N \to A$ such that $w'|(N \setminus Y) = w|(N \setminus Y)$ and $\underline{A} \models (N:\Phi)[w']$. ■■■

7.2.5 REMARKS. (i): A broad discussion of the properties of existence equations and the motivation of this kind of logic in general is contained in the survey article [B 82] of BURMEISTER; therefore -- because of lack of space -- we do not go here into further details. Let us only recall that we have **hidden existential quantifiers** in the E-equations: Let us consider every K_φ-ary operation $\varphi^{\underline{A}}$ of output sort s of some partial algebra $(\underline{A},v_A)$ as a $(K_\varphi \cup \{K_\varphi\})$-ary relation, say $\varphi^* := \mathrm{graph}(\varphi^{\underline{A}})$, on (A,v_A), and set $\underline{A}^* := ((A,(\varphi^*)_{\varphi\in\Omega}),v_A)$; then we have for any assignment $w:M \to A$ that $\underline{A} = (M:(x_k|k \in K_\varphi)\varphi \overset{e}{=} (x_k|k \in K_\varphi)\varphi)[w]$ iff for some variable $y \in (s)v_X^{-1} \setminus \{x_k|k \in K_\varphi\}$, $\underline{A}^* \models (M:\exists^*\{y\}\varphi^*((x_k|k \in K_\varphi),y))[w]$. Thus the satisfaction of the TE-expression for $\underline{A}$ is equivalent to the satisfaction of an existential formula with respect to $\underline{A}^*$. As sketched in [B 82], 2.18, this can be extended to E-equations between arbitrary terms, where we get quite complex existential expressions.

(ii): Let us further observe that $\underline{A} \models (\emptyset:\exists x\ x \overset{e}{=} x)$ is equivalent with the fact that, for $s := xv_X$, $A_s \neq \emptyset$, i.e. the phylum of $\underline{A}$ of the sort s of x is non-empty. Thus our language L is not only strong enough to "talk" about evaluability and non-evaluability of terms, but also about emptiness and non-emptiness of phyla of models: Namely $\underline{A} \models (\emptyset:\exists x\ x \overset{e}{=} x)$ iff there exists an ($\underline{S}$-) assignment $w:\{x\} \to A$ such that $\underline{A} \models (\{x\}:x \overset{e}{=} x)[w]$; but since this formula is trivially satisfied by every assignment, there only remains as non-trivial consequence that w exists, and this is equivalent to the fact that A_s is non-empty.

7.3 Coincidence and Substitution Theorem

7.3.1. For lack of space we do not present here many details about the language which we have just introduced. For the finitary homogeneous case it is discussed to a great extent -- including a completeness theorem -- in [B 82], while for the finitary heterogeneous case a similar discussion, which even contains proofs, is presented in [Ka 82] by H.KAPHENGST for a language almost identical with the above one.

In order to provide a better feeling for the properties of such a language for heterogeneous partial algebras, we only briefly discuss the Coincidence Theorem and the Theorem of Substitution, which are of great importance, but differ in our case quite a bit from the usual ones, since we now have to observe evaluability of terms and compatibility with the sort mappings (cf. also section 6 for the properties of E-equations). A reader who is not so familiar with logic is referred e.g. to H.HERMES [He 63].

7.3.2 COINCIDENCE THEOREM. Let $\underline{A} \in \mathbf{Alg}(\underline{S})$ be any partial algebra, let $(M:\Phi)$ and $(N:\Phi)$ be any L-formulas belonging to the same basic formula Φ. If $v:M \to A$ and $w:N \to A$ are any two ($\underline{S}$-) assignments such that $v|fvar(\Phi) = w|fvar(\Phi)$, then $\underline{A} \models (M:\Phi)[v]$ iff $\underline{A} \models (N:\Phi)[w]$.

PROOF. We use induction on the structure of Φ for all possible M, N, v, w:
For atomic formulas Φ, i.e. for E-equations the statement follows from 5.5.5.(ii). -- We omit the cases of negation and c-restricted conjunction, which are quite simple, and assume that $\Phi = \forall^* Y\Psi$, $\#Y < q$ in X, and assume that $\underline{A} \models (M:\forall^* Y\Psi)[v]$, i.e. that for all $v':M \cup Y \to A$ satisfying $v'|(M \setminus Y) = v|(M \setminus Y)$ we have $\underline{A} \models (M \cup Y:\Psi)[v']$. By symmetry we now only have to show that $\underline{A} \models (N:\forall^* Y\Psi)[w]$ for (every) $w:N \to A$ with $w|fvar(\Phi) = v|fvar(\Phi)$:
First case: There is no assignment $v':M \cup Y \to A$ (compatible with the sort mappings!), which means that there is some $y \in Y$, say of sort $s \in S$, such that $A_s = \emptyset$ -- namely this cannot occur for $m \in M$, since v maps M in a compatible way into A. But then there also is no (compatible) $w':N \cup Y \to A$, and trivially $\underline{A} \models (N:\forall^* Y\Psi)[w]$.
Second case: There is some compatible $v':M \cup Y \to A$; then there also is some compatible $w':N \cup Y \to A$, and in both cases it can then be arranged that the restrictions to $M \setminus Y$ (respectively $N \setminus Y$) coincide with the corresponding restrictions of v respectively w. Now, given such a w', i.e. $w'|(N \setminus Y) = w|(N \setminus Y)$, we consider $v':M \cup Y \to A$, where $v'|(Y \setminus fvar(\forall^* Y\Psi)) := w'|(Y \setminus fvar(\forall^* Y\Psi))$, and the restriction of v' to the rest of M is the same as the corresponding restriction of v. Then -- by the construction of v' and the assumptions on v and w -- it is obvious that $v'|(M \setminus Y) = v|(M \setminus Y)$ as well as $v'|fvar(\Psi) = w'|fvar(\Psi)$. The first equality implies $\underline{A} \models (M \cup Y:\Psi)[v']$, while the second one implies together with the induction hypothesis for v' and $M \cup Y$ on one side and w' and $N \cup Y$ on the other side that $\underline{A} \models (N \cup Y:\Psi)[w']$. Since w' was arbitrary (except for the necessary restrictions), we get $\underline{A} \models (N:\forall^* Y\Psi)[w]$. ■■■

7.3.3 REMARK. It should be stressed once more that we cannot derive in the **heterogeneous case** from the Coincidence Theorem that $\underline{A} \models (M:\Phi)$ implies $\underline{A} \models (fvar(\Phi):\Phi)$, as it can be done in the homogeneous case! But observing that the proof of 7.3.2 heavily uses the simultaneous existence of assignments, we have at least the following result which can be strengthened, as the corresponding conditions for E-equations suggest.

7.3.4 COROLLARY. If $\underline{A} \in \mathbf{Alg}(\underline{S})$, and if $(M:\Phi)$ and $(N:\Phi)$ are L-formulas for the same basic formula Φ such that $[M]v_X \subseteq [N]v_X$, then $\underline{A} \models (M:\Phi)$ implies $\underline{A} \models (N:\Phi)$. ■■■

7.3.5 THEOREM of SUBSTITUTION. Let $\underline{A}$ be any partial algebra, and let and $(N:\Phi)$ be an L-formula. Moreover, let $w:M \to A$ be any assignment and $f:N \to dom\,\tilde{w}$ a mapping (both compatible with the sorts!), such that f is admissible for $(N:\Phi)$. Then

$$\underline{A} \models (N:\Phi)[f\tilde{w}] \quad \text{iff} \quad \underline{A} \models (M:(\Phi)f)[w] .$$

In particular, if $M = N$, then we get the usual form for some $m \in M$:

$$\underline{A} \models (M:\Phi)[w_m^{mf\tilde{w}}] \quad \text{iff} \quad \underline{A} \models (M:(\Phi)_m^{mf})[w] ,$$

where, for each $n \in M$, $nw_m^{mf\widetilde{w}} := nw$ if $m \neq n$, and $nw_m^{mf\widetilde{w}} := mf\widetilde{w}$ if $m = n$; and we get $(\Phi)_m^{mf}$ from Φ by replacing each free occurrence of m in Φ by mf.

PROOF. The proof is by induction on the structure of Φ:
(a): Let $\Phi = t \stackrel{e}{=} t'$. Then
$\underline{A} \models (M:t \stackrel{e}{=} t')[f\widetilde{w}]$ iff $(t,t') \in \ker(f\widetilde{w})^{\sim}$, iff $(t,t') \in \ker(((\widetilde{f})_{\underline{dom}\ \widetilde{w}})\cdot\widetilde{w})$, iff $(t\widetilde{f},t'\widetilde{f}) \in \ker \widetilde{w}$, iff $(t\overline{f},t'\overline{f}) \in \ker \widetilde{w}$, iff $\underline{A} \models (M:t\overline{f} \stackrel{e}{=} t'\overline{f})[w]$, iff $\underline{A} \models (M:(\Phi)f)[w]$.
Since the cases, when $\Phi = \neg\Psi$ and $\Phi = \Lambda^*F$, are straightforward -- using 7.1.5 and the induction hypothesis --, we only consider the case
(b): Let $\Phi = \forall^*Y\Psi$. Then let $M'' := \mathrm{fvar}(\Phi)$ and $f'' := f|M''$; therefore $M'' \cap Y = \emptyset$, and the Coincidence Theorem yields $\underline{A} \models (N:\Phi)[f\widetilde{w}]$ iff $\underline{A} \models (M'':\Phi)[f''\widetilde{w}]$. Since it is not too difficult to prove (say also by induction on the structure of formulas using 7.1.5 and the Coincidence Theorem) that
(*) $f|\mathrm{fvar}(\Phi) = f''|\mathrm{fvar}(\Phi)$ implies $(\Phi)f = (\Phi)f''$,
it is sufficient to prove
(**) $\underline{A} \models (M'':\Phi)[f''\widetilde{w}]$ iff $\underline{A} \models (M:(\Phi)f'')[w]$.
We therefore introduce a mapping $f' := f''_Y:M'' \cup Y \to \mathrm{dom}\ \widetilde{w}$ by setting $xf' := xf = xf''$, if $x \in M'' = \mathrm{fvar}(\Phi)$, and else $xf' := x$ (cf. 7.1.5.(ii).(d)). As in 7.1.6 we conclude that f' is admissible for $(M'' \cup Y:\Psi)$. -- Let us first assume that $\underline{A} \models (M'':\Phi)[f''\widetilde{w}]$, i.e. $\underline{A} \models (M'' \cup Y:\Psi)[u]$ for all $u:M'' \cup Y \to A$ satisfying $u|M'' = f''\widetilde{w}$. Let $v:M \cup Y \to A$ be such that $v|(M \setminus Y) = w|(M \setminus Y)$. Observe that for all $x \in M''$ $xf'\widetilde{v} = xf''\widetilde{v} = xf(w|(M \setminus Y))^{\sim} = xf''\widetilde{w}$; thus $\underline{A} \models (M'' \cup Y:\Psi)[f'\widetilde{v}]$. The induction hypothesis then yields $\underline{A} \models (M \cup Y:(\Psi)f')[v]$. Hence $\underline{A} \models (M:(\forall^*Y\Psi)f'')[w]$. -- Conversely, let us assume that $\underline{A} \models (M:(\forall^*Y\Psi)f'')[w]$, and let $u:M'' \cup Y \to A$ satisfy $u|M'' = f''\widetilde{w}$. Then define $v:M \cup Y \to A$ as follows: $yv := yu\ (= yf'\widetilde{v})$ for all $y \in Y$, and $xv := xw$ for all $x \in (M \setminus Y)$. Hence we get $xf'\widetilde{v} = x\widetilde{v} = xu$ for all $x \in Y$, and $xf'\widetilde{v} = xf''\widetilde{w} = xu$, if $x \in M''$ (observe that $Y \cap [M'']f'' = \emptyset$), i.e. $f'\widetilde{v} = u$. Thus -- observing (*) for f' and f'' as well as the Coincidence Theorem -- $\underline{A} \models (M \cup Y:(\Psi)f')[v]$ implies by induction hypothesis $\underline{A} \models (M'' \cup Y:\Psi)[f'\widetilde{v}]$, i.e. $\underline{A} \models (M'' \cup Y:\Psi)[u]$. The choice of u then implies $\underline{A} \models (M'':\forall^*Y\Psi)[f''w]$, completing the proof. ■■■

7.3.6 COROLLARY (Theorem of Substitution for Validity).
Let $\underline{A} \in \mathbf{Alg}(\underline{S})$, $M,N \subseteq X$ c-small subsets, $(N:\Phi)$ an L-formula and $f:N \to \mathrm{dom}\ \mathbf{Eq}_M(\{\underline{A}\})$ a mapping admissible for $(N:\Phi)$. Then $\underline{A} \models (N:\Phi)$ implies $\underline{A} \models (M:(\Phi)f)$.

PROOF. Let $w:M \to A$ be any assignment. Then $f\widetilde{w}$ is also an assignment, since $\mathrm{dom}\ \mathbf{Eq}_M(\{\underline{A}\}) \subseteq \mathrm{dom}\ \widetilde{w}$. Since f is admissible for $(N:\Phi)$, $\underline{A} \models (N:\Phi)$ implies $\underline{A} \models (N:\Phi)[f\widetilde{w}]$, and therefore -- by 7.3.5 -- $\underline{A} \models (M:(\Phi)f)[w]$. Hence $\underline{A} \models (M:(\Phi)f)$.
■■■

7.3.7 DEFINITION. Let $\mathbf{K} \subseteq \mathbf{Alg}(\underline{S})$ be any class of partial algebras and $F \subseteq L_{\mathbf{c},\mathbf{q}}(\underline{S},X^{`}) =: L$ any set of L-formulas. We then define
$\mathbf{K} \models F$ iff $\underline{A} \models \Sigma$ for every $\underline{A} \in \mathbf{K}$ and for every $\Sigma \in F$
(for one-element sets the brackets are omitted). With this notation we get:
$\mathbf{Th}(\mathbf{K}) := \{\Sigma \in L \mid \mathbf{K} \models \Sigma\}$ and $\mathbf{Mod}(F) := \{\underline{A} \in \mathbf{Alg}(\underline{S}) \mid \underline{A} \models F\}$.

Th(K) is called the theory of K, and **Mod**(F) is called the model class of F. -- K is said to be axiomatizable (by F) iff K = **Mod**(F) for some suitable set $F \subseteq L$.

7.3.8 REMARKS. (i): Since " $\models$ " is a binary relation between **Alg**($\underline{S}$) and L, **Th** and **Mod** are just the closure operators corresponding to the related Galois connection, and they satisfy the same properties as observed for **Eeq** and **Mod** (cf. 6.1.3).

(ii): As we have already seen in section 6 it is sometimes useful not to consider the whole language, but only some special part of it, for instance all E-equations and later on all elementary implications, and then the operator " **Th** " will be replaced by the corresponding operator.

(iii): In the rest of this book we shall mainly be concerned with so-called open (or universal) formulas. Within the scope of this book it is not possible to continue the discussion of the general language which we shall use here mainly as a background for the description of phenomena and not as a mathematical theory to be studied in its own right. We would only like to mention some further facts:

If S is finite and if the similarity type is finitary, then it is not difficult to convince onéself in connection with the results of model theory for relational systems -- by replacing partial operations by their graphs (cf.7.2.5.(i)) and the phyla by disjoint unary relations, whose domains cover the whole set, and by requiring by appropriate axioms that the i-th component of some relation has to be of the appropriate sort -- that axiomatizable classes of such partial ($\underline{S}$-) algebras are exactly those classes which are closed with respect to the formation of ultraproducts, ultraroots and isomorphic copies -- $\underline{A} \in$ **Alg**($\underline{S}$) is an ultraroot of some K-algebra, iff some ultrapower of $\underline{A}$ belongs to K (i.e. we have the Theorem of Keisler and Shelah). We have the strong feeling that this is no longer true, when the set S of sorts is infinite.

Moreover -- still assuming S to be finite and τ to be finitary --, the above encoding also shows that the Compactness Theorem, Łos' Lemma (= Ultraproduct Theorem), Skolem's Theorems etc. also hold for finitary partial algebras. Thus our language L is just a suitable shorthand so that we do not have to carry along troublesome axioms. -- Moreover, it should be easily realized that it would be almost impossible to get our results in sections 5 and 6 on terms, if we had to represent them by existential formulas as indicated in 7.2.5.

(iv): The scope of the above language was mainly motivated by compatibility with Part II and with the category theoretical encoding of formulas by ANDRÉKA, NÉMETI and SAIN (as quoted at the beginning of this chapter), which we shall heavily use in the case of open formulas in this and the next chapter.

8 Elementary implications, universal Horn formulas

8.1 Some basic kinds of universal Horn formulas

8.1.1. Usually the equational theory of algebraic structures is still considered as part of universal algebra, while the theory of elementary implications is often already counted as belonging to model theory. In connection with partial algebras the E-equational theory is not of the same importance as equational theory is for total algebras, but elementary implications, i.e. formulas of the form $(M:(\wedge^* E \Rightarrow t \overset{e}{=} t'))$, where E is any set of E-equations, gain fundamental importance, e.g. in connection with the description of the domains of partial algebras (see e.g. Part II).

For instance the class **Cat*** of all small categories considered as homogeneous partial algebras $(M;\cdot,Dom^*,Cod^*)$ -- the domain M being the set of all morphisms -- of similarity type (2,1,1) can be axiomatized by the following axioms (universal quantifiers and some parenthesis are omitted):

(C1*) $(x\cdot y \overset{e}{=} x\cdot y \wedge y\cdot z \overset{e}{=} y\cdot z) \Rightarrow x\cdot(y\cdot z) \overset{e}{=} (x\cdot y)\cdot z$,

(C2a*) $x\cdot y \overset{e}{=} x\cdot y \Rightarrow Dom^*(x\cdot y) \overset{e}{=} Dom^*(x)$,

(C2b*) $x\cdot y \overset{e}{=} x\cdot y \Rightarrow Cod^*(x\cdot y) \overset{e}{=} Cod^*(y)$,

(C3*) $x\cdot y \overset{e}{=} x\cdot y \Rightarrow Cod^*(x) \overset{e}{=} Dom^*(y)$,

(C4a*) $Dom^*(x)\cdot x \overset{e}{=} x$,

(C4b*) $x\cdot Cod^*(x) \overset{e}{=} x$.

-- If we consider (small) categories as two-sorted partial algebras -- getting the class **Cat** -- of type $\tau = (2,1,1,1)$, where $\Omega = (\cdot,Dom,Cod,\mathbf{1})$, and $\underline{S}$ is given by $S:= \{Ob,Mor\}$, $graph(\cdot^{\underline{S}}):= \{((Mor,Mor),Mor)\}$, $graph(Dom^{\underline{S}}):= \{((Mor),Ob)\} =: graph(Cod^{\underline{S}})$ and $graph(\mathbf{1}^{\underline{S}}):= \{((Ob),Mor)\}$, where the argument of **1** is written as usual as an index, then we can formulate the axioms as $(fvar(\Phi):\Phi)$, where Φ is one of the following formulas (m, m' and m" being of sort Mor and o being of sort Ob):

(C1) $(m\cdot m' \overset{e}{=} m\cdot m' \wedge m'\cdot m'' \overset{e}{=} m'\cdot m'') \Rightarrow m\cdot(m'\cdot m'') \overset{e}{=} (m\cdot m')\cdot m''$;

(C2a) $m\cdot m' \overset{e}{=} m\cdot m' \Rightarrow Dom(m\cdot m') \overset{e}{=} Dom(m)$,

(C2b) $m\cdot m' \overset{e}{=} m\cdot m' \Rightarrow Cod(m\cdot m') \overset{e}{=} Cod(m')$,

(C3) $m\cdot m' \overset{e}{=} m\cdot m' \Rightarrow Cod(m) \overset{e}{=} Dom(m')$,

(C4a) $\mathbf{1}_o\cdot m \overset{e}{=} \mathbf{1}_o\cdot m \Rightarrow \mathbf{1}_o\cdot m \overset{e}{=} m$,

(C4b) $m\cdot \mathbf{1}_o \overset{e}{=} m\cdot \mathbf{1}_o \Rightarrow m\cdot \mathbf{1}_o \overset{e}{=} m$,

(C5a) $Dom(\mathbf{1}_o) \overset{e}{=} o$,

(C5b) $Cod(\mathbf{1}_o) \overset{e}{=} o$,

(C6a) $\mathbf{1}_{Dom(m)}\cdot m \overset{e}{=} m$,

(C6b) $m\cdot \mathbf{1}_{Cod(m)} \overset{e}{=} m$. --

Notice that in both cases the elementary implications which occur as axioms only contain TE-expressions in the premise (cf. 6.1.2.(ii)), that TE-expressions are trivially satisfied in all total algebras, and in this case all the above axioms behave like E-equations. In each case the "definability condition"

$(\{m,m'\}:Dom(m') \overset{e}{=} Cod(m) \Rightarrow m\cdot m' \overset{e}{=} m\cdot m')$ (respectively $Dom^*(y) \overset{e}{=} Cod^*(x) \Rightarrow x\cdot y \overset{e}{=} x\cdot y$) is a consequence of the other axioms. It is also easy to see that the homomorphic image of a small category need not be a small category: Let the category $\underline{\mathbf{C}}$ have exactly four objects, say o_1, o_2, o_3 and o_4 and besides the images of $Dom^{\underline{\mathbf{C}}}$ and $Cod^{\underline{\mathbf{C}}}$ only two further morphisms $m:o_1 \rightarrow o_2$ and $m':o_4 \rightarrow o_3$ (using the usual notation for morphisms, indicating e.g. that $Dom^{\underline{\mathbf{C}}}(m) = o_1$ and $Cod^{\underline{\mathbf{C}}}(m) = o_2$); when we identify o_2 with o_4 and take the quotient with respect to the induced congruence, say R, then we get a partial algebra, which is not a (small) category, since [m]R and [m']R would be composable in the quotient, if this were a category, and they are not.

Since we know that E-varieties are closed with respect to homomorphic images, we may conclude that neither **Cat*** nor **Cat** are axiomatizable by E-equations.

Moreover, almost all the examples in Part II (by H.REICHEL) are not definable by E-

equations, but only by elementary implications. -- In general it will not even be sufficient to use elementary implications of the form $(M:\Lambda^*F \Rightarrow t \overset{e}{=} t')$, where F is a set of TE-expressions (we shall call such axioms ECE-equations).
Hopefully having motivated the reader a little bit by these observations we now make the above concepts precise by introducing those kinds of open formulas which will be most important for further considerations -- the reader should always keep in mind that we usually deal with partial S-algebras without explicitely mentioning this.

8.1.2 DEFINITION. Let $(M:\Phi)$ be an L-formula (cf. section 7); then $(M:\Phi)$ (possibly with some cardinality restrictions as in section 7) is called

-- TE-expression (i.e. term existence expression), iff $\Phi = t \overset{e}{=} t$ for some $t \in F(M,\mathbf{TAlg}(S))$;
-- E-equation (i.e. existence equation), iff $\Phi = t \overset{e}{=} t'$ for some $t,t' \in F(M,\mathbf{TAlg}(S))$;
-- ECE-equation (i.e. existentially conditioned existence equation), iff $\Phi = (\Lambda^*F \Rightarrow t \overset{e}{=} t')$, where F is some set of TE-expressions with variables in M and $t,t' \in F(M,\mathbf{TAlg}(S))$;
-- QE-equation (i.e. quasi-existence equation), iff $\Phi = (\Lambda^*F \Rightarrow t \overset{e}{=} t')$, where F is a set of E-equations and $t \overset{e}{=} t'$ a single E-equation with variables in M;
-- elementary implication, if $\Phi = (\Lambda^*F \Rightarrow \Lambda^*F')$, where F and F' are sets of E-equations with variables in M;
-- NTE-expression (i.e. negation of a TE-expression), iff $\Phi = \neg(t \overset{e}{=} t)$ for some term t with variables in M;
-- NE-equation (i.e. negation of an E-equation), iff $\Phi = \neg(t \overset{e}{=} t')$ for some terms t,t' with variables in M;
-- ECH-formula (i.e. existentially conditioned universal Horn formula), iff either $\Phi = \neg\Lambda^*F$ with F a set of TE-expressions, or Φ is an ECE-equation, both with variables in M;
-- SECH-formula (i.e. special existentially conditioned universal Horn formula), iff either $\Phi = \neg(t \overset{e}{=} t)$ or Φ is an ECE-equation, each with variables in M;
-- H-formula (i.e. universal Horn formula), iff either $\Phi = \neg\Lambda^*F$, where F is a set of E-equations, or Φ is a QE-equation, each with variables in M.

If $\mathbf{K} \subseteq \mathbf{Alg}(S)$ is any class of partial algebras, then
TE(K), **Eeq(K)**, **ECEeq(K)**, **QEeq(K)**, **Impl(K)**, **NTE(K)**, **NEeq(K)**, **ECH(K)**, **SECH(K)** and **Horn(K)** (cf. ECHEeq(K)), SHEeq(K), respectively Heq(K) in [B 82]), respectively, are the corresponding sets of L-formulas valid in **K** -- when the operators carry an index M, then we want to indicate that we only want to consider such axioms with M as corresponding set of variables --; the model classes of a set of the corresponding axioms are called TE-, E-, ECE-, QE-, Impl-, NTE-, NE-, ECH-, SECH-, and Horn-varieties, respectively, while TE, Eeq, ECEeq, QEeq, Impl, NTE, NEeq, ECH, SECH and Horn designate the corresponding sets of formulas of L, respectively.

8.1.3 REMARKS. **(i)**: It should be observed that the hep-varieties of H.REICHEL in Part II are very special QE-varieties.
(ii): For a long time it has been quite confusing among those dealing with partial algebras that validity of identities "could be defined in so different ways", and that

all these definitions coincided in the case of total algebras with the usual definition of validity of equations (cf. the papers of SŁOMIŃSKI, KERKHOFF, HÖFT, POYTHRESS, JOHN, EDWARDS, KOCHEN & SPECKER in the references). Therefore it should be noticed that all ECE-equations and all conjunctions of them with the same conclusion reduce to mere equations in the case of total algebras, since then all TE-expressions are trivially satisfied. In each case of a "notion of validity" (except for the one called "strong validity" by SŁOMIŃSKI [Sł 68], where satisfaction refers to the congruence generated by ker $\tilde{w}$ ($w{:}M \to \underline{A}$) in the completion $\underline{F}(M,\mathbf{TAlg}(\underline{S}))$ of $\underline{\mathrm{dom}}\ \tilde{w}$) one actually deals with validity with respect to a set of ECE-equations (or special conjunctions of them) with special prescribed form of the premise. Let us briefly discuss the most frequently used concepts (only for the homogeneous case, whence we omit the reference set of variables):

(a) "Weak validity" of $t \overset{e}{=} t'$: "If t and t' both are interpreted, then their interpretations are equal." -- This is just validity of the ECE-equation $(t \overset{e}{=} t \wedge t' \overset{e}{=} t') \Rightarrow t \overset{e}{=} t'$ ("weak equation": $t \overset{w}{=} t'$).

(b) "Validity" (in SŁOMIŃSKI [Sł 68]: in recent literature of computer science this is often called "strong validity") of $t \overset{e}{=} t'$: "If either t or t' is interpreted, then both terms are equally interpreted." -- This is described by validity of the formula $(t \overset{e}{=} t \Rightarrow t \overset{e}{=} t') \wedge (t' \overset{e}{=} t' \Rightarrow t \overset{e}{=} t')$ ("strong equation": $t \overset{s}{=} t'$).

(c) "Evans' validity" of $t \overset{e}{=} t'$ (cf. T.EVANS [Ev 51]): "If t is interpreted, and if all strict subterms of t' are interpreted, then t' is interpreted with the same value as t, and the same holds for t and t' interchanged." -- This is the description of the following conjunction of ECE-equations:

$$(\Lambda^*\{t \overset{e}{=} t,\ q \overset{e}{=} q \mid q \in ({\downarrow}t' \setminus \{t'\})\} \Rightarrow t \overset{e}{=} t') \wedge$$
$$\wedge\ (\Lambda^*\{t' \overset{e}{=} t',\ p \overset{e}{=} p \mid p \in ({\downarrow}t \setminus \{t\})\} \Rightarrow t \overset{e}{=} t') .$$ ("Evans' equation" $t \overset{E}{=} t'$)

Notice that $\underline{A} \models t \overset{s}{=} t'$ iff $t^{\underline{A}} = t'^{\underline{A}}$, i.e. "strong equations" are valid in $\underline{A}$, iff the corresponding induced term operations are equal. And validity of "Evans' equations" is connected with the embeddability as relative subalgebras (one has to take all elements of $\underline{A}$ as fundamental constants into the type getting, say, $\underline{A}_A$; if $\underline{B}_A$ is any partial algebra similarly extended with the constants for A, then one has to consider all "Evans' equations" valid in $\underline{B}_A$ which contain no free variable; if all these are also valid in $\underline{A}_A$, then $\underline{A}$ is fully embeddable into $\underline{B}$). "Weak equations" are e.g. closely related to quantum logic (cf. KOCHEN and SPECKER [KoSp 68]).

(iii): It can be said that R.KERKHOFF [Ke 70] and V.POYTHRESS [Po 73] deal in their papers with specific SECH-formulas, namely with NTE-expressions and "weak equations", since they consider "weak equations" where the terms can only be taken from a given subset of the term algebra, while all other terms are never allowed to be interpreted.

The following definition is of great importance for model theoretic investigations:

8.1.4 DEFINITION. A set F of L-formulas is <u>preserved by some algebraic operator</u>, say **X** (see 4.6 for examples of operators) iff, for every class $\mathbf{K} \subseteq \mathbf{Alg}(\underline{S})$ and for every subset $F' \subseteq F$, $\mathbf{K} \subseteq \mathbf{Mod}(F')$ implies $\mathbf{X}(\mathbf{K}) \subseteq \mathbf{Mod}(F')$, i.e. if $\underline{A}$ is obtained from **K**-algebras by using **X** (say products, homomorphic images, etc.) and if $\mathbf{K} \models F'$, then $\underline{A} \models F'$.

8.1.5. Notice that the First Birkhoff Theorem (cf. also 6.2.1) for E-equations especially tells us that E-equations are preserved by homomorphic images, subalgebras and direct as well as $\mathbf{d}^{\&}$-reduced products.

8.2 Algebraic encoding of universal Horn formulas

8.2.1. Before we further discuss preservation of formulas by algebraic operators, we present an algebraic description of those formulas which will be of interest for us. Among others B.BANASCHEWSKI and H.HERRLICH used this method in [BaH 76] for total algebras, while, independently, R.JOHN in [J 75a] and H.ANDRÉKA and I.NÉMETI in [AN 76], [AN 78] and [AN79a] independently applied it to partial algebras (and in [AN 78] and [AN 79a] even to arbitrary categories and for arbitrary formulas, see also Hien & SAIN [HiSa 83]). -- We shall deal in this book only with Horn formulas.

8.2.2 DEFINITION. Let $(M:\Phi)$ be any elementary implication, $\Phi = (\Lambda^*F \Rightarrow \Lambda^*G)$, where $F,G \subseteq Eeq_M$ (cf. 6.1.1). Then we associate with $(M:\Phi)$ four partial algebras and an epimorphism between two of them in the following way (observe the sort homomorphisms, which are always added in a natural way -- for the definition of $\underline{M}_d$ see 1.1.4):

$\underline{T}_p(M:\Phi) := \underline{M}_d + \downarrow(\Lambda^*F)$, the partial algebra of the <u>premise terms</u> in Φ,

$\underline{Pr}(M:\Phi) := \underline{T}_p(M:\Phi)/Cong_{\underline{T}_p(M:\Phi)}(F)$, the <u>premise algebra</u> for $(M:\Phi)$,

$\underline{T}_c(M:\Phi) := \underline{M}_d + \downarrow\Phi$, the partial algebra of all subterms of terms in Φ,

$\underline{Co}(M:\Phi) := \underline{T}_c(M:\Phi)/Cong_{\underline{T}_c(M:\Phi)}(F \cup G)$, the <u>conclusion algebra</u> for $(M:\Phi)$;

i.e. we have the commutative diagram in Figure 8.1, where $nat_p(M:\Phi)$ and $nat_c(M:\Phi)$ are the corresponding quotient homomorphisms.

The induced homomorphism $e(M:\Phi):\underline{Pr}(M:\Phi) \to \underline{Co}(M:\Phi)$ then exists according to the Diagram Completion Lemma and the fact that $\ker(nat_p(M:\Phi)) \subseteq \subseteq \ker(id \cdot nat_c(M:\Phi))$ by construction. $e(M:\Phi)$ is called the <u>encoding epimorphism</u> for $(M:\Phi)$.

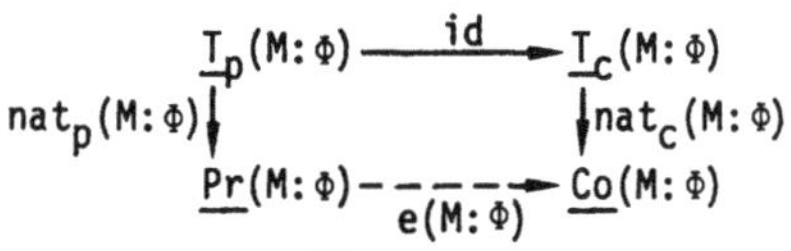

Figure 8.1.

8.2.3 PROPOSITION. Let $(M:\Phi) = (M:\Lambda^*F \Rightarrow \Lambda^*G)$ be any elementary implication with encoding epimorphism $e(M:\Phi):\underline{Pr}(M:\Phi) \to \underline{Co}(M:\Phi)$, let $\underline{A} \in \mathbf{Alg}(\underline{S})$, and let $w:M \to A$ be any assignment.

(i) $e(M:\Phi)$ is in fact an epimorphism.

(ii) $\underline{A} \models (M:\Lambda^*F)[w]$ iff there exists a homomorphism $f_w:\underline{Pr}(M:\Phi) \to \underline{A}$ satisfying $id_M \cdot nat_p(M:\Phi) \cdot f_w = w$.

(iii) $\underline{A} \models (M:\Phi)[w]$, iff: whenever w induces a homomorphism $f_w:\underline{Pr}(M:\Phi) \to \underline{A}$ satisfying $id_M \cdot nat_p(M:\Phi) \cdot f_w = w$, then there exists a homomorphism $g_w:\underline{Co}(M:\Phi) \to \underline{A}$ such that $e(M:\Phi) \cdot g_w = f_w$ (cf. Figure 8.2).

(iv) If $\underline{A} \in \mathbf{Alg}(\underline{S})$, then $\underline{A} \models (M:\Phi)$ iff $\underline{A} \in Inj(\{e(M:\Phi)\})$ (cf. 5.1.4).

-- In the homogeneous case we may replace M by $fvar(\Phi)$, if $fvar(\Phi)$ is non-empty, or else by any one-element set, if M is non-empty.

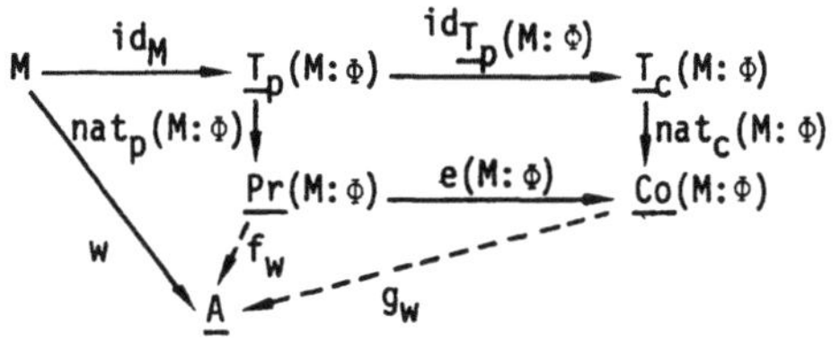

Figure 8.2.

PROOF. Since M generates both the domain and the codomain of $e(M:\Phi)$ which is induced over id_M, (i) is obvious. -- (iv) is a consequence of (iii), and the last remark follows from the Coincidence Theorem. -- Let us first prove (ii): Assume that f_w exists, satisfying $id_M \cdot nat_p(M:\Phi) \cdot f_w = w$; then $T_p(M:\Phi) = M \cup \downarrow(\Lambda^*F) \subseteq dom\ \tilde{w}$ and $F \subseteq ker\ \tilde{w}$, since $F \subseteq ker(nat_p(M:\Phi)) \subseteq ker(nat_p(M:\Phi) \cdot f_w) \subseteq ker\ \tilde{w}$, i.e. $\underline{A} \models \Lambda^*F[w]$. Conversely, if $\underline{A} \models \Lambda^*F[w]$, i.e. if $F \subseteq ker\ \tilde{w}$, then the Diagram Completion Theorem 5.8.2 for epimorphisms implies the existence of f_w . ∎

Ad (iii): If $\underline{A} \models (M:\Lambda^*F)[w]$ is **not true**, then f_w does not exist, but $(M:\Phi)$ is trivially satisfied by w in $\underline{A}$, and conversely. Hence the case of $\underline{A} \models (M:\Lambda^*F)[w]$, i.e. of the existence of f_w, is indeed the only interesting one. Then $\underline{A} \models (M:\Phi)[w]$ iff $(F \cup G) \subseteq ker\ \tilde{w}$, and by (ii) this is equivalent to the existence of $g_w:\underline{Co}(M:\Phi) \to \underline{A}$ satisfying $id_M \cdot id_{\underline{T}_p(M:\Phi)} \cdot nat_c(M:\Phi) \cdot g_w = w = id_M \cdot nat_p(M:\Phi) \cdot f_w$. Commutativity of the upper right hand square in Figure 8.2 yields $id_M \cdot nat_p(M:\Phi) \cdot f_w = id_M \cdot id_{\underline{T}_p(M:\Phi)} \cdot nat_c(M:\Phi) \cdot g_w = id_M \cdot nat_p(M:\Phi) \cdot e(M:\Phi) \cdot g_w$. Since $id_M \cdot nat_p(M:\Phi)$ is an epimorphism, this is equivalent to $f_w = e(M:\Phi) \cdot g_w$. ∎∎∎

8.2.4 COROLLARY 1. If F is any set of E-equations with variables in M, and if $\underline{T}_p := \underline{M}_d + \underline{\downarrow}(\Lambda^*F)$, then $e_F := id_M \cdot nat_{Cong_{\underline{T}_p}(F)} : M \to \underline{T}_p/Cong_{\underline{T}_p}(F)$ encodes $(M:\Lambda^*F)$. ∎∎∎

8.2.5 COROLLARY 2. Each epimorphism $e:\underline{P} \to \underline{C}$ represents an elementary implication $(M:\Lambda^*F \Rightarrow \Lambda^*G)$ (for some suitable cardinality **c** restricting the lengths of the conjunctions), namely if e is given, then let M be some generating subset of $\underline{P}$, let $i:M \to P$ be the identity as a mapping out of $\underline{F}(M,\mathbf{TAlg}(\underline{S}))$, and let F and $F \cup G$ be some generating subsets of the closed congruence relations $ker\ \tilde{i}$ and $ker(ie)^\sim$, respectively, such that, for $\underline{T}_p := \underline{M}_d + \underline{\downarrow}F$ and $\underline{T}_c := \underline{M}_d + \underline{\downarrow}(F \cup G)$: $\underline{P} \cong \underline{T}_p/Cong_{\underline{T}_p}(F)$ and $\underline{C} \cong \underline{T}_c/Cong_{\underline{T}_c}(F \cup G)$. -- If a cardinality restriction by **c** is given, then e represents a **c**-restricted implication iff M, F and G can be chosen such that $\#M, \#F, \#G < \mathbf{c}$. ∎∎∎

Since $(M:\neg\Lambda^*F)$ can be considered as an implication with an always false conclusion:

8.2.6 COROLLARY 3. If $F \subseteq Eeq_M$, $\underline{A} \in \mathbf{Alg}(\underline{S})$ and $w:M \to A$, then $\underline{A} \models (M:\neg\Lambda^*F)[w]$ iff there exists in the diagram corresponding to Figure 8.2 for $\Phi := \Lambda^*F$ no homomorphism $f_w:\underline{Pr}(M:\Lambda^*F) \to \underline{A}$ (see 8.2.4 for $\underline{Pr}(M:\Lambda^*F)$) satisfying $id_M \cdot nat_p(M:\Lambda^*F) \cdot f_w = w$; and $\underline{A} \models (M:\neg\Lambda^*F)$ iff $\underline{Pr}(M:\Lambda^*F)$ does not allow any homomorphism into $\underline{A}$. ∎∎∎

8.2.7 REMARK. It should be observed that one has for every $\underline{A} \in \mathbf{Alg}(\underline{S})$, for every $w:M \to A$, for every basic L-formula Φ, and for every non-empty set F of basic L-formulas that

$\underline{A} \models (M:\Phi \Rightarrow \Lambda^*F)[w]$ iff $\underline{A} \models (M:\Lambda^*\{\Phi \Rightarrow \Psi | \Psi \in F\})[w]$, iff $\underline{A} \models (M:\Phi \Rightarrow \Psi)[w]$ for every $\Psi \in F$.

Hence, if $(M:\Phi \Rightarrow \Lambda^*F)$ is just an elementary implication, then this can always be replaced by a set of QE-equations, i.e. the language of all QE-equations is as powerful as the one of all elementary implications, although the latter is much more comprehensive. -- If we had allowed an empty set F of E-equations in the conclusion, then we would get a formula which would be always true (cf. 8.2.3), and we would still have the same implicational model theory as before.

8.3 Preservation properties of algebraic operators -- first visit

8.3.1. In Definition 4.6.2 we have presented a list of operators on **Alg**($\underline{S}$) which are of interest for us, in Definition 8.1.2 we have given a list of special kinds (sets) of formulas, and in both cases we have a partial order related to set theoretical inclusion. We now want to investigate a little bit more closely the relationship via preservation between these two lists, and this order will be useful, as e.g. the following observation shows:

8.3.2 LEMMA. Let **Q** and **Q'** be two algebraic operators on **Alg**($\underline{S}$) such that $Q \subseteq Q'$, and let F and F' be two sets of L-formulas such that $F \subseteq F'$. Then:

(i) If **Q'** preserves F' (cf. 8.1.4), then **Q** preserves F.

(ii) If **Q** does not preserve F, then **Q'** does not preserve F'.

PROOF. Assume that **Q'** preserves F', and let $K \subseteq$ **Alg**($\underline{S}$) such that $K \models F''$ for any subset F" of F. Since F" is also a subset of F', we get $Q(K) \subseteq Q'(K) \subseteq$ **Mod**(F") by our assumptions. -- (ii) is just the contrapositive of (i). ∎∎∎

8.3.3 REMARK. From 4.6.4 and 8.1.2 we get the Hasse diagrams for the corresponding partial orders as shown in Figure 8.3. For some of the sets or classes of formulas the formulas are restricted in size by some cardinality $c \geq d^{\&}$ as in the definition of L, and this is indicated by an upper index. For reduced products the upper index **k** ($\geq$ **d***) indicates that the filters under consideration are at least **k**-complete. Notice that filters containing only the index set are complete for all cardinals. Assume $k \leq k'$. Observe, too, that E-equations can be considered as special ECE-equations with an empty premise.

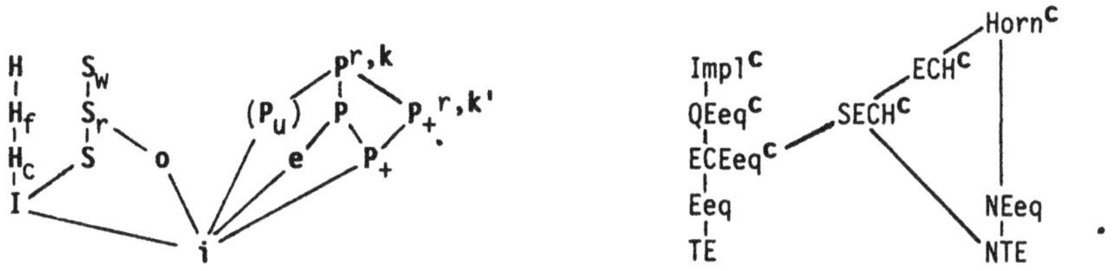

Algebraic operators sets of formulas

Figure 8.3.

8.3.4 THEOREM. Let the sets of formulas defined in 8.1.2 be taken in $L_{c,q}(\underline{S},X`) =: L$, and consider the operators from 4.6.2. Then Table 8.1 presents the preservation properties of these operators with respect to the indicated sets of formulas ("+" means preservation, "-" means no preservation, "*" means preservation only in the case of $d^{\&}, c \leq k$, "**" means $d^* \leq k$). In the case of ultraproducts #S as well as all arities and formulas have to be finite. In connection with L, **o** preserves no formulas of the indicated kinds, while in the case, when the assignments always concern the whole set X of variables, then **o** preserves all Horn formulas.

PROOF. At this place we only consider the case of products, the rest of the proof of 8.3.4 will be postponed until section 11, since it will follow more easily from later results. -- Preservation of all elementary implications: Let $e: \underline{P} \rightarrow \underline{C}$ be the encoding epimorphism of an elementary implication, and let $(\underline{A}:= X(\underline{A}_i|i \in I),(pr_i|i \in I))$ be a product such that in each $\underline{A}_i$ the implication encoded by e is valid, and let $p:\underline{P} \to \underline{A}$ be any homomorphism. If $I = \emptyset$, then $\underline{A} \cong \underline{S}$ preserves all E-equations and therefore also all elementary implications; but note that it then does not preserve negations of TE-expressions, since it is total, and therefore we get all the negative entries in Table 8.1. Thus assume $I \neq \emptyset$; then, for each $i \in I$, $p \cdot pr_i$ induces (by assumption) a homomorphism $q_i:\underline{C} \to \underline{A}_i$ satisfying $e \cdot q_i = p \cdot pr_i$. $(q_i|i \in I)$ then induces a homomorphism $q:\underline{C} \to \underline{A}$ satisfying $q \cdot pr_i = q_i$ for each $i \in I$; therefore $p = e \cdot q$, showing that e is valid in $\underline{A}$. If $I \neq \emptyset$ and $(M:\Phi)$ is an NTE-formula encoded by the partial algebra $\underline{C}'$, such that in all $\underline{A}_i$ $(M:\Phi)$ is valid, then a homomorphism from $\underline{C}'$ into $\underline{A}$ would contradict this assumption; thus $(M:\Phi)$ is also valid in $\underline{A}$ (cf. 8.2.6). Because of 8.3.2 the entries in Table 8.1 for products (and for the operator **e**) are proved. ∎

	TE	Eeq	$ECEeq^c$	$QEeq^c$	$Impl^c$	NTE	NEeq	ECH^c	$SECH^c$	$Horn^c$
H	+	+	-	-	-	-	-	-	-	-
$\mathbf{H_f}$	+	+	-	-	-	-	-	-	-	-
$\mathbf{H_c}$	+	+	+	-	-	+	-	+	+	-
$\mathbf{S_w}$	-	-	-	-	-	+	+	-	-	-
$\mathbf{S_r}$	-	-	-	-	-	+	+	-	-	-
S	+	+	+	+	+	+	+	+	+	+
P	+	+	+	+	+	-	-	-	-	-
$\mathbf{P_+}$	+	+	+	+	+	+	+	+	+	+
$\mathbf{P^{r,k}}$	**	**	*	*	*	-	-	-	-	-
$\mathbf{P_+^{r,k}}$	**	**	*	*	*	**	**	*	*	*
e	+	+	+	+	+	-	-	-	-	-
I	+	+	+	+	+	+	+	+	+	+
(o)	-	-	-	-	-	-	-	-	-	-
$\mathbf{(P_u)}$	+	+	+	+	+	+	+	+	+	+

Preservation (via validity) of formulas by algebraic operators
(for explanations see Theorem 8.3.4)
Table 8.1.

Also without proofs at this place we mention here the following results whose proofs (see section 13.3) will follow from the main theorem of section 12.

8.3.5 THEOREM. Let $\mathbf{K} \subseteq \mathbf{Alg}(\underline{S})$ be any class of partial algebras, and let $\mathbf{M_c}$ be an additional algebraic operator, defined by

$\mathbf{M_c}(\mathbf{K}) := \{\underline{A} \in \mathbf{Alg}(\underline{S}) |$there exist $\underline{B} \in \mathbf{K}$ and a closed and surjective homomorphism $f:\underline{A} \to \underline{B}\}$,

i.e. the class of all **closed preimages** of **K**-algebras. -- Then we have for $\mathbf{k} := \max(\mathbf{d}^{\&},\mathbf{c})$:

(i) $\mathbf{ModTE}(\mathbf{K}) = \mathbf{HM_cP}(\mathbf{K})$;

(ii) $\mathbf{ModEeq}(\mathbf{K}) = \mathbf{HSP}(\mathbf{K})$;

(iii) $\mathbf{ModECEeq^c}(\mathbf{K}) = \mathbf{HSP^{r,k}}(\mathbf{K})$;

(iv) $\mathbf{ModQEeq^c}(\mathbf{K}) = \mathbf{SP^{r,k}}(\mathbf{K})$;

(v) $\mathbf{ModECH^c}(\mathbf{K}) = \mathbf{H_cSP_+^{r,k}}(\mathbf{K})$;

(vi) $\mathbf{ModSECH^c}(K) = \begin{cases} H_c SP_+^{r,k}(K)\ , \text{ if } \mathbf{NTE}(K) = \emptyset\ , \\ H_c SP_+^{r,k}(K)\ , \text{ else;} \end{cases}$ $\Big\}$ only, if each assignment interprets all of X;

(vii) $\mathbf{ModHorn^c}(K) = SP_+^{r,k}(K)$.

8.3.6. We shall see in later sections that especially within the framework of ECE-equations even finer distinctions can be made than in the above result.

8.4 The diagonal-fill-in property -- motivation

8.4.1 OBSERVATIONS. Let **Q**, **T** be classes of homomorphisms between partial ($\underline{S}$-) algebras, and at the same time, for any class $\mathbf{K} \subseteq \mathbf{Alg}(\underline{S})$, let

$\mathbf{Q}(\mathbf{K}) := \{\underline{B} \in \mathbf{Alg}(\underline{S})\,|\,$there exists $f:\underline{A} \to \underline{B}$, $f \in \mathbf{Q}$, for some $\underline{A} \in \mathbf{K}\}$

be the class of all **Q**-images of **K**-algebras, and

$\mathbf{T}(\mathbf{K}) := \{\underline{A} \in \mathbf{Alg}(\underline{S})\,|\,$there exists $f:\underline{A} \to \underline{B}$, $f \in \mathbf{T}$, for some $\underline{B} \in \mathbf{K}\}$

be the class of all **T**-preimages of **K**-algebras.

Moreover, let $(M:\Phi)$ be any elementary implication encoded by the epimorphism $e:\underline{P} \to \underline{C}$, and let $f:\underline{A} \to \underline{B}$ be any homomorphism.

(i): Let f be a **T**-homomorphism, and assume $\underline{B} \in \mathrm{Inj}(\{e\})$ (cf. 5.1.4). We want to know, when $\underline{A} \in \mathrm{Inj}(\{e\})$ is satisfied, i.e. when $(M:\Phi)$ respectively e "is retractable against **T**-homomorphism" ("hereditary with respect to '**T**-subobjects'"), since such a situation occurred in connection with E-equations and **S** instead of **T**, and since we will often meet similar situations below. -- Therefore, let $p:\underline{P} \to \underline{A}$ be any homomorphism. Then $\underline{B} \models (M:\Phi)[pf]$ (e.g. $\underline{B} \in \mathrm{Inj}(\{e\})$) implies the existence of a homomorphism $q:\underline{C} \to \underline{B}$ such that $eq = pf$, i.e. we get a commutative square as the fully drawn one in Figure 8.4.

Figure 8.4.

What we need is a homomorphism $d:\underline{C} \to \underline{A}$ satisfying $ed = p$. Since e is an epimorphism, we may conclude in the case of the existence of such a d from $eq = pf = edf$ that $q = df$, and that d is unique. Hence, with our assumptions, $\underline{A} \in \mathrm{Inj}(\{e\})$, iff each such square as in Figure 8.4 has a unique "diagonal-fill-in" d making both triangles commutative.

(ii): Let us now assume $f \in \mathbf{Q}$ and $\underline{A} \in \mathrm{Inj}(\{e\})$ (i.e. $\underline{A} \models (M:\Phi)$), and this time we ask when we can conclude that also $\underline{B} \models (M:\Phi)$. Thus assume $r:\underline{P} \to \underline{B}$ . .If we want to apply $\underline{A} \in \mathrm{Inj}(\{e\})$, we need that for each such $r:\underline{P} \to \underline{B}$ there exists a (not necessarily unique) homomorphism $p := p_r:\underline{P} \to \underline{A}$ such that $pf = r$. Namely under such circumstances we get a homomorphism $d:\underline{C} \to \underline{A}$ satisfying $ed = p$; and this implies $edf = pf = r$, i.e. df is then the homomorphism we have looked for in connection with r. Again a diagram similar to the one in Figure 8.4 appears. But this time we need a special property for $\underline{A}$, namely that it be "e-projective" (see below).

These two observations motivate to consider the following two definitions (see e.g. [N 82] or [NSa 77]; [NSa 77] contains several historical notes on the subject):

8.4.2 DEFINITION. We say that any two homomorphisms $e:\underline{P} \to \underline{C}$ and $f:\underline{A} \to \underline{B}$ have the (<u>unique</u>) <u>diagonal-fill-in-property</u> (in symbols: Difip(e,f)) iff for every $p:\underline{P} \to \underline{A}$ and $q:\underline{C} \to \underline{B}$ satisfying $eq = pf$ (i.e. making the fully drawn part in Figure 8.4 commuta-

tive) there exists a (unique) homomorphism $d:\underline{C} \to \underline{A}$ such that $ed = p$ and $df = q$.
As is well known (cf. section 24), any such relation induces a Galois connection (Galois correspondence), here between Hom(**Alg**($\underline{S}$)) and Hom(**Alg**($\underline{S}$)), with two closure operators, which in this connection are often called Λ and Λ^{op}, i.e. for any $\mathbf{M} \subseteq$ Hom(**Alg**($\underline{S}$)) we have:

$$\Lambda(\mathbf{M}) := \{b \in \mathrm{Hom}(\mathbf{Alg}(\underline{S})) \mid \mathrm{Difip}(a,b) \text{ for all } a \in \mathbf{M}\},$$
$$\Lambda^{op}(\mathbf{M}) := \{a \in \mathrm{Hom}(\mathbf{Alg}(\underline{S})) \mid \mathrm{Difip}(a,b) \text{ for all } b \in \mathbf{M}\}.$$

8.4.3 DEFINITION. Let $\underline{P} \in$ **Alg**($\underline{S}$), and $f:\underline{A} \to \underline{B}$. $\underline{P}$ is called f-projective, iff for every homomorphism $r:\underline{P} \to \underline{B}$ there exists a homomorphism $q:\underline{P} \to \underline{A}$ satisfying $qf = r$. If $\mathbf{M} \subseteq$ Hom(**Alg**($\underline{S}$)), and if $\underline{P}$ is f-projective for each $f \in \mathbf{M}$, then we say that $\underline{P}$ is **M**-projective. By Pj(**M**) we denote the class of all **M**-projective partial ($\underline{S}$-) algebras.

8.4.4. Returning to our observations in 8.4.1 we realize that in connection with the "preservation of elementary implications by '**Q**-homomorphic images'" (case (ii)), we will have to study the class Pj(**Q**) of all **Q**-projective objects.
And in connection with the first observation we can now say that all formulas of some given set F of elementary implications are preserved by the operator T (see 8.4.1), iff for the set, say Ep(F), of encoding epimorphisms of all F-formulas, i.e. $\mathrm{Ep}(F) := \{e(M:\Phi):\underline{\mathrm{Pr}}(M:\Phi) \to \underline{\mathrm{Co}}(M:\Phi) \mid (M:\Phi) \in F\}$, we have $\mathbf{T} \subseteq \Lambda(\mathrm{Ep}(F))$ -- this time for the **class T**.
We shall prove in section 10 that for every class **E** of epimorphisms in **Alg**($\underline{S}$) the "pair" (of classes) $(\Lambda^{op}(\Lambda(\mathbf{E})), \Lambda^{op}(\mathbf{E}))$ is a so-called **factorization system**; and our observations show that factorization systems form the appropriate background for an (algebraic) treatment of the preservation of implications by any kind of subobjects (see also the next section for another motivation; and see the Remark in [AN 79], p. 148). And in section 12 we shall see how this concept relates to the meta-theorem of ANDRÉKA, NÉMETI and SAIN about the axiomatizability of classes of partial algebras by elementary implications.

9 Preservation (reflection) of formulas by mappings

The universal algebraic theory of total algebras usually gets along with homomorphisms, where injectivity, surjectivity or being an epimorphism are the only additional relevant properties. We have already seen that we need for partial algebras in addition at least closedness and fullness. But this is only the top of an iceberg, as we shall see with more examples at the end of this section. Many users of partial algebras complain as much about the multitude of properties for homomorphisms between partial algebras (usually calling them "different concepts of homomorphisms") as the multitude of "notions of validity" of equations (see 8.1.3). As in the latter case the main reason for the richness of properties of homomorphisms between partial algebras is the "newly gained importance" of term existence expressions (respectively of their negations), which play only a trivial rôle in the case of total algebras. And concepts which are usually considered part of model theory have to be used in order to organize this wealth of properties, as we want to show in this section. We could already realize in section 8 -- and we shall get new evidence for it here and in the next section -- that the category theoretical concept of a factorization system has a model theoretic aspect in connection with implications, and it is a very useful tool also in connection with the concepts described in this section. -- One should know in advance that the concepts of preservation (respectively reflection) of a formula by a mapping (homomorphism) are not actually entirely different ones, but are like two sides of the same coin.

9.1 Preservation and reflection

9.1.1 DEFINITION. Let $\underline{A},\underline{B} \in$ **Alg**($\underline{S}$) be arbitrary partial algebras, $(M:\Phi)$ any L-formula and $f:A \to B$ any mapping (not necessarily a homomorphism, but always compatible with the sort mappings).

(i): We say that f <u>preserves</u> $(M:\Phi)$ <u>with respect to</u> $\underline{A}$ and $\underline{B}$, iff for every assignment $w:M \to A$ we have:

$$\underline{A} \models (M:\Phi)[w] \quad \text{implies} \quad \underline{B} \models (M:\Phi)[wf] .$$

(ii): We say that f <u>reflects</u> $(M:\Phi)$ <u>with respect to</u> $\underline{A}$ and $\underline{B}$, iff for every assignment $w:M \to A$ we have:

$$\underline{B} \models (M:\Phi)[wf] \quad \text{implies} \quad \underline{A} \models (M:\Phi)[w] .$$

If we assume -- which we shall usually do -- that f is a homomorphism, then we shall usually omit the addendum "with respect to $\underline{A}$ and $\underline{B}$", since then the structures of $\underline{A}$ and $\underline{B}$ are already involved. -- If F is any set of formulas, then f preserves (reflects) F iff f preserves (reflects) each formula in F. In a similar way we say that a set F of formulas is preserved (reflected) by a class **M** of mappings, if this is done by each mapping.

9.1.2 REMARKS. (i): It is immediate from Definition 2.1.1 that **homomorphisms** between $\underline{A}$ and $\underline{B}$ are exactly those mappings from A into B which preserve all formulas $(M_\varphi: (x_k|k \in K_\varphi)\varphi \stackrel{e}{=} y)$ $(\varphi \in \Omega)$, where $M_\varphi := \{x_k, y | k \in K_\varphi\}$, $y \notin \{x_k | k \in K_\varphi\}$, and where $k,k' \in K_\varphi$ and $k \neq k'$ imply $x_k \neq x_{k'}$ (in this case we shall say that the sequence $\underline{x} := (x_k|k \in K_\varphi)$ is <u>injective</u> or <u>without repetitions</u>). -- It is then easy to realize that homomorphisms also preserve all atomic formulas (E-equations); namely, if $(M:t \stackrel{e}{=} t')$ is given, if $f:\underline{A} \to \underline{B}$ is any homomorphism, and if $w:M \to A$ is any assignment such that $\underline{A} \models (M:t \stackrel{e}{=} t')[w]$, then $t\tilde{w} = t'\tilde{w}$, and this implies $t(wf)^{\sim} = t'(wf)^{\sim}$ (cf. 5.7.5), i.e. $\underline{B} \models (M:t \stackrel{e}{=} t')[wf]$. Moreover, homomorphisms trivially preserve all formulas of the form $(M:\neg(x \stackrel{e}{=} x))$ $(x \in M \subseteq X)$, since we always have $\underline{A} \models (M:x \stackrel{e}{=} x)$; therefore $(M:\neg(x \stackrel{e}{=} x))$ is never satisfied, neither in $\underline{A}$, nor in $\underline{B}$. -- While in the case of total algebras we can extend this observation to all NTE-expressions $(M:\neg(t \stackrel{e}{=} t))$, we shall see below that preservation of all NTE-expressions by $f:\underline{A} \to \underline{B}$ is equivalent to closedness of f.

As we have mentioned before, the "individuality" of TE-expressions and their negations in connection with partial algebras is the clue for the basic difference between total and partial algebras and the manifold of partial-algebra-concepts, into which almost any total-algebra-concept splits.

(ii): For the above observation, and for what follows about closed homomorphisms, it is important that we are able to restrict assignments to the set of free variables, as long as the empty partial algebra or heterogeneous partial algebras with empty phyla are allowed. Namely if we had only allowed assignments for all of X into A, and if A had some empty phylum, then preservation or reflection of a formula would mean nothing in connection with $\underline{A}$ for any formula, since there would be no assignments into A at all.

In order to show that there is not much difference between the concepts of preservation and reflection of formulas by mappings, we state:

9.1.3 PROPOSITION. Let $\underline{A}$ and $\underline{B}$ be any similar partial algebras, $f:A \to B$ any mapping, and $(M:\Phi)$ any formula. Then with respect to $\underline{A}$ and $\underline{B}$ f preserves $(M:\Phi)$ iff f reflects $(M:\neg\Phi)$.

PROOF. Let $w:M \to A$. If f preserves $(M:\Phi)$, and if $\underline{B} \models (M:\neg\Phi)[wf]$, then $\underline{A} \models (M:\Phi)[w]$ cannot hold, i.e. f must reflect $(M:\neg\Phi)$. The converse argument is similar. ■■■

9.1.4. We shall in general use either reflection or preservation in such a way that we may always deal with the more simple kind of formulas. Moreover, we shall usually assume in what follows that the mappings under consideration preserve all E-equations, i.e. are homomorphisms.

9.1.5 EXAMPLES. (i): An ($\underline{S}$-) mapping $f:A \to B$ is **injective** iff, for every $a,b \in A$ (of the same sort), $a \neq b$ implies $af \neq bf$, i.e. iff f preserves $(\{x,y\}:\neg(x \overset{e}{=} y))$ (f reflects $(\{x,y\}:x \overset{e}{=} y)$), for any distinct variables $x,y \in X$ of the same sort. -- If we only consider those formulas $(\{x,y\}:x \overset{e}{=} y)$, where $xv_X = yv_X \in S' \subseteq S$ for a proper subset, then reflection of these formulas means injectivity of $f|[S']v_A^{-1}$.

(ii) Recall that a homomorphism $f:\underline{A} \to \underline{B}$ is **closed** iff, for every $\phi \in \Omega$ and for every $\underline{a} \in A^{K_\phi}$: $\underline{a}f \in \operatorname{dom} \phi^{\underline{B}}$ implies $\underline{a} \in \operatorname{dom} \phi^{\underline{A}}$, i.e. iff f reflects all formulas $([K_\phi]\underline{x}:\underline{x}\phi \overset{e}{=} \underline{x}\phi)$ ($\phi \in \Omega$, $\underline{x} \in X^{K_\phi}$ preserving sorts); it suffices to require that f reflects such a formula "for each $\phi \in \Omega$ for at least one such sequence without repetitions". We shall see below that this will be equivalent to the fact that f **reflects all TE-expressions.**

Further examples will follow later. But observe that full homomorphisms cannot be defined by reflection of formulas, since the original sequence (assignment) might be changed.

9.1.6. The following lemma is only formulated with respect to reflection, but because of 9.1.3 one gets the corresponding results for preservation by interchanging "reflects" and "preserves" at each occurrence, and if a special kind of formula is referred to, then it has to be replaced by its negation.

9.1.7 LEMMA. Let $\underline{A}$, $\underline{B}$ and $\underline{C}$ be any similar partial algebras, $f:A \to B$ and $g:B \to C$ any mappings, and $\Sigma = (M:\Phi)$, $\Sigma_i = (M_i:\Phi_i)$ $(i \in I)$ any formulas ($\#I < \mathbf{c}$, see 7.1). Then:

(i) If both f and g reflect Σ, then so does fg.

(ii) If fg reflects Σ, and if g preserves Σ, then f reflects Σ.

(iii) If fg reflects Σ, and if f is surjective and preserves Σ, then g reflects Σ.

(iv) If f reflects Σ_i for each $i \in I$, then f reflects $(\bigcup\{M_i | i \in I\}:\wedge^*\{\Phi_i | i \in I\})$ as well as $(\bigcup\{M_i \mid i \in I\}:\forall^*\{\phi_i \mid i \in I\})$.

(v) If $0 < \#Y < \mathbf{q}$, $Y \subseteq X$, and if f reflects $(M:\Phi)$, then f reflects $(M \setminus Y:\forall^*Y\Phi)$, and if f is surjective, then it also reflects $(M \setminus Y:\exists^*Y\Phi)$.

(vi) If f reflects $(M:\Phi)$, and if $M \subseteq N \subseteq X$, then f reflects $(N:\Phi)$; and if $fvar(\Phi) \subseteq M' \subseteq M$, and if there exists an assignment, say v, from M into A, then f also reflects $(M':\Phi)$.

(vii) Let $N \subseteq X$, let $q:M \to F(N,\mathbf{TAlg}(\underline{S}))$ be a mapping which is admissible for $(M:\Phi)$, let f reflect $(M:\Phi)$ and $(N:mq \overset{e}{=} mq)$ for each $m \in M$, let Φ be an E-equation, and let there exist an assignment $w:M \to A$. Then f reflects $(N:(\Phi)q)$ (cf. 7.1.5).

PROOF. Since the proofs of (i), (ii) and (iii) are relatively straightforward, we leave them as exercises -- observe in (iii) that surjectivity of f allows an assignment into B to be factorized by one into A via f.

Ad(vi): We only prove the second case (the proof of the first part should be obvious): Let $v:M \to A$ be given; let $w:M' \to A$ be any assignment, and assume $\underline{B} \models (M':\Phi)[wf]$. Then $\underline{B} \models (M:\Phi)[w'f]$ -- where $w'|M' := w$, and $w'|(M \setminus M') := v|(M \setminus M')$ --, because of the Coincidence Theorem. Since f reflects $(M:\Phi)$, we have $\underline{A} \models (M:\Phi)[w']$, thus $\underline{A} \models (M':\Phi)[w]$, again by the Coincidence Theorem. ∎

(iv) is now an immediate consequence of (vi) and the definition of $\underline{A} \models (M:\wedge^*\{\Phi_i \mid i \in I\})$ (cf. 7.2.2) . ∎

Ad (v): We only prove the case of the existential quantifier: Let f be surjective, and let $w:M \setminus Y \to A$ be any assignment such that $\underline{B} \models (M \setminus Y:\exists^*Y\Phi)[wf]$. Then there exists an assignment $u:M \cup Y \to B$ such that $u|(M \setminus Y) = wf$ and $\underline{B} \models (M \cup Y:\Phi)[u]$; thus $\underline{B} \models (M:\Phi)[u]$ by the Coincidence Theorem. Since f is surjective, there exists an assignment $w':M \cup Y \to A$ such that $u|Y = w'f|Y$ and $w'|(M \setminus Y) = w$; hence $u = w'f$. Now $\underline{A} \models (M:\Phi)[w'|M]$, since f reflects $(M:\Phi)$; hence $\underline{A} \models (M:\exists^*Y\Phi)[w]$. ∎

Ad (vii): Let $v:N \to A$ be an assignment such that $\underline{B} \models (N:(\Phi)q)[vf]$, let $\Phi = t \overset{e}{=} t'$, and let $t_m := mq$ $(m \in M)$. (vi) and the fact that $\underline{B} \models (N:(t_m|m \in M)t \overset{e}{=} (t_m|m \in M)t')[vf]$ -- i.e. $(t\tilde{q})(vf)^\sim = (t'\tilde{q})(vf)^\sim$ -- imply $\underline{B} \models (N:t_m \overset{e}{=} t_m)[vf]$. Moreover, the assumption that f reflects $(N:t_m \overset{e}{=} t_m)$ yields $\underline{A} \models (N:t_m \overset{e}{=} t_m)[v]$ $(m \in M)$. Hence we may define $u:M \to A$ by $mu := t_m\tilde{v} = mq\tilde{v}$ $(m \in M)$. Therefore $(t\tilde{q})(vf)^\sim = ((vf)(N:t_m)^{\underline{B}}|m \in M)(M:t)^{\underline{B}} = ((uf)(M:m)^{\underline{B}}|m \in M)(M:t)^{\underline{B}} = (uf)(M:t)^{\underline{B}}$; similarly $t'(uf)^\sim = (uf)(M:t')^{\underline{B}} = (t'\tilde{q})(vf)^\sim = (t\tilde{q})(vf)^\sim = (uf)(M:t)^{\underline{B}}$. Thus $\underline{B} \models (M:t \overset{e}{=} t')[uf]$. Since f reflects $(M:t \overset{e}{=} t')$, we get $\underline{A} \models (M:t \overset{e}{=} t')[u]$, i.e. $ut^{\underline{A}} = ut'^{\underline{A}}$, and therefore $(q\tilde{v})(M:t)^{\underline{A}} = (q\tilde{v})(M:t')^{\underline{A}}$; and this shows that $\underline{A} \models (M:(t \overset{e}{=} t')q)[v]$. ∎∎∎

9.1.8 COROLLARY 1. Let $f:\underline{A} \to \underline{B}$ be any homomorphism and $(M:\Phi)$ any formula.

(i) If $\underline{A} \models (M:\Phi)$ and if f is surjective and preserves $(M:\Phi)$, then $\underline{B} \models (M:\Phi)$.

(ii) If $\underline{B} \models (M:\Phi)$ and if f reflects $(M:\Phi)$, then $\underline{A} \models (M:\Phi)$. ∎∎∎

9.1.9 COROLLARY 2. Let $f:\underline{A} \to \underline{B}$ and $g:\underline{B} \to \underline{C}$ be any homomorphisms and $(M:t \overset{e}{=} t')$ any $\underline{E}$-equation.

(i) If f and g both reflect $(M:t \overset{e}{=} t')$, then so does fg.

(ii) If fg reflects $(M:t \overset{e}{=} t')$, then so does f.

(iii) If f is surjective and fg reflects $(M:t \overset{e}{=} t')$, then g reflects $(M:t \overset{e}{=} t')$.

PROOF. Observe that homomorphisms preserve all E-equations. ■■■

9.1.10 PROPOSITION. Closed homomorphisms reflect all TE-expressions.

PROOF. (By algebraic induction:) Let $f:\underline{A} \to \underline{B}$ be closed. Then f trivially reflects all $(M:x \overset{e}{=} x)$ $(x \in M \subseteq X)$. Thus assume $t := (t_k|k \in K_\phi)\phi \in F(M,\mathbf{TAlg}(\underline{S}))$ and that f reflects $(M:t_k \overset{e}{=} t_k)$ $(k \in K_\phi)$. By assumption f reflects all $(fvar(\underline{x}\phi):\underline{x}\phi \overset{e}{=} \underline{x}\phi)$, where $\underline{x} := (x_k|k \in K_\phi)$ is any injective sequence of variables (of correct sorts!) -- since f is closed. Observe that $q:x_k \mapsto t_k$ $(k \in K_\phi)$ is an admissible mapping from $fvar(\underline{x}\phi)$ into $\underline{F}(M,\mathbf{TAlg}(\underline{S}))$, and let $w:M \to A$ be an assignment, for which $\underline{B} \models (M:t \overset{e}{=} t)[wf]$. Then $\underline{B} \models (M:t_k \overset{e}{=} t_k)[wf]$ for each $k \in K_\phi$. Now 9.1.7.(vii) implies that f also reflects $(M:t \overset{e}{=} t)$, since $t \overset{e}{=} t = (\underline{x}\phi \overset{e}{=} \underline{x}\phi)q$. ■■■

Of fundamental importance for later applications is the following

9.1.11 THEOREM. Let $f:\underline{A} \to \underline{B}$ be any homomorphism and
$\Sigma = (M:\Lambda^*\{t_i \overset{e}{=} t_i'|i \in I\} \Rightarrow \Lambda^*\{p_j \overset{e}{=} p_j'|j \in J\})$
be any elementary implication with encoding epimorphism $e:\underline{P} \to \underline{C}$. Then f reflects Σ, iff the pair (e,f) has the diagonal-fill-in property: Difip(e,f) (cf. 8.4.2).

PROOF. The above statement follows with no great effort from 8.4.1.(i) and 8.2.3. ■■■

9.2 Examples for the reflection of formulas by homomorphisms

9.2.1 DEFINITION. Let $f:\underline{A} \to \underline{B}$ be any homomorphism. f is called <u>initial</u> (an <u>initial homomorphism</u>) iff, for every **mapping** $g:C \to A$ starting from some partial algebra $\underline{C}$, g is a homomorphism from $\underline{C}$ into $\underline{A}$ iff gf is a homomorphism. -- In the dual situation, i.e. if, for every $\underline{D} \in \mathbf{Alg}(\underline{S})$ and for every **mapping** $h:B \to D$, h is a homomorphism from $\underline{B}$ into $\underline{D}$ iff fh is a homomorphism, then f is called <u>final</u> (a <u>final homomorphism</u>).

9.2.2 REMARK. It is easy to realize that every full and injective homomorphism is initial, and that every full and surjective homomorphism is final -- but the converses are not true. ■

9.2.3 PROPOSITION. For any homomorphism $f:\underline{A} \to \underline{B}$ the following statements are equivalent:

(i) f is initial.

(ii) For every $a \in \mathbf{U}\{[A^{K_\phi}]\phi^{\underline{A}}|\phi \in \Omega\}$ one has $\#([\{af\}]f^{-1}) = 1$, and if $\underline{a} \in A^{K_\phi}$ such that $\underline{a}f \in dom\ \phi^{\underline{B}}$ and $\underline{a}f\phi^{\underline{B}} \in [A]f$, then $\underline{a} \in dom\ \phi^{\underline{A}}$.

(iii) f reflects all formulas of the form $(\{y\} \cup [K_\phi]\underline{x}:\underline{x}\phi \overset{e}{=} y)$ (where we may assume that $\underline{x}$ is injective -- respecting the sorts! --, and $y \notin [K_\phi]\underline{x}$).

PROOF. Let us first observe that in Definition 9.2.1 of initiality the implication "g is a homomorphism implies that gf is a homomorphism" is trivial; thus we have only to prove the other implication.

"(i) implies (ii)" (by contraposition): Let $a \in \mathbf{U}\{[A^{K_\phi}]\phi^{\underline{A}}|\phi \in \Omega\}$, say $a = \underline{b}\phi'^{\underline{A}}$, and let $cf = af$ for some $c \neq a$. Define $C := \{c\} \cup [K_{\phi'}]\underline{b}$, $graph(\phi'^{\underline{C}}) := \{(\underline{b},c)\}$, $graph(\phi^{\underline{C}}) := \emptyset$, else. It is easy to see that $id_C:\underline{C} \to \underline{A}$ is no homomorphism, while $id_C \cdot f$

is one. Hence f is not initial. Similarly, if $\underline{b} \in A^{K_\phi} \setminus \mathrm{dom}\ \phi^{\underline{A}}$, but $(\underline{b}f)\phi^{\underline{B}} = cf$, then define $\underline{C}$ as above, and we get the same result. ∎

"(ii) implies (iii)": Let (ii) be satisfied, let $(N:\underline{x}\phi \stackrel{e}{=} y)$ be given as in (iii) -- $N:= \{y\} \cup [K_\phi]\underline{x}$ --, and let $w:N \to A$ be an assignment such that $\underline{B} \models (N:\underline{x}\phi \stackrel{e}{=} y)[wf]$, i.e. $(\underline{x}wf)\phi^{\underline{B}} = ywf$. Then the assumptions of (ii) are seen to be satisfied, and (ii) implies $\underline{x}w \in \mathrm{dom}\ \phi^{\underline{A}}$ and $[\{ywf\}]f^{-1} = \{yw\}$; therefore necessarily $\underline{x}w\phi^{\underline{A}} = yw$, i.e. $\underline{A} = (N:\underline{x}\phi \stackrel{e}{=} y)[w]$, i.e. f reflects $(N:\underline{x}\phi \stackrel{e}{=} y)$. ∎

"(iii) implies (i)": Assume (iii), and let $g:C \to A$ be any mapping such that gf is a homomorphism $gf:\underline{C} \to \underline{B}$. Then gf preserves all formulas $(N:\underline{x}\phi \stackrel{e}{=} y)$ and f reflects them; hence Lemma 9.1.7.(ii) implies together with Remark 9.1.2.(i) that g is a homomorphism, i.e. f is initial. ∎∎∎

9.2.4 REMARKS. (i): In analogy to the definition of initial homomorphisms one may call homomorphisms $f:\underline{A} \to \underline{B}$ <u>relatively closed</u>, when they satisfy "for every $\phi \in \Omega$ and for every $\underline{a} \in A^{K_\phi}$: $\underline{a}f \in \mathrm{dom}\ \phi^{\underline{B}}$ and $\underline{a}f\phi^{\underline{B}} \in [A]f$ imply $\underline{a} \in \mathrm{dom}\ \phi^{\underline{A}}$ ", since $f':\underline{A} \to \underline{C}$ is closed, where graph(f'):= graph(f) and $\underline{C}$ is the relative subalgebra of $\underline{B}$ with carrier set [A]f. It is easily realized that relatively closed homomorphisms cannot be described by reflection of formulas, since composition of relatively closed homomorphisms need not be relatively closed (cf. 9.1.7.(i)), as the example in Figure 9.1 shows (for a homogeneous example of type $\tau:= (1)$), where f and g are relatively closed -- g is even closed --, but fg is not relatively closed. Notice that all initial homomorphisms are relatively closed.

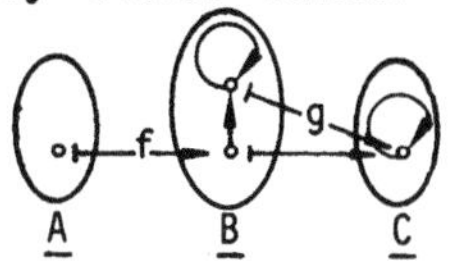

Figure 9.1.

(ii): Proposition 9.2.3 shows that "a homomorphism is injective and full iff it is injective and initial". Thus the injective and full homomorphisms can be defined via reflection of special formulas, namely $(\{x,y\}:x \stackrel{e}{=} y)$ and $(\{y\} \cup [K_\phi]\underline{x}:\underline{x}\phi \stackrel{e}{=} y)$. But full homomorphisms alone cannot be defined by reflection of formulas, since this class is not closed with respect to composition. The class of full and surjective homomorphisms is closed with respect to composition, but we do not know of any set of formulas which could define this class via reflection. However, in the next section we shall see that this latter class is connected with a class defined by the reflection of some formulas through a "factorization system".

9.2.5 DEFINITION. Let P be any set of terms (together with their set of variables), and let $F_P:= \{(M_t:t \stackrel{e}{=} t) \mid X \cap \downarrow t \subseteq M_t \subseteq F(M,\mathbf{TAlg}(\underline{S}))\ ,\ (M_t:t) \in P\}$ be any set of TE-expressions. We say that a homomorphism $f:\underline{A} \to \underline{B}$ is <u>P-reflecting</u> iff f reflects all formulas from F_P.

9.2.6 REMARKS. (i): If in 9.2.5 $P = \emptyset$ or $P \subseteq \{(M:x) \mid x \in M \subseteq X\}$, then P-reflecting homomorphisms are just homomorphisms, but if $P = T_{loc}{}^{\mathbf{c}} := \bigcup\{\{M\} \times F(M,\mathbf{TAlg}(\underline{S})) \mid M \subseteq X\ ,\ \#M < \mathbf{c}\}$ -- $\mathbf{c}$ being taken from 7.1.1 --, or at least if $\{\underline{x}\phi \mid \phi \in \Omega\ ,\ \underline{x}$ an injective sequence in $X^{K_\phi}\} \subseteq P$, then all P-reflecting homomorphisms are closed (therefore we have sometimes already called them "P-closed"). Thus the concept of P-reflecting homomorphisms is a common generalization of the concepts of homomorphisms and of closed homomorphisms.

(ii): If a homomorphism reflects $(M:(t_k \mid k \in K_\phi)\phi \stackrel{e}{=} (t_k \mid k \in K_\phi)\phi)$, then it need not reflect $(M:t_k \stackrel{e}{=} t_k)$ for any $k \in K_\phi$; for instance consider $\tau := (1)$, $\Omega := \{\phi\}$, $A := \{a\}$, $B := \{b,b'\}$, $graph(\phi^{\underline{A}}) := \emptyset$, $graph(\phi^{\underline{B}}) := \{((b),b')\}$, and $graph(f) := \{(a,b)\}$. Then f trivially reflects $(\{x\}:x\phi\phi \stackrel{e}{=} x\phi\phi)$, but it does not reflect $(\{x\}:x\phi \stackrel{e}{=} x\phi)$. Thus, if we want P to include all subterms of terms occurring in P (where the sets of variables are not changed!), then we say that we consider $\downarrow$P-reflecting homomorphisms. This distinction will gain some importance in later chapters, since $\downarrow$P-reflecting homomorphisms have the "nicer" theory.

In a similar way as P-reflecting homomorphisms generalize closed homomorphisms and homomorphisms, initial and "relatively normal" homomorphisms can be generalized as follows:

9.2.7 DEFINITION. Let P be any set of terms,
$\mathfrak{C}_P := \{(M \cup \{y\}:t \stackrel{e}{=} y) \mid (M:t) \in P,\ y \in X \setminus fvar(t),\ t\tilde{v}_X = yv_X\}$. Then a homomorphism $f:\underline{A} \to \underline{B}$ is called <u>P-initial</u> (or <u>P-normal</u>) iff f reflects all formulas from $\mathfrak{C}_P$. If $P = T_{loc}{}^{\mathbf{c}}$ (see 9.2.6.(i)), then we call a P-normal homomorphism <u>relatively normal</u>.

9.2.8 REMARK. It is easy to see from 9.2.3 that initial homomorphisms are P-initial (= P-normal) ones, when we choose
$P := \{(\{y\} \cup [K_\phi]\underline{x}:\underline{x}\phi \stackrel{e}{=} y) \mid y \in X \setminus [K_\phi]\underline{x},\ \underline{x}\phi\tilde{v}_X = yv_X,\ \underline{x} \in X^{K_\phi}$ respecting the sort mappings, $\phi \in \Omega\}$.
Recalling the concept of normal relative subalgebras from 5.3.2, where we have met them in connection with free completions, we get the following motivation for the notation "relatively normal":

9.2.9 PROPOSITION. Let $\underline{A}$ be a relative subalgebra of $\underline{B}$ such that A generates $\underline{B}$. Then $\underline{B}$ is a normal extension of $\underline{A}$ iff $id_A:\underline{A} \to \underline{B}$ is a relatively normal homomorphism.

PROOF. First assume that $\underline{B}$ is an A-generated normal extension of $\underline{A}$, and let $\Sigma = (M \cup \{y\}:t \stackrel{e}{=} y)$ be any of the formulas considered in 9.2.7. If $t \in X$, then Σ is reflected by id_A, since id_A is injective. Thus let us assume that $t = (t_k \mid k \in K_\phi)\phi$, and also that $\{(M \cup \{z_k\}:t_k \stackrel{e}{=} z_k) \mid k \in K_\phi\}$ is reflected by id_A, where we assume the z_k to be mutually distinct and not contained in $M \cup \{y\}$. Let $w:M \cup \{y\} \to A$ be any assignment such that $\underline{B} \models (M \cup \{y\}:t \stackrel{e}{=} y)[w \cdot id_A]$. Then this is only possible, when $(w|M)(M:t)^{\underline{B}} \in A$, i.e. $(w|M)((M:t_k)^{\underline{B}} \mid k \in K_\phi)\phi^{\underline{B}} \in A$. Thus normality of $\underline{A}$ in $\underline{B}$ implies $(w|M)(M:t_k)^{\underline{B}} \in \underline{A}$ $(k \in K_\phi)$, and $(w|M)((M:t_k)^{\underline{B}} \mid k \in K_\phi)\phi^{\underline{B}} = (w|M)(M:t)^{\underline{A}} = yw$, i.e. id_A reflects Σ. ∎
Conversely, let id_A be relatively normal and $\mathbf{C}_B A = B$. Let $\underline{b} \in dom\ \phi^{\underline{B}}$, $\underline{b}\phi^{\underline{B}} =: a \in A$. By assumption we can find sets $N \subseteq A$, $M \subseteq X$, $\#M, \#N < \mathbf{c}$, a bijection $w:M \to N$ and terms $t_k \in F(M,\mathbf{TAlg}(\underline{S}))$ such that $(k)\underline{b} = w(M:t_k)^{\underline{B}}$ $(k \in K_\phi)$. Hence we get for $t := (t_k \mid k \in K_\phi)\phi$ and $w':M \cup \{y\} \to A$ with $w'|M := w$, $yw' := a$ (where $y \in X \setminus M$, $yv_X = t\tilde{v}_X$) that $\underline{B} \models (M \cup \{y\}:t \stackrel{e}{=} y)[w' \cdot id_A]$, and therefore $\underline{A} \models (M \cup \{y\}:t \stackrel{e}{=} y)[w']$, i.e. $w(M:t)^{\underline{A}} = yw' = a = (w(M:t_k)^{\underline{A}} \mid k \in K_\phi)\phi^{\underline{A}}$. Thus $w(M:t_k)^{\underline{A}} = w(M:t_k)^{\underline{B}} = b_k \in A$ for all $k \in K_\phi$ and $\underline{b}(M:t)^{\underline{B}} = \underline{b}(M:t)^{\underline{A}}$, i.e. $\underline{B}$ is a normal extension of $\underline{A}$. ∎∎∎

9.2.10. It is still an open **problem** for us, whether one can describe normality of an extension in general by reflection of formulas. -- It should be observed that a relatively normal homomorphism is always injective.

Let us now recall the following definition:

9.2.11 DEFINITION. A mapping $f:A \to B$ is an <u>elementary embedding</u> of $\underline{A}$ into $\underline{B}$ iff, for every L-formula $(M:\Phi)$ and for every assignment $w:M \to A$, one has $\underline{A} \models (M:\Phi)[w]$ iff $\underline{B} \models (M:\Phi)[wf]$.
f is an <u>L-pure embedding</u> of $\underline{A}$ into $\underline{B}$ if the above is true whenever Φ is of the form $\Phi = \exists^* Y(\Lambda^* F)$, where F is a set of E-equations, $\#F < \mathbf{c}$, $\#Y < \mathbf{q}$.
Hence we have because of 9.1.3:

9.2.12 PROPOSITION. Let $\underline{A},\underline{B} \in \mathbf{Alg}(\underline{S})$, and let $f:A \to B$ be any mapping. Then the following statements are equivalent:

(i) f is an elementary embedding of $\underline{A}$ into $\underline{B}$.

(ii) f preserves every L-formula $(M:\Phi)$ with respect to $\underline{A}$ and $\underline{B}$.

(iii) f reflects every L-formula $(M:\Phi)$ with respect to $\underline{A}$ and $\underline{B}$.

If we have a finitary language, then it is well known that these are equivalent to

(iv) For every formula $(\{x\}:\Phi)$ (i.e. $\#fvar(\Phi) \le 1$) f reflects $(\emptyset:\exists x\Phi)$ with respect to $\underline{A}$ and $\underline{B}$. ■■■

9.2.13. A similar result with respect to (i), (ii) and (iii) above can be formulated for L-pure embeddings.

9.2.14. There exists an approach by Z.PAWLAK (cf. [BtRaSk 75]) to describe computers basically as partial algebras with one unary partial operation, i.e. as $(A;\phi^{\underline{A}})$. The morphisms used in this connection are called "machine morphisms" and are defined as homomorphisms $f:\underline{A} \to \underline{B}$ such that $a \notin \mathrm{dom}\,\phi^{\underline{A}}$ with $a \in [A]\phi$ implies $af \notin \mathrm{dom}\,\phi^{\underline{B}}$, i.e. f has to preserve the formulas $(\{x,y\}:x\phi \overset{e}{=} y \wedge \neg(y\phi \overset{e}{=} y\phi))$ or, equivalently, f has to reflect the formulas $(\{x,y\}:x\phi \overset{e}{=} y \Rightarrow y\phi \overset{e}{=} y\phi)$.

No.	Property of the homomorphism	Type of the reflected formula
1.	injective (monomorphism)	$x \overset{e}{=} y$ $(x,y \in X,\ xv_X = yv_X)$
2.	closed	$\underline{x}\phi \overset{e}{=} \underline{x}\phi$ ($\phi \in \Omega$, $\underline{x} \in X^{K_\phi}$ injective, $\underline{x}\cdot v_X = v_{K_\phi}$)
	closed	$t \overset{e}{=} t$ (t arbitrary term)
3.	injective and closed	$x \overset{e}{=} y$, $\underline{x}\phi \overset{e}{=} \underline{x}\phi$ (see 1. and 2.)
4.	initial	$\underline{x}\phi \overset{e}{=} y$ (see 2., $y \notin [K_\phi]\underline{x}$)
5.	injective and initial (inj.& full)	$x \overset{e}{=} y$, $\underline{x}\phi \overset{e}{=} y$ (see 1. and 4.)
6.	relatively normal	$t \overset{e}{=} y$ (t arbitrary term, $y \notin fvar(t)$)
7.	P-reflecting	$(M:t \overset{e}{=} t)$ (for all $(M:t) \in P$; $P \subseteq T_{loc}$)
8.	P-initial = P-normal	$(M \cup \{y\}:t \overset{e}{=} y)$ (for all $(M:t) \in P$, $y \notin fvar(t)$)
9.	Pawlak's machine morphism	$x\phi \overset{e}{=} y \Rightarrow y\phi \overset{e}{=} y\phi$
10.	elementary embedding	$(M:\Phi)$ (for all L-formulas)
11.	L-pure embedding	$(M:\exists^* Y(\Lambda^* F))$ ($F \subseteq Eeq^*_X$, $\#F < \mathbf{c}$, $\#Y < \mathbf{q}$)

Some properties of homomorphisms definable by reflection of formulas:
Table 9.1. (see 9.2.15)

9.2.15 TABLE. We collect in Table 9.1 the examples obtained so far about properties of homomorphisms definable by reflection of formulas: If the set of variables is just the set of free variables of the body of the formula, this set is omitted. Arbitrary combinations of these properties are possible.-- Recall that $f:\underline{A} \to \underline{B}$ is a homomorphism, iff f preserves all E-equations (atomic formulas), iff f preserves all formulas $([K_\phi]\underline{x} \cup \{y\}:\underline{x}\phi \stackrel{e}{=} y)$, where $\underline{x} \in X^{K_\phi}$ is an injective sequence compatible with the sort mappings, $y \in X \setminus [K_\phi]\underline{x}$, $yv_X = (\underline{x}\phi)\tilde{v}_X$.

9.2.16 OBSERVATIONS concerning additional concepts.

(i): We have already observed that fullness cannot be described by reflection of formulas, nor do we know sets of formulas describing surjectivity, finality and related properties by their reflection. For some properties we shall find tools related to reflection in the next section in connection with factorization systems, but fullness will then only be describable by some process of "relativization". -- Nevertheless one can find concepts describable in a model theoretic way to characterize the properties mentioned above:

We say that a mapping $f:A \to B$ <u>weakly reflects</u> a formula $(M:\Phi)$ with respect to $\underline{A}$ and $\underline{B}$ iff, for every assignment $w:M \to A$, $\underline{B} \models (M:\Phi)[wf]$ implies the existence of an assignment $w':M \to A$ satisfying $w'f = wf$ and $\underline{A} \models (M:\Phi)[w']$.

And we say that f <u>retracts</u> $(M:\Phi)$ with respect to $\underline{A}$ and $\underline{B}$, iff for every assignment $u:M \to B$ (!), for which $\underline{B} \models (M:\Phi)[u]$, there exists an assignment $u':M \to A$ such that $u = u'f$ and $\underline{A} \models (M:\Phi)[u']$.

(ii): It is now easy to realize that a homomorphism $f:\underline{A} \to \underline{B}$ is **full**, iff f weakly reflects all formulas of the form $(\{y\} \cup [K_\phi]\underline{x}:\underline{x}\phi \stackrel{e}{=} y)$ $(\phi \in \Omega)$, where $y \in X \setminus [K_\phi]\underline{x}$, $yv_X = (\underline{x}\phi)\tilde{v}_X$, and $\underline{x}$ is an injective K_ϕ-sequence in X respecting the sort mappings.

Thus we may already conclude that properties of homomorphisms defined by weak reflection are in general not inherited by the composition of such homomorphisms.

(iii): The simplest example for retraction is the class of surjective homomorphisms, since obviously a mapping $f:A \to B$ is surjective iff it retracts $(\{x\}:x \stackrel{e}{=} x)$ for all $x \in X$. -- Similarly one easily sees that $f:\underline{A} \to \underline{B}$ is a full and surjective homomorphism iff f retracts all formulas of the form $(\{x\}:x \stackrel{e}{=} x)$ and $([K_\phi]\underline{x}:\underline{x}\phi \stackrel{e}{=} \underline{x}\phi)$, $x \in X$ and $\underline{x}$ an injective K_ϕ-sequence respecting the sort mappings. -- In addition **final** homomorphisms can be characterized by retraction, since (as we mention without proof) for $f:\underline{A} \to \underline{B}$ the following statements are equivalent:

(a) f is final.

(b) f is full, and for all $\phi \in \Omega$ and $\underline{b} \in B^{K_\phi}$, $\underline{b} \in \mathrm{dom}\, \phi^{\underline{B}}$ implies $\{\underline{b}\phi^{\underline{B}}\} \cup [K_\phi]\underline{b} \subseteq [A]f$.

(c) f retracts all formulas of the form $([K_\phi]\underline{x}:\underline{x}\phi \stackrel{e}{=} \underline{x}\phi)$, where $\phi \in \Omega$ and $\underline{x}$ is any injective, sort respecting K_ϕ-sequence in X.

-- Hence a homomorphism is full and surjective, iff it is final and surjective.

(iv): Finally let us show that the properties of full homomorphisms formulated in 2.4.5.(iii) follow from properties of the concept of weak reflection, which always tacitly refer to the given structures:

Let $\underline{A},\underline{B},\underline{C} \in \mathbf{Alg}(\underline{S})$, $f:A \to B$ and $g:B \to C$ be mappings (compatible with the sort mappings), and let $\Sigma = (M:\Phi)$ be some L-formula. Then we have

(a) If f is surjective and weakly reflects Σ, then f retracts Σ.

(b) If f and g both weakly reflect Σ, and if f is surjective or g is injective, then fg weakly reflects Σ.

(c) If fg weakly reflects Σ, and if g is injective and preserves Σ, then f weakly reflects Σ.

(d) If fg weakly reflects Σ, and if f is surjective and preserves Σ, then g weakly reflects Σ.

PROOF. (a) follows directly from the definitions in (i) if f is surjective. ∎

Ad (b): Let f and g both weakly reflect Σ, and let $w:M \to A$ be any assignment such that $\underline{C} \models (M:\Phi)[wfg]$. If g is injective, then $\underline{B} \models (M:\Phi)[wf]$, since g weakly reflects Σ, and there is no assignment $u:M \to B$ different from wf satisfying $wfg = ug$. Since f weakly reflects Σ, there now exists $v:M \to A$ satisfying $vf = wf$ (and $vfg = wfg$) and $\underline{A} \models (M:\Phi)[v]$. -- If g is not injective, but f is surjective, then we first have $u:M \to B$ satisfying $ug = wfg$ and $\underline{B} \models (M:\Phi)[u]$, and since f is surjective and weakly reflects Σ, we have $v:M \to A$ such that $vf = u = wf$ -- whence $vfg = wfg$ -- and v can be chosen such that $\underline{A} \models (M:\Phi)[v]$. ∎

Ad (c): Let $w:M \to A$ be such that $\underline{B} \models (M:\Phi)[wf]$. Hence -- g preserving Σ -- $\underline{C} \models (M:\Phi)[wfg]$. Since fg weakly reflects Σ, we have $w':M \to A$ such that $w'fg = wfg$ and $\underline{A} \models (M:\Phi)[w']$. Since g is injective, we may conclude $wf = w'f$, showing that f weakly reflects Σ. ∎

Ad (d): Let $u:M \to B$ satisfy $\underline{C} \models (M:\Phi)[ug]$. Since f is surjective, there is $w:M \to A$ with $wf = u$. Since fg weakly reflects Σ, we have $w':M \to A$ satisfying $w'fg = wfg$ and $\underline{A} \models (M:\Phi)[w']$. Since f preserves Σ, we get $\underline{B} \models (M:\Phi)[w'f]$, showing that g weakly reflects Σ. ∎∎∎

10 Factorization systems of partial algebras

As we have already indicated in sections 8 and 9, the "diagonal-fill-in property" (Difip) (cf. 8.4.2) has a fundamental importance in connection with the reflection of elementary implications by homomorphisms and with algebraic operators connected with elementary implications. Since elementary implications are encoded by epimorphisms, the diagonal-fill-in is even unique in all cases considered so far, and this allows factorization systems to enter the scene, as we shall see below. Actually we shall not meet examples, where the diagonal-fill-in will not be unique; therefore we shall treat in this presentation only the unique case, although the more general case in many instances yields the same results.

We start this section in a more general category theoretical context, which will show that the concepts treated here can also be applied in greater generality, say in "nice" subclasses of **Alg**($\underline{S}$), especially in so-called "full (epireflective) subcategories". Actually the concept of factorization systems was to our knowledge first considered in more detail in general topology, since, owing to lack of more than one interesting factorization system , this concept was not then of much interest in the study of total algebras.

For partial algebras the situation is different because of the wealth of interesting factorization systems, as we shall see below. The general theory of partial algebras much resembles the theory of (Hausdorff) point set topology, while total algebras correspond to compact topological spaces (e.g. the free completion $\underline{F}(\underline{A},\mathbf{TAlg}(\underline{S}))$ of $\underline{A}$ is comparable to the Stone-Čech-compactification of a Hausdorff space).

10.1 The Galois connection induced by the unique-diagonal-fill-in property

10.1.1. Without any significant changes Definition 8.4.2 can be applied to arbitrary categories $\underline{\mathbf{C}} = (\mathrm{Ob}(\underline{\mathbf{C}}) \cup \mathrm{Mor}(\underline{\mathbf{C}});\cdot,\mathrm{Dom},\mathrm{Cod},\mathbf{1})$, we do not repeat it here. Then the operators Λ and Λ^{op} defined in 8.4.2 have the following properties (cf. G.E.STRECKER [S 72]):

10.1.2 THEOREM. Let $\mathbf{H} \subseteq \mathrm{Mor}(\underline{\mathbf{C}})$ be any class of morphisms of some given category $\underline{\mathbf{C}}$ (say $\mathrm{Ob}(\underline{\mathbf{C}}) = \mathbf{Alg}(\underline{S})$, $\mathrm{Mor}(\underline{\mathbf{C}}) = \mathrm{Hom}(\mathbf{Alg}(\underline{S}))$). Then:

(i) $\Lambda(\mathbf{H})$ contains all isomorphisms from $\underline{\mathbf{C}}$.

(ii) $\Lambda(\mathbf{H})\cdot\Lambda(\mathbf{H}) := \{h\cdot h' \mid h,h' \in \Lambda(\mathbf{H})\} \subseteq \Lambda(\mathbf{H})$, i.e. $\Lambda(\mathbf{H})$ is closed with respect to composition.

(iii) $\Lambda(\mathbf{H}) \cap \mathbf{H} \subseteq \mathrm{Iso}(\underline{\mathbf{C}})$, i.e. $\Lambda(\mathbf{H})$ and $\mathbf{H}$ have only isomorphisms in common; and if $\mathrm{Iso}(\underline{\mathbf{C}}) \subseteq \mathbf{H}$, then $\Lambda(\mathbf{H}) \cap \mathbf{H} = \mathrm{Iso}(\underline{\mathbf{C}})$.

(iv) If $\mathbf{H'} \subseteq \mathbf{H}$, then $\Lambda(\mathbf{H}) \subseteq \Lambda(\mathbf{H'})$, $\mathbf{H} \subseteq \Lambda^{op}\Lambda(\mathbf{H})$, and $\Lambda\Lambda^{op}\Lambda(\mathbf{H}) = \Lambda(\mathbf{H})$.

(v) If $f\cdot g \in \Lambda(\mathbf{H})$, and if g is a monomorphism (i.e. if $g \in \mathrm{Mono}(\underline{\mathbf{C}})$), then $f \in \Lambda(\mathbf{H})$, i.e. $\Lambda(\mathbf{H})$ is "<u>right cancellable with respect to monomorphisms</u>".

(vi) If $\mathbf{H} \subseteq \mathrm{Epi}(\underline{\mathbf{C}})$ is a class of epimorphisms of $\underline{\mathbf{C}}$, then $\Lambda(\mathbf{H})$ is right cancellable: $f\cdot g \in \Lambda(\mathbf{H})$ implies $f \in \Lambda(\mathbf{H})$.

(vii) $\Lambda(\mathbf{H})$ is closed with respect to pullbacks, i.e. if $(P;p,q)$ is a pullback of $(f,g;C)$ (in particular $p\cdot f = q\cdot g$, cf. 4.2.2) with $f \in \Lambda(\mathbf{H})$, then $q \in \Lambda(\mathbf{H})$.

(viii) $\Lambda(\mathbf{H})$ is closed with respect to multiple pullbacks, i.e. if $(P;(p_i \mid i \in I))$ is a multiple pullback of $((f_i \mid i \in I);C)$ (cf. 4.2.2), and if for some $i' \in I$ $\{f_i \mid i \in I \setminus \{i'\}\} \subseteq \Lambda(\mathbf{H})$, then $p_{i'} \in \Lambda(\mathbf{H})$ (#$I \geq 2$, I a set) -- in particular: if $\{f_i \mid i \in I\} \subseteq \Lambda(\mathbf{H})$, then $\{p_i \mid i \in I\} \subseteq \Lambda(\mathbf{H})$.

(ix) $\Lambda(\mathbf{H})$ is closed with respect to products, i.e. if $(f_i : A_i \to B_i \mid i \in I)$ is a family of $\underline{\mathbf{C}}$-morphisms (where I is any set), if $\{f_i \mid i \in I\} \subseteq \Lambda(\mathbf{H})$, and if $f : \mathbf{X}(A_i \mid i \in I) \to \mathbf{X}(B_i \mid i \in I)$ is -- in the case when these products exist -- the induced product morphism, then $f \in \Lambda(\mathbf{H})$.

(x) If $\mathbf{H} \subseteq \mathrm{Epi}(\underline{\mathbf{C}})$, then $\Lambda(\mathbf{H})$ is closed with respect to induced product morphisms with respect to non-empty index sets; i.e. if $(f_i : B \to A_i \mid i \in I)$ is a source of $\underline{\mathbf{C}}$-morphisms (I a non-empty set) such that $\{f_i \mid i \in I\} \subseteq \Lambda(\mathbf{H})$, if the product $(\mathbf{X}(A_i \mid i \in I);(p_i \mid i \in I))$ exists and $f : B \to \mathbf{X}(A_i \mid i \in I)$ is the induced product morphism, then $f \in \Lambda(\mathbf{H})$.

Dually one has for $\mathbf{S} \subseteq \mathrm{Mor}(\underline{\mathbf{C}})$:

$(i)^{op}$ $\Lambda^{op}(\mathbf{S})$ contains all isomorphisms from $\underline{\mathbf{C}}$.

$(ii)^{op}$ $\Lambda^{op}(\mathbf{S})\cdot\Lambda^{op}(\mathbf{S}) \subseteq \Lambda^{op}(\mathbf{S})$.

$(iii)^{op}$ $\Lambda^{op}(\mathbf{S}) \cap \mathbf{S} \subseteq \mathrm{Iso}(\underline{\mathbf{C}})$; and if $\mathrm{Iso}(\underline{\mathbf{C}}) \subseteq \mathbf{S}$, then $\Lambda^{op}(\mathbf{S}) \cap \mathbf{S} = \mathrm{Iso}(\underline{\mathbf{C}})$.

$(iv)^{op}$ If $\mathbf{S'} \subseteq \mathbf{S}$, then $\Lambda^{op}(\mathbf{S}) \subseteq \Lambda^{op}(\mathbf{S'})$, $\mathbf{S} \subseteq \Lambda\Lambda^{op}(\mathbf{S})$, and $\Lambda^{op}\Lambda\Lambda^{op}(\mathbf{S}) = \Lambda^{op}(\mathbf{S})$.

$(v)^{op}$ If $fg \in \Lambda^{op}(\mathbf{S})$, and if f is an epimorphism, then $g \in \Lambda^{op}(\mathbf{S})$, i.e. $\Lambda^{op}(\mathbf{S})$ is "<u>left cancellable with respect to epimorphisms</u>".

$(vi)^{op}$ If $\mathbf{S} \subseteq \mathrm{Mono}(\underline{\mathbf{C}})$, then $\Lambda^{op}(\mathbf{S})$ is left cancellable: $f\cdot g \in \Lambda^{op}(\mathbf{S})$ implies $g \in \Lambda^{op}(\mathbf{S})$.

$(vii)^{op}$ $\Lambda^{op}(\mathbf{S})$ is closed with respect to pushouts.

(viii)^{op} $\Lambda^{op}(\mathbf{S})$ is closed with respect to multiple pushouts.

(ix)^{op} $\Lambda^{op}(\mathbf{S})$ is closed with respect to coproducts.

(x)^{op} If $\mathbf{S} \subseteq \text{Mono}(\underline{\mathbf{C}})$, then $\Lambda^{op}(\mathbf{S})$ is closed with respect to induced coproduct morphisms with respect to non-empty index sets.

Because of the importance of these concepts and results in later sections (since we cannot assume the reader to be much familiar with category theoretical argumentation, and since we do not know any place containing detailed proofs of these statements) we sketch the proofs of these facts in connection with the operator Λ, the dual statements then just follow by "reversing the arrows". The reader is advised to draw in each case the corresponding diagram for which we do not have enough space (cf. Fig. 8.4).

PROOF. Ad (i): Since isomorphisms are invertible, the statement is trivial. ∎

Ad (ii): Let $s, s' \in \Lambda(\mathbf{H})$ such that $s \cdot s'$ exists. Let $h \in \mathbf{H}$, $p, q \in \text{Mor}(\underline{\mathbf{C}})$ such that $h \cdot q = p \cdot s \cdot s'$, then we first have a unique $d' \in \text{Mor}(\underline{\mathbf{C}})$ such that $h \cdot d' = p \cdot s$ and $d' \cdot s' = q$, since $s' \in \Lambda(\mathbf{H})$; and next we have a unique $d \in \text{Mor}(\underline{\mathbf{C}})$ such that $h \cdot d = p$ and $d \cdot s = d'$; therefore $d \cdot s \cdot s' = d' \cdot s' = q$, showing that we have $\text{Difip}(h, s \cdot s')$, i.e. $s \cdot s' \in \Lambda(\mathbf{H})$. ∎

Ad (iii): $f \in \Lambda(\mathbf{H}) \cap \mathbf{H}$ implies $\text{Difip}(f,f)$, and choosing in Figure 8.4 $p := \mathbf{1}$ and $q := \mathbf{1}$ (the identities for the corresponding objects), then the diagonal-fill-in d satisfies $f \cdot d = \mathbf{1}$ and $d \cdot f = \mathbf{1}$, i.e. $f \in \text{Iso}(\underline{\mathbf{C}})$. If $\text{Iso}(\underline{\mathbf{C}}) \subseteq \mathbf{H}$, then, by (i), $\mathbf{H} \cap \Lambda(\mathbf{H}) = \text{Iso}(\underline{\mathbf{C}})$. ∎

Ad (iv): These are just properties of the closure operators related to a Galois connection. ∎

Ad (v)/(vi): Assume $h \in \mathbf{H}$, $f \cdot g \in \Lambda(\mathbf{H})$, and let morphisms p, q be given such that $p \cdot f = h \cdot q$. Since then also $p \cdot f \cdot g = h \cdot q \cdot g$, we have a unique d satisfying $h \cdot d = p$ and $d \cdot f \cdot g = q \cdot g$. If g is a monomorphism, then the last equation implies $d \cdot f = q$; if h is an epimorphism, then $h \cdot q = p \cdot f = h \cdot d \cdot f$ implies $q = d \cdot f$, i.e. $\text{Difip}(h,f)$ in both cases. ∎

Ad (vii)/(viii): Let $(P;(p_i|i \in I))$ be a multiple pullback of $((f_i|i \in I);C)$, and assume for some $i' \in I$ that $\{f_i|i \in I \setminus \{i'\}\} \subseteq \Lambda(\mathbf{H})$ $(\#I \geq 2$, I a set); moreover, let $h \in \mathbf{H}$, and let p and q be morphisms, such that $h \cdot q = p \cdot p_{i'}$. Then we have for each $j \in I \setminus \{i'\}$ that $h \cdot q \cdot f_{i'} = p \cdot p_{i'} \cdot f_{i'} = p \cdot p_j \cdot f_j$. Thus $(f_j \in \Lambda(\mathbf{H}))$ there is a unique morphism d_j satisfying $h \cdot d_j = p \cdot p_j$ and $d_j \cdot f_j = q \cdot f_{i'}$. Since $(P;(p_i|i \in I))$ is a multiple pullback, there is a unique d such that $d \cdot p_j = d_j$ and $d \cdot p_{i'} = q$ $(j \in I \setminus \{i'\})$. Since also $h \cdot d_j = p \cdot p_j = h \cdot d \cdot p_j$ $(j \in I \setminus \{i'\})$ and $h \cdot q = p \cdot p_{i'} = h \cdot d \cdot p_{i'}$, the morphisms p and $h \cdot d$ can be considered as the unique morphism induced by the family $(h \cdot d_j | j \in I)$, where $d_{i'} := q$. Thus $h \cdot d = p$, and we have $\text{Difip}(h, p_{i'})$. ∎

Ad (ix): The argumentation is almost identical with the one for (viii); therefore we leave it as an exercise. ∎

Ad (x): Assume $\mathbf{H} \subseteq \text{Epi}(\underline{\mathbf{C}})$, and let $(f_i: A \to B_i | i \in I)$ be a source of $\Lambda(\mathbf{H})$-morphisms, where I is any set, let the product $(B := \mathbf{X}(B_i|i \in I);(q_i|i \in I))$ exist, and let $f: A \to B$ be the induced product morphism; moreover, let $h \in \mathbf{H}$, an let p and q be morphisms, such that $h \cdot q = p \cdot f$. Then $h \cdot q \cdot q_i = p \cdot f \cdot q_i = p \cdot f_i$ for each $i \in I$. Since each f_i belongs to $\Lambda(\mathbf{H})$, we get for each $i \in I$ a unique morphism d_i such that $h \cdot d_i = p$ and $d_i \cdot f_i = q \cdot q_i = d_i \cdot f \cdot q_i$. Since h is an epimorphism, the first family of equations implies $d_i = d_j$ for all $i,j \in I$, and since $(B;(q_i|i \in I))$ is a product, the last equation of the second family implies $q = d\cdot f$; thus: Difip(h,f). ∎∎∎

Before we discuss examples in **Alg**($\underline{S}$), we first investigate the relationship of "Difip" to factorization systems.

10.2 Factorization systems

10.2.1 DEFINITION. Let $\underline{\mathbf{C}}$ be any category, and let $\mathbf{H},\mathbf{S} \subseteq Mor(\underline{\mathbf{C}})$ be any classes of morphisms. Then one says that $(\mathbf{H},\mathbf{S})$ is a factorization system -- or that $\underline{\mathbf{C}}$ is a bicategory with respect to $\mathbf{H}$ and $\mathbf{S}$ --, iff the following axioms hold:

(FS1) $\mathbf{H}\cdot\mathbf{S} = Mor(\underline{\mathbf{C}})$, i.e. each $f \in Mor(\underline{\mathbf{C}})$ has an $(\mathbf{H},\mathbf{S})$-factorization $(h,s) \in \mathbf{H} \times \mathbf{S}$ such that $h\cdot s = f$.

(FS2) $Iso(\underline{\mathbf{C}}) \subseteq \mathbf{H} \cap \mathbf{S}$.

(FS3) $\mathbf{H}\cdot\mathbf{H} \subseteq \mathbf{H}$ and $\mathbf{S}\cdot\mathbf{S} \subseteq \mathbf{S}$.

(FS4) The $(\mathbf{H},\mathbf{S})$-factorization of any $\underline{\mathbf{C}}$-morphism is unique up to unique isomorphism, i.e. whenever there are $h,h' \in \mathbf{H}$ and $s,s' \in \mathbf{S}$ such that $h\cdot s = h'\cdot s'$, then there exists a **unique isomorphism** $d \in Iso(\underline{\mathbf{C}})$ such that $h\cdot d = h'$ and $d\cdot s' = s$.

If $(\mathbf{H},\mathbf{S})$ is a factorization system, then we call $\mathbf{H}$ its left factor and $\mathbf{S}$ its right factor.

10.2.2 PROPOSITION. Let $(\mathbf{H},\mathbf{S})$ be a factorization system in the category $\underline{\mathbf{C}}$. Then:

(i) $\mathbf{H} \cap \mathbf{S} = Iso(\underline{\mathbf{C}})$.

(ii) $\mathbf{H}\cdot\mathbf{H} = \mathbf{H}$ and $\mathbf{S}\cdot\mathbf{S} = \mathbf{S}$.

(iii) If $p,p':A \to B$ are $\underline{\mathbf{C}}$-morphisms such that $h\cdot p = h\cdot p'$ and $p\cdot s = p'\cdot s$ for some $h \in \mathbf{H}$ and $s \in \mathbf{S}$, then $p = p'$.

(iv) $\mathbf{H} \times \mathbf{S} \subseteq$ Difip.

(v) If $h\cdot h', h \in \mathbf{H}$, then $h' \in \mathbf{H}$; if $s\cdot s', s' \in \mathbf{S}$, then $s \in \mathbf{S}$.

PROOF. Ad (i): Let $f:A \to B$, $f \in \mathbf{H} \cap \mathbf{S}$. Since $Iso(\underline{\mathbf{C}}) \subseteq \mathbf{H} \cap \mathbf{S}$ and $1_A, 1_B \in Iso(\underline{\mathbf{C}})$, $(1_A,f)$ and $(f,1_B)$ are $(\mathbf{H},\mathbf{S})$-factorizations of f; thus there exists by (FS4) a unique isomorphism $i:B \to A$ such that $f\cdot i = 1_A$ and $i\cdot f = 1_B$, i.e. f is an isomorphism. Combined with (FS2) we get $Iso(\underline{\mathbf{C}}) = \mathbf{H} \cap \mathbf{S}$. ∎

Ad (ii): Because of (FS1) and (FS2), (FS3) implies $\mathbf{H} \subseteq \mathbf{H}\cdot Iso(\underline{\mathbf{C}}) \subseteq \mathbf{H}\cdot\mathbf{H} \subseteq \mathbf{H}$, and the same for S; therefore (ii) follows. ∎

Ad (iii): Let the assumptions of (iii) be satisfied, and let (h^*,s^*) and (h',s') be $(\mathbf{H},\mathbf{S})$-factorizations of p respectively p'. By (FS3) $h\cdot h^*, h\cdot h' \in \mathbf{H}$ and $s^*\cdot s, s'\cdot s \in \mathbf{S}$, and the assumptions imply $(h\cdot h^*)\cdot s^* = (h\cdot h')\cdot s'$, $h^*\cdot(s^*\cdot s) = h'\cdot(s'\cdot s)$ and $(h\cdot h^*)\cdot(s^*\cdot s) = (h\cdot h')\cdot(s'\cdot s)$; hence (FS4) implies the existence of unique isomorphisms d, d' and d" such that

(a) $(h\cdot h^*)\cdot d = h\cdot h'$ and $d\cdot s' = s^*$;

(b) $h^*\cdot d' = h'$ and $d'\cdot(s'\cdot s) = s^*\cdot s$;

(c) $(h\cdot h^*)\cdot d'' = h\cdot h'$ and $d''\cdot(s'\cdot s) = s^*\cdot s$.

From (a) and (b) we get

(d) $(h\cdot h^*)\cdot d = h\cdot h'$ and $d\cdot(s'\cdot s) = s^*\cdot s$ as well as

(e) $(h\cdot h^*)\cdot d' = h\cdot h'$ and $d'\cdot(s'\cdot s) = s^*\cdot s$.
The uniqueness of the isomorphisms in each case yields that (d) and (c) imply $d'' = d$, while (e) and (c) imply $d' = d''$. Hence (a) and (b) imply $d\cdot s' = s^*$ and $h^*\cdot d = h'$, thus $p = h^*\cdot s^* = (h'\cdot d^{-1})\cdot(d\cdot s') = h'\cdot s' = p'$. ∎
Ad (iv): Assume $h \in \mathbf{H}$, $s \in \mathbf{S}$, let $p,q \in \mathrm{Mor}(\underline{\mathbf{C}})$ such that $h\cdot q = p\cdot s$, and let (h',s') and (h'',s'') be $(\mathbf{H},\mathbf{S})$-factorizations of p respectively q. Since now $(h\cdot h'')\cdot s'' = h'\cdot(s'\cdot s)$, (FS3) and (FS4) imply the existence of a unique isomorphism i satisfying $(h\cdot h'')\cdot i = h'$ and $i\cdot(s'\cdot s) = s''$. Define $d := h''\cdot i\cdot s'$; then $h\cdot d = h\cdot h''\cdot i\cdot s' = h'\cdot s' = p$ and $d\cdot s = h''\cdot i\cdot s'\cdot s = h''\cdot s'' = q$, i.e. d is a diagonal-fill-in for h and s. If we have another diagonal-fill-in d' satisfying $h\cdot d' = p$ and $d'\cdot s = q$, then (iii) implies $d = d'$, showing the uniqueness, i.e. Difip(h,s). ∎
Ad (v): (v) follows from (iii) (cf. 10.1.2.(v), $(v)^{op}$) and is left as an exercise. ∎∎∎

10.2.3 REMARK. 10.2.2.(iii) shows that even without requiring $\mathbf{H} \subseteq \mathrm{Epi}(\underline{\mathbf{C}})$ or $\mathbf{S} \subseteq \mathrm{Mono}(\underline{\mathbf{C}})$ -- which would immediately imply the uniqueness of the diagonal-fill-in in 10.2.2.(iv) -- this uniqueness can be proved, when $(\mathbf{H},\mathbf{S})$ is a factorization system. Moreover, the pairs $(h,s) \in \mathbf{H} \times \mathbf{S}$ share properties of epimorphisms and monomorphisms. In this book we shall have in most cases $\mathbf{H} \subseteq \mathrm{Epi}(\mathbf{Alg}(\underline{S}))$, but the factorization system (final homomorphisms, bijective homomorphisms) (see 10.3) shows that this is not necessary, while the factorization system ($\mathbf{TAlg}(\underline{S})$-extendable epimorphisms,closed homomorphisms) (see 10.3) shows that the right factor need not consist of monomorphisms. A.PASZTOR has in [Pa 79], Beispiel 1.12 and Beispiel 1.13, examples of factorization systems in categories $\underline{\mathbf{C}}$, where neither the left factor is contained in $\mathrm{Epi}(\underline{\mathbf{C}})$, nor is the right factor contained in $\mathrm{Mono}(\underline{\mathbf{C}})$.
From I.NÉMETI, I.SAIN [NSa 77] we take the following

10.2.4 PROPOSITION. Let $\underline{\mathbf{C}}$ be any category, $\mathbf{H},\mathbf{S} \subseteq \mathrm{Mor}(\underline{\mathbf{C}})$. Then the following statements are equivalent:
(i) $(\mathbf{H},\mathbf{S})$ is a factorization system in $\underline{\mathbf{C}}$.
(ii) $\mathbf{H}$ and $\mathbf{S}$ satisfy (FS1), (FS5) and (FS6), where
(FS5) $\mathbf{H}\cdot\mathrm{Iso}(\underline{\mathbf{C}}) \subseteq \mathbf{H}$ and $\mathrm{Iso}(\underline{\mathbf{C}})\cdot\mathbf{S} \subseteq \mathbf{S}$.
(FS6) $\mathbf{H} \times \mathbf{S} \subseteq \mathrm{Difip}$ (i.e. each pair $(h,s) \in \mathbf{H} \times \mathbf{S}$ has the unique diagonal-fill-in property).

PROOF. Let $(\mathbf{H},\mathbf{S})$ be a factorization system, then (FS2) and (FS3) obviously imply (FS5), and (FS6) has been proved in 10.2.2.(iv). -- Conversely, let us assume that (FS1), (FS5) and (FS6) are satisfied. First assume $h',h'' \in \mathbf{H}$, and let (h,s) be an $(\mathbf{H},\mathbf{S})$-factorization of $h'\cdot h''$; let $p := h$ and $q := \mathbf{1}_{\mathrm{Cod}(h'')}$. Applying (FS6) twice yields diagonal-fill-ins d' and $d'' =: d$ such that first $h'\cdot d' = p$ (= h) and $d'\cdot s = h''\cdot q = h''$, while in the second case $h''\cdot d = d'$ and $d\cdot s = q = \mathbf{1}_{\mathrm{Cod}(h'')}$. From the last equation it easily follows that s is an epimorphism. Now $h\cdot(s\cdot d) = (h'\cdot h'')\cdot d = h = h\cdot\mathbf{1}_{\mathrm{Dom}(s)}$, and $(s\cdot d)\cdot s = s\cdot(d\cdot s) = s = \mathbf{1}_{\mathrm{Dom}(s)}\cdot s$. Since $s\cdot d$ and $\mathbf{1}_{\mathrm{Dom}(s)}$ can be considered as diagonal-fill-ins of the same diagram, the uniqueness in (FS6) implies $s\cdot d = \mathbf{1}_{\mathrm{Dom}(s)}$. Altogether we have shown that s is an isomorphism inverse to d; therefore (FS5) implies $h'\cdot h'' \in \mathbf{H}\cdot\mathrm{Iso}(\underline{\mathbf{C}}) \subseteq \mathbf{H}$. The argument for $\mathbf{S}\cdot\mathbf{S} \subseteq \mathbf{S}$ is similar. ∎
In order to prove (FS2) let (h,s) be an $(\mathbf{H},\mathbf{S})$-factorization of $i \in \mathrm{Iso}(\underline{\mathbf{C}})$, i.e. $h\cdot s = i$; thus $h\cdot(s\cdot i^{-1}) = \mathbf{1}_{\mathrm{Dom}(h)}$ and $(i^{-1}\cdot h)\cdot s = \mathbf{1}_{\mathrm{Cod}(s)}$. Then one easily realizes

$h\cdot(s\cdot i^{-1}\cdot h) = h = h\cdot 1_{Cod(h)}$, and $(s\cdot i^{-1}\cdot h)\cdot s = 1_{Cod(h)}\cdot s$; as above, (FS6) implies $s\cdot i^{-1}\cdot h = 1_{Cod(h)}$. This shows that h and s are isomorphisms; therefore $i = h\cdot s \in \mathbf{H}\cdot Iso(\underline{\mathfrak{C}}) \cap Iso(\underline{\mathfrak{C}})\cdot\mathbf{S} \subseteq \mathbf{H} \cap \mathbf{S}$ (by (FS5)), hence (FS2). ∎

In order to prove (FS4) let $f \in Mor(\underline{\mathfrak{C}})$, and let (h,s) and (h',s') be two (**H**,**S**)-factorizations of f: $f = h\cdot s = h'\cdot s'$. Difip(h,s') implies for (p,q):= (h',s) the existence of a unique d satisfying $h\cdot d = h'$ and $d\cdot s' = s$, while a similar choice for Difip(h',s) yields a morphisms d' such that $h'\cdot d' = h$ and $d'\cdot s = s'$; thus $h = (h\cdot d)\cdot d' = h\cdot 1_{Cod(h)}$ and $d\cdot(d'\cdot s) = s = 1_{Cod(h)}\cdot s$, and (FS6) implies $d\cdot d' = 1_{Cod(h)}$, and in a similar way one can show that $d'\cdot d = 1_{Cod(h')}$. Hence d and d' are isomorphisms uniquely determined because of (FS6), and (FS4) is satisfied. ∎∎∎

That the ties between the unique diagonal-fill-in-property Difip and factorization systems are even closer is shown by the following result (cf. [NSa 77]):

10.2.5 PROPOSITION. If (**H**,**S**) is a factorization system in a category $\underline{\mathfrak{C}}$, then $\Lambda(\mathbf{H}) = \mathbf{S}$ and $\Lambda^{op}(\mathbf{S}) = \mathbf{H}$ (uniqueness of the diagonal-fill-in need not even be required here!).

PROOF. Assume $f \in \Lambda(\mathbf{H})$, let $f = h\cdot s$, $(h,s) \in \mathbf{H} \times \mathbf{S}$. Because of Difip(h,f) there exists d such that $h\cdot d = 1_{Dom(f)}$ and $d\cdot f = s$. Hence $h\cdot(d\cdot h) = (h\cdot d)\cdot h = 1_{Dom(f)}\cdot h = h = h\cdot 1_{Cod(h)}$, and $(d\cdot h)\cdot s = d\cdot(h\cdot s) = d\cdot f = s = 1_{Cod(h)}\cdot s$. 10.2.2.(iii) then implies $d\cdot h = 1_{Cod(h)}$, showing that h is an isomorphism, i.e. $f = h\cdot s \in Iso(\underline{\mathfrak{C}})\cdot\mathbf{S} \subseteq \mathbf{S}$ because of (FS5). Hence $\Lambda(\mathbf{H}) \subseteq \mathbf{S}$. (FS6) implies $\mathbf{S} \subseteq \Lambda(\mathbf{H})$; therefore equality holds. $\Lambda^{op}(\mathbf{S}) = \mathbf{H}$ is proved in a similar (dual) way. ∎∎∎

From H.HERRLICH and G.E.STRECKER [HS 73], exercise 34K, we take the following

10.2.6 DEFINITION. Let $\underline{\mathfrak{C}}$ be any category.

(i): Every monomorphism $m \in Mono(\underline{\mathfrak{C}}) \cap \Lambda(Epi(\underline{\mathfrak{C}}))$ is called a <u>strong monomorphism</u> of $\underline{\mathfrak{C}}$, every epimorphism $e \in Epi(\underline{\mathfrak{C}}) \cap \Lambda^{op}(Mono(\underline{\mathfrak{C}}))$ is called a <u>strong epimorphism</u> of $\underline{\mathfrak{C}}$.

(ii): $\underline{\mathfrak{C}}$ is called <u>well-powered</u>, iff for every object $A \in Ob(\underline{\mathfrak{C}})$ there exists a **set**, say $Sub(A) \subseteq \{(B,m) \mid m:B \to A,\ m \in Mono(\underline{\mathfrak{C}})\}$ such that for every $\underline{\mathfrak{C}}$-monomorphism $n:C \to A$ there exist $(B,m) \in Sub(A)$ and an isomorphism $i:C \to B$ such that $i\cdot m = n$. -- $\underline{\mathfrak{C}}$ is called <u>co-(well-powered)</u>, if $\underline{\mathfrak{C}}$ has the analogous (dual) property for epimorphisms.

10.2.7 EXAMPLE. For every similarity type τ and for every (heterogeneous sort algebra) $\underline{S} \in \mathbf{Alg}(\tau)$ the category $(\mathbf{Alg}(\underline{S}) \cup Hom(\mathbf{Alg}(\underline{S}));\cdot,Dom,Cod,\mathbf{1})$ is well-powered and co-(well-powered). Namely for every $\underline{A} \in \mathbf{Alg}(\underline{S})$

$\{(\underline{B},id_{BA}:B \to A) \mid \underline{B}$ is a weak subalgebra of $\underline{A}\}$

is such a representative set for monomorphisms into $\underline{A}$; and

$\{(nat_R|A:\underline{A} \to \underline{C}/R,\underline{C}/R) \mid R$ is a closed, A-initial congruence in $\underline{F}(\underline{A},\mathbf{TAlg}(\underline{S}))$ with domain $\underline{C}\}$

is such a representative system for epimorphisms (cf. 5.8). ∎

As far as the strong epimorphisms and the strong monomorphisms in $\mathbf{Alg}(\underline{S})$ are concerned, we have the following

10.2.8 PROPOSITION. In $\mathbf{Alg}(\underline{S})$ we have:

(i) $\Lambda(Epi(\mathbf{Alg}(\underline{S}))) = \{m \mid m$ is a closed and injective homomorphism$\}$,

i.e. here the strong monomorphisms are exactly the closed and injective homomorphisms.

(ii) $\Lambda^{op}(\text{Mono}(\mathbf{Alg}(\underline{S}))) = \{e \mid e \text{ is a full and surjective homomorphism}\}$,
i.e. the strong epimorphisms are exactly the full and surjective homomorphims.

(iii) $(\text{Epi}(\mathbf{Alg}(\underline{S})),\Lambda(\text{Epi}(\mathbf{Alg}(\underline{S}))))$ is a factorization system in $\mathbf{Alg}(\underline{S})$.

(iv) $(\Lambda^{op}(\text{Mono}(\mathbf{Alg}(\underline{S}))),\text{Mono}(\mathbf{Alg}(\underline{S})))$ is a factorization system in $\mathbf{Alg}(\underline{S})$.

(v) In $\mathbf{TAlg}(\underline{S})$ -- the full subcategory of $\mathbf{Alg}(\underline{S})$ of all total $\underline{S}$-algebras -- the two factorization systems corresponding to (iii) and (iv) coincide.

PROOF. Because of 10.2.5 it is sufficient for (i) to (iv) to prove

(a) ($\text{Epi}(\mathbf{Alg}(\underline{S}))$, all closed and injective homomorphisms) is a factorization system of $\mathbf{Alg}(\underline{S})$;

(b) (all full and surjective homomorphisms, $\text{Mono}(\mathbf{Alg}(\underline{S}))$) is a factorization system of $\mathbf{Alg}(\underline{S})$.

Because of Theorem 10.1.2 and the general properties of epimorphisms and monomorphisms in both cases only (FS1) and (FS4) have to be proved, but in connection with (a) one easily realizes that every homomorphism $f:\underline{A} \to \underline{B}$ can be factored -- uniquely up to isomorphism -- into $f = f'\cdot i$, where $f':\underline{A} \to \mathbf{C}_{\underline{B}}([A]f)$, $\text{graph}(f') = \text{graph}(f)$, and i is the closed inclusion homomorphism from the $[A]f$-generated subalgebra of $\underline{B}$ into $\underline{B}$. Concerning (b), (FS1) and (FS4) follow from the First Homomorphism Theorem 2.7.4. ∎

Since in $\mathbf{TAlg}(\underline{S})$ all $\mathbf{TAlg}(\underline{S})$-homomorphisms are closed, all $\mathbf{TAlg}(\underline{S})$-epimorphisms are full and surjective and all $\mathbf{TAlg}(\underline{S})$-monomorphisms are closed and injective, we have (v). ∎∎∎

10.2.9. Proposition 10.2.8 may give an idea, how useful the following general result may be for the case of partial algebras, while in the case of total algebras -- at least as long as the left factor is required to consist of some class of epimorphisms and the right factor has to consist of some class of monomorphisms -- only one "relevant" factorization system exists. Notice that, trivially, $(\text{Mor}(\underline{\mathbf{C}}),\text{Iso}(\underline{\mathbf{C}}))$ and $(\text{Iso}(\underline{\mathbf{C}}),\text{Mor}(\underline{\mathbf{C}}))$ are always factorization systems.

10.2.10 THEOREM. Let $\underline{\mathbf{C}}$ be any co-(well-powered) category, which has multiple pushouts (i.e. for every set I and for every source $(A;(f_i \mid i \in I))$ a multiple pushout exists), and if ($\text{Epi}(\underline{\mathbf{C}})$,all strong $\underline{\mathbf{C}}$-monomorphisms) is a factorization system in $\underline{\mathbf{C}}$, then for any class $\mathbf{E} \subseteq \text{Epi}(\underline{\mathbf{C}})$, $(\Lambda^{op}\Lambda(\mathbf{E}),\Lambda(\mathbf{E}))$ is a factorization system of $\underline{\mathbf{C}}$, and $\Lambda^{op}\Lambda(\mathbf{E}) \subseteq \text{Epi}(\underline{\mathbf{C}})$.

-- A dual statement holds also for monomorphisms.

PROOF. Assume $\mathbf{E} \subseteq \text{Epi}(\underline{\mathbf{C}})$, let $f:A \to B$ be any $\underline{\mathbf{C}}$-morphism, and let $(A \xrightarrow{e_i} D_i \xrightarrow{m_i} B \mid i \in I)$ be a "representative system" for all factorizations of f: $e_i \in \Lambda^{op}\Lambda(\mathbf{E})$, $e_i\cdot m_i = f$ for each $i \in I$, and if $A \xrightarrow{e} D \xrightarrow{m} B$ is a factorization of f ($e\cdot m = f$, $e \in \Lambda^{op}\Lambda(\mathbf{E})$), then there exist $i \in I$ and an isomorphism $d:D_i \to D$ such that $e_i\cdot d = e$ and $d\cdot m = m_i$. Since $\mathbf{E} \subseteq \text{Epi}(\underline{\mathbf{C}})$, we have $\Lambda(\mathbf{E}) \supseteq \Lambda(\text{Epi}(\underline{\mathbf{C}}))$, i.e. $\Lambda^{op}\Lambda(\mathbf{E}) \subseteq \Lambda^{op}\Lambda(\text{Epi}(\underline{\mathbf{C}})) = \text{Epi}(\underline{\mathbf{C}})$ by assumption; therefore $\Lambda^{op}\Lambda(\mathbf{E})$ consists only of $\underline{\mathbf{C}}$-epimorphisms. Thus, and since $\underline{\mathbf{C}}$ is assumed to be co-(well-powered), we may assume that I is a set. Let $((j_i \mid i \in I);C)$ be a multiple pushout of $(A;(e_i \mid i \in I))$. Notice that $I \neq \emptyset$, since $1_A\cdot f$ is a factorization of f with $1_A \in \Lambda^{op}\Lambda(\mathbf{E})$ (cf. (FS2)). Let $m:C \to B$ be the $\underline{\mathbf{C}}$-morphism induced with respect to the family $(m_i:D_i \to B \mid i \in I)$ by the properties of a multiple pushout. Hence $j_i\cdot m = m_i$ $(i \in I)$. Since $I \neq \emptyset$, and since it can always be achieved that $\#I \geq 2$ (e.g. by counting a factorization twice), we may conclude from 10.1.2.(viii)^{op} that $j_i \in \Lambda^{op}\Lambda(\mathbf{E})$ for each $i \in I$, i.e. $e := e_i \cdot j_i \in \Lambda^{op}\Lambda(\mathbf{E})$. Hence there exist $i' \in I$ and an isomorphism $d: D_{i'} \to C$ such that $e_{i'} \cdot d = e$ and $d \cdot m = m_{i'}$. Without loss of generality we may assume that $C = D_{i'}$ and $d = \mathbf{1}_C$.

We now want to show that $m \in \Lambda(\mathbf{E}) = \Lambda\Lambda^{op}\Lambda(\mathbf{E})$ (the idea for this part is due to A.Pasztor (oral communication)): Let $v \in \Lambda^{op}\Lambda(\mathbf{E})$ and $p,q \in Mor(\underline{\mathbf{C}})$ such that $v \cdot q = p \cdot m$ (see Figure 10.1), let $(r,s;P)$ be a pushout of $(X := Dom(v);p,v)$, and let $u: P \to B$ be the morphism induced by $(m,q;B)$; thus $r \cdot u = m$, and $s \cdot u = q$. Because of 10.2.(vii)^{op} $r \in \Lambda^{op}\Lambda(\mathbf{E})$, hence $e \cdot r \in \Lambda^{op}\Lambda(\mathbf{E})$, and therefore we may assume as above that there is $i'' \in I$ such that $P = D_{i''}$, $e \cdot r = e_{i''}$ and $u = m_{i''}$. In particular we have $j_{i''}: P \to C$ such that $e \cdot r \cdot j_{i''} = e$; therefore $r \cdot j_{i''} = \mathbf{1}_C$, since e is an epimorphism.

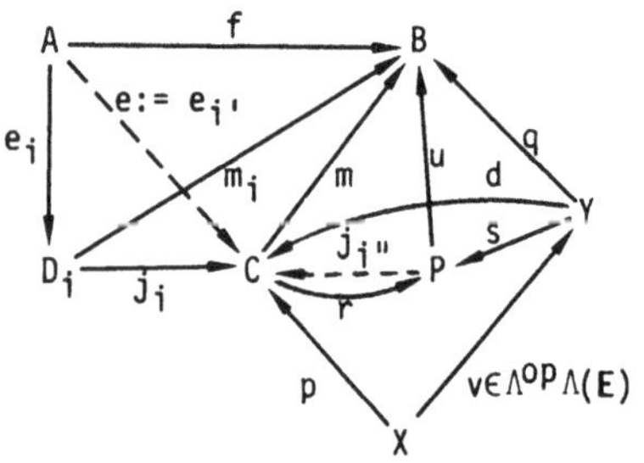

Figure 10.1.

Define $d := s \cdot j_{i''}: Y \to C$, where $Y := Cod(v)$. Then we have $v \cdot d = v \cdot (s \cdot j_{i''}) = p \cdot r \cdot j_{i''} = p$, and $v \cdot d \cdot m = p \cdot r \cdot u = p \cdot m = v \cdot q$. Since v is an epimorphism, this implies $d \cdot m = q$ and also the uniqueness of d, i.e. d is a unique diagonal-fill-in, and we have Difip(v,m); therefore $m \in \Lambda(\mathbf{E})$. This shows that $(\Lambda^{op}\Lambda(\mathbf{E}), \Lambda(\mathbf{E}))$ satisfies (FS1). (FS5) follows from Theorem 10.1.2, and by definition $\Lambda^{op}\Lambda(\mathbf{E}) \times \Lambda(\mathbf{E}) \subseteq$ Difip, i.e. we have (FS6). Hence $(\Lambda^{op}\Lambda(\mathbf{E}), \Lambda(\mathbf{E}))$ is indeed a factorization system. ∎∎∎

10.2.11 REMARK. Since $\mathbf{Alg}(\underline{S})$ is co-(well-powered) and has multiple pushouts, and since $(Epi(\mathbf{Alg}(\underline{S}))$, all strong monomorphisms) is a factorization system in $\mathbf{Alg}(\underline{S})$, all the classes of homomorphisms -- in particular those in Table 9.1 of section 9.2 -- defined by reflection of some set of implications form the right factor of some factorization system.

10.2.12. In order that we do not become too much involved here with more special category theoretical concepts we investigate the following specialization of Theorem 10.2.10 only for the category of partial $\underline{S}$-algebras -- any subclass closed with respect to subalgebras and direct products would do, too. Let us recall now from Definition 5.1.1 the concept of **K**-extendable homomorphisms, and let us define for $\mathbf{K} \subseteq \mathbf{Alg}(\underline{S})$: $Ext(\mathbf{K}) := \{e \in Epi(\mathbf{Alg}(\underline{S})) \mid e \text{ is } \mathbf{K}\text{-extendable}\}$.

10.2.13 REMARKS. (i): 5.1.6 implies: $Ext(\mathbf{K}) = Ext(\mathbf{SP}(\mathbf{K}))$ for every $\mathbf{K} \subseteq \mathbf{Alg}(\underline{S})$.
(ii): Let $\mathbf{K} \subseteq \mathbf{Alg}(\underline{S})$, $\mathbf{K} = \mathbf{SP}(\mathbf{K})$, let $\underline{A} \in Inj(Ext(\mathbf{K}))$, and let $(r_{\underline{A},\mathbf{K}}, \underline{F}(\underline{A},\mathbf{K}))$ be its **K**-universal solution. Since then $r_{\underline{A},\mathbf{K}} \in Ext(\mathbf{K})$, $\underline{A} \in Inj(Ext(\mathbf{K}))$ implies that there exists $d: \underline{F}(\underline{A},\mathbf{K}) \to \underline{A}$ such that $r_{\underline{A},\mathbf{K}} \cdot d = id_A$. Since $r_{\underline{A},\mathbf{K}}$ is an epimorphism, we additionally have from $r_{\underline{A},\mathbf{K}} \cdot d \cdot r_{\underline{A},\mathbf{K}} = id_A \cdot r_{\underline{A},\mathbf{K}} = r_{\underline{A},\mathbf{K}} \cdot id_{F(\underline{A},\mathbf{K})}$ that $d \cdot r_{\underline{A},\mathbf{K}} = id_{F(\underline{A},\mathbf{K})}$, showing that $r_{\underline{A},\mathbf{K}}$ is an isomorphism; thus $\underline{A} \in \mathbf{K}$. Thus every quasiprimi-

tive class is uniquely determined as Ext(**K**) for some $\mathbf{K} \subseteq \mathbf{Alg}(\underline{S})$. In particular:

10.2.14 PROPOSITION. For every subclass $\mathbf{K} \subseteq \mathbf{Alg}(\underline{S})$: Inj(Ext(**K**)) = **SP**(**K**) . ■■■

10.2.15 THEOREM. For every subclass $\mathbf{K} \subseteq \mathbf{Alg}(\underline{S})$ one has:

(i) If $\underline{A},\underline{B} \in$ **SP**(**K**) and $f:\underline{A} \to \underline{B}$, then $f \in \Lambda(\mathrm{Ext}(\mathbf{K}))$.

(ii) $(\mathrm{Ext}(\mathbf{K}),\Lambda(\mathrm{Ext}(\mathbf{K})))$ is a factorization system.

PROOF. Ad (i): Let $e \in \mathrm{Ext}(\mathbf{K})$, $e:\underline{P} \to \underline{C}$, $\underline{A},\underline{B} \in$ **SP**(**K**) , $f:\underline{A} \to \underline{B}$, and let p, q be given such that pf = eq . Since $\underline{A} \in$ **SP**(**K**) and $e \in \mathrm{Ext}(\mathbf{K})$, there exists $d:\underline{C} \to \underline{A}$ such that ed = p . Since e is an epimorphism, d is unique and edf = pf = eq implies df = q , showing that Difip(e,f) and $f \in \Lambda(\mathrm{Ext}(\mathbf{K}))$. ■

Ad (ii): Since $\mathbf{Alg}(\underline{S})$ satisfies the assumptions of Theorem 10.2.10, we only have to show that $\Lambda^{op}\Lambda(\mathrm{Ext}(\mathbf{K})) \subseteq \mathrm{Ext}(\mathbf{K})$. Therefore let $e \in \Lambda^{op}\Lambda(\mathrm{Ext}(\mathbf{K}))$, $e:\underline{P} \to \underline{C}$, let $\underline{A} \in \mathbf{K}$ and let $f:\underline{P} \to \underline{A}$ be any homomorphism. Consider the **K**-universal solutions $(r_{\underline{P},\mathbf{K}},\underline{F}(\underline{P},\mathbf{K})) =: (r_{\underline{P}},\underline{F}_{\underline{P}})$ and $(r_{\underline{C},\mathbf{K}},\underline{F}(\underline{C},\mathbf{K})) =: (r_{\underline{C}},\underline{F}_{\underline{C}})$. Then f induces a homomorphism $\overline{f}:\underline{F}_{\underline{P}} \to \underline{A}$ such that $f = r_{\underline{P}}\cdot\overline{f}$, and $e\cdot r_{\underline{C}}$ induces a homomorphism $q:\underline{F}_{\underline{P}} \to \underline{F}_{\underline{C}}$ satisfying $r_{\underline{P}}\cdot q = e\cdot r_{\underline{C}}$. Because of (i), $q \in \Lambda(\mathrm{Ext}(\mathbf{K}))$; thus there exists a unique diagonal-fill-in $d:\underline{C} \to \underline{F}_{\underline{P}}$ satisfying $e\cdot d = r_{\underline{P}}$; hence $e\cdot d\cdot\overline{f} = r_{\underline{P}}\cdot\overline{f} = f$, showing that $e \in \mathrm{Ext}(\mathbf{K})$. ■■■

Following H.HERRLICH [H 72], we define

10.2.16 DEFINITION. Let $\mathbf{K} \subseteq \mathbf{Alg}(\underline{S})$. Then the elements of $\Lambda(\mathrm{Ext}(\mathbf{K}))$ are called **K**-perfect homomorphisms.

10.2.17 REMARK. We would like to recall from category theory the following concepts:

Let $\mathbf{A} \subseteq \mathrm{Ob}(\underline{\mathbf{C}})$ for some category $\underline{\mathbf{C}}$. Then **A** is called a reflective subclass of $\underline{\mathbf{C}}$, iff

(a) for every $B \in \mathrm{Ob}(\underline{\mathbf{C}})$ there exist an object $F(B,\mathbf{A}) \in \mathbf{A}$ and a morphism $r_{B,\mathbf{A}}:B \to F(B,\mathbf{A})$ such that for every $A \in \mathbf{A}$ and for every $f:B \to A$ there exists a unique $\overline{f}:F(B,\mathbf{A}) \to A$ satisfying $f = r_{B,\mathbf{A}}\cdot\overline{f}$; and

(b) for every $f:B \to C$ in $\mathrm{Mor}(\underline{\mathbf{C}})$ there exists a morphism $F(f,\mathbf{A}):F(B,\mathbf{A}) \to F(C,\mathbf{A})$ such that $f\cdot r_{C,\mathbf{A}} = r_{B,\mathbf{A}}\cdot F(f,\mathbf{A})$, $F(1_C,\mathbf{A}) = 1_{F(C,\mathbf{A})}$ and $F(f\cdot g,\mathbf{A}) = F(f,\mathbf{A})\cdot F(g,\mathbf{A})$.

This means that $F(.,\mathbf{A})$ is a **functor** from $\underline{\mathbf{C}}$ into the full subcategory of $\underline{\mathbf{C}}$ with object class **A** and **all** $\underline{\mathbf{C}}$-morphisms between **A**-objects. In category theoretical terminology $F(.,\mathbf{A})$ is called a left adjoint functor for the inclusion functor from **A** into $\underline{\mathbf{C}}$.

A reflective subcategory **A** is called epireflective, if, for each $B \in \mathbf{A}$, $r_{B,\mathbf{A}}$ is an epimorphism of $\underline{\mathbf{C}}$; it is called monoreflective, if each $r_{B,\mathbf{A}}$ is a monomorphism; and **H**-reflective -- where $(\mathbf{H},\Lambda(\mathbf{H}))$ is a factorization system in $\underline{\mathbf{C}}$ --, if each $r_{B,\mathbf{A}}$ belongs to **H**. In particular, for $\mathbf{K} \subseteq \mathbf{Alg}(\underline{S})$ the class Inj(Ext(**K**)), i.e. each quasiprimitive subclass, yields always an epireflective full subcategory of $\mathbf{Alg}(\underline{S})$.

It is proved in category theory that each monoreflective full subcategory of $\underline{\mathbf{C}}$ is epireflective (cf. [HS 73], Proposition 36.3). The epimorphisms of $\underline{\mathbf{C}}$ which are also monomorphisms are called bimorphisms. In $\mathbf{TAlg}(\underline{S})$ all bimorphisms are isomorphisms, since closed and injective epimorphisms are isomorphisms. Thus in $\mathbf{TAlg}(\underline{S})$ the only "abstract" monoreflective subclass ("abstract" meaning to be closed with respect to isomorphic copies) in $\mathrm{Ob}(\mathbf{TAlg}(\underline{S}))$ is $\mathbf{TAlg}(\underline{S})$ itself, while in $\mathbf{Alg}(\underline{S})$ there are usually abundantly many monoreflective subcategories, namely all quasiprimitive subclasses **A** containing **TAlg**($\underline{S}$), since then, for every $\underline{B} \in$ **Alg**($\underline{S}$), $\underline{F}(\underline{B},\mathbf{TAlg}(\underline{S})) \in \mathbf{A}$, and $r_{\underline{B},\mathbf{A}}$ has to be a left factor of $id_B : \underline{B} \to \underline{F}(\underline{B},\mathbf{TAlg}(\underline{S}))$; therefore $r_{\underline{B},\mathbf{A}}$ is injective.

10.3 Examples of factorization systems for partial algebras

10.3.1. In what follows we want to describe some -- in our opinion important -- factorization systems in the category **Alg**($\underline{S}$) of partial algebras. In order to get a formal description of factorization systems connected with P-reflecting homomorphisms for any set P of terms we have to introduce some generalization of the concept of a term and a term operation. As far as we know, this concept has been first introduced by A.Obtułowicz in connection with Lawvere style investigations of the category **Alg**(τ) of all partial algebras of some given type τ. We shall consider this concept later in section 15 in more detail; here we only need the definition and some special items.

10.3.2 DEFINITION. (i): Let $M \subseteq X$, $\#M < \mathbf{d}^{\&}$, and let $\{t\} \cup D \subseteq F(M,\mathbf{TAlg}(\underline{S}))$. Then $(M:t|D)$ is called a <u>conditioned term</u> (briefly: c-term; more precisely we should say "conditioned $\underline{S}$-term") with set M of variables.

(ii): If $\underline{A} \in$ **Alg**($\underline{S}$) and $(M:t|D)$ is a conditioned term, then the induced <u>c-term operation</u> $(M:t|D)^{\underline{A}}$ is defined by:

$dom(M:t|D)^{\underline{A}} := \bigcap\{dom(M:p)^{\underline{A}} | p \in \{t\} \cup D\} = \{\underline{a} \in A^M | \underline{a}(M:p)^{\underline{A}}$ exists for every $p \in \{t\} \cup D\}$,

and if $\underline{a} \in dom(M:t|D)^{\underline{A}}$, then $\underline{a}(M:t|D)^{\underline{A}} := \underline{a}(M:t)^{\underline{A}}$.

(iii): Let $P \subseteq T_{loc}$ (see 6.1.1) be any set of terms, then we recall that $\downarrow P := \{(M:t) | t \in \downarrow p$ for some $(M:p) \in P\}$. If $\underline{A} \in$ **Alg**($\underline{S}$), then let us define

$$\langle \underline{A},P\rangle := (A,((M:t|\{p\})^{\underline{A}} | t \in \downarrow p \text{ and } (M:p) \in P)).$$

If necessary, $\langle \underline{A},P\rangle$ is connected with S by v_A, since v_A is compatible with $\underline{S}$-term operations and therefore also with conditioned $\underline{S}$-term operations.

10.3.3 EXAMPLES. (i): Let $(I \dot{\cup} Z \dot{\cup} O;\delta,\mu)$ be a partial Mealy automaton, i.e. for $S := \{i,z,o\}$ we have $graph(\delta^{\underline{S}}) = \{((i,z),z)\}$, $graph(\mu^{\underline{S}}) = \{((i,z),o)\}$, where I is the set of **input elements**, Z the set of **states**, and O the set of **output elements**, and i,z,o are the corresponding "sorts" in S. Then, sometimes, in connection with the process of reduction only terms having their value in O may be of interest, like $(\{i'',i',z'\}:(i'',(i',z')\delta)\mu)$ (i', i" are of sort i, z' is of sort z). In this connection a term like $(\{i'',i',z'\}:(i',z')\delta)$ might only be of interest, when it occurs as a subterm with value in O, i.e. e.g. in the form $(\{i'',i',z'\}:(i',z')\delta|\{(i'',(i',z')\delta)\mu\})$.

(ii): Let $\underline{P}$ be the smallest homogeneous partial Peano algebra on $\{a,b,c\}$ of type $\tau := (2,2)$, in which $xx\phi yy\phi\phi' \stackrel{e}{=} xx\phi yy\phi\phi'$ is valid, and in which $(a,b)\phi^{\underline{P}}$, $(a,c)\phi^{\underline{P}}$ and $(b,c)\phi^{\underline{P}}$ exist, and let $\underline{Q}$ be the relative subalgebra of $\underline{P}$ generated by $\{a,b,c\}$ by the conditioned term operations $(\{x,y\}:xy\phi|\{xy\phi xy\phi\phi'\})^{\underline{P}}$ and $(\{x,y\}:xy\phi xy\phi\phi')^{\underline{P}}$. Then $\underline{Q}$ is the smallest partial Peano algebra on $\{a,b,c\}$, in which $xx\phi xx\phi\phi' \stackrel{e}{=} xx\phi xx\phi\phi'$ is valid; in particular $(a,b)\phi^{\underline{P}}$, $(a,c)\phi^{\underline{P}}$ and $(b,c)\phi^{\underline{P}}$ do not belong to Q, since none of these elements occurs in a sequence in dom $\phi'^{\underline{P}}$.

As a further preparation for the description of the factorization system connected with P-reflecting homomorphisms let us observe that we have as an immediate consequence of

Corollary 5.8.4 of the Epimorphism Theorem (see also [BJPa 78]):

10.3.4 PROPOSITION. Let $\mathbf{M}_c$ be the class of all closed homomorphisms in $\mathbf{Alg}(\underline{S})$. Then

(i) $(\mathrm{Ext}(\mathbf{TAlg}(\underline{S})), \mathbf{M}_c)$ is a factorization system, i.e. the closed homomorphisms are exactly the $\mathbf{TAlg}(\underline{S})$-perfect homomorphisms.

(ii) An epimorphism $e:\underline{A} \to \underline{B}$ is $\mathbf{TAlg}(\underline{S})$-extendable (i.e. belongs to $\mathrm{Ext}(\mathbf{TAlg}(\underline{S}))$), iff $e = e' \cdot i$, where i is an isomorphism, and e' is a full and injective epimorphism from $\underline{A}$ into an A-generated relative subalgebra of $\underline{F}(\underline{A},\mathbf{TAlg}(\underline{S}))$ such that $\mathrm{graph}(e') = \Delta_A$. ■■■

10.3.5 PROPOSITION. Let $P \subseteq T_{loc}$ (see 6.1.1) be any set of terms, and let $\mathbf{M}_P \subseteq \mathrm{Hom}(\mathbf{Alg}(\underline{S}))$ designate the class of all P-reflecting homomorphisms of $\mathbf{Alg}(\underline{S})$; then:

(i) $(\Lambda^{op}(\mathbf{M}_P), \mathbf{M}_P)$ is a factorization system in $\mathbf{Alg}(\underline{S})$.

(ii) A homomorphism $e:\underline{A} \to \underline{B}$ belongs to $\Lambda^{op}(\mathbf{M}_P)$ iff

(*) e is a full and injective epimorphism (i.e. full bimorphism) such that there exists a full and injective homomorphism $j_{\underline{B}}:\underline{B} \to \underline{F}(\underline{A},\mathbf{TAlg}(\underline{S}))$ satisfying $e \cdot j_{\underline{B}} = id_A$; and if $\underline{C}$ is the relative subalgebra of $\underline{F}(\underline{A},\mathbf{TAlg}(\underline{S}))$ on $[B]j_{\underline{B}}$, then $C_{<\underline{C},P>}A = C$.

(iii) $\Lambda^{op}(\mathbf{M}_P) \subseteq \mathrm{Ext}(\mathbf{Mod}(\{(M:t \overset{e}{=} t) \mid (M:t) \in P\}))$, and if $P = \downarrow P$, then equality holds.

PROOF. Ad (i): For each $(M:t) \in P$ consider $e(M:t):\underline{M}_d \to \underline{\downarrow t\ U\ M}$, the encoding epimorphism of $(M:t \overset{e}{=} t)$ (cf. 8.2.2 and 8.2.4). Then $\mathbf{M}_P = \Lambda(\{e(M:t) \mid (M:t) \in P\})$, implying that $\Lambda^{op}(\mathbf{M}_P) \subseteq \mathrm{Epi}(\mathbf{Alg}(\underline{S}))$ and that $(\Lambda^{op}(\mathbf{M}_P), \mathbf{M}_P)$ is a factorization system because of Theorem 10.2.10. ■

For the proofs of the other two items we only sketch the ideas and leave details as exercises.

Ad (ii): Knowing already that we deal with the factors of a factorization system, we have to see how we can describe in a way unique up to isomorphism a decomposition of any given homomorphism, say $f:\underline{A} \to \underline{C}$, in an epimorphism $e:\underline{A} \to \underline{B}$ followed by a P-reflecting homomorphism, say $g:\underline{B} \to \underline{C}$. Having already 10.3.4 at hand, and knowing that $\mathbf{M}_P$ contains $\mathbf{M}_c$, we get $\Lambda^{op}(\mathbf{M}_P) \subseteq \mathrm{Ext}(\mathbf{TAlg}(\underline{S}))$. Since the classes involved are closed with respect to isomorphisms, we may assume that $\underline{B}$ is an A-generated relative subalgebra of $\underline{F}(\underline{A},\mathbf{TAlg}(\underline{S}))$, and $\mathrm{graph}(e) = \Delta_A$. In order to make g P-reflecting we have to take to A as many elements of $\underline{F}(\underline{A},\mathbf{TAlg}(\underline{S}))$ as are required by the reflection of TE-expressions induced by P and allowed by $\underline{C}$. This idea leads to the formal description in (ii). ■

Ad (iii): For any set P of terms let $F_P := \{(M:t \overset{e}{=} t) \mid (M:t) \in P\}$. Then it is obvious (cf. section 6) that $\mathbf{Mod}(F_P) = \mathbf{Mod}(F_{\downarrow P})$. It is also easy to realize that $\{e(M:t) \mid (M:t) \in \downarrow P\} \subseteq \mathrm{Ext}(\mathbf{Mod}(F_P))$; therefore $\Lambda^{op}(\mathbf{M}_P) \subseteq \Lambda^{op}(\mathbf{M}_{\downarrow P}) \subseteq \mathrm{Ext}(\mathbf{Mod}(F_P))$. Hence it remains to show (cf. 10.2.15) that each $\mathbf{Mod}(F_P)$-extendable epimorphism, say $e:\underline{A} \to \underline{B}$ -- where we can assume $\underline{B}$ to be an A-generated relative subalgebra of $\underline{F}(\underline{A},\mathbf{TAlg}(\underline{S}))$ -- belongs to $\Lambda^{op}(\mathbf{M}_{\downarrow P})$: But in order that e be $\mathbf{Mod}(F_P)$-extendable, $id_A:\underline{A} \to \underline{F}(\underline{A},\mathbf{Mod}(F_P))$ -- observe that $\mathbf{Mod}(F_P)$ is monoreflective, since it contains $\mathbf{TAlg}(\underline{S})$ -- has to be extendable to $\underline{B}$, thus $\underline{B}$ has to be isomorphic (over id_A) to a relative subalgebra of $\underline{F}(\underline{A},\mathbf{Mod}(F_P))$. On the other hand it is not too difficult to realize because of $\downarrow P$ being closed with respect to subterms that $\underline{B}$ is generated by A with respect to the term operations induced by the elements of $\downarrow P$, showing in connection with (ii) that e belongs to $\Lambda^{op}(\mathbf{M}_{\downarrow P})$. ■■■

10.3.6. Proposition 10.3.5 yields an explicit description of a usually relatively large set of factorization systems, and because of 9.2.6.(ii) not all of them are connected with quasivarieties, namely usually not in the case, when P is not closed with respect to subterms. Two similarly large families will follow below, but first we consider one single very important factorization system, which has been studied e.g. in the thesis of A.PASZTOR [Pa 79] with respect to subcategories of $\mathbf{Alg}(\underline{S})$.

10.3.7 PROPOSITION. Let $\mathbf{H}_w$ be the class of all surjective and $\mathbf{S}_i$ the class of all injective and initial homomorphisms. Then $(\mathbf{H}_w, \mathbf{S}_i)$ is a factorization system in $\mathbf{Alg}(\underline{S})$.

PROOF. This is easy to realize, when one observes that every homomorphism $f:\underline{A} \to \underline{B}$ can be decomposed into a surjective homomorphism $f':\underline{A} \to \underline{B}'$ -- graph(f'):= graph(f) , $\underline{B}'$ being the relative subalgebra of $\underline{B}$ with carrier set [A]f -- followed by the full and injective (= initial and injective) embedding of $\underline{B}'$ into $\underline{B}$, and that this factorization is unique up to unique isomorphism. ■■■

10.3.8 LEMMA. Let (E,M) and (H,S) be factorization systems of $\mathbf{Alg}(\underline{S})$ such that the following properties are satisfied:

(a) **M** consists of all homomorphisms, which reflect some given set F of formulas.

(b) Each h in **H** is surjective and preserves all formulas from F.

(c) Either **E** consists of epimorphisms, or **S** consists of monomorphisms.

Then $(\mathbf{E}\cdot(\mathbf{H} \cap \mathbf{M}), \mathbf{S} \cap \mathbf{M}) = (\mathbf{E}\cdot\mathbf{H}, \mathbf{S} \cap \mathbf{M})$ is a factorization system in $\mathbf{Alg}(\underline{S})$.

PROOF. We first prove that the conditions of the Lemma imply $\Lambda^{op}(\mathbf{M} \cap \mathbf{S}) = \mathbf{E}\cdot(\mathbf{H} \cap \mathbf{M})$: For this purpose assume $k \in \Lambda^{op}(\mathbf{M} \cap \mathbf{S})$, let (e,m) be its (E,M)-factorization, and let (h,s) be an (H,S)-factorization of m (see Figure 10.2). Lemma 9.1.7.(iii) implies that $s \in \mathbf{M}$, because hs belongs to **M**, h is surjective and preserves F. Since $\mathbf{E} \subseteq \mathrm{Epi}(\mathbf{Alg}(\underline{S}))$ or $\mathbf{S} \subseteq \mathrm{Mono}(\mathbf{Alg}(\underline{S}))$, either 10.1.2.(vi) or 10.1.2.(v) imply together with $hs \in \mathbf{M}$ that $h \in \mathbf{M}$.

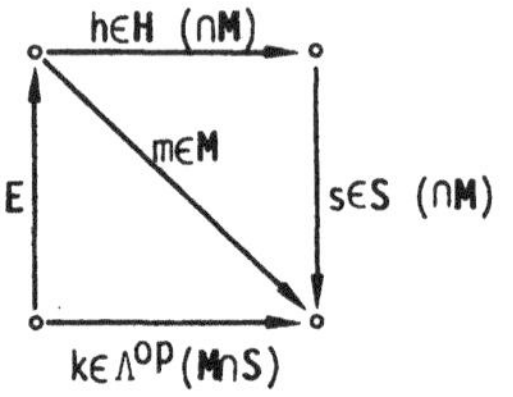

Figure 10.2.

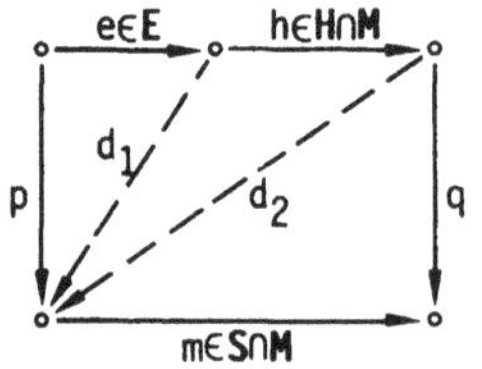

Figure 10.3.

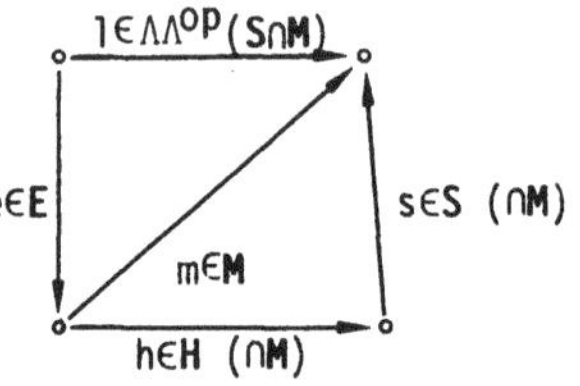

Figure 10.4.

Thus $h \in \mathbf{H} \cap \mathbf{M}$, $s \in \mathbf{S} \cap \mathbf{M}$: Moreover, since $k = ehs \in \Lambda^{op}(\mathbf{S} \cap \mathbf{M})$, and since either $eh \in \mathrm{Epi}(\mathbf{Alg}(\underline{S}))$ or $\mathbf{S} \cap \mathbf{M} \subseteq \mathrm{Mono}(\mathbf{Alg}(\underline{S}))$ (see (c)), $10.1.2.(v)^{op}$ or $10.1.2.(vi)^{op}$ imply $s \in \Lambda^{op}(\mathbf{S} \cap \mathbf{M})$. Hence the axioms of factorization systems easily imply that s is an isomorphism, and $hs \in \mathbf{H} \cap \mathbf{M}$, i.e. $k = e(hs) \in \mathbf{E}\cdot(\mathbf{H} \cap \mathbf{M})$.

Moreover, $\mathbf{E}\cdot(\mathbf{H} \cap \mathbf{M}) \subseteq \Lambda^{op}(\mathbf{S} \cap \mathbf{M})$: To prove this consider a commutative diagram like the one in Figure 10.3 (without the broken lines). Then there exists a unique d_1 such that $ed_1 = p$ and $d_1m = hq$, since $m \in \mathbf{M}$. Since also $m \in \mathbf{S}$ and $h \in \mathbf{H}$ (!), there exists a unique d_2 such that $hd_2 = d_1$ and $d_2m = q$. Now $ehd_2 = ed_1 = p$ showing that

d_2 is the desired diagonal-fill-in (uniqueness easily follows from the assumptions). Altogether we get $\mathbf{E}\cdot(\mathbf{H}\cap\mathbf{M}) = \Lambda^{op}(\mathbf{S}\cap\mathbf{M})$ (= $\mathbf{E}\cdot\mathbf{H}$, since we have seen that we did not need in the last argument that h had to belong to **M**). We still have to show that $\Lambda\Lambda^{op}(\mathbf{S}\cap\mathbf{M}) \subseteq \mathbf{S}\cap\mathbf{M}$ (the other inclusion being obvious): Let $l \in \Lambda\Lambda^{op}(\mathbf{S}\cap\mathbf{M})$; then we decompose l as shown in Figure 10.4 and argue in a similar way as for k getting that $eh \in \Lambda(\mathbf{E}\cdot(\mathbf{H}\cap\mathbf{M})) \cap \mathbf{E}\cdot(\mathbf{H}\cap\mathbf{M}) \subseteq \mathrm{Iso}(\mathbf{Alg}(\underline{S}))$, showing that $ehs \in \mathbf{S}\cap\mathbf{M}$. ■■■

10.3.9 COROLLARY. Let P be any set of terms for $\mathbf{Alg}(\underline{S})$, and let us define $\mathbf{S}_w := \mathrm{Mono}(\mathbf{Alg}(\underline{S}))$, $\mathbf{H}_f :=$ class of all full and surjective $\mathbf{Alg}(\underline{S})$-homomorphisms. Then

$$(\Lambda^{op}(\mathbf{M}_P)\cdot(\mathbf{H}_f\cap\mathbf{M}_P),\ \mathbf{M}_P\cap\mathbf{S}_w) \quad\text{and}\quad (\Lambda^{op}(\mathbf{M}_P)\cdot(\mathbf{H}_w\cap\mathbf{M}_P),\ \mathbf{M}_P\cap\mathbf{S}_i)$$

are factorization systems in $\mathbf{Alg}(\underline{S})$.

PROOF. The assumptions of Lemma 10.3.8 are satisfied for $(\Lambda^{op}(\mathbf{M}_P), \mathbf{M}_P)$ and $(\mathbf{H}_f, \mathbf{S}_w)$ (respectively $(\mathbf{H}_w, \mathbf{S}_i)$). ■■■

10.3.10 REMARK. When we remember that in Lemma 10.3.8 we have $\mathbf{E}\cdot(\mathbf{H}\cap\mathbf{M}) = \mathbf{E}\cdot\mathbf{H}$, and when we define -- also for later use -- $\mathbf{H}_P := \mathbf{M}_P\cap\mathbf{H}_w$, $\mathbf{H}_{fP} := \mathbf{M}_P\cap\mathbf{H}_f$, $\mathbf{S}_P := \mathbf{M}_P\cap\mathbf{S}_w$, $\mathbf{S}_{iP} := \mathbf{M}_P\cap\mathbf{S}_i$, then we can formulate the results of 10.3.8 and 10.3.9 also in the shorter form: For every set P of terms $(\Lambda^{op}(\mathbf{M}_P), \mathbf{M}_P)$, $(\Lambda^{op}(\mathbf{M}_P)\cdot\mathbf{H}_{fP}, \mathbf{S}_P)$, and $(\Lambda^{op}(\mathbf{M}_P)\cdot\mathbf{H}_P, \mathbf{S}_{iP})$, are factorization systems.

10.3.11 Some further examples of factorization systems.

We briefly present some further examples of factorization systems in $\mathbf{Alg}(\underline{S})$, which might be of interest, but which will not be much used later in this book. In this connection we recall that because of Theorem 10.2.10 all classes of homomorphisms which are defined by the reflection of a given set of elementary implications are right factors of factorization systems, where the left factor consists of a class of epimorphisms. In what follows we want to give a brief description of some such left factors.

(i): Let $\mathbf{M}_i$ be the class of all **initial** homomorphisms in $\mathbf{Alg}(\underline{S})$ (see 9.2), let $f:\underline{A}\to\underline{B}$ be a homomorphism, and let (s,j) be an $(\mathbf{H}_w, \mathbf{S}_i)$-factorization of f (see 10.3.7). If (e',i') is a $(\Lambda^{op}(\mathbf{M}_i), \mathbf{M}_i)$-factorization of s, then (e',i'j) is a $(\Lambda^{op}(\mathbf{M}_i), \mathbf{M}_i)$-factorization of f, and we can restrict our considerations to the case, when f is surjective:

Let $(id_A, \tilde{f})$ be an $(\mathrm{Ext}(\mathbf{TAlg}(\underline{S})), \mathbf{M}_c)$-factorization of f (cf.10.3.4), observe that every initial homomorphism is relatively closed, and every surjective and initial homomorphism is closed. Assume that (e,i) is the $(\Lambda^{op}(\mathbf{M}_i), \mathbf{M}_i)$-factorization we look for, where $e:\underline{A}\to\underline{C}$. Since for surjective f also i is surjective and -- being initial -- also closed, $\underline{C}$ has to be a closed homomorphic image of $\underline{\mathrm{dom}}\ \tilde{f}$ (see 5.8.2.(vi)) in such a way that for $a \in \mathbf{U}\{[C^{K_\phi}]\phi^{\underline{C}} \mid \phi\in\Omega\}$ one has $\#((ai)i^{-1}) = 1$. Since i is closed, we get $[\mathbf{U}\{[C^{K_\phi}]\phi^{\underline{C}} \mid \phi\in\Omega\}]i = \mathbf{U}\{[B^{K_\phi}]\phi^{\underline{B}} \mid \phi\in\Omega\}$. Hence let

$R := \{(a,b) \mid a,b \in \mathrm{dom}\ \tilde{f},\ a\tilde{f} = b\tilde{f} \in \mathbf{U}\{[B^{K_\phi}]\phi^{\underline{B}} \mid \phi\in\Omega\}\}$. Then R is a closed congruence on $\underline{\mathrm{dom}}\ \tilde{f}$, and it is not too difficult to see that $\underline{C} \cong (\underline{\mathrm{dom}}\ \tilde{f})/R$. These observations show -- perhaps together with some additional calculations -- that $\Lambda^{op}(\mathbf{M}_i)$ consists of all surjective (!) epimorphisms $e:\underline{A}\to\underline{C}$ of $\mathbf{Alg}(\underline{S})$ satisfying that $c \in C \setminus \mathbf{U}\{[C^{K_\phi}]\phi^{\underline{C}} \mid \phi\in\Omega\}$ implies $\#((c)e^{-1}) = 1$. ■

(ii): Relatively normal homomorphisms are defined to reflect all formulas $(\mathrm{fvar}(t) \cup \{y\}:t \overset{e}{=} y)$, where t is any term and $y \notin \mathrm{fvar}(t)$ any variable of correct sort; in particular $(\{x,y\}:x \overset{e}{=} y)$, for $x \neq y$ any variables of the same sort, is such a formula; thus relatively normal homomorphisms are injective; moreover, they are initial. Let $\mathbf{M}_n$ be the class of all relatively normal **Alg**($\underline{S}$)-homomorphisms. We claim that $\Lambda^{op}(\mathbf{M}_n)$ consists of all epimorphisms $e:\underline{A} \to \underline{C}$ such that for every $c \in C$ there exist a term $(M:t)$, a sequence $\underline{c} \in \mathrm{dom}(M:t)^{\underline{C}}$ and an $m \in M$ such that $(m)\underline{c} = c$ and $\underline{c}(M:t)^{\underline{C}} \in [A]e$ -- equivalently: $e \in \Lambda^{op}(\mathbf{M}_n)$ iff $e:\underline{A} \to \underline{C}$ is an epimorphism such that $\downarrow([A]e) = C$ with respect to the algebraic quasiorder on $\underline{C}$. ∎

(iii): We have not yet seen the necessity to work out a detailed description of the left factors of **P-initial** (= **P-normal**) **homomorphisms** respectively **injective and P-normal homomorphisms.**

(iv) Final homomorphisms form the left factor of a factorization system in **Alg**($\underline{S}$) such that this left factor does not consist only of epimorphisms. It is not difficult to realize that the corresponding right factor consists of the class of all bijective homomorphisms in **Alg**($\underline{S}$). ∎

No.	Left factor: Class of all ...	Right factor: Class of all ..
1	full and surj. homs ($\mathbf{H}_f$)	inj. homs ($\mathbf{S}_w$)
2	**TAlg**($\underline{S}$)-ext. homs ($\underline{A} \twoheadrightarrow \underline{B}$, $\underline{B} \cong$ over id_A to some A-initial segment of $\underline{F}(\underline{A},$**TAlg**$(\underline{S}))$)	closed homs ($\mathbf{M}_c$)
3	epis ($\mathbf{E}_w$)	closed monos ($\mathbf{S}_c$)
4	surj. homs $f:\underline{A} \twoheadrightarrow \underline{B}$, where $a \in B \setminus \bigcup\{[B^{K_\phi}]\phi \mid \phi \in \Omega\}$ implies $\#(af^{-1})=1$	initial homs ($\mathbf{M}_i$)
5	surj. homs ($\mathbf{H}_w$)	initial monos ($\mathbf{S}_i = \mathbf{S}_f$)
6	epis $f:\underline{A} \twoheadrightarrow \underline{B}$, where for $b \in \underline{B}$ there exist $(M:t)$, $\underline{b} \in \mathrm{dom}(M:t)^{\underline{B}}$, $m \in M$: $b=(m)\underline{b}$ and $\underline{b}(M:t)^{\underline{B}} \in [A]f$	relatively normal homs ($\mathbf{M}_n$)
7	epis $f:\underline{A} \twoheadrightarrow \underline{B}$, where $\underline{B}$ is isom. over id_A to a P-generated A-initial segment of $\underline{F}(\underline{A},$**TAlg**$(\underline{S}))$	P-refl. homs ($\mathbf{M}_P$)
7'	**Mod**(F_P)-ext. epis	↓P-refl. homs ($\mathbf{M}_{\downarrow P}$)
8	-- not yet described in detail (problem) --	P-initial (= P-normal) homs
9	$e \in \Lambda^{op}(\mathbf{M}_P)\cdot\mathbf{H}_f$	P-refl. monos ($\mathbf{S}_P$)
10	$e \in \Lambda^{op}(\mathbf{M}_P)\cdot\mathbf{H}_w$	P-refl. initial homs ($\mathbf{M}_{iP}$)
11	final homs ($\mathbf{M}_{f_i}$)	bijective homs ($\mathbf{H}_b$)

Factorization systems in **Alg**($\underline{S}$)

Table 10.1. (see 10.3.12)

10.3.12. We conclude this Chapter by presenting in Table 10.1 a survey on some factorization systems in **Alg**($\underline{S}$) described earlier in this section (the first eight numbers correspond to those in Table 9.1) and in Figure 10.5 the "closure system" on Hom(**Alg**($\underline{S}$)) generated by the most interesting classes of homomorphisms considered so far (or which will be considered in sections 12 and 13). In addition to the notation already introduced, let us use the following one: $\mathbf{E}_w$ designates the class of all epimorphisms, **I** the class of all isomorphisms, **M** (with some index) stands for the class of all homo-

morphisms with the additional property indicated in the index, similarly, **H** (with some index) designates a class of surjective and **S** (with some index) a class of injective homomorphisms with the additional properties shown in the index. The additional properties are abbreviated by: "f" for "full", "i" for "initial", "c" for "closed", "f_i" for "final", "b" for "bijective", and "P" for "P-reflecting" -- it should be observed that "P" is a parameter, and that in connection with Figure 10.5 we have separating examples for the cases when the similarity type concerns at least two at least unary operation symbols, say ϕ and ϕ', such that for some sequence $\underline{x}$ of variables $\underline{x}\phi$ belongs to P, while for all admissible sequences $\underline{t}$ of terms, $\underline{t}\phi'$ does not belong to P (for more details see [BWo 82a]; the final version of Figure 10.5 is due to R.Wille (see also [Wi 84] for the general principles of drawing such "concept lattices")).
Observe that already $\mathbf{E}_W$, $\mathbf{M}_c$, $\mathbf{M}_f$, $\mathbf{M}_i$, $\mathbf{M}_{f_i}$, $\mathbf{M}_P$, $\mathbf{H}_W$ and $\mathbf{S}_W$ generate this closure system.

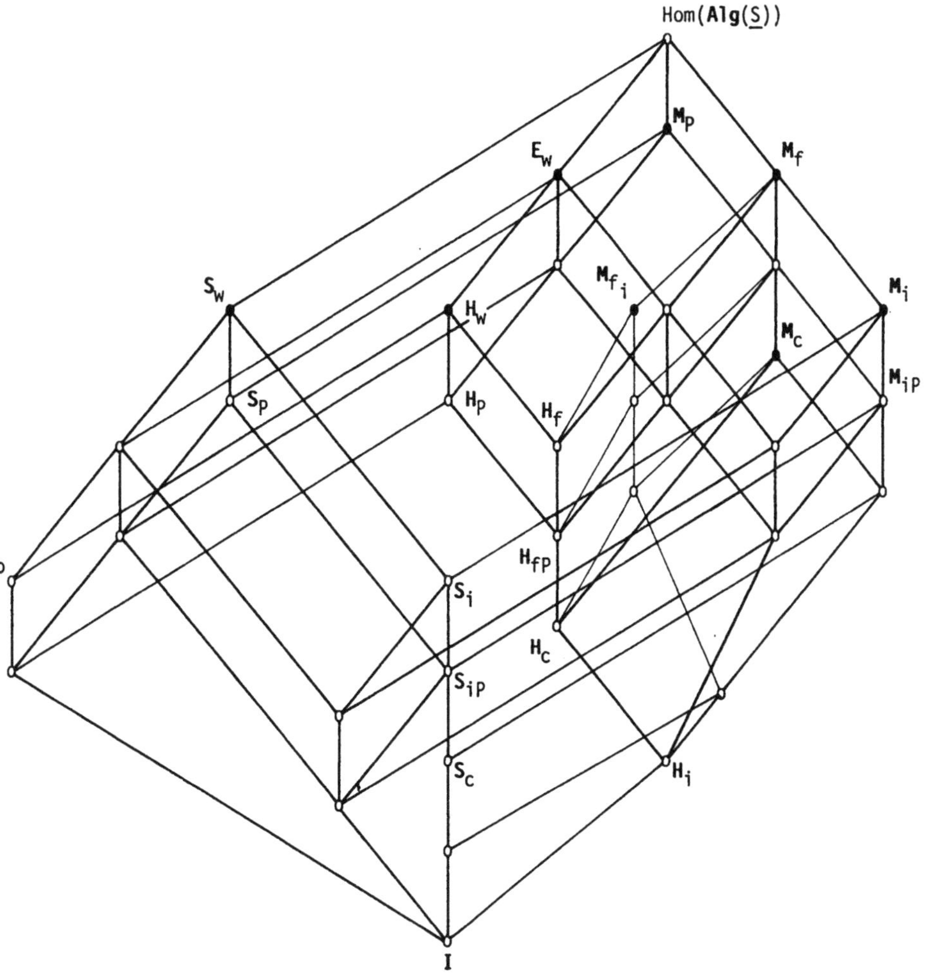

Figure 10.5 (see 10.3.12).

CHAPTER III

BIRKHOFF TYPE RESULTS FOR ELEMENTARY IMPLICATIONS

As can already be realized from the examples in Part II (and also from those in Chapter II of this part), the rôle of equations as most important axioms in the total case is taken over by elementary implications in the case of partial algebras, in particular by ECE-equations. Now H.ANDRÉKA, I.NÉMETI and I.SAIN have shown in [AN 76], [AN 82] and [NSa 77] that category theoretical methods help to provide a meta-theorem which yields Birkhoff type theorems for a wealth of special kinds of elementary implications. This meta-theorem will be the core of this chapter and is presented and proved in section 12, while the rest of this chapter is then dedicated to some of its applications within the category **Alg**($\underline{S}$) of partial algebras.

In **section 11** we first revisit direct limits and introduce some concepts which shall be needed in connection with the main result. In particular morphisms between directed systems are introduced and investigated, and it is investigated how properties of homomorphisms are inherited by the colimiting cocone (respectively, the induced morphisms) between direct limits. A very important feature of direct limits is the fact that they provide the possibility of measuring "smallness". At the end we prove Theorem 8.3.4.

In **section 12** we formulate and prove the Meta-Birkhoff Theorem for elementary implications, and a first attempt is made to test some concepts in the theory of partial algebras as to whether they could be used in connection with this theorem.

In **section 13** we provide as many examples as possible within the category **Alg**($\underline{S}$) for the main theorem of section 12, i.e. of characterizations of classes of partial algebras axiomatizable by special elementary implications.

Finally, in **section 14** we present descriptions of closed classes of some special kinds of elementary implications.

11 Direct limits revisited

In this section we continue the study of direct limits started in subsection 4.4, and we especially investigate how they behave in connection with model theoretic concepts introduced in the meantime. We shall use again the notation introduced in subsection 4.4, and the reader is advised to reread this subsection.

11.1 Morphisms between directed systems

11.1.1 DEFINITION. Let $\underline{C}$ be any category, $\underline{I} := (I;\leq)$ and $\underline{J} := (J;\leq)$ be directed sets, and $\mathbf{A}_{\underline{I}} := (A_i, f_{il}: A_i \to A_l \mid i,l \in I\ ,\ i \leq l)$ and $\mathbf{B}_J := (B_j, g_{jl}: B_j \to B_l \mid j,l \in J\ ,\ j \leq l)$

be two directed systems in $\underline{\mathfrak{C}}$. Then a pair $(m,(m_i|i \in I))$ is called a morphism between directed systems, here from $\mathfrak{A}_{\underline{I}}$ into $\mathfrak{B}_{\underline{J}}$ iff:

(D1) $m:(I;\leq) \to (J;\leq)$ is a monotone mapping, i.e. $i \leq i'$ implies $im \leq i'm$.

(D2) For every $i \in I$, $m_i:A_i \to B_{(i)m}$ is a $\underline{\mathfrak{C}}$-morphism.

(D3) For any $i,i' \in I$ with $i \leq i'$ one has $f_{ii'}\cdot m_{i'} = m_i\cdot g_{(i)m(i')m}$ (i.e. a compatibility condition between the two systems).

-- Such a morphism is called cofinal, if $[I]m$ is cofinal in $\underline{J}$, i.e. if for every $j \in J$ there is $i \in I$ such that $j \leq (i)m$.

-- If $\underline{I} = \underline{J}$ and $m = id_I$, then we call $(id_I,(m_i|i \in I))$ a special morphism.

-- If $I \subseteq J$, $\leq_{\underline{I}} = \leq_{\underline{J}} \cap (I \times I)$ and $m_i = id_{A_i}$ for each $i \in I$, then $(id_I,(id_{A_i}|i \in I)$ is called an inclusion morphism between the given directed systems.

11.1.2. Notice that for a directed system $\mathfrak{A}_I$ (with direct limit $((f_i|i \in I),A)$) and a sink $((m_i:A_i \to B)_{i\in I},B)$, this family can be considered as a special morphism from $\mathfrak{A}_{\underline{I}}$ into $\mathfrak{B}_{\underline{I}} := (B_i := B,\ g_{ij} := id_B|i \leq j \in I)$. This observation will often tacitly be used.

11.1.3 LEMMA. The class of all morphisms (cofinal morphisms, respectively special morphisms) between directed systems (for any fixed poset $\underline{I}$, if necessary) is closed with respect to composition and therefore forms, together with the corresponding directed families, a category. In particular, for directed systems $\mathfrak{A}_{\underline{I}}$, $\mathfrak{B}_{\underline{J}}$ and $\mathfrak{C}_{\underline{K}}$ and morphisms $m^* = (m,(m_i|i \in I)):\mathfrak{A}_{\underline{I}} \to \mathfrak{B}_{\underline{J}}$ and $n^* = (n,(n_j|j \in J)):\mathfrak{B}_{\underline{J}} \to \mathfrak{C}_{\underline{K}}$ we define their composition as $m^*\cdot n^* := (mn,(m_i\cdot n_{(i)m}|i \in I)):\mathfrak{A}_{\underline{I}} \to \mathfrak{C}_{\underline{K}}$.

PROOF. (D1) and (D2) are easily verified. For (D3) (draw a diagram!) we compute:
$f_{ii'}\cdot m_{i'}\cdot n_{(i')m} = m_i\cdot g_{(i)m(i')m}\cdot n_{(i')m} = m_i\cdot n_{(i)m}\cdot h_{(i)mn(i')mn}$.
If $m = n = id_I$ and $\underline{I} = \underline{J} = \underline{K}$, then $mn = id_I$. -- If m^* and n^* are cofinal, and if we consider $k \in K$, then there are $j \in J$, $i \in I$ such that $k \leq (j)n$, $j \leq (i)m$, i.e. $k \leq (j)n \leq (i)mn$, showing that $m^*\cdot n^*$ is cofinal. ■■■

11.1.4 PROPOSITION. Let $\mathfrak{A}_{\underline{I}}$, $\mathfrak{B}_{\underline{J}}$ and $\mathfrak{C}_{\underline{K}}$ be directed systems in **Alg**($\underline{S}$), and let $m^*:\mathfrak{A}_{\underline{I}} \to \mathfrak{B}_{\underline{J}}$ and $n^*:\mathfrak{B}_{\underline{J}} \to \mathfrak{C}_{\underline{K}}$ be morphisms between them. Moreover, let $((f_i|i \in I),\underline{A})$, $((g_j|j \in J),\underline{B})$ and $((h_k|k \in K),\underline{C})$ be direct limits of $\mathfrak{A}_{\underline{I}}$, $\mathfrak{B}_{\underline{J}}$ and $\mathfrak{C}_{\underline{K}}$, respectively. Then:

(i) m^*, n^* and $m^*\cdot n^*$ induce unique homomorphisms $\overline{m}:\underline{A} \to \underline{B}$, $\overline{n}:\underline{B} \to \underline{C}$ and $\overline{p}:\underline{A} \to \underline{C}$, respectively, such that $f_i\overline{m} = m_i g_{(i)m}$ $(i \in I)$, $g_j\overline{n} = n_j h_{(j)n}$ $(j \in J)$, and $f_i\overline{p} = m_i n_{(i)m} h_{(i)mn}$ $(i \in I)$, respectively. Moreover: $\overline{m}\cdot\overline{n} = \overline{p}$.

(ii) If m^* is cofinal, and if all m_i $(i \in I)$ are isomorphisms, then $\overline{m}$ is an isomorphism. If, in addition, $I \subseteq J$, $m_i = id_{A_i}$ and $f_{ij} = g_{ij}$ $(i \leq j \in I)$, then one may choose $((f_i|i \in I),\underline{A}) := ((g_i|i \in I),\underline{B})$.

(iii) Let $\mathbf{S} \subseteq \mathrm{Hom}(\mathbf{Alg}(\underline{S}))$ such that $(\Lambda^{op}(\mathbf{S}),\Lambda\Lambda^{op}(\mathbf{S}))$ is a factorization system, let m^* be cofinal, and assume $\{m_i|i \in I\} \subseteq \Lambda^{op}(\mathbf{S})$, then $\overline{m} \in \Lambda^{op}(\mathbf{S})$ (it is sufficient that, for some $i_0 \in I$, $\{m_i|i_0 \leq i \in I\} \subseteq \Lambda^{op}(\mathbf{S})$).

(iv) Let $\mathbf{S} \subseteq \mathrm{Hom}(\mathbf{Alg}(\underline{S}))$ such that $(\Lambda^{op}(\mathbf{S}),\Lambda\Lambda^{op}(\mathbf{S}))$ is a factorization system, let $\{f_{ii'}|i \leq i' \in I\} \subseteq \Lambda^{op}(\mathbf{S})$, then $\{f_i|i \in I\} \subseteq \Lambda^{op}(\mathbf{S})$.

PROOF. Ad (i): For each $i \in I$ we have $m_i g_{(i)m}:\underline{A}_i \to \underline{B}$ such that, for $i \leq j \in I$, one has $f_{ij} m_j g_{(j)m} = m_i g_{(i)m(j)m} g_{(j)m} = m_i g_{(i)m}$; from (DL3) we therefore infer the existence of a unique $\overline{m}:\underline{A} \to \underline{B}$ such that, for each $i \in I$, one has $f_i \overline{m} = m_i g_{(j)m}$. In the same way we get $\overline{n}$ and $\overline{p}$ with respect to n^* and $m^* \cdot n^*$. Since $f_i \overline{p} = m_i n_{(i)m} h_{(i)mn} = m_i g_{(i)m} \overline{n} = f_i \overline{mn}$ for each $i \in I$ the uniqueness of the induced morphism in (DL3) implies $\overline{p} = \overline{mn}$. ∎

Ad (ii): Let m^* be cofinal, let each m_i be an isomorphism and let $\overline{m}:\underline{A} \to \underline{B}$ be the induced homomorphism according to (i). We observe that, for each $j \in J$, we have some $i \in I$ satisfying $(i)m \geq j$; therefore $q_j := g_{j(i)m} \cdot m_i^{-1} \cdot f_i : \underline{B}_j \to \underline{A}$ is a homomorphism which is independent from the choice of i: if we have $i' \in I$ such that $(i')m \geq j$, then there is $l \in I$, $i, i' \leq l$, and we get

$$q_j = g_{j(i)m} \cdot m_i^{-1} \cdot f_i = g_{j(i)m} \cdot m_i^{-1} \cdot f_{il} \cdot f_l = g_{j(i)m} \cdot g_{(i)m(l)m} \cdot m_l^{-1} \cdot f_l = g_{j(l)m} \cdot m_l^{-1} \cdot f_l = \ldots = g_{j(i')m} \cdot m_{i'}^{-1} \cdot f_{i'}.$$

In the same way we get, for $j \leq j'$ in J and for $l \in I$ satisfying $j' \leq (l)m$: $g_{jj'} \cdot q_{j'} = g_{jj'} \cdot g_{j'(l)m} \cdot m_l^{-1} \cdot f_l = g_{j(l)m} \cdot m_l^{-1} \cdot f_l = q_j$. Thus $(q_j | j \in J)$ induces a homomorphism $\overline{q}:\underline{B} \to \underline{A}$. Now it is not difficult to conclude (because of the uniqueness of the induced homomorphisms) that $\overline{m} \cdot \overline{q} = id_A$ and $\overline{q} \cdot \overline{m} = id_B$ (since $m_i m_i^{-1} = id_{A_i}$ and $g_j \cdot \overline{q} \cdot \overline{m} = g_{j(i)m} \cdot m_i^{-1} \cdot f_i \cdot \overline{m} = g_{j(i)m} \cdot m_i^{-1} \cdot m_i \cdot g_{(i)m} = id_{B_j} \cdot g_j = g_j \cdot id_B$). ∎

Ad (iii): Assume $\{m_i | i \in I\} \subseteq \Lambda^{op}(\mathbf{S})$, let $s \in \mathbf{S}$, $s:\underline{U} \to \underline{V}$, $a, b \in \mathrm{Hom}(\mathbf{Alg}(\underline{S}))$ such that $as = \overline{m}b$. Then (draw a diagram!) for each $i \in I$: $m_i \cdot g_{(i)m} \cdot b = f_i \cdot \overline{m} \cdot b = f_i \cdot a \cdot s$; because of $\mathrm{Difip}(m_i, s)$ there exists a unique $d_{(i)m}:B_{(i)m} \to \underline{U}$ such that $m_i \cdot d_{(i)m} = f_i \cdot a$ and $d_{(i)m} \cdot s = g_{(i)m} \cdot b$. Defining, for arbitrary $j \in J$, $d_j:\underline{B}_j \to \underline{U}$ by $d_j := g_{j(i)m} \cdot d_{(i)m}$ for some $i \in I$, it is not difficult to realize that $((d_j | j \in J), \underline{U})$ is compatible with $\underline{\mathbf{B}}_J$, therefore inducing a unique homomorphism $d:\underline{B} \to \underline{U}$ such that for each $j \in J$ one has $g_j d = d_j$. Since for each $i \in I$ we have $f_i \cdot a = m_i \cdot g_{(i)m} \cdot d = f_i \cdot \overline{m} \cdot d$, the uniqueness property in (DL3) implies $a = \overline{m} \cdot d$. Similarly we get, for each $j \in J$, $g_j \cdot b = d_j \cdot s = g_j \cdot d \cdot s$, and therefore, because of (DL3) and (ii), $b = d \cdot s$ showing that $\overline{m} \in \Lambda^{op}(\mathbf{S})$. ∎

Ad (iv): Assume now that $\{f_{ii'} | i \leq i' \in I\} \subseteq \Lambda^{op}(\mathbf{S})$, let $i \in I$, $s \in \mathbf{S}$, $s:\underline{U} \to \underline{V}$, and $a, b \in \mathrm{Hom}(\mathbf{Alg}(\underline{S}))$ such that $f_i b = as$. Observe that $I_i := \{i' \in I | i \leq i'\}$ is cofinal in $\underline{I}$. For each $i' \in I_i$ we have $as = f_i b = f_{ii'} f_{i'} b$ and $f_{ii'} \in \Lambda^{op}(\mathbf{S})$. Therefore there exists a unique $d_{i'}:\underline{A}_{i'} \to \underline{U}$ such that $f_{ii'} d_{i'} = a$ and $d_{i'} s = f_{i'} b$. Let $i'' \geq i' \geq i$; then $f_{ii'} f_{i'i''} d_{i''} = a = f_{ii'} d_{i'}$, and $d_{i'} s = f_{i'} b = f_{i'i''} f_{i''} b = f_{i'i''} d_{i''} s$. 10.2.2.(iii) then implies $f_{i'i''} d_{i''} = d_{i'}$, and we therefore get a unique $d:\underline{A} \to \underline{U}$ such that $f_{i'} d = d_{i'}$ for each $i' \in I_i$. Since $d_i = a$, this yields $f_i d = d_i$. For each $i' \in I_i$ we have, in addition, $f_{i'} ds = d_{i'} s = f_{i'} b$. Therefore the uniqueness in (DL3) implies $ds = b$, showing that $f_i \in \Lambda^{op}(\mathbf{S})$. ∎∎∎

From 11.1.4.(iv) and 10.3 (see Table 10.1) we infer the

11.1.5 COROLLARY. Let $(\underline{A}_i, (f_{ij} | i \leq j \in I))$ be a directed system with direct limit $((f_i | i \in I), \underline{A})$. Assume that each f_{ij} is surjective, full and surjective, respectively an epimorphism (or has any other property described on the left hand side of Table 10.1). Then each f_i $(i \in I)$ has the same property. ∎∎∎

Similarly 11.1.4.(iii) implies together with 10.3

11.1.6 COROLLARY. Let $(m,(m_i|i \in I)):\mathbb{A}_{\underline{I}} \to \mathbb{B}_{\underline{J}}$ be a cofinal morphism between directed systems, and let $\bar{m}$ be the induced homomorphism between the corresponding direct limits. Then one has: If each m_i $(i \in I)$ is surjective, full and surjective, respectively an epimorphism (or has any other property described on the left hand side of Table 10.1), then $\bar{m}$ has the same property. ■■■

11.2 A device to measure "smallness"

Results similar to those of Proposition 11.1.4 and its corollaries can also be obtained with respect to the operator "Λ", but not in the same generality. As a preparation we need the following definition from BANASCHEWSKI and HERRLICH (see [BaH 76], ANDRÉKA and NÉMETI [AN 76] and other papers):

11.2.1 DEFINITION. Let $B \in Ob(\underline{\mathbb{C}})$ for some category $\underline{\mathbb{C}}$, and let $\mathbf{k}$ be any infinite regular cardinal. Then B is called $\mathbf{k}$-<u>small</u> iff

(*) for every $\mathbf{k}$-directed system $\mathbb{A}_{\underline{I}} = (A_i, f_{ij}|i \leq j \in I)$ in $\underline{\mathbb{C}}$ with $\mathbf{k}$-direct limit $((f_i|i \in I),A)$, and for every $\underline{\mathbb{C}}$-morphism $g:B \to A$ there exist some $i \in I$ and $h:B \to A_i$ such that $hf_i = g$.

B is called <u>strongly</u> $\mathbf{k}$-<u>small</u> iff one has in addition to (*):

(**) For any $\mathbf{k}$-directed system $\mathbb{A}_{\underline{I}}$ with $\mathbf{k}$-direct limit as in (*) and for any two $\underline{\mathbb{C}}$-morphisms $h,h':B \to A_i$ satisfying $hf_i = h'f_i$ there exists $j \geq i$ in I such that already $hf_{ij} = h'f_{ij}$.

11.2.2 PROPOSITION. Let $\mathbf{k} \geq \mathbf{d^*}$ and $\underline{B} \in \mathbf{Alg}(\underline{S})$. Then the following statements are equivalent:

(i) $\underline{B}$ is $\mathbf{k}$-small.

(ii) $\underline{B}$ is strongly $\mathbf{k}$-small.

(iii) $\#B < \mathbf{k}$ and $\#(\mathbf{U}(\{\phi\} \times \mathrm{dom}\, \phi^{\underline{B}}|\phi \in \Omega)) < \mathbf{k}$.

PROOF. "(iii) ⇒ (ii)": Let (iii) be satisfied, let $\mathbb{A}_{\underline{I}}$ be a $\mathbf{k}$-directed system with direct limit $((f_i|i \in I),A)$. Let $g:\underline{B} \to \underline{A}$ be any homomorphism. Because of the $\mathbf{d^*}$-directedness of $\underline{I}$ $(\mathbf{d^*} \leq \mathbf{k})$ we know by 4.4.4 that there exists, for every $b \in B$, some $(b)i \in I$ such that $bg = a_b' f_{(b)i}$ for some $a_b' \in A_{(b)i}$. By the $\mathbf{k}$-directedness of $\underline{I}$ we get some j in I such that -- having chosen (b)i and a_b' for each $b \in B$ -- we have $(b)i \leq j$ and $bg = a_b' f_{(b)ij} f_j$. Define $h':B \to A_j$ by $bh' := a_b' f_{(b)ij}$.
Let $\underline{b} \in \mathrm{dom}\, \phi^{\underline{B}}$, then $\underline{b}g \in \mathrm{dom}\, \phi^{\underline{A}}$, and 4.4.4.(ii) implies the existence of $(\underline{b})i \in I$ such that there is $\underline{a_b} \in \mathrm{dom}\, \phi^{\underline{A}(\underline{b})i}$ satisfying $\underline{a_b} \cdot f_{(\underline{b})i} = \underline{b} \cdot g$. Since $\underline{I}$ is $\mathbf{k}$-directed for an infinite regular cardinal $\mathbf{k}$ there exists, because of the assumptions about $\underline{B}$, some $i \in I$ such that $j,(\underline{b})i \leq i$ for each $\phi \in \Omega$ and for each $\underline{b} \in \mathrm{dom}\, \phi^{\underline{B}}$. Then it is easy to realize that $h := h'f_{ji}:\underline{B} \to \underline{A}_i$ is a homomorphism which satisfies $hf_i = g$. This shows that $\underline{B}$ is $\mathbf{k}$-small.
-- Now assume that $p,q:\underline{B} \to \underline{A}_i$ are two homomorphisms for some $i \in I$ such that $pf_i = qf_i$. Then a similar argument as above shows the existence of some $j \geq i$ in I such that $bpf_{ij} = bqf_{ij}$ for all $b \in B$, i.e. $pf_{ij} = qf_{ij}$, showing that $\underline{B}$ is even

strongly **k**-small. ∎

"(ii) ⇒ (i)" is obvious. ∎

"(i) ⇒ (iii)" (by contraposition): Assume that $\underline{B}$ violates (iii). Then (by 4.4.7) $\underline{B} = \varinjlim \mathfrak{A}_{I_3(\underline{B},B)}$ where each $\underline{C}$ in $I_3(\underline{B},B)$ satisfies (iii), and $\underline{B}$ cannot be isomorphic to any such $\underline{C}$. By choosing $g := id_B : \underline{B} \to \varinjlim \mathfrak{A}_{I_3(\underline{B},B)} = \underline{B}$ we realize that there cannot exist any homomorphism $h:\underline{B} \to \underline{C}$ for any $\underline{C} \in I_3(\underline{B},B)$ such that $h \cdot id_C = id_B = g$. ∎∎∎

11.2.3 REMARKS. (i): Observe that in 11.2.2 the arguments proving "(iii) ⇒ (ii) ⇒ (i)" remain true in general for arbitrary infinite regular cardinals, and therefore the objects described in 11.2.2.(iii) are always strongly **k**-small in **Alg**($\underline{S}$). But the argument which we have used in proving "(i) ⇒ (iii)" fails in general (see 4.4.5). We did not check, whether 11.2.2 remains true also for $\#\omega_0 \leq \mathbf{k} < \mathbf{d^*}$, since we shall not need this later. Thus we leave it as a **problem** to the reader.

(ii): In the case of total (heterogeneous) algebras the classes of **k**-small and strongly **k**-small algebras do not coincide. Without proof we state that for $\mathbf{k} \geq \mathbf{d^*}$ an algebra $\underline{A} \in \mathbf{TAlg}(\underline{S})$ is

(a) strongly **k**-small in **TAlg**($\underline{S}$), iff $\#A \leq \mathbf{k}$; and it is

(b) **k**-small, iff $\underline{A}$ is isomorphic to the quotient $\underline{T}(M,\mathbf{TAlg}(\underline{S}))/R$ of a **k**-small generated Peano algebra ($\#M < \mathbf{k}$) with respect to a **k**-small generated congruence R. ∎

(iii): From 11.2.2 it easily follows that full homomorphic images of (strongly) **k**-small partial algebras ($\mathbf{k} \geq \mathbf{d^*}$) are also (strongly) **k**-small.

(iv): For historical remarks -- also for the following concept -- see [NSa 77].

11.2.4 DEFINITION. A category $\underline{\mathfrak{C}}$ is called <u>strongly</u> **k**-<u>algebroidal</u> (see [BaH 76]), iff each $\underline{\mathfrak{C}}$-object is the direct limit of strongly **k**-small objects.

11.2.5. 4.4.7 and 11.2.3 together show that **Alg**($\underline{S}$) is strongly **k**-algebroidal for every regular cardinal $\mathbf{k} \geq \mathbf{d^*}$.

11.3 Direct limits and the operator "Λ"

Now we are able to prove an analogue to Proposition 11.1.4.(iii) and (iv):

11.3.1 PROPOSITION. Let $\mathfrak{A}_{\underline{I}} = (\underline{A}_i, f_{ik} | i \leq k \in I)$ and $\mathfrak{B}_{\underline{J}} = (\underline{B}_j, g_{jl} | j \leq l \in J)$ be **k**-directed systems ($\mathbf{k} \geq \mathbf{d^*}$, regular), and let $((f_i | i \in I), \underline{A})$ respectively $((g_j | j \in J), \underline{B})$ be their **k**-direct limits. Moreover, let $m^* = (m, (m_i | i \in I)) : \mathfrak{A}_{\underline{I}} \to \mathfrak{B}_{\underline{J}}$ be a cofinal morphism between directed systems. Finally assume $\mathbf{E} \subseteq \mathrm{Hom}(\mathbf{Alg}(\underline{S}))$ such that domain and codomain of each **E**-morphism are strongly **k**-small, and that $(\Lambda^{op}\Lambda(\mathbf{E}), \Lambda(\mathbf{E}))$ is a factorization system. Then we obtain:

(i) If each m_i belongs to $\Lambda(\mathbf{E})$ (for all $i \geq i_0$ in I for some i_0), then so does the colimit morphism $\overline{m}:\underline{A} \to \underline{B}$.

(ii) If $\{f_{ik} | i \leq k \in I\} \subseteq \Lambda(\mathbf{E})$, then $\{f_i | i \in I\} \subseteq \Lambda(\mathbf{E})$.

PROOF. Ad (i): Let $e:\underline{X} \to \underline{Y}$ be any **E**-morphism and $a,b \in \mathrm{Hom}(\mathbf{Alg}(\underline{S}))$ such that $a\overline{m} = eb$. Since $\underline{X}$ and $\underline{Y}$ are strongly **k**-small, there exist $i \in I$, $j \in J$, $a':\underline{X} \to \underline{A}_i$, $b':\underline{Y} \to \underline{B}_j$ such that $a = a'f_i$ and $b = b'g_j$. Since [I]m is cofinal in $\underline{J}$, since $\underline{I}$ and

$\underline{J}$ are directed, and since $\underline{X}$ and $\underline{Y}$ are strongly **k**-small, there exists $k \in I$ such that $(i)m,j \leq (k)m$ and $a'f_{ik}m_k = b'g_{j(k)m}$. $m_k \in \Lambda(\mathbf{E})$ implies the existence of a unique $d_k:\underline{Y} \to \underline{A}_k$ such that $a'f_{ik} = ed_k$ and $d_km_k = b'g_{j(k)m}$. Therefore, $a = a'f_{ik}f_k = e(d_kf_k)$ and $(d_kf_k)\overline{m} = d_km_kg_{(k)m} = b$, showing that $d := d_kf_k$ is the diagonal-fill-in we have looked for. -- 10.2.2.(iii) and 10.2.5 then imply the uniqueness of d. ∎

Ad (ii): Let $i \in I$ be arbitrary, consider $e \in \mathbf{E}$ as in the proof of (i) and $a,b \in \mathrm{Hom}(\mathbf{Alg}(\underline{S}))$ such that $af_i = eb$. The facts that $\underline{Y}$ is (strongly) **k**-small and $\mathcal{A}_{\underline{I}}$ is **k**-directed imply the existence of $k \in I$ (w.l.o.g. $k \geq i$) and $b':\underline{Y} \to \underline{A}_k$ such that $b'f_k = b$. Since $\underline{X}$ is strongly **k**-small there exists $j \geq k$ such that $af_{ij} = eb'f_{kj}$. W.l.o.g. $k = j$, i.e. $af_{ik} = eb'$. Since $f_{ik} \in \Lambda(\mathbf{E})$, there exists a unique $d:\underline{Y} \to \underline{A}_i$ such that $ed = a$ and $df_{ik} = b'$. Then $df_i = df_{ik}f_k = b'f_k = b$. Hence $\mathrm{Difip}(e,f_i)$. -- As in (i) we can infer the uniqueness of d. ∎∎∎

From section 9 (cf. Table 9.1) and section 10 (see Table 10.1) we can infer the

11.3.2 COROLLARY 1. Let $(\underline{A}_i, f_{ij} | i \leq j \in I)$ be a **d***-directed system in **Alg**($\underline{S}$), and let $((f_i | i \in I), \underline{A})$ be its direct limit. If all f_{ij} are injective, closed, initial, relatively normal, P-reflecting, respectively P-initial (for some set P of terms), then each f_i has the same property. ∎∎∎

11.3.3 COROLLARY 2. Let $(m,(m_i | i \in I)):\mathcal{A}_{\underline{I}} \to \mathcal{B}_{\underline{J}}$ be a cofinal morphism between **k**-directed systems, and let $\overline{m}:\varinjlim \mathcal{A}_{\underline{I}} \to \varinjlim \mathcal{B}_{\underline{J}}$ be the induced homomorphism. If each m_i is injective, closed, initial, relatively normal, P-reflecting, respectively P-initial (for some set P of terms), then $\overline{m}$ has the same property. ∎∎∎

11.3.4. The results of 11.1.4 and 11.3.1 can also be applied in connection with properties, which are not necessarily directly related to a factorization system, but only so-called "relativized" properties like "fullness" or "relative closedness":

-- $f:\underline{A} \to \underline{B}$ is full, iff for every $(\mathbf{H}_f, \mathbf{S}_w)$-factorization (s,m) of f (i.e. $f = sm$) the monomorphism m is initial;
-- $f:\underline{A} \to \underline{B}$ is relatively closed, iff for every $(\mathbf{H}_w, \mathbf{S}_i)$-factorization (s,m) of f the epimorphism s is closed;
-- $f:\underline{A} \to \underline{B}$ is relatively P-reflecting, iff for every $(\mathbf{H}_w, \mathbf{S}_i)$-factorization (s,m) of f the epimorphism s is P-reflecting.

Thus the following results will show that all these properties are preserved in the sense of 11.3.2 as well as in the sense of 11.3.3 by **k**-direct limits ($\mathbf{k} \geq \mathbf{d^*}$) of **k**-directed systems.

11.3.5 LEMMA. With the same general assumptions and notation on $\mathcal{A}_{\underline{I}}$, $\mathcal{B}_{\underline{J}}$ and m^* as in 11.3.1 let $(\mathbf{H},\mathbf{S}) := (\Lambda^{op}\Lambda(\mathbf{E}), \Lambda(\mathbf{E}))$ be any factorization system as in 11.3.1. Moreover, for each $i \in I$ let (h_i, s_i) be an $(\mathbf{H},\mathbf{S})$-factorization of m_i, $h_i:\underline{A}_i \to \underline{E}_i$, $s_i:\underline{E}_i \to \underline{B}_i$. Then there exists for $i \leq j \in I$ a homomorphism $e_{ij}:\underline{E}_i \to \underline{E}_j$ such that, for $i \leq j \leq k \in I$, one has $h_ie_{ij} = f_{ij}h_j$, $s_ig_{(i)m(j)m} = e_{ij}s_j$, $e_{ii} = id_{\underline{E}_i}$ and $e_{ij}e_{jk} = e_{ik}$, i.e. $\mathcal{E}_{\underline{I}} := (\underline{E}_i, e_{ij} | i \leq j \in I)$ is a **k**-directed system, and $h^* := (id_I, (h_i | i \in I)):\mathcal{A}_{\underline{I}} \to \mathcal{E}_{\underline{I}}$ and $s^* := (m, (s_i | i \in I)):\mathcal{E}_{\underline{I}} \to \mathcal{B}_{\underline{J}}$ are cofinal morphisms between **k**-directed systems, and the induced homomorphisms $\overline{h}:\varinjlim \mathcal{A}_{\underline{I}} \to \varinjlim \mathcal{E}_{\underline{I}}$ and

$\overline{s}:\underrightarrow{\lim}\, \mathcal{E}_{\underline{I}} \to \underrightarrow{\lim}\, \mathcal{B}_{\underline{J}}$ belong to **H**, respectively to **S** -- the assumptions concerning the existence of **E** are only needed for " $\overline{s} \in S$ " .

PROOF. Since for $i \le j \in I$ we have $h_i(s_i g_{(i)m(j)m}) = m_i g_{(i)m(j)m} = f_{ij} m_j = (f_{ij} h_j) s_j$, where $h_i \in \mathbf{H}$ and $s_j \in \mathbf{S}$, there exists a unique diagonal-fill-in $e_{ij}:\underline{E}_i \to \underline{E}_j$ such that the usual commutativity relations hold. The uniqueness of diagonal-fill-ins also easily implies $e_{ij}e_{jk} = e_{ik}$ for $i \le j \le k \in I$. The special assumptions on (**H**,**S**) imply together with 11.1.4.(iii) respectively 11.3.1.(i) that $\overline{h} \in \mathbf{H}$ and $\overline{s} \in \mathbf{S}$. ■■■

11.3.6. Since $(\mathbf{H}_w, \mathbf{S}_i)$ and $(\mathbf{H}_f, \mathbf{S}_w)$ satisfy the assumptions on (**H**,**S**) in Lemma 11.3.5, it is now easy to realize that **k**-direct limits (**k** ≥ **d***) of **k**-directed systems preserve fullness, relative closedness or P-reflection, since they preserve surjectivity (together with fullness), injectivity, initialness, closedness and P-reflection. ■■■

11.4 Sections, retractions and direct limits

The following two properties of morphisms are often useful in proving that some morphism is in fact an isomorphism:

11.4.1 DEFINITION. Let $\underline{\mathbf{C}}$ be any category, $A,B \in \mathrm{Ob}(\underline{\mathbf{C}})$, $s:A \to B$, $r:B \to A$ $\underline{\mathbf{C}}$-morphisms. s is called a <u>section</u> iff there is $r':B \to A$ such that $s \cdot r' = \mathbf{1}_A$, and r is called a <u>retraction</u> iff there is $s':A \to B$ such that $s' \cdot r = \mathbf{1}_A$.
If r is a retraction or s a section, then A is called a <u>retract</u> of B.

11.4.2. It should be observed that every section is a monomorphism, and that every retraction is an epimorphism. Moreover, in 11.4.1 r' is a retraction and s' is a section, too. ■

The simple category theoretical proof of the following statement is left as an exercise (see [HS 73], 6.7 and 6.15):

11.4.3 PROPOSITION. Let $\underline{\mathbf{C}}$ be any category, $f \in \mathrm{Mor}(\underline{\mathbf{C}})$. Then the following statements are equivalent:
(i) f is an isomorphism.
(ii) f is a section and an epimorphism of $\underline{\mathbf{C}}$.
(iii) f is a retraction and a monomorphism of $\underline{\mathbf{C}}$. ■■■

11.4.4 PROPOSITION. In **Alg**($\underline{S}$) every section is injective and closed, and every retraction is full and surjective.

PROOF. Let $s:\underline{A} \to \underline{B}$ be a section with corresponding retraction $r:\underline{B} \to \underline{A}$. Since $sr = id_A$ is closed and bijective, s is closed and injective, and r is surjective. Since the restriction of r to <u>[A]s</u> is still closed and surjective, it is easy to realize that r is full. ■■■

11.4.5 PROPOSITION. Let $m^*:\mathcal{A}_{\underline{I}} \to \mathcal{B}_{\underline{I}}$ and $n^*:\mathcal{B}_{\underline{I}} \to \mathcal{A}_{\underline{I}}$ be special morphisms between directed systems, and let $\overline{m}:\underrightarrow{\lim}\, \mathcal{A}_{\underline{I}} \to \underrightarrow{\lim}\, \mathcal{B}_{\underline{I}}$ respectively $\overline{n}:\underrightarrow{\lim}\, \mathcal{B}_{\underline{I}} \to \underrightarrow{\lim}\, \mathcal{A}_{\underline{I}}$ be the induced colimit homomorphisms. Assume that each m_i is a section and n_i its corresponding retraction. Then $\overline{m}$ is a section and $\overline{n}$ its corresponding retraction.

PROOF. Since $m_i n_i = id_{A_i}$, we get $\overline{m}\cdot\overline{n} = id_.$ by 11.1.4.(i) and (ii). ∎

11.5 Some further concepts connected with direct limits

In this subsection we describe some category theoretical concepts and their partial algebraic interpretations in **Alg**($\underline{S}$), which will be needed for the meta theorem of H.ANDRÉKA, I.NÉMETI and I.SAIN on implicationally definable classes, which we shall present in the next section. In 11.2.5 we have already realized that **Alg**($\underline{S}$) is a strongly **d***-algebroidal category.

11.5.1 DEFINITION. Let $\underline{\mathbf{C}}$ be any category, **k** any infinite cardinal. A class S of $\underline{\mathbf{C}}$-morphisms is called <u>**k**-inductive</u> iff any **k**-direct limit of "S-subobjects" is again an "S-subobject". I.e.: S is **k**-inductive iff for any **k**-directed system $(A_i, f_{ij} | i \le j \in I)$ with direct limit $((f_i | i \in I), A)$ and for every compatible sink $((m_i : A_i \to B | i \in I), B)$ with $\{m_i | i \in I\} \subseteq S$ one has $\overline{m} \in S$ for the induced $\underline{\mathbf{C}}$-morphism $\overline{m} : A \to B$.

11.5.2 REMARK and EXAMPLES. Propositions 11.1.4.(iii) and 11.3.1.(i) tell us that in **Alg**($\underline{S}$) we have: For every factorization system (**E**,**M**) of **Alg**($\underline{S}$), **E** is **k**-inductive for every infinite regular cardinal **k**. Moreover, **M** is **k**-inductive for **k** ≥ **d***, if **M** = Λ(**E'**), where **E'** ⊆ **E**, and for each **E'**-morphism, say $e:\underline{X} \to \underline{Y}$, $\underline{X}$ and $\underline{Y}$ are strongly **k**-small.
In particular, the classes of all epimorphisms, all surjective homomorphisms, respectively of all full and surjective homomorphisms are **k**-inductive for every infinite regular cardinal **k**; and the classes of all injective homomorphisms, of all closed, closed and injective, initial, P-reflecting, injective and initial, injective and P-reflecting respectively relatively normal homomorphisms are **d***-inductive and **k**-inductive for every regular cardinal **k** ≥ **d*** .

11.5.3 DEFINITION. Let $\underline{\mathbf{C}}$ be any category and **k** any infinite regular cardinal. A class E of $\underline{\mathbf{C}}$-morphisms is called <u>**k**-dual algebroidal</u> (cf. [AN 82], p.16f) iff every E-morphism $e:A \to B$ -- for which A is strongly **k**-small -- is a direct limit of E-morphisms f_{ij} for which $Dom(f_{ij})$ as well as $Cod(f_{ij})$ are strongly **k**-small. More precisely: **E** is **k**-dual algebroidal iff, for every **E**-morphism $e:A \to B$ with strongly **k**-small domain A, there exists a directed system $(A_i, f_{ij} | i \le j \in I)$ with direct limit $((f_i | i \in I), B)$ such that there is $i' \in I$ satisfying $e = f_{i'}$, and for $i \le j \in I$ the domain as well as the codomain of f_{ij} are strongly **k**-small, and $f_{ij} \in$ **E** .

11.5.4 REMARK. Notice that the above concept is a generalization of the fact that every (E-) congruence relation is a directed union of finitely (or **d***-small) generated (E-) congruence relations (cf. 4.4.11). -- If **E** consists of all full and surjective homomorphisms, then 4.4.11, 11.2.2 and 11.2.3.(iii) show that **E** is **d***-dual algebroidal. And in combination with 4.4.7 we may conclude that many classes of epimorphisms in **Alg**($\underline{S}$) are **d***-dual algebroidal, as we shall see in section 13.1.

11.6 Proof of Theorem 8.3.4

The reader should recall the contents of subsection 8.3 and observe, that the statements of Theorem 8.3.4 concerning the operators **P**, $\mathbf{P}_+$ and e have already been proved in 8.3.

11.6.1. The operators **H** (= $\mathbf{H}_w$) and $\mathbf{H}_f$: That (full) homomorphic images preserve E-equations, has been shown in 6.2.1.(i). That ECE-equations are in general not preserved is shown by the example sketched in Figure 11.1 for $\tau = (1)$ and $F = (\{x\}:x\phi\phi \stackrel{e}{=} x\phi\phi \Rightarrow x\phi\phi\phi \stackrel{e}{=} x\phi\phi\phi)$. ∎

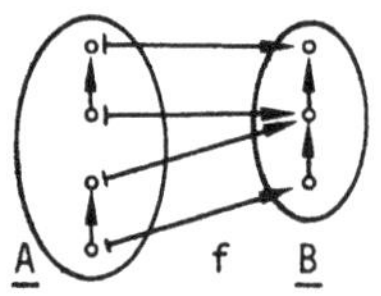

Figure 11.1.

The homomorphism of Fig. 11.1 shows that negations of TE-expressions -- here $(\{x\}:\neg(x\phi\phi \stackrel{e}{=} x\phi\phi))$ -- are not preserved by (full) homomorphic images. ∎

11.6.2. That there is some sort of ECE-equations which is preserved by full homomorphic images will be discussed in section 13 (e.g. in 13.3).

11.6.3. The operator $\mathbf{H}_c$: By definition closed homomorphisms preserve all negations of TE-expressions (since they reflect all TE-expressions). Therefore also arbitrary disjunctions of such formulas are preserved (cf. 9.1.7.(iv)). Thus, if $f:\underline{A} \to \underline{B}$ is a closed and surjective homomorphism, if $\underline{A} \models (M:V^*\{\neg(t_i \stackrel{e}{=} t_i)|i \in I\})$, and if $w:M \to B$ is any assignment, then the surjectivity of f implies that we can choose $w':M \to A$ (always compatible with the sort mappings!) such that $w = w'f$. Since f preserves our formula and $\underline{A} \models (M:V^*\{\neg(t_i \stackrel{e}{=} t_i)|i \in I\})[w']$, we get
$\underline{B} \models (M:V^*\{\neg(t_i \stackrel{e}{=} t_i)|i \in I\})[w'f]$ $(w = w'f)$.
Moreover, we show below that surjective and closed homomorphisms preserve ECE-equations, and therefore the same argument as above shows that $\mathbf{H}_c$ preserves ECE-equations: Assume that $\underline{A} \models (M:\Lambda^*\{t_i \stackrel{e}{=} t_i|i \in I\} \Rightarrow t \stackrel{e}{=} t')$, and let $w:M \to B$ be an assignment, such that $\underline{B} \models (M:\Lambda^*\{t_i \stackrel{e}{=} t_i|i \in I\})[w]$. As above we have $w':M \to A$ such that $w'f = w$. Since f reflects $(M:\Lambda^*\{t_i \stackrel{e}{=} t_i|i \in I\})$, we get $\underline{A} \models (M:\Lambda^*\{t_i \stackrel{e}{=} t_i|i \in I\})[w']$ and therefore $\underline{A} \models (M:t \stackrel{e}{=} t')[w']$. Thus $\underline{B} \models (M:t \stackrel{e}{=} t')[w'f]$, since f preserves E-equations.
Finally we have to show that closed homomorphic images preserve neither negations of E-equations nor QE-equations. For this purpose consider, for $\tau := (1)$, the formulas $F:= (\{x\}:\neg(x\phi \stackrel{e}{=} x\phi\phi))$ and $G:= (\{x\}:x\phi \stackrel{e}{=} x\phi\phi \Rightarrow x\phi \stackrel{e}{=} x)$, and consider Figure 11.2, where $\underline{A}:= (\mathbf{N};\phi)$ is the set of natural numbers with successor operation ϕ. Then F and G are valid in $\underline{A}$, but not in $\underline{B}$. ∎

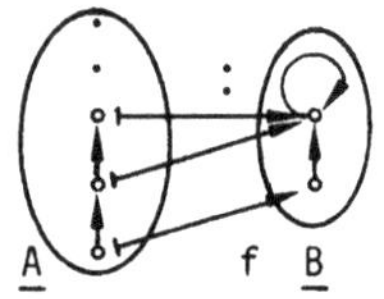

Figure 11.2.

11.6.4. The operators $\mathbf{S}_w$ and $\mathbf{S}_r$: Let $f:\underline{A} \to \underline{B}$ be an injective homomorphism. Since f preserves atomic formulas, it reflects negations of them. Thus, if in $\underline{B}$ the negation of an E-equation is valid, then this is also valid in $\underline{A}$. Now consider $F:= (\{x\}:x\phi \stackrel{e}{=} x\phi)$, $(\tau:= (1))$, then Figure 11.3 shows that (weak) relative subalgebras do not preserve TE-expressions. ∎

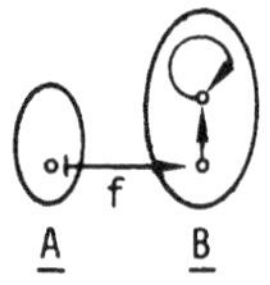

Figure 11.3.

11.6.5. The operators **S** (= $\mathbf{S}_c$) and **I**: Since closed monomorphisms reflect all elementary implications (i.e. epimorphisms) and therefore also their disjunctions, 9.1.8.(ii) implies that subalgebras preserve all elementary implications. Moreover, they reflect negations of E-equations and therefore also their disjunctions. Thus 9.1.8.(ii) implies that subalgebras preserve all universal Horn formulas. The same trivially holds for isomorphic copies. ∎

In preparation for the next step we prove the following

11.6.6 LEMMA. Let $F := (M : \wedge^*\{t_k \stackrel{e}{=} t'_k \mid k \in K\} \Rightarrow \wedge^*\{\bar{t}_l \stackrel{e}{=} \bar{t}'_l \mid l \in L\})$ be any elementary implication with $\#K, \#L < \mathbf{k}$ for some infinite regular cardinal $\mathbf{k} \geq \mathbf{d}^{\&}$, and let $e : \underline{P} \to \underline{C}$ be its encoding epimorphism. Let $\mathbf{A}_I := (\underline{A}_i, f_{ij} \mid i \leq j \in I)$ be a $\mathbf{k}$-directed system with $\mathbf{k}$-direct limit $((f_i \mid i \in I), \underline{A})$, and assume that $\underline{A}_i \models F$ for all $i \in I$. Then $\underline{A} \models F$.

PROOF. Observe that the assumptions on F imply that $\underline{P}$ and $\underline{C}$ are strongly $\mathbf{k}$-small, that each f_{ij} trivially preserves and reflects F, and that each f_i reflects F, because F is valid in $\underline{A}_i$. Now, assume that we have $g : \underline{P} \to \underline{A}$. Since $\underline{P}$ is strongly $\mathbf{k}$-small, there exist $i \in I$ and $g' : \underline{P} \to \underline{A}_i$ such that $g = g'f_i$. Since F is valid in $\underline{A}_i$, there exists $h' : \underline{C} \to \underline{A}_i$ such that $g' = eh'$. Thus $eh'f_i = g'f_i = g$. This shows that $\underline{A} \models F$. ∎∎∎

11.6.7. The operators $\mathbf{P}^{r,\mathbf{k}}$, $\mathbf{P}_+{}^{r,\mathbf{k}}$, $\mathbf{P}_u$ and $\underrightarrow{\lim}^{\mathbf{k}}$: Since the sort algebra $\underline{S}$ always belongs to $\mathbf{P}^{r,\mathbf{k}}(\mathbf{K})$ for any class $\mathbf{K}$, $\mathbf{P}^{r,\mathbf{k}}$ cannot preserve negations of TE-expressions. -- If $\mathbf{k} \geq \mathbf{d}^{\&}, \mathbf{c}$ ($\mathbf{c}$ being the cardinal connected with the language L, see 7.1.1) is an infinite regular cardinal, then Lemma 11.6.6 and the definition of $\mathbf{k}$-reduced products in 4.5.3 show that the operators $\mathbf{P}^{r,\mathbf{k}}$, $\mathbf{P}_+{}^{r,\mathbf{k}}$ and $\underrightarrow{\lim}^{\mathbf{k}}$ preserve elementary implications (restricted by $\mathbf{c}$). Thus it remains to show that $\mathbf{P}_+{}^{r,\mathbf{k}}$ preserves formulas of the kind $F := (M : \neg(\wedge^*\{t_k \stackrel{e}{=} t'_k \mid k \in K\}))$ ($\#K < \mathbf{c}$): Let $\underline{P}$ be the relative subalgebra of $\underline{F}(M, \mathbf{TAlg}(\underline{S}))$ with carrier set $M \cup \downarrow F$. Then we recall that validity of F in $\underline{A}$ means that there is no homomorphism from $\underline{P}$ into $\underline{A}$. Let $(\underline{A}_i \mid i \in I)$ be a non-empty family of partial algebras such that F is valid in each $\underline{A}_i$, and let $\mathbf{G}$ be a proper $\mathbf{k}$-complete filter on I. Since $\underline{P}$ is strongly $\mathbf{k}$-small (even strongly $\mathbf{c}$-small), there cannot exist any homomorphism from $\underline{P}$ into $\mathbf{X}^{\mathbf{k}}(\underline{A}_i \mid i \in I)/\mathbf{G}$, since this would imply the existence of a homomorphism from $\underline{P}$ into some $\underline{A}_i$. Hence in this $\mathbf{k}$-reduced product F is also valid. -- Since ultrafilters are proper filters, the index set has to be non-empty; thus $\mathbf{P}_u$ is contained in the operator $\mathbf{P}_+{}^{r,\#\omega_0}$, and like this operator it preserves all kinds of finite formulas under consideration. ∎∎∎

11.6.8 REMARKS. (i): The statement about ultraproducts also follows from the fact that the Ultraproduct Theorem or Łos' Lemma holds in all categories (cf. [AN 78]): A finitary and finite formula F holds in an ultraproduct $\underline{A} := \mathbf{X}^{\#\omega_0}(\underline{A}_i \mid i \in I)/\mathbf{G}$ iff $\{i \in I \mid \underline{A}_i \models F\} \in \mathbf{G}$.

(ii): We would also like to mention without proof that formulas of the kind $(M : \wedge^*\{t_k \stackrel{e}{=} t'_k \mid k \in K\} \Rightarrow \vee^*\{\bar{t}_l \stackrel{e}{=} \bar{t}'_l \mid l \in L\})$ for an at least two-element set L are only preserved by the operators $\mathbf{S}_c$, $\mathbf{I}$, $\mathbf{e}$ and $\mathbf{P}_u$.

12 A Meta Birkhoff Type Theorem

12.1 $HSP^{r,k}$-cones

12.1.1. In a series of papers (see [AN 76], [AN 82] or [NSa 77]) H.ANDRÉKA, I.NÉMETI and I.SAIN have presented category theoretical axiomatizability results which yield Birkhoff type results of the form $\mathbf{Mod}K_{HSF}(K) = HSF(K)$, where H is an operator of "H-images", S an operator of "S-preimages" for classes **H** and **S** of morphisms, respectively, and F is an operator of some special kind of reduced products (for some class $\mathcal{F}$ of filters). We do not present here the most general result from [AN 82], but one which is relatively close to it. And we shall present it in a category theoretical form, since in this way it can relatively easily be applied to "nice" subclasses of $\mathbf{Alg}(\underline{S})$ like E-, ECE- or QE-varieties, for example. Finally we prove an even weaker version. One should observe that almost all the classes of objects appearing in PART II are QE-varieties.
In 8.4 and also somehow in 11.6 we have tried to motivate, how one can describe axioms which are preserved by such operators **H**, S and **F**, respectively. We do not claim that those axioms are the only ones which do the job, but the Main Theorem in subsection 12.2 will show that under certain conditions these axioms are sufficient. Thus we ask the reader to recall the concepts of H-projective objects, of Pj(**H**) from 8.4.3, and also of the operator Λ^{op} from 8.4.2 or from section 10. The following definition now presents an encoding of formulas which, as we shall show to some extent in 12.2, are then preserved by the operators they have been connected with (see e.g. Table 8.1).

12.1.2 DEFINITION. (i): Let $\underline{\mathbf{C}}$ be any category, let **H** and **S** be classes of $\underline{\mathbf{C}}$-morphisms, let **k** be any infinite regular cardinal or "∞" (i.e. no cardinal bound at all), and let I be any set: A family $(D,(f_i|i \in I))$ $(f_i:D \to Cod(f_i)$, $i \in I)$ is called a <u>cone</u>; let **Cone** designate the class of all set indexed cones in $\underline{\mathbf{C}}$; and in particular, let us define:

$K_{\mathbf{H}} := \{(D,(f_i|i \in I))|D \in Pj(\mathbf{H})\}$, (**H**-<u>cones</u>);

$K_{\mathbf{S}} := \{(D,(f_i|i \in I))|f_i \in \Lambda^{op}(\mathbf{S})$ for all $i \in I\}$, (**S**-<u>cones</u>);

$K_{\mathbf{P}^{r,k}} := \{(D,(f_i|i \in I))|\#I = 1$, D is strongly **k**-small$\}$, ($\mathbf{P}^{r,k}$-<u>cones</u>);

$K_{\mathbf{P}_+^{r,k}} := \{(D,(f_i|i \in I))|\#I \le 1$, D is strongly **k**-small$\}$, ($\mathbf{P}_+^{r,k}$-<u>cones</u>);

$K_{\mathbf{e}} := \{(D,(f_i|i \in I))|I \neq \emptyset\}$, (**e**-<u>cones</u>);

$K_{\mathbf{P}_u} := \{(D,(f_i:D \to D_i|i \in I))|D$ and D_i are strongly ω_0-small, I finite$\}$, ($\mathbf{P}_u$-<u>cones</u>).

If we designate strongly **k**-small objects in diagrams by " $\overset{\mathbf{k}}{\blacksquare}$ " and **S**-morphisms by " $\overset{\mathbf{S}}{\longrightarrow}$ ", we can sketch the situation in Figure 12.1. (If nothing is said about I, then the set I may be empty; for $I = \emptyset$ the cone starting with D is written as (D,) .)

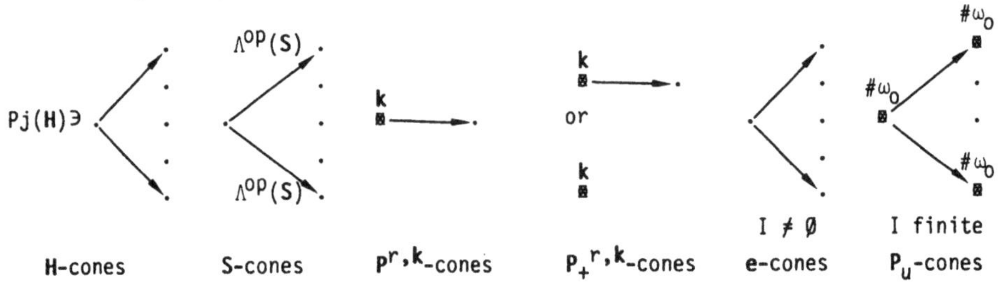

Figure 12.1 (see 12.1.2.(i))

(ii): Let $Y \subseteq \{\mathbf{H},\mathbf{S},\mathbf{P},\mathbf{P}^{r,\mathbf{k}},\mathbf{P}_+{}^{r,\mathbf{k}},\mathbf{e},\mathbf{P}_u\}$, and let X be the string formed from the operator symbols in Y (usually written in the order $\mathbf{HSPP}^{r,\mathbf{k}}\mathbf{P}_+{}^{r,\mathbf{k}}\mathbf{eP}_u$), then:

$$K_X := \bigcap\{K_Q \mid Q \in Y\} .$$

(iii): Let $c := (D,(f_i \mid i \in I)) \in$ **Cone**, and $A \in Ob(\underline{\mathbf{C}})$. We say that c <u>holds</u> (<u>is valid</u>) in A and that A <u>is injective for</u> c (in symbols: $A \models c$), iff for every $\underline{\mathbf{C}}$-morphism $q:D \to A$ there exist some $i \in I$ and $f:Cod(f_i) \to A$ such that $q = f_i \cdot f$ (see Figure 12.2); (D,) holds in A iff there is no morphism from D into A.

If $\mathbf{K} \subseteq Ob(\underline{\mathbf{C}})$ and $c \in K \subseteq$ **Cone**, then we define

$\mathbf{K} \models c$ iff $A \models c$ for all $A \in \mathbf{K}$;

$A \models K$ iff $A \models c$ for all $c \in K$;

$\mathbf{K} \models K$ iff $A \models K$ for all $A \in \mathbf{K}$;

and

$Inj(K) := \{A \in Ob(\underline{\mathbf{C}}) \mid A \models K\}$;

$K_X(\mathbf{K}) := \{c \in K_X \mid \mathbf{K} \models c\}$.

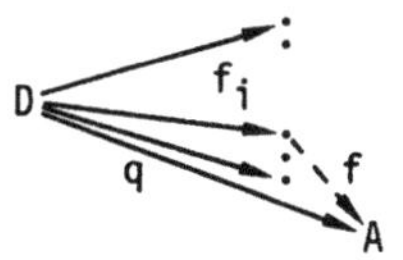

Figure 12.2.

12.1.3 REMARK. "Inj" and "K_X" are -- for each string X corresponding to a subset of $\{\mathbf{H},\mathbf{S},\mathbf{P},\mathbf{P}^{r,\mathbf{k}},\mathbf{P}_+{}^{r,\mathbf{k}},\mathbf{e},\mathbf{P}_u\}$ for any given infinite regular cardinal **k** (including ∞) -- the operators of a Galois correspondence which defines a language $(K_X, Ob(\underline{\mathbf{C}}), \models)$ in the sense of abstract model theory (see [NSa 77]). In the case that we consider the category $\mathbf{Alg}(\underline{S})$ of partial algebras, each cone, say $(D,(d_i \mid i \in I))$, represents an implication $(M:\Lambda^*\{t_l \overset{e}{=} t'_l \mid l \in L\} \Rightarrow V^*\{\Lambda^*\{\bar{t}_{ij} \overset{e}{=} \bar{t}'_{ij} \mid j \in J_i\} \mid i \in I\})$, where in the case that I is empty we replace the conclusion by a statement which is always false, and the whole formula then becomes equivalent to $(M:\neg(\Lambda^*\{t_l \overset{e}{=} t'_l \mid l \in L\}))$. In the proof and the applications of the main theorem we shall restrict our considerations to the cases where I is an at most one-element set, but the general theory goes further.

Observe, too, that for $\mathbf{k} = \infty$ we get $\mathbf{P}^{r,\infty} = \mathbf{P}$, $\mathbf{P}_+{}^{r,\infty} = \mathbf{P}_+$.

12.2 A Meta Birkhoff Theorem

12.2.1 DEFINITION. Let $\underline{\mathbf{C}}$ be any category and $\mathbf{H} \subseteq Mor(\underline{\mathbf{C}})$. We recall from 8.4.3 that a $\underline{\mathbf{C}}$-object A is called **H**-projective, iff for every **H**-morphism, say $h:B \to C$, and for every $f:A \to C$ there is a $g:A \to B$ such that $gh = f$. When $Pj(\mathbf{H})$ designates the class of all **H**-projective $\underline{\mathbf{C}}$-objects, then we say that <u>$\underline{\mathbf{C}}$ has enough **H**-projectives</u>, iff for every $C \in Ob(\underline{\mathbf{C}})$ there are $P \in Pj(\mathbf{H})$ and $h:P \to C$ in **H**, i.e. every $\underline{\mathbf{C}}$-object is the H-image of some **H**-projective object.

12.2.2. Observe that all partial Peano algebras are $\mathbf{H}_f$-projective for the class $\mathbf{H}_f$ of all full and surjective homomorphisms, and that each partial algebra is a full and surjective homomorphic image of some suitable partial Peano algebra.

The main theorem is hidden in the following

12.2.3 LEMMA. Let **k** be an infinite regular cardinal or ∞ (i.e. there may be no bound on the cardinalities), and let $\underline{\mathbf{C}}$ be a strongly **k**-algebroidal category (see 11.2.4) in which direct products and direct limits of **k**-directed systems exist. Moreover, let **H** and **S** be classes of $\underline{\mathbf{C}}$-morphisms, for which we consider the following properties:

(1) $\underline{\mathbf{C}}$ has enough **H**-projectives (see 12.2.1).

(2) Every **H**-projective $P \in Pj(\mathbf{H})$ is the **k**-direct limit of strongly **k**-small **H**-projective objects.

(3) $(\Lambda^{op}(\mathbf{S}),\mathbf{S})$ is a factorization system such that $\Lambda^{op}(\mathbf{S}) \subseteq \mathbf{Epi}(\underline{C})$.

(4) If $fg \in \mathbf{H}$ and $f \in \Lambda^{op}(\mathbf{S})$, then $g \in \mathbf{H}$, i.e. **H** is left cancellable with respect to $\Lambda^{op}(\mathbf{S})$-morphisms.

(5) $\underline{C}$ is $\Lambda^{op}(\mathbf{S})$-co-(well-powered) (i.e. from each $\underline{C}$-object there start up to isomorphism only a set of $\Lambda^{op}(\mathbf{S})$-objects).

(6) $\Lambda^{op}(\mathbf{S})$ is **k**-dual algebroidal (see 11.5.3).

(7) **S** is **k**-inductive (see 11.5.1).

Let $\mathbf{K} \subseteq Ob(\underline{C})$ be any class of $\underline{C}$-objects, and recall that $\mathbf{H}(\mathbf{K})$ designates the class of all **H**-images of **K**-objects, while $\mathbf{S}(\mathbf{K})$ designates the class of all **S**-preimages of **K**-objects. Then we have the following statements:

(i) Without any assumptions on **H** and **S** we have for any string **Z** of operators from $\{\mathbf{H},\mathbf{S},\mathbf{P}_+{}^{r,\mathbf{k}},\mathbf{P}^{r,\mathbf{k}},\mathbf{e},\mathbf{P}_u\}$

$$\mathbf{Z}(\mathbf{K}) \subseteq InjK_{\mathbf{Z}}(\mathbf{K}) .$$

(ii) (1), (3), (4) and (5) imply

$$InjK_{\mathbf{HS}}(\mathbf{K}) = \mathbf{HS}(\mathbf{K}) \text{ , and}$$
$$InjK_{\mathbf{HSe}}(\mathbf{K}) = \mathbf{HSe}(\mathbf{K}) .$$

(iii) (1) through (7) imply

$$InjK_{\mathbf{HSP}_+{}^{r,\mathbf{k}}}(\mathbf{K}) = \mathbf{HSP}_+{}^{r,\mathbf{k}}(\mathbf{K}) \text{ , and}$$
$$InjK_{\mathbf{HSP}^{r,\mathbf{k}}}(\mathbf{K}) = \mathbf{HSP}^{r,\mathbf{k}}(\mathbf{K}) .$$

(iv) If $\mathbf{k} = \infty$, then (1), (3), (4) and (5) imply

$$InjK_{\mathbf{HSP}_+}(\mathbf{K}) = \mathbf{HSP}_+(\mathbf{K}) \text{ , and}$$
$$InjK_{\mathbf{HSP}}(\mathbf{K}) = \mathbf{HSP}(\mathbf{K}) .$$

(v) (1) through (7) imply

$$InjK_{\mathbf{HSP}_u}(\mathbf{K}) = \mathbf{HSP}_u(\mathbf{K}) \text{ , and}$$
$$InjK_{\mathbf{HSP}_u\mathbf{e}}(\mathbf{K}) = \mathbf{HSP}_u\mathbf{e}(\mathbf{K}) .$$

12.2.4. The rest of this subsection will be dedicated to the proof of this Lemma, or at least to the proofs of (i), (iii) and (iv) in it, and also in connection with (i) we restrict our considerations to the operators **HSF**, where F is either $\mathbf{P}_+{}^{r,\mathbf{k}}$ or $\mathbf{P}^{r,\mathbf{k}}$. The statements in (ii) and (v) will not be used later, and therefore we refer the interested reader to their proofs in [NSa 77] and [AN 82]. But first we collect the statements (by slightly weakening them) in the following

12.2.5 COROLLARY (META BIRKHOFF THEOREM). With the notation and all assumptions from 12.2.3 one has for any operator $\mathbf{F} \in \{\mathbf{i},\mathbf{P}^{r,\mathbf{k}},\mathbf{P}_+{}^{r,\mathbf{k}},\mathbf{e},\mathbf{P}_u,\mathbf{eP}_u\}$ (cf. 4.6.2),

$$InjK_{\mathbf{HSF}}(\mathbf{K}) = \mathbf{HSF}(\mathbf{K}) . \blacksquare\blacksquare\blacksquare$$

12.2.6. Before we go into the details of the proof of 12.2.3, we already briefly comment on them:

(i): Part (i) of the proof (see 12.2.7) will be closely connected with the proof of Theorem 8.3.4 (see Table 8.1 and also subsection 11.6).

(ii): Let us turn to the proof of the other inclusion in 12.2.3.(iii) -- which will be the main part of the following proof, since we do not prove all the stated facts -- (compare Figure 12.3): While in the proof of the original Birkhoff Theorem (for equations) one works with the Peano algebra freely generated by the carrier set of the algebra $\underline{A}$ under consideration -- i.e. which satisfies all specified formulas valid in **K** --, one has to replace the Peano algebra in the general case by an **H**-projective object, say Q, together with an **H**-morphism from Q to the object A (**H** takes the place of the class of all quotient mappings). Inductivity of the Peano algebra is replaced by the assumption that the **H**-projective object Q is the direct limit of a **k**-directed system consisting of strongly **k**-small **H**-projective objects Q_j -- in order to get the premises of the implications to be considered down to the desired size. For each such "premise" the possible conclusions are collected from the $(\Lambda^{op}(S),S)$-factorizations of morphisms from the Q_j into **K**. Since we want to form products of the possible conclusions and necessary **K**-objects, we need the $\Lambda^{op}(S)$-co-(well-poweredness). With the sets of necessary "conclusions" respectively necessary **K**-objects two directed systems are formed, the latter yielding a **k**-reduced product, say K, of **K**-objects, the other one yielding a direct limit, say R, which is an "S-subobject" of K and allows an **H**-morphism into A. Verification of this last statement is especially involved and needs an extra Lemma (12.2.14), where we need that $\Lambda^{op}(S)$-morphisms starting with strongly **k**-small objects can be approximated as direct limits by **k**-directed systems of $\Lambda^{op}(S)$-morphisms between strongly **k**-small objects -- i.e. admissible implications -- (needing strong **k**-algebroidalness of $\underline{\mathbf{C}}$ and **k**-dual algebroidalness of $\Lambda^{op}(S)$).

12.2.7 PROOF of 12.2.3.(i). Since we restrict the proof to operators **HSF** with $F \in \{P_+^{r,k}, P^{r,k}\}$, the corresponding cones $c \in K_{HSF}$ have the form $c = (D,d:D \to D')$ or $c = (D^*,)$ (see 12.1.2.(i), the latter is only admissible, if $F = P_+^{r,k}$), where D and D* are strongly **k**-small **H**-projective objects and $d \in \Lambda^{op}(S)$. Now, assume $\mathbf{K} \subseteq Ob(\underline{\mathbf{C}})$, $A \in \mathbf{HSF}(\mathbf{K})$ and $c \in K_{HSF}$. Thus there are an **H**-morphism $h:B \to A$, and an S-morphism $s:B \to X^k(A_i|i \in I)/f$, where, for some set I, $\{A_i|i \in I\} \subseteq \mathbf{K}$, and **f** is a **k**-complete filter on I (observe that ∞-completeness of **f** means that **f** is a principal filter on I, and $P^{r,\infty} = P$, $P_+^{r,\infty} = P_+$); the directed system connected with the above reduced product is denoted as in 4.5 by $(A_F,p_{FG}|F \supseteq G \in \mathbf{f})$, where $A_F := X(A_i|i \in F)$ and p_{FG} is the projection from A_F onto A_G.

(a): **f** contains the empty set: Then $X^k(A_i|i \in I)/f =: T$ is just the terminal element in $\underline{\mathbf{C}}$, i.e. each $\underline{\mathbf{C}}$-object allows exactly one morphism into T. We then have only cones of the form $c = (D,d:D \to D')$. Assume $f:D \to A$; since D is **H**-projective and $h \in \mathbf{H}$, there is $f':D \to B$ such that $f'h = f$. Because of the nature of T there is a (unique) $q:D' \to T$ such that $dq = f's$. Since $s \in S$ and $d \in \Lambda^{op}(S)$, there is a unique $d':D' \to B$ such that $dd' = f'$ (and $d's = q$). Thus $dd'h = f'h = f$, showing that $A \models c$, since f was arbitrarily chosen; hence $A \in InjK_{HSF}(\mathbf{K})$.

(b): **f** is a proper filter, $I \neq \emptyset$: Let either $c = (D,d:D \to D')$ or $c' = (D,)$ be in $K_{HSF}(\mathbf{K})$. In either case assume $f:D \to A$, and as in (a) we get $f':D \to B$ satisfying $f'h = f$. Since D is strongly **k**-small and since **f** is **k**-complete, there are $G \in \mathbf{f}$ and $f_G:D \to A_G$ such that $f_Gp_G = f's$. If we have started with c', then $f_Gp_{Gi}:D \to A_i$ for some $i \in G$ contradicts $A_i \in \mathbf{K}$ and $c' \in K_{HSF}(\mathbf{K})$. Thus in this case there cannot be any morphism from D to A, i.e. $A \models c'$. -- If we consider c, then c holds in every A_i $(i \in G)$, and therefore also in A_G, and we get $g:D' \to A_G$ such that $dg = f_G$. Now, $f's = f_Gp_G = dgp_G$, $s \in S$ and $d \in \Lambda^{op}(S)$. Thus there exists $d':D' \to B$ such that $dd' = f'$ (and $d's = gp_G$). As in (a) this completes the proof. ∎

12.2.8 Sketch of the proof of 12.2.3.(iii) and (iv). From what we have said earlier, (iv) is a special case of (iii) for $\mathbf{k} = \infty$. Assume that we have already proved $\mathbf{HSP}_+{}^{r,\mathbf{k}}(\mathbf{K}) = \mathrm{Inj}\mathbf{K}_{\mathbf{HSP}_+{}^{r,\mathbf{k}}}(\mathbf{K})$. Then $\mathbf{HSP}^{r,\mathbf{k}}(\mathbf{K}) = \mathbf{HSP}_+{}^{r,\mathbf{k}}(\mathbf{K}\cup\{T\}) = \mathrm{Inj}\mathbf{K}_{\mathbf{HSP}_+{}^{r,\mathbf{k}}}(\mathbf{K}\cup\{T\})$, where T designates the terminal object of $\underline{\mathbf{C}}$, representable as a product with empty index set. Since all cones $(D,d:D \to D')$ hold in T, while no cone $(D,)$ does, we have $\mathbf{K}_{\mathbf{HSP}_+{}^{r,\mathbf{k}}}(\mathbf{K} \cup \{T\}) = \mathbf{K}_{\mathbf{HSP}^{r,\mathbf{k}}}(\mathbf{K})$; therefore (iii) (and (iv)) follow in the case of $F = \mathbf{P}^{r,\mathbf{k}}$, too.

Thus -- because of (i) -- it remains only to show $\mathrm{Inj}\mathbf{K}_{\mathbf{HSP}_+{}^{r,\mathbf{k}}}(\mathbf{K}) \subseteq \mathbf{HSP}_+{}^{r,\mathbf{k}}(\mathbf{K})$. This proof follows to a great extent the one in [ADN 80] for the statement given here in 8.3.5.(iv), and which is based on a proof of A.I.MAL'CEV in [Mal 73] for quasivarieties (compare also the proof of the First Birkhoff Theorem 6.2.5) -- in [AN 82] a proof for a more general result can be found.

We have to show that $A \in \mathrm{Inj}\mathbf{K}_{\mathbf{HSP}_+{}^{r,\mathbf{k}}}(\mathbf{K})$ implies $A \in \mathbf{HSP}_+{}^{r,\mathbf{k}}(\mathbf{K})$:

Thus assume $A \in \mathrm{Inj}\mathbf{K}_{\mathbf{HSP}_+{}^{r,\mathbf{k}}}(\mathbf{K})$. Since we have assumed in (1) that $\underline{\mathbf{C}}$ has enough $\mathbf{H}$-projectives, there exist $Q \in \mathrm{Pj}(\mathbf{H})$ and an $\mathbf{H}$-morphism $h:Q \to A$. Because of assumption (2) we may choose a $\mathbf{k}$-directed system, say $\mathfrak{Q}_J := (Q_j, q_{jj'} \mid j \leq j' \in J)$ with direct limit $((q_j \mid j \in J), Q)$ (i.e. representing Q) such that each Q_j is a strongly $\mathbf{k}$-small $\mathbf{H}$-projective object; and in order to make some notation simpler, we assume that $Q_j \neq Q_{j'}$, whenever $j \neq j'$ in J.

Then the main line of the rest of the proof is as follows:

(a): For each $j \in J$ we choose a **set** $\mathbf{K}_j \subseteq \mathbf{K}$ and a **set** $\mathbf{E}_j \subseteq \Lambda^{op}(\mathbf{S})$ such that each $\mathbf{E}_j$-morphism starts at Q_j and in some way $\mathbf{E}_j$ and $\mathbf{K}_j$ represent all morphisms from Q_j into some $\mathbf{K}$-object with respect to $\Lambda^{op}(\mathbf{S})$:

If $K \in \mathbf{K}$, $f:Q_j \to K$ and if (e,s) is a $(\Lambda^{op}(\mathbf{S}),\mathbf{S})$-factorization of f, then there are $K' \in \mathbf{K}_j$ and $f':Q_j \to K'$ such that f' has a $(\Lambda^{op}(\mathbf{S}),\mathbf{S})$-factorization (e',s') satisfying

(a1) $e' \in \mathbf{E}_j$, and

(a2) There is an isomorphism $i:\mathrm{Cod}(e') \to \mathrm{Cod}(e)$ such that $e'i = e$.

CLAIM (a): The choices in (a) above are possible, and $\mathbf{E}_j \neq \emptyset$ and $\mathbf{K}_j \neq \emptyset$, in particular there exists a morphism from Q_j into some $\mathbf{K}$-object for each $j \in J$. (See 12.2.10.)

(b): Now, we define the set $\mathbf{K}_A := \bigcup\{\mathbf{K}_j \mid j \in J\}$ which is relevant for the morphisms starting from A into some $\mathbf{K}$-object (in the equational case this was the set of all representatives of subalgebras of $\mathbf{K}$-algebras which were generated by homomorphic images of A, cf. $\mathbf{Rsuba}(\underline{A};;\mathbf{K})$ in 5.11). Define $I := \bigcup\{\mathrm{Mor}(Q_j,K) \mid j \in J, \text{ and } K \in \mathbf{K}_A\}$, which is also a set. For each $j \in J$ we define $I_j := \bigcup\{\mathrm{Mor}(Q_{j'},K) \mid j \leq j' \in J, K \in \mathbf{K}_A\}$.

CLAIM (b1): $\{I_j \mid j \in J\}$ is the base of a $\mathbf{k}$-complete proper filter $\mathbf{f} := \{L \subseteq I \mid \text{there is } j \in J \text{ with } I_j \subseteq L\}$ on I. (See 12.2.11.)

For what follows compare Figure 12.3 on page 168: For each $j \in J$ define $K_j := \mathsf{X}(\mathrm{Cod}(f) \mid f \in I_j)$, and let $f_j:Q_j \to K_j$ be the (product) morphism induced by the source $(q_{jj'}f \mid j \leq j' \in J, f \in \mathrm{Mor}(Q_{j'},K), K \in \mathbf{K}_A)$. Moreover, let

$D_j := \mathsf{X}(\mathrm{Cod}(e_f) \mid f \in \mathrm{Mor}(Q_{j'},K), j \leq j' \in J, K \in \mathbf{K}_A, e_f \in \mathbf{E}_{j'},$
$(e_f,s_f) \text{ is a fixed } (\Lambda^{op}(\mathbf{S}),\mathbf{S})\text{-factorization of } q_{jj'}f)$.

Let $e^*_j:Q_j \to D_j$ be the (product) morphism induced by the source $(e_f \mid f \in I_j)$, and let

$s^*_j := \mathsf{X}(s_f \mid f \in I_j): D_j \to K_j$ be the product in the sense of 4.1.3.(d) of the family of morphisms $s_f \in \mathbf{S}$ occurring in the definition of D_j. Since $\mathbf{S} = \Lambda\Lambda^{op}(\mathbf{S})$ is closed with respect to products (see 10.1.2.(ix)), we get $s^*_j \in \mathbf{S}$ $(j \in J)$.
Let (e_j, s'_j) be a $(\Lambda^{op}(\mathbf{S}),\mathbf{S})$-factorization of e^*_j, such that $R_j := \mathrm{Cod}(e_j) = \mathrm{Dom}(s'_j)$. Then $s_j := s'_j s^*_j$ is an $\mathbf{S}$-morphism, and (e_j, s_j) is a $(\Lambda^{op}(\mathbf{S}),\mathbf{S})$-factorization of f_j.
Since $I_l \subseteq I_j$ for $j \le l \in J$, we have a projection morphism (or restriction morphism) $k_{jl}: K_j \to K_l$ induced by the family $(p^{K_j}{}_f: K_j \to \mathrm{Cod}(f) \mid f \in I_l)$ of projections.

CLAIM (b2): $k_{jj} = \mathbf{1}_{K_j}$ and $k_{jl}k_{lm} = k_{jm}$ for $j \le l \le m \in J$, i.e. $\mathfrak{K}_{\underline{J}} := (K_j, k_{jl} \mid j \le l \in J)$ is a $\mathbf{k}$-directed system cofinal in $\mathfrak{K}_{(\mathbf{f};\supseteq)} := (K_F, k_{FG} \mid F \supseteq G \in \mathbf{f})$ -- let $((k_j \mid j \in J), K)$ be its direct limit, which is obviously isomorphic to the reduced product $\mathsf{X}(\mathrm{Cod}(f) \mid f \in I)/\mathbf{f}$ (cf. 11.1.4.(ii)), i.e. $K \in \mathbf{P}_+{}^{r,\mathbf{k}}(\mathbf{K})$. (See 12.2.12.)

CLAIM (b3): $f_j k_{jl} = q_{jl} f_l$ for every $j \le l \in J$, i.e. $f^* := (\mathrm{id}_J, (f_j \mid j \in J))$ is a cofinal morphism between the $\mathbf{k}$-directed systems $\mathfrak{Q}_{\underline{J}}$ and $\mathfrak{K}_{(\mathbf{f},\supseteq)}$ -- let $\overline{f}: Q \to K$ be the induced colimit morphism. (See 12.2.13.)

(c): Claim (b3) implies for $j \le l \in J$, when we decompose f_j and f_l: $e_j(s_j k_{jl}) = (q_{jl} e_l) s_l$. Observing that $e_j \in \Lambda^{op}(\mathbf{S})$ and $s_l \in \mathbf{S}$, we get from Lemma 11.3.5 for $j \le l \in J$ a unique morphism $r_{jl}: R_j \to R_l$ (recall $R_j = \mathrm{Cod}(e_l)$) such that $e_j r_{jl} = q_{jl} e_l$ and $r_{jl} s_l = s_j k_{jl}$. And in particular $(R_j, r_{jl} \mid j \le l \in J) =: \mathfrak{R}_{\underline{J}}$ is a $\mathbf{k}$-directed system -- with colimit, say, $((r_j \mid j \in J), R)$ -- and $e^* := (\mathrm{id}_J, (e_j \mid j \in J)): \mathfrak{Q}_{\underline{J}} \to \mathfrak{R}_{\underline{J}}$ as well as $s^* := (\mathrm{id}_J, (s_j \mid j \in J)): \mathfrak{R}_{\underline{J}} \to \mathfrak{K}_{\underline{J}}$ are special morphisms between $\mathbf{k}$-directed systems -- with induced morphisms, say, $\overline{e}: Q \to R$ and $\overline{s}: R \to K$, respectively. Obviously $\overline{e}\cdot\overline{s} = \overline{f}$.

CLAIM (c1): $\overline{e} \in \Lambda^{op}(\mathbf{S})$ and $\overline{s} \in \mathbf{S}$, in particular $R \in \mathbf{SP}_+{}^{r,\mathbf{k}}(\mathbf{K})$. (See 12.2.15.)

(d): Because of Claim (c1) it remains to prove that there exists an $\mathbf{H}$-morphism from R to A. Since for each $j \in J$ the morphism e_j belongs to $\Lambda^{op}(\mathbf{S})$ and starts at a strongly $\mathbf{k}$-small $\mathbf{H}$-projective object; thus (Q_j, e_j) is an $\mathbf{HSP}_+{}^{r,\mathbf{k}}$-cone.

CLAIM (d1): For each $j \in J$ we have $e_j \in K_{\mathbf{HSP}_+{}^{r,\mathbf{k}}(\mathbf{K})}$. (See 12.2.16.)

Thus, because of (3), e_j is a $\mathbf{K}$-extendable epimorphism. Since $A \in \mathrm{Inj}K_{\mathbf{HSP}_+{}^{r,\mathbf{k}}(\mathbf{K})}$, we get for each $j \in J$ a unique $g_j: R_j \to A$ satisfying $q_j h = e_j g_j$.

CLAIM (d2): $(g_j \mid j \in J)$ is compatible with $\mathfrak{R}_{\underline{J}}$, i.e. for each $j \le l \in J$ we have $r_{jl} g_l = g_j$. (See 12.2.17.)

Hence we obtain a unique morphism $\overline{g}: R \to A$ such that $r_j \overline{g} = g_j$ for each $j \in J$.

CLAIM (d3): $\overline{e}\cdot\overline{g} = h$. (See 12.2.18.)

Since $\overline{e} \in \Lambda^{op}(\mathbf{S})$ and $h \in \mathbf{H}$, assumption (3) together with Claim (d3) yields $\overline{g} \in \mathbf{H}$, finally showing that $A \in \mathbf{HSP}_+{}^{r,\mathbf{k}}(\mathbf{K})$.

This completes the sketch of the proof of 12.2.3.(iii) and (iv) -- observe that, for $\mathbf{k} = \infty$, J above is a one-element set. ∎

12.2.9. We now prove the claims stated in 12.2.8. At some instances we shall refer to the fact that we have proved results before for the category $\mathbf{Alg}(\underline{S})$, which are actually true in general for arbitrary categories or at least for categories satisfying the assumptions of 12.2.3. We shall not stress the corresponding references more explicitly.

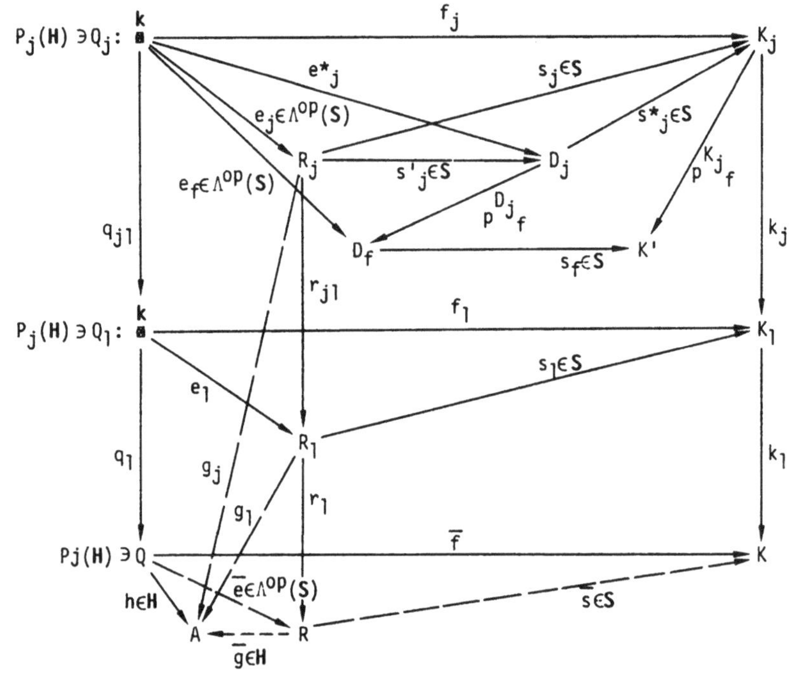

Figure 12.3.

12.2.10 Proof of Claim (a): Since by 12.2.3.(5) $\underline{\mathbf{C}}$ is $\Lambda^{op}(\mathbf{S})$-co-(well-powered), there exists for each $j \in J$ a **set** $\mathbf{E'}_j \subseteq \Lambda^{op}(\mathbf{S})$ such that each $\Lambda^{op}(\mathbf{S})$-morphism starting from Q_j is represented up to isomorphism in $\mathbf{E'}_j$. Let

$$\mathbf{E}_j := \{e \in \mathbf{E'}_j \mid \text{there are } s \in \mathbf{S} \text{ and } K \in \mathbf{K} \text{ such that } es:Q_j \to K\},$$

i.e. $\mathbf{E}_j$ consists of all $\mathbf{E'}_j$-morphisms which are the first factor of a $(\Lambda^{op}(\mathbf{S}),\mathbf{S})$-factorization of a morphism from Q_j into some $\mathbf{K}$-object. For each $e \in \mathbf{E}_j$ choose some $K_e \in \mathbf{K}$ such that there is $s \in \mathbf{S}$ satisfying $es:Q_j \to K_e$. Define $\mathbf{K}_j := \{K_e \mid e \in \mathbf{E}_j\}$. Obviously $\mathbf{K}_j \neq \emptyset$, if $\mathbf{E}_j \neq \emptyset$, and the latter is true, if there is at least one morphism from Q_j into some $\mathbf{K}$-object: Assume that there is no such morphism; then $(Q_j,) \in K_{\mathbf{HSP}_+{}^r,\mathbf{k}}(\mathbf{K})$, but this contradicts the facts that $q_j h:Q_j \to A$ and $A \in \mathrm{Inj}K_{\mathbf{HSP}_+{}^r,\mathbf{k}}(\mathbf{K})$. Thus we have for each $j \in J$ at least one morphism from Q_j into some $\mathbf{K}$-object, $\mathbf{E}_j$ as well as $\mathbf{K}_j$ are non-empty, and they are sets by construction. In particular, if there is some $A \in \mathrm{Inj}K_{\mathbf{HSP}_+{}^r,\mathbf{k}}(\mathbf{K})$, then $\mathbf{K}$ has to be non-empty. ∎

12.2.11 Proof of Claim (b1): Consider $J' \subseteq J$, $\#J' < \mathbf{k}$, then $\mathbf{k}$-directedness of $\underline{J}$ implies the existence of an upper bound of J', say j. Thus $I_j \subseteq I_{j'}$ $(j' \in J')$. Since $I_j \neq \emptyset$ by Claim (a), this easily implies that $\mathbf{f}$ is a proper $\mathbf{k}$-complete filter. ∎

12.2.12 Proof of Claim (b2): This is a simple consequence of the definition of k_{jl} as a restriction morphism between products and the fact that for $j \leq l \leq m \leq j'$ and $f \in \mathrm{Mor}(Q_{j'},K')$ for some $K' \in \mathbf{K}_A$ we have $k_{jl}k_{lm}p^{K_m}{}_f = k_{jl}p^{K_l}{}_f = p^{K_j}{}_f = k_{jm}p^{K_m}{}_f$. ∎

12.2.13 Proof of Claim (b3): Again by the definition of K_j and f_j in 12.2.8.(b), we have, for $j \leq 1 \in J$ and $f \in I_1$, i.e. for $f \in Mor(Q_{j'},K')$ for some $j' \geq 1$ and $K' \in K_A$, that $q_{j1}f_1p^{K_1}{}_f = q_{j1}q_{1j'}f = q_{jj'}f = f_jp^{K_j}{}_f = f_jk_{j1}p^{K_1}{}_f$, and the statement follows. ∎

For the proof of Claim (c1) we need an auxiliary lemma (cf. [NSa 77], Lemma 11(b)):

12.2.14 LEMMA. If $\underline{C}$ is a strongly **k**-algebroidal category -- **k** an infinite regular cardinal --, S a class of $\underline{C}$-morphisms satisfying conditions (3), (5), (6) and (7) from 12.2.3, then $S = \Lambda(\Lambda^{op}(S)^{\mathbf{k}}_{\mathbf{k}})$,

where $\Lambda^{op}(S)^{\mathbf{k}}_{\mathbf{k}} := \{e \in \Lambda^{op}(S) | Dom(e)$ and $Cod(e)$ are strongly **k**-small$\}$.

PROOF. Since $(\Lambda^{op}(S),S)$ is a factorization system, 10.2.5 implies $S = \Lambda\Lambda^{op}(S)$. Therefore it is sufficient (cf. 10.1.2.(iv)) to prove $\Lambda(\Lambda^{op}(S)^{\mathbf{k}}_{\mathbf{k}}) \subseteq \Lambda(\Lambda^{op}(S))$:

Thus, assume that $s \in \Lambda(\Lambda^{op}(S)^{\mathbf{k}}_{\mathbf{k}})$, let $h \in \Lambda^{op}(S)$, $h:A \to B$, $s:C \to D$, and let $p,q \in Mor(\underline{C})$ such that $hq = ps$. We have to show the existence of a unique diagonal-fill-in $d:B \to C$, $hd = p$, $ds = q$.

Since $\underline{C}$ is strongly **k**-algebroidal, A is the direct limit, say $((h_i|i \in I),A)$ of a **k**-directed system, say $\underline{A_I} := (A_i,h_{ij}|i \leq j \in I)$. For each $i \in I$ let (t_i,r_i) -- $t_i:A_i \to E_i$ in $\Lambda^{op}(S)$, $r_i:E_i \to B$ in **S** -- be a $(\Lambda^{op}(S),S)$-factorization of h_ih: $t_ir_i = h_ih = (h_{ij}t_j)r_j$. $t_i \in \Lambda^{op}(S)$ and $r_j \in S$ imply the existence of a unique diagonal-fill-in $e_{ij}:E_i \to E_j$, and the uniqueness of diagonal-fill-ins (or the fact that t_i is an epimorphism) easily implies $e_{ij}e_{j1} = e_{i1}$ for each $i \leq j \leq 1 \in I$ (cf.11.3.5). Hence $\underline{E_I} := (E_i,e_{ij}|i \leq j \in I)$ is a **k**-directed system with colimiting cocone, say $((e_i|i \in I),E)$.

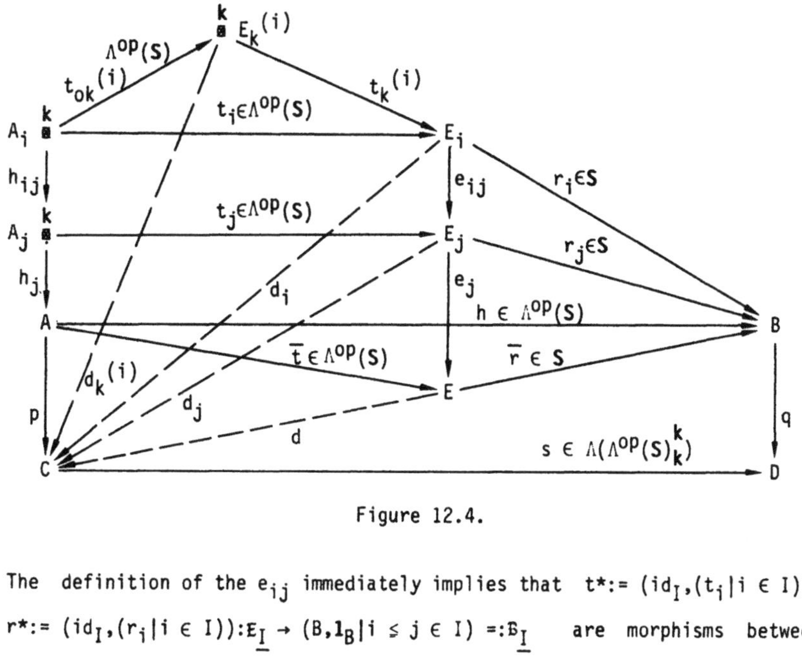

Figure 12.4.

The definition of the e_{ij} immediately implies that $t^* := (id_I,(t_i|i \in I)):\underline{A_I} \to \underline{E_I}$ and $r^* := (id_I,(r_i|i \in I)):\underline{E_I} \to (B,1_B|i \leq j \in I) =: \underline{B_I}$ are morphisms between **k**-directed systems. Let $\bar{t}:A \to E$ and $\bar{r}:E \to B$ be the induced colimit morphisms (consider the commutative diagram in Figure 12.4). Obviously $\bar{t}\cdot\bar{r} = h$; moreover, $\bar{r} \in \mathbf{S}$, since $\mathbf{S}$ is $\mathbf{k}$-inductive, and $\bar{t} \in \Lambda^{op}(S)$ by 11.1.4.(iii). In addition, since $h = \bar{t}\cdot\bar{r}$ is a $\Lambda^{op}(S)$-morphism, too, the uniqueness of such factorizations implies that $\bar{r}$ is an isomorphism. It is therefore now sufficient to construct a diagonal-fill-in, say d, starting from E instead of B (since then $p = \bar{t}d = h(\bar{r}^{-1}d)$ and $q = \bar{r}^{-1}\bar{r}q = (\bar{r}^{-1}d)s$).

In order to obtain d, we use that $\Lambda^{op}(S)$ is $\mathbf{k}$-dual algebroidal:

For each $t_i:A_i \to E_i$ we have a $\mathbf{k}$-directed system $\mathcal{E}_{\underline{I}_i}{}^{(i)} := (E_k{}^{(i)}, t_{kl}{}^{(i)} | k \le l \in I_i)$ with colimiting cocone $((t_k{}^{(i)}|k \in I_i), E_i)$ such that each $E_k{}^{(i)}$ is strongly $\mathbf{k}$-small, $t_i = t_o{}^{(i)}$ for some $o \in I_i$, and $t_{kl}{}^{(i)} \in \Lambda^{op}(S)$ for any $k \le l \in I_i$ and $i \in I$. Observe that, for $o \le k \in I_i$: $t_{ok}{}^{(i)}(t_k{}^{(i)}r_iq) = h_ihq = (h_ip)s$ and $s \in \Lambda(\Lambda^{op}(S)_{\mathbf{k}}^{\mathbf{k}})$. Thus there exists for each $i \in I$ and each $k \in I_i$ a $d_k{}^{(i)}:E_k{}^{(i)} \to C$ such that $t_{ok}{}^{(i)}d_k{}^{(i)} = h_ip$ and $d_k{}^{(i)}s = t_k{}^{(i)}r_iq = t_k{}^{(i)}e_i\bar{r}q$. Since each $t_{ok}{}^{(i)}$ is an epimorphism, it is easy to realize that $t_{kl}{}^{(i)}d_l{}^{(i)} = d_k{}^{(i)}$ for $k \le l \in I_i$, $i \in I$. Therefore there exists a unique colimit morphism $d_i:E_i \to C$ such that $t_k{}^{(i)}d_i = d_k{}^{(i)}$ $(k \in I_i)$, and in particular $t_id_i = t_o{}^{(i)}d_i = t_{ok}{}^{(i)}t_k{}^{(i)}d_i = t_{ok}{}^{(i)}d_k{}^{(i)} = h_ip$.

Now, for $i \le j \in I$ the facts that t_i is an epimorphism and $t_ie_{ij}d_j = h_{ij}t_jd_j = h_{ij}h_jp = h_ip = t_id_i$ imply that $e_{ij}d_j = d_i$, i.e. the family $(d_i|i \in I)$ is compatible with $\mathcal{A}_{\underline{I}}$ and therefore induces a morphism $d:E \to C$ satisfying $e_id = d_i$ $(i \in I)$. The uniqueness of induced colimit morphisms and the facts that $(t_id_i:A_i \to C|i \in I)$ is a family compatible with $\mathcal{A}_{\underline{I}}$ and that $t_id_i = h_ip = t_ie_id = h_i\bar{t}d$ implies $\bar{t}d = p$. Since $\bar{t}$ is an epimorphism, $\bar{t}ds = ps = hq = \bar{t}\bar{r}q$ implies $ds = \bar{r}q$, i.e. we have got the morphism d which we have looked for. ∎

12.2.15 Proof of Claim (c1): The fact that $\bar{e} \in \Lambda^{op}(S)$ follows from 11.1.4.(iii), while " $\bar{s} \in \mathbf{S}$ " is now an immediate consequence of Lemma 12.2.14 and Proposition 11.3.1.(i). From Claim (b2) then $R \in \mathbf{SP}_+{}^{r,\mathbf{k}}(\mathbf{K})$ follows. ∎

12.2.16 Proof of Claim (d1): Assume $g:Q_j \to K$ for some $K \in \mathbf{K}$. Then -- by the assumptions on $\mathbf{E}_j$ and $\mathbf{K}_j$ -- there exists a $(\Lambda^{op}(S),S)$-factorization, say (e_g,s_g), of g such that $e_g \in \mathbf{E}_j$, and there are $K' \in \mathbf{K}_j$ and $f:Q_j \to K'$ such that f has a $(\Lambda^{op}(S),S)$-factorization (e_g,s') (see (a)); in particular $e_g = e_f$ and $f \in I_j$. Thus we have (see Figure 12.3) $g = e_gs_g = e_fs_g = e_js'_jp^{D_j}{}_fs_g$, showing that g factors through e_j, i.e. $e_j \in K_{\mathbf{HSP}_+{}^{r,\mathbf{k}}(\mathbf{K})}$. ∎

12.2.17 Proof of Claim (d2): Since each e_j is an epimorphism, this follows from $e_jr_{jl}g_l = q_{jl}e_lg_l = q_{jl}q_lh = q_jh = e_jg_j$ (see Figure 12.3). ∎

12.2.18 Proof of Claim (d3): Since $e_jg_j = q_jh$ for each $j \in J$, the uniqueness of induced colimit morphisms implies, as often before, that indeed $\bar{e}\cdot\bar{g} = h$. ∎

12.2.19 Conclusion of the proof of 12.2.3 (iii) and (iv): Now all claims of 12.2.8 have been proved, and we get $\mathrm{Inj}K_{\mathbf{HSP}_+{}^{r,\mathbf{k}}(\mathbf{K})} \subseteq \mathbf{HSP}_+{}^{r,\mathbf{k}}(\mathbf{K})$, which is what we had to show.

As already mentioned the missing parts of the proof of 12.2.3 can be found in [NSa 77] and [AN 82] (in the latter paper the statement is even more general than in 12.2.3) or the reader may construct proofs by himself along the line of the above one. ■■■

12.2.20 REMARK. We have just presented the approach of ANDRÉKA, NÉMETI and SAIN to handle the abundance of axiomatizability results for Horn formulas. One should observe that R.JOHN -- in [J 75a] for the general case and in [J 75] for special kinds of ECE-equations -- has got similar results by combining the operators **S** and **F** to one, called **M**, and **M** is based on a class M of special cocones ("sinks"). His approach appears perhaps to be more elegant, but we have the feeling that the one in part presented here follows more closely the lines of existing proofs (cf. the proof of BIRKHOFF in [Bi 35] for equational classes of universal algebras or the one of MAL'CEV in [Mal 73] for quasivarieties), and the special axiomatizability results can be read more easily from it. For further historical remarks see the paper [NSa 77].

12.3 "Compatible" classes H and S in Alg($\underline{S}$)

12.3.1 REMARK and DEFINITION. The rest of this section as well as the next section are now dedicated to the investigation of examples of the Meta Birkhoff Theorem in the category **Alg**($\underline{S}$). In particular we shall study families of classes **H** and **S** of homomorphisms satisfying the conditions (1) through (7) of Lemma 12.2.3. That **Alg**($\underline{S}$) is strongly **k**-algebroidal for each infinite regular cardinal $\mathbf{k} \geq \mathbf{d^*}$ has already been shown in 11.2.5. Moreover, products as well as direct limits of (**k**-) directed systems exist in **Alg**($\underline{S}$). Thus **Alg**($\underline{S}$) satisfies the general assumptions about $\underline{\mathbf{C}}$ of 12.2.3. We also know already (see Table 10.1) that **Alg**($\underline{S}$) has in general infinite families of factorization systems.

In this subsection we want to investigate possible combinations of classes **H** and **S** of homomorphisms (respectively of the corresponding operators) satisfying the "compatibility-condition" 12.2.3.(4), and we shall call such pairs of classes compatible. Observe that 12.2.3.(4) is the only condition in 12.2.3 which connects **H** and **S**.

As far as **H** is concerned, we shall always assume that it be a class of surjective homomorphisms -- because of condition 12.2.3.(1) that there should always be enough **H**-projectives --, although surjectivity is not necessary for the existence of **H**-projectives in general: with respect to the class $\mathbf{F_i}$ of all **final homomorphisms** (see 9.2.1 and 9.2.16.(iii)) it is not too difficult to show -- by using 9.2.16.(iii).(c) -- the following:

(*): A partial Peano algebra $\underline{P}$ on some set X is $\mathbf{F_i}$-projective iff $P = X \cup B_{\underline{P}}{}^1 X$ such that for each $x \in X$ there are exactly one $\phi \in \Omega$, exactly one $\underline{a} \in \mathrm{dom}\,\phi^{\underline{P}}$ and exactly one $k \in K_\phi$ such that $x = (k)\underline{a}$

(i.e. the domains of the operations $\phi^{\underline{P}}$ consist of disjoint injective sequences -- also disjoint for different operations --, which cover X). Thus such a $\underline{P}$ can only be discrete when it is empty, and therefore no non-empty discrete partial algebra can be the final homomorphic image of an $\mathbf{F_i}$-projective partial algebra.

12.3.2 LEMMA. For any strongly **k**-algebroidal category $\underline{\mathbf{C}}$ and any class **S** of $\underline{\mathbf{C}}$-morphisms satisfying $\Lambda^{op}(\mathbf{S}) \subseteq \mathrm{Epi}(\underline{\mathbf{C}})$ the class $\mathbf{H} := \mathrm{Iso}(\underline{\mathbf{C}})$ of all isomorphisms in $\underline{\mathbf{C}}$ is compatible with S and also satisfies 12.2.3.(1) and (2).

PROOF. It is obvious that every $\underline{\mathbf{C}}$-object is Iso($\underline{\mathbf{C}}$)-projective, and strong $\mathbf{k}$-algebroidalness of $\underline{\mathbf{C}}$ implies 12.2.3.(2). Let $h = fg \in \mathrm{Iso}(\underline{\mathbf{C}})$, $f \in \Lambda^{op}(S) \subseteq \mathrm{Epi}(\underline{\mathbf{C}})$, then we have $fgh^{-1} = \mathbf{1}$, and therefore f is a section as well as an epimorphism, i.e. an isomorphism (see 11.4.3). Thus $g = f^{-1}h \in \mathrm{Iso}(\underline{\mathbf{C}})$. ■■■

12.3.3 COROLLARY. Let the assumptions of Lemma 12.2.3 be satisfied for the category $\underline{\mathbf{C}}$, for $\mathbf{H},\mathbf{S} \subseteq \mathrm{Mor}(\underline{\mathbf{C}})$, and for some infinite regular cardinal $\mathbf{k}$ ("*" means that $\mathbf{k} = \#\omega_0$). Then one has for every class $\mathbf{K} \subseteq \mathrm{Ob}(\underline{\mathbf{C}})$:

(i) $\mathrm{InjK}_{\mathbf{S}}(\mathbf{K}) = \mathbf{S}(\mathbf{K})$;

(ii) $\mathrm{InjK}_{\mathbf{Se}}(\mathbf{K}) = \mathbf{Se}(\mathbf{K})$;

(iii)* $\mathrm{InjK}_{\mathbf{SP}_u}(\mathbf{K}) = \mathbf{SP}_u(\mathbf{K})$;

(iv)* $\mathrm{InjK}_{\mathbf{SP}_u\mathbf{e}}(\mathbf{K}) = \mathbf{SP}_u\mathbf{e}(\mathbf{K})$;

(v) $\mathrm{InjK}_{\mathbf{SP}_+{}^{r,\mathbf{k}}}(\mathbf{K}) = \mathbf{SP}_+{}^{r,\mathbf{k}}(\mathbf{K})$;

(vi) $\mathrm{InjK}_{\mathbf{SP}^{r,\mathbf{k}}}(\mathbf{K}) = \mathbf{SP}^{r,\mathbf{k}}(\mathbf{K})$;

(vii) $\mathrm{InjK}_{\mathbf{HS}}(\mathbf{K}) = \mathbf{HS}(\mathbf{K})$;

(viii) $\mathrm{InjK}_{\mathbf{HSe}}(\mathbf{K}) = \mathbf{HSe}(\mathbf{K})$;

(ix)* $\mathrm{InjK}_{\mathbf{HSP}_u}(\mathbf{K}) = \mathbf{HSP}_u(\mathbf{K})$;

(x)* $\mathrm{InjK}_{\mathbf{HSP}_u\mathbf{e}}(\mathbf{K}) = \mathbf{HSP}_u\mathbf{e}(\mathbf{K})$;

(xi) $\mathrm{InjK}_{\mathbf{HSP}_+{}^{r,\mathbf{k}}}(\mathbf{K}) = \mathbf{HSP}_+{}^{r,\mathbf{k}}(\mathbf{K})$;

(xii) $\mathrm{InjK}_{\mathbf{HSP}^{r,\mathbf{k}}}(\mathbf{K}) = \mathbf{HSP}^{r,\mathbf{k}}(\mathbf{K})$;

(xiii) $\mathrm{InjK}_{\mathbf{SP}_+}(\mathbf{K}) = \mathbf{SP}_+(\mathbf{K})$;

(xiv) $\mathrm{InjK}_{\mathbf{SP}}(\mathbf{K}) = \mathbf{SP}(\mathbf{K})$;

(xv) $\mathrm{InjK}_{\mathbf{HSP}_+}(\mathbf{K}) = \mathbf{HSP}_+(\mathbf{K})$;

(xvi) $\mathrm{InjK}_{\mathbf{HSP}}(\mathbf{K}) = \mathbf{HSP}(\mathbf{K})$.

PROOF. In the cases (i) through (vi), (xiii) and (xiv) we have chosen $\mathbf{H} = \mathrm{Iso}(\underline{\mathbf{C}})$ (see 12.3.2), in the cases (iii), (iv), (ix) and (x) the assumptions of 12.2.3 have to be satisfied for $\mathbf{k} = \#\omega_0$, and in the cases (xiii) to (xvi) one has to choose $\mathbf{k} = \infty$; but then, by assumption, the conditions of 12.2.3 are satisfied. ■■■

12.3.4. In 12.3.3 we have listed separately all possibilities for the operator $\mathbf{F}$ in 12.2.5, and we have distinguished between arbitrary "admissible" operators $\mathbf{H}$ -- satisfying 12.2.3.(i) and (ii), and for which we still have to investigate compatibility with "admissible" operators S -- and the case $\mathbf{H} = \mathbf{I}$ (isomorphisms), where every "admissible" S -- i.e. such S satisfying 12.2.3.(3), (5), (6) and (7) -- can be chosen.

12.3.5 LEMMA. Let $\mathbf{S} \subseteq \mathrm{Hom}(\mathbf{Alg}(\underline{S}))$, and let $\mathbf{H} =: \mathbf{H}_F \subseteq \mathbf{H}_W$ be the class of all surjective homomorphisms reflecting some given set F of formulas, such that F is preserved by $\Lambda^{op}(\mathbf{S})$, and assume that $fg \in \mathbf{H}_F$ and $f \in \Lambda^{op}(\mathbf{S})$ together imply surjectivity of f. Then $\mathbf{H}_F$ is compatible with $\mathbf{S}$, i.e. $\mathbf{H}_F$ is left cancellable with respect to $\Lambda^{op}(\mathbf{S})$.

PROOF. Since surjectivity of fg implies surjectivity of g, this is a consequence of

9.1.7.(ii). ∎∎∎

Since -- by 10.3.7 and 10.1.2.(iv) -- $\Lambda^{op}(S) \subseteq \mathbf{H}_W$ iff **S** contains the class of all full and injective homomorphisms, and since for closed $h = fg$, where f is an epimorphism, f is closed and surjective, we get:

12.3.6 COROLLARY. Let $\mathbf{S} \subseteq \mathrm{Hom}(\mathbf{Alg}(\underline{S}))$, and let $\mathbf{H} \subseteq \mathbf{H}_W$ be the class of all surjective homomorphisms reflecting some given set F of E-equations.

(i) Then **H** is left cancellable with respect to $\Lambda^{op}(\mathbf{S})$, whenever $\mathbf{S} = \Lambda\Lambda^{op}(\mathbf{S})$ contains the class $\mathbf{S}_i$ of all full and injective homomorphisms.

(ii) If, in addition, each h in **H** is closed, then **H** is left cancellable with respect to $\Lambda^{op}(\mathbf{S})$, whenever $\Lambda^{op}(\mathbf{S}) \subseteq \mathrm{Epi}(\mathbf{Alg}(\underline{S}))$. ∎∎∎

12.3.7 PROPOSITION. *If* **H** *is the class either of all full and surjective or of all surjective homomorphisms in* $\mathbf{Alg}(\underline{S})$, *then* **H** *is left cancellable for any class* **S** *of homomorphisms.*

PROOF. If fg is surjective, then so is g; and if, in addition, fg is full, then g is full -- see 2.4.5.(iii).(d). ∎∎∎

Combining 12.3.7 with 12.3.6 (or with 12.3.5) we get the

12.3.8 COROLLARY. If $\mathbf{S} \subseteq \mathrm{Hom}(\mathbf{Alg}(\underline{S}))$, and if **H** is the class of all full and surjective homomorphisms reflecting some given set F of E-equations, then **H** is compatible with **S**, whenever $\Lambda\Lambda^{op}(\mathbf{S})$ contains the class of all full and injective homomorphisms. ∎∎∎

12.3.9 LEMMA. Let **H** be the class of all surjective -- respectively of all full and surjective -- homomorphisms reflecting some set F of E-equations; and let **S** be the class (a): of all, (b): of all injective, (c): of all initial, respectively (d): of all injective and initial homomorphisms in each case reflecting some set F' of E-equations. Then $F' \subseteq F$ implies that **H** and **S** are compatible.

PROOF. We already know that under the conditions of the Lemma $fg \in \mathbf{H}$ implies that g is (full and) surjective. Thus we only have to show that g reflects F: Lemma 10.3.8 tells us that for $\mathbf{S} = \mathbf{M} \cap \mathbf{N}$ with **M** being the class of all F'-reflecting homomorphisms and **N** being one of the classes in 12.3.9.(a) through (d), and therefore satisfying $\Lambda^{op}(\mathbf{N}) \subseteq \mathbf{H}_W$, one has $\Lambda^{op}(\mathbf{S}) = \Lambda^{op}(\mathbf{M}) \cdot (\Lambda^{op}(\mathbf{N}) \cap \mathbf{M})$. Therefore f can be represented as $f = uv$, where $u \in \Lambda^{op}(\mathbf{M})$ and $v \in \Lambda^{op}(\mathbf{N}) \cap \mathbf{M} \subseteq \mathbf{H}_W$. Since vg preserves all E-equations, and since uvg reflects F ($\supseteq$ F'), 9.1.7.(ii) implies that u reflects F and therefore F'. Thus $u \in \mathbf{M} \cap \Lambda^{op}(\mathbf{M}) \subseteq \mathrm{Iso}(\mathbf{Alg}(\underline{S}))$. Hence $f = uv$ is surjective and preserves all E-equations, and 9.1.7.(iii) implies that g reflects F, showing that $g \in \mathbf{H}$. ∎∎∎

12.3.10 COROLLARY. For some special instances of the operators **H** and **S** the information about compatibility derived from the results in 12.3.2 through 12.3.9 is collected in Table 12.1. No entry at some place means only that this case has not yet been investigated, while an entry "m" ("m*") says that compatibility of the corresponding pair (**H**,**S**) follows from 12.3.m (under the condition $P' \subseteq P$, where P and P' are any sets of terms). As abbreviations we use the notation corresponding to the one in 4.6.2 and in 10.3.12, and in addition to the one in 10.3.12 we use "P-i" for P-initialness (for any set P of terms). ■

H \ S	S_i	S_w	M_i	I	S_c	M_c	$M_{p'}$	$S_{p'}$	$M_{ip'}$	$S_{ip'}$	M_n	$M_{p'-i}$	$S_{p'-i}$
I	2	2	2	2	2	2	2	2	2	2	2	2	2
H_c	6	6	6	6	6	6	6	6	6	6	6	6	6
H_f	8	8	8	7	7	7	7	7	7	7	7	7	7
H_w	6	6	6	7	7	7	7	7	7	7	7	7	7
H_p	6	6	6				9*	9*	9*	9*			
H_{fp}	8	8	8				9*	9*	9*	9*			
H_i	6	6	6										
H_{p-i}	6	6	6									9*	9*
H_b	6	6	6										

Table 12.1. Compatibility of the operators **H** and **S** (see 12.3.10)

12.3.11 REMARK. The compatibilities listed in Table 12.1 do not mean that we actually already have examples for Lemma 12.2.3, since all the other conditions of this Lemma have not yet been checked; this will be done in detail in the next section. We only know from Table 10.1 that for the classes S listed in Table 12.1 $(\Lambda^{op}(S), S)$ is a factorization system satisfying $\Lambda^{op}(S) \subseteq \mathrm{Epi}(\mathbf{Alg}(\underline{S}))$, and that, for $\mathbf{k} \geq \mathbf{d^*}$, S from Table 12.1 is **k**-inductive (see 11.3.1). Thus we have assumptions 12.2.3.(3) and (7). Moreover, co-(well-poweredness) of $\mathbf{Alg}(\underline{S})$ with respect to epimorphisms (see the Epimorphism Theorem 5.8.3) and the fact that $\Lambda^{op}(S) \subseteq \mathrm{Epi}(\mathbf{Alg}(\underline{S}))$ imply that $\mathbf{Alg}(\underline{S})$ is $\Lambda^{op}(S)$-co-(well-powered); therefore also assumption 12.2.3.(5) is satisfied in Table 12.1. Since for $\mathbf{k} \geq \mathbf{d^*}$ all the basic assumptions about the category $\underline{\mathbf{C}}$ in 12.2.3 are satisfied in $\mathbf{Alg}(\underline{S})$, it remains only to investigate conditions 12.2.3.(1), (2) and (6) for the "pairings" of Table 12.1. This will be done in section 13.
It should be observed that still other combinations of properties of homomorphisms could be included in Table 12.1, which therefore only presents some arbitrary choice. But even now we have already 33 pairs with one and 10 pairs with two parameters which usually may run through infinitely many distinct values.

13 Construction of examples for the Meta Birkhoff Theorem

In this section we continue the discussion of examples for the Meta Birkhoff Theorem 12.2.5, which we have already begun in 12.3. Our starting point is Table 12.1 of compatible operators; we shall deal only with the operators already considered in this table, and we shall also use its abbreviations. Because of Remark 12.3.11 it remains only to investigate conditions (1), (2) and (6) of Lemma 12.2.3.
Actually we shall not discuss all the operators considered in Table 12.1, but only those about which we already know a bit more (e.g. $\Lambda^{op}(\mathbf{M}_{p'-i})$ has not yet been investigated); -- the investigation of the others and possibly further operators is left as an exercise or **problem**. For some other operators we only state the results and leave the details of the proofs as exercises.

13.1 k-dual algebroidalness of $\Lambda^{op}(S)$ for some classes S of Alg($\underline{S}$)-morphisms

An immediate consequence of 11.2.2 is

13.1.1 LEMMA. Let $f:\underline{A} \to \underline{B}$ be a full and surjective homomorphism, and let $\underline{A}$ be strongly **k**-small for some regular infinite cardinal $\mathbf{k} \geq \mathbf{d^*}$. Then $\underline{B}$ is strongly **k**-small. ■■■

13.1.2 COROLLARY. The classes $\mathbf{H_f}$ (= $\Lambda^{op}(\mathbf{S_w})$) and $\mathbf{H_c}$ are **k**-dual algebroidal (see 11.5.3) for every regular cardinal $\mathbf{k} \geq \mathbf{d^*}$. ■■■

13.1.3 LEMMA. The classes $\mathbf{E_w}$ = Epi(**Alg**($\underline{S}$)) (= $\Lambda^{op}(\mathbf{S_c})$) and $\mathbf{H_w}$ (= $\Lambda^{op}(\mathbf{S_r})$) are **k**-dual algebroidal for each regular infinite cardinal $\mathbf{k} \geq \mathbf{d^*}$.

PROOF. Let $e:\underline{A} \to \underline{B}$ be any epimorphism, $\underline{A} \xrightarrow{s} \underline{C} \xrightarrow{m} \underline{B}$ its ($\mathbf{H_f}$,$\mathbf{S_w}$)-factorization, where we may assume that $\underline{C}$ is a weak subalgebra of $\underline{B}$. Then $B = \mathbf{C}_{\underline{B}}C$. Let $\underline{A}$ be strongly **k**-small, then so is $\underline{C}$ because of 13.1.1. From 4.4.7 and 11.2.2 we know that $\underline{B}$ is the **k**-direct limit (union!) of all weak subalgebras $\underline{D}$ of $\underline{B}$, which are strongly **k**-small and generated by some subset of C (i.e. $\underline{D} \in I_3(\underline{B},C)$). But then $\underline{B}$ is also the **k**-direct limit of the system of all $\underline{D}$ U* $\underline{C}$ (see 3.2.2 for "U*"), where $\underline{D} \in I_3(\underline{B},C)$ with the inclusion homomorphisms; and each $\underline{D}$ U* $\underline{C}$ is strongly **k**-small. It is easy to check that for $\underline{D} \subseteq \underline{D}' \in I_3(\underline{B},C)$ the inclusion mapping $id_{C\cup D}:\underline{D}$ U* $\underline{C} \to \underline{D}'$ U* $\underline{C}$ as well as the homomorphism $f_{\underline{AD}}:=f:\underline{A} \to \underline{D}$ U* $\underline{C}$ are epimorphisms. Thus, adding $\underline{A}$ as a least element to $I_3(\underline{B},C)$ and the $f_{\underline{AD}}$ as corresponding homomorphisms, we get $f_{\underline{A}} = f:\underline{A} \to \underline{B}$, and therefore the desired "approximation" of f. The argument for $\mathbf{H_w}$ is similar. ■■■

13.1.4 LEMMA. Let $\Lambda^{op}(S) \subseteq$ Epi(**Alg**($\underline{S}$)) for some factorization system ($\Lambda^{op}(S)$,S), and let $\mathbf{k} \geq \mathbf{d^*}$ be a regular cardinal. Assume that we have for $e:\underline{A} \to \underline{B}$ in $\Lambda^{op}(S)$ with $\underline{A}$ strongly **k**-small:
Let $\underline{C}$ be the weak subalgebra of $\underline{B}$ with carrier set [A]e and the structure induced from $\underline{A}$ by e; and if $\underline{D}$ is any C-generated strongly **k**-small weak subalgebra of $\underline{B}$ containing $\underline{C}$, then let there be a strongly **k**-small weak subalgebra $\underline{D}'$ of $\underline{B}$ containing $\underline{D}$ such that $e:\underline{A} \to \underline{D}'$ belongs to $\Lambda^{op}(S)$.
Then $\Lambda^{op}(S)$ is **k**-dual algebroidal.

PROOF. Let $e:\underline{A} \to \underline{B}$ be in $\Lambda^{op}(S)$. With the assumptions of 13.1.4 set
$I_5(\underline{B},C):= \{\underline{D} | \underline{D}$ is a strongly **k**-small C-generated weak subalgebra of $\underline{B}$ and $e:\underline{A} \to \underline{D}$ is in $\Lambda^{op}(S)\}$.
Then the assumptions of 13.1.4 say that $I_5(\underline{B},C)$ is cofinal in $I_3(\underline{B},C)$; thus $\underline{B}$ is the direct limit of $I_5(\underline{B},C)$. Adding to $I_5(\underline{B},C)$ the inclusion mappings, $\underline{A}$ and the epimorphisms from $\underline{A}$ into $\underline{D} \in I_5(\underline{B},C)$ with the same graph as e, then 10.1.2.(v)^{op} implies that all morphisms of this **k**-directed system belong to $\Lambda^{op}(S)$ showing that $\Lambda^{op}(S)$ is **k**-dual algebroidal. ■■■

In view of 10.3.8 and 10.3.9 the following Lemma will be useful:

13.1.5 LEMMA. Let P be any set of TE-expressions, and let $\mathbf{k} \geq \mathbf{d^*}$ be any regular cardinal. Then $\Lambda^{op}(\mathbf{S_P})$ and $\Lambda^{op}(\mathbf{S_{jP}})$ are **k**-dual algebroidal.

SKETCH OF PROOF. We only prove the statement for S_{ip}, the proof of the other one is similar. Combining the results in 10.3.5 and 10.3.9 we may say:

$g:\underline{C} \to \underline{D}$ belongs to $\Lambda^{op}(S_{ip})$ iff $G := \text{dom}\,\tilde{g}$ and $\tilde{g}$ have the following properties:

(*) If $G_P := \mathbf{C}_{<\underline{G},P>}C$, then each element of $G \setminus G_P$ is identified by $\ker \tilde{g}$ with some element of G_P.

Now consider $f:\underline{A} \to \underline{B}$ in $\Lambda^{op}(S_{ip}) = \Lambda^{op}(\mathbf{M}_P)\cdot \mathbf{H}_W =: \mathbf{E}$; let $\underline{F} := \text{dom}\,\tilde{f}$, $F_P := C_{<\underline{F},P>}A$. Let J be the set of all pairs $(\underline{D},R)$ such that $\underline{D}$ is a strongly $\mathbf{k}$-small $\underline{A}$-generated relative subalgebra of $\underline{F}$, $R \in \text{Cong}(\underline{D})$, $R \subseteq \ker f$, $\#R < \mathbf{k}$ and each element of $D \setminus D_P$ is R-equivalent with some element of D_P.

For elements of J define $(\underline{D},R) \leq (\underline{D}',R')$ iff $\underline{D}$ is a relative subalgebra of $\underline{D}'$ and $R \subseteq R'$. Then it is easy to realize that $(J;\leq)$ is a $\mathbf{k}$-directed poset.

If $(\underline{D},R) \leq (\underline{D}',R')$, then there are unique homomorphisms $f_{RR'}:\underline{D}/R \to \underline{D}'/R'$ satisfying $\text{nat}_R \cdot f_{RR'} = \text{nat}_{R'}|D$, and $f_R:\underline{D} \to \underline{B}$ satisfying $\text{nat}_R \cdot f_R = \tilde{f}|D$. Moreover, $\underline{D}/R$ and $\underline{D}'/R'$ are strongly $\mathbf{k}$-small. From (*) we may infer that each $f_{RR'}$ and each f_R belong to $\mathbf{E}$ and that $(\underline{D}/R, f_{RR'} | (\underline{D},R) \leq (\underline{D}',R') \text{ in } J)$ is a $\mathbf{k}$-directed system with $\mathbf{k}$-direct limit $((f_R | (\underline{D},R) \in J), \underline{B})$, $(\underline{A},\Delta_A) \in J$ and $f_{\Delta_A} = f$. This shows that $\Lambda^{op}(S_{ip})$ is $\mathbf{k}$-dual algebroidal. ■■■

Combining these results and using the notation from 12.3.10 we get the

13.1.6 PROPOSITION. Let $\mathbf{k} \geq \mathbf{d^*}$ be a regular cardinal. The classes $\Lambda^{op}(\mathbf{S})$ are $\mathbf{k}$-dual algebroidal, where $\mathbf{S}$ is one of the classes $\mathbf{I}$, S_c, S_i, S_w, $S_{P'}$, $S_{iP'}$, $\mathbf{M}_c$, $\mathbf{M}_i$, $\mathbf{M}_{P'}$, $\mathbf{M}_n$.

PROOF. Since the treatment of $\mathbf{I}$ is trivial because of $\mathbf{k}$-algebroidalness of $\mathbf{Alg}(\underline{S})$, it remains to treat $S_{P'}$, $S_{iP'}$, $\mathbf{M}_c$, $\mathbf{M}_i$, $\mathbf{M}_{P'}$, $\mathbf{M}_n$. The statements for $\mathbf{M}_c$ and $\mathbf{M}_{P'}$ can easily be derived from 13.1.4 and the descriptions of their left factors in the corresponding factorization system. Since $\Lambda^{op}(S_{P'}) = \Lambda^{op}(\mathbf{M}_{P'})\cdot\Lambda^{op}(S_w)$ and $\Lambda^{op}(S_{iP'}) = \Lambda^{op}(\mathbf{M}_{P'})\cdot\Lambda^{op}(S_i)$, their $\mathbf{k}$-dual algebroidalness follows from 13.1.5. -- The proofs for $\mathbf{M}_i$ and $\mathbf{M}_n$ are left as exercises. ■■■

13.1.7 REMARKS and PROBLEMS. **(i):** The details of the proofs of the cases, where $\mathbf{S}$ in 13.1.6 is $\mathbf{M}_i$ or $\mathbf{M}_n$ have not been given, since they are quite technical using the characterizations of $\Lambda^{op}(\mathbf{M}_i)$ respectively of $\Lambda^{op}(\mathbf{M}_n)$, and we do not yet know how important they will be. The cases of $\mathbf{M}_{iP'}$, $\mathbf{M}_{P'-i}$ and $S_{P'-i}$ have not yet been investigated by us. The positive solution of one of the following two problems would help a lot.

(ii): All the results in 13.1.6 would be trivial consequences if the following **conjecture** (or even a stronger one) could be proved:

CONJECTURE: Let $F \subseteq L_{\mathbf{d^*},\mathbf{d^*}}$ be any set of formulas for $\mathbf{Alg}(\underline{S})$, let $\mathbf{M}_F$ be the class of all homomorphisms reflecting all formulas from F. Then $\Lambda^{op}(\mathbf{M}_F)$ is $\mathbf{k}$-dual algebroidal for every regular cardinal $\mathbf{k} \geq \mathbf{d^*}$.

We have no idea whether this conjecture is true. But it should be observed that $\Lambda^{op}(\mathbf{M}_F)$ is "generated" by a set E_F of epimorphisms between strongly $\mathbf{k}$-small objects as $\Lambda^{op}(\mathbf{M}_F) = \Lambda^{op}(\Lambda(E_F))$.

(iii): The solution of the following **problem** would probably be of great help for the above one:

PROBLEM: Let $\underline{\mathfrak{C}}$ be any ("nice") category and $E \subseteq \text{Epi}(\underline{\mathfrak{C}})$ be any class of epimorphisms. Is there a complete set of rules inducing a concept of derivability of morphisms from a given class such that a $\underline{C}$-morphism e is derivable from **E** iff $e \in \Lambda^{op}(\Lambda(\mathbf{E}))$? Perhaps such results are known in category theory, but we have not yet come upon them, and some experts whom we asked did not know about them, either.

13.2 H-projective partial $\underline{S}$-algebras

The results of this subsection are motivated by conditions 12.2.3.(1) and (2) and the wish to apply this meta theorem in as many instances as possible. We recall that we have already agreed in 12.3 to choose the class **H** of 12.2.3 as a class of surjective homomorphisms. Again we use the notation of 12.3.10.

13.2.1 REMARK. Condition 12.2.3.(2) that every **H**-projective object of $\mathbf{Alg}(\underline{S})$ has to be the direct limit of a **k**-directed system of strongly **k**-small **H**-projective objects is obviously satisfied for $\mathbf{k} \geq \mathbf{d^*}$ as long as we know that each weak subalgebra generated by a subset of a fixed generating set of an **H**-projective object is **H**-projective, too. 12.2.3.(2) then follows from 11.2.2 and 4.4.7. Thus our main objective is to characterize **H**-projective objects for different classes **H**.

13.2.2 LEMMA. $Pj(\mathbf{H}_w)$ is the class of all discrete partial algebras.

PROOF. That each discrete partial algebra is projective with respect to surjective homomorphisms is a consequence of the Axiom of Choice. If $\underline{A}$ is not discrete, then $id_A:\underline{A} \to \underline{A}$ cannot be "lifted" against $id_A:\underline{A}_d \to \underline{A}$ in $\mathbf{H}_w$. ■■■

13.2.3 COROLLARY. $\mathbf{H}_w$ satisfies conditions 12.2.3.(1) and (2). ■■■

13.2.4 LEMMA. (i): $Pj(\mathbf{H}_f)$ consists of all partial algebras $\underline{P}$ satisfying

(*) $\underline{P}$ is a partial Peano algebra on some Peano basis, say X.

(**) For all $\varphi,\varphi' \in \Omega$, for all $\underline{a} \in \operatorname{dom} \varphi^{\underline{P}}$ and $\underline{b} \in \operatorname{dom} \varphi'^{\underline{P}}$, and for all $k \in K_\varphi$, $l \in K_{\varphi'}$:

If $\varphi \neq \varphi'$, or if $\varphi = \varphi'$, but $\underline{a} \neq \underline{b}$, or if $k \neq l$, then $(k)\underline{a} \neq (l)\underline{b}$, and we always have $(k)\underline{a},(l)\underline{b} \in X$.

I.e. $\underline{P}$ is a partial Peano algebra on some set X such that $P = X \cup \mathbf{B}_{\underline{P}}{}^1X = \mathbf{D}_{\underline{P}}{}^1X$, and each element of X occurs at most once in any sequence belonging to some of the fundamental partial operations (thus these argument sequences are injective and mutually disjoint).

(ii): $\mathbf{H}_f$ satisfies conditions 12.2.3.(1) and (2) for every regular cardinal $\mathbf{k} \geq \mathbf{d^*}$.

PROOF. Let $f:\underline{A} \to \underline{B}$ be any full and surjective homomorphism, let $\underline{P}$ be a partial Peano algebra on X satisfying (*) and (**), and let $p:\underline{P} \to \underline{B}$ be any homomorphism. Let $x \in X$. If there are $\varphi \in \Omega$, $\underline{a} \in \operatorname{dom} \varphi^{\underline{P}}$ and $k \in K_\varphi$ such that $x = (k)\underline{a}$, then φ, $\underline{a}$ and k are uniquely determined by x. We have $\underline{ap} \in \operatorname{dom} \varphi^{\underline{B}}$, and since f is full and surjective, there is $\underline{c} \in \operatorname{dom} \varphi^{\underline{A}}$ such that $\underline{cf} = \underline{ap}$, and we choose one such $\underline{c}$ and define $(l)\underline{a}q := (l)\underline{c}$ for each $l \in K_\varphi$ (in particular the q-value of x determines the q-value of any other argument of $\underline{a}$). If x is not in the domain of any such sequence, then we define xq arbitrarily in $(xp)f^{-1}$. For each element $a = \underline{a}\varphi^{\underline{P}}$ of P we define (recursively) $aq := (\underline{a}q)\varphi^{\underline{A}}$. Obviously $q:\underline{P} \to \underline{A}$ satisfies $qf = p$. ■

It is easily seen that any partial algebra $\underline{D}$ can be represented as a quotient of a partial Peano algebra $\underline{P}$ satisfying (*) and (**): Choose for some $o \notin \Omega$:
$X := (\{o\} \times A) \cup \bigcup(\{\phi\} \times \{\underline{a}\} \times K_\phi | \phi \in \Omega$ and $\underline{a} \in \mathrm{dom}\, \phi^{\underline{A}})$;
$P \setminus X :\!= \bigcup(\{\phi\} \times \mathrm{dom}\, \phi^{\underline{A}} | \phi \in \Omega)$;
$\mathrm{graph}(\phi^{\underline{P}}) := \{(((\phi,\underline{a},k)|k \in K_\phi),(\phi,\underline{a}))|\underline{a} \in \mathrm{dom}\, \phi^{\underline{A}}\ ,\ \phi \in \Omega\}$;

$$p:P \to A : \quad (o,a) \mapsto a\ ,\qquad (\phi,\underline{a},k) \mapsto (k)\underline{a}\ ,\qquad (\phi,\underline{a}) \mapsto (\underline{a})\phi^{\underline{A}}\ .$$

It is easy to realize that $\underline{P}$ satisfies (*) and (**) and that $p:\underline{P} \to \underline{A}$ is a full and surjective homomorphism). ∎

If $\underline{Q}$ is any partial algebra which does not satisfy (*) and (**) for any generating subset, then there is a congruence R on a partial Peano algebra $\underline{P}$ which satisfies (*) and (**) such that $\underline{P}/R$ is isomorphic to $\underline{Q}$. Then $id_{P/R}$ cannot be lifted against nat_R, i.e. $\underline{Q}$ cannot belong to $Pj(\mathbf{H}_f)$. ∎∎∎

13.2.5 REMARK. It is easy to see that all partial algebras are Iso($\mathbf{Alg}(\underline{S})$)-projective, and that 12.2.3.(1) and (2) are satisfied for any regular cardinal $\mathbf{k} \geq \mathbf{d^*}$ in connection with the operator **I**. ∎

13.2.6 REMARK. Observe that for the class $\mathbf{H}_b$ of all bijective homomorphisms of $\mathbf{Alg}(\underline{S})$ we have $Pj(\mathbf{H}_b) = Pj(\mathbf{H}_w)$ (since we allow the Axiom of Choice), and therefore we know from 13.2.3 that conditions 12.2.3.(1) and (2) are satisfied for $\mathbf{H}_b$ and regular cardinals $\mathbf{k} \geq \mathbf{d^*}$.

But this means that switching between $\mathbf{H}_b$ and $\mathbf{H}_w$ in the cases of **S** as $\mathbf{S}_r$, $\mathbf{S}_w$ or $\mathbf{M}_i$ -- where 12.2.3 is applicable to $\mathbf{H}_b$ as well as to $\mathbf{H}_w$ -- does not change the corresponding sets of formulas. Therefore we get from Table 12.1 that

$$\mathbf{H}_w\mathbf{S} = \mathbf{H}_b\mathbf{S}\ ,$$
$$\mathbf{H}_w\mathbf{S}\mathbf{P}^{r,\mathbf{k}} = \mathbf{H}_b\mathbf{S}\mathbf{P}^{r,\mathbf{k}}\ ,$$
$$\mathbf{H}_w\mathbf{S}\mathbf{P}_+{}^{r,\mathbf{k}} = \mathbf{H}_b\mathbf{S}\mathbf{P}_+{}^{r,\mathbf{k}}\ ,$$

where **S** is one of the operators $\mathbf{S}_r$, $\mathbf{S}_w$ or $\mathbf{M}_i$.
However, it should also be observed that $\mathbf{H}_w$ is additionally compatible with other operators **S**, while we did not investigate their compatibility with $\mathbf{H}_b$. ∎

Let us add some more general results

13.2.7 LEMMA. Let **H** and **H'** be classes of surjective homomorphisms; then

$$Pj(\mathbf{H} \cap \mathbf{H'}) \supseteq Pj(\mathbf{H}) \cup Pj(\mathbf{H'}) \cup \{\underline{A}+\underline{B} | \underline{A} \in Pj(\mathbf{H})\ ,\ \underline{B} \in Pj(\mathbf{H'})\}\ .$$

PROOF. Obviously $Pj(\mathbf{H} \cap \mathbf{H'}) \supseteq Pj(\mathbf{H}) \cup Pj(\mathbf{H'})$. Let $(j_1,j_2;\underline{A_1}+\underline{A_2} =: \underline{B})$ be a coproduct such that $\underline{A_1} \in Pj(\mathbf{H})$ and $\underline{A_2} \in Pj(\mathbf{H'})$, and let $f:\underline{E} \to \underline{F}$ be any $(\mathbf{H} \cap \mathbf{H'})$-morphism, $g:\underline{B} \to \underline{F}$. By assumption there are $h_i:\underline{A_i} \to \underline{E}$ such that $h_i f = j_i g$ $(i \in \{1,2\})$. Since $(j_1,j_2;\underline{B})$ is a coproduct, there is $h:\underline{B} \to \underline{E}$ such that $j_i h = h_i$. Therefore $j_i g = h_i f = j_i hf$ $(i \in \{1,2\})$. Thus $hf = g$. ∎∎∎

13.2.8 LEMMA. Let $\underline{P_1}$ and $\underline{P_2}$ be partial Peano algebras, and let $(j_1,j_2;\underline{P_1}+\underline{P_2})$ be their coproduct. Then j_1 and j_2 are injective. ∎∎∎

13.2.9 LEMMA. (i): Let **H** be a class of surjective homomorphisms and $(\Lambda^{op}(\mathbf{S}),\mathbf{S})$ be a factorization system. Moreover, let $e:\underline{Q} \to \underline{P}$ be in $\Lambda^{op}(\mathbf{S})$ with **H**-projective $\underline{Q}$. Then $\underline{P}$ is **E**-projective for $\mathbf{E} := \mathbf{H} \cap \mathbf{S}$.

(ii): If, in addition, $\mathbf{Alg}(\underline{S})$ has enough **H**-projectives, if **H** is the left factor of a factorization system (it suffices that **H** is left cancellable with respect to epimorphisms), and if $\Lambda^{op}(\mathbf{S}) \subseteq \mathrm{Epi}(\mathbf{Alg}(\underline{S}))$, then $\mathbf{Alg}(\underline{S})$ has enough **E**-projective objects.

(iii): With the assumptions of (i) and (ii) we can also state that each **E**-projective object $\underline{P}$ is isomorphic to a subalgebra of a $\Lambda^{op}(\mathbf{S})$-homomorphic image of an **H**-projective object.

PROOF. Take the assumptions and notation of (i), let $f:\underline{C} \to \underline{D}$, $f \in \mathbf{E}$, and let $g:\underline{P} \to \underline{D}$. Since $\underline{Q}$ is **H**-projective, there is $h:\underline{Q} \to \underline{C}$ such that $hf = eg$. Now the assumptions on e and f imply the existence of a diagonal-fill-in $d:\underline{P} \to \underline{C}$ satisfying in particular $df = g$, showing that $\underline{P}$ is **E**-projective. ∎

Now let the assumptions on **H** and **S** of (ii) also be satisfied, and let $\underline{A} \in \mathbf{Alg}(\underline{S})$. By assumption there exist $\underline{Q} \in \mathrm{Pj}(\mathbf{H})$ and an **H**-morphism $q:\underline{Q} \to \underline{A}$. Let $q = gs$ be a $(\Lambda^{op}(\mathbf{S}),\mathbf{S})$-factorization of q, $g:\underline{Q} \to \underline{P}$. By the first part $\underline{P} \in \mathrm{Pj}(\mathbf{H} \cap \mathbf{S})$. Since g is an epimorphism, and since **H** is left cancellable with respect to epimorphisms, we have $s \in \mathbf{H}$ ($\cap$ **S**), i.e. $\underline{A}$ is an ($\mathbf{H} \cap \mathbf{S}$)-homomorphic image of an ($\mathbf{H} \cap \mathbf{S}$)-projective object. ∎

Ad (iii): Let $\underline{P} \in \mathrm{Pj}(\mathbf{E})$; since **H** has enough **H**-projectives, there are $\underline{Q} \in \mathrm{Pj}(\mathbf{H})$ and an **H**-morphism $f:\underline{Q} \to \underline{P}$. Let $f = ts$ be a $(\Lambda^{op}(\mathbf{S}),\mathbf{S})$-factorization of f, $t:\underline{Q} \to \underline{Z}$. Because of (i) $\underline{Z}$ is **E**-projective. Since **H** is left cancellable with respect to epimorphisms, s belongs to **E**, and therefore $id_P:\underline{P} \to \underline{P}$ can be lifted against s to a homomorphism $u:\underline{P} \to \underline{Z}$ satisfying $us = id_P$. Thus u is a section and therefore closed (see 11.4.4), i.e. $\underline{P}$ is isomorphic to a subalgebra of $\underline{Z}$. ∎∎∎

13.2.10 PROPOSITION. Let **H** be one of the classes $\mathbf{H}_w$ or $\mathbf{H}_f$, let P be any set of terms, and let **S** be the class of all P-reflecting homomorphisms. Then $\mathbf{E} := \mathbf{H} \cap \mathbf{S}$ satisfies 12.2.3.(1) and (2). In particular each **E**-projective object can be obtained in the way described in 13.2.9. More precisely -- using 10.3.2 and 13.2.4 -- we get

$\mathrm{Pj}(\mathbf{H}_P) = \{\underline{Q} \mid \underline{Q}$ is a partial Peano algebra on some set X and $Q = \mathbf{C}_{<\underline{Q},P>}X\}$.

$\mathrm{Pj}(\mathbf{H}_{fP}) = \{\underline{Q} \mid \underline{Q}$ is a partial Peano algebra on some set X, and $\underline{Q}$ contains some X-generated relative subalgebra $\underline{B} \in \mathrm{Pj}(\mathbf{H}_f)$ such that $Q = \mathbf{C}_{<\underline{Q},P>}B\}$.

PROOF. Let **E** be either $\mathbf{H}_P$ or $\mathbf{H}_{fP}$ and **H** the corresponding class $\mathbf{H}_w$ respectively $\mathbf{H}_f$. Using 10.3.5 the property in the description of Pj(**E**) applies at least to those objects which are $\Lambda^{op}(\mathbf{M}_P)$-homomorphic images of **H**-projective objects. By 13.2.9.(iii) each **E**-projective object is a subalgebra of such a partial algebra, and it is easily seen (take the preimage of [P]u with respect to t in the proof of 13.2.9.(iii)) that it also has to have the corresponding property. Moreover, the fact that $\mathbf{H}_w$ and $\mathbf{H}_f$ have property 12.2.3.(2) and the above description of Pj(**E**) imply that **E** has property 12.2.3.(2). ∎∎∎

13.2.11 COROLLARY. $\mathrm{Pj}(\mathbf{H}_c)$ is the class of all partial Peano algebras, and $\mathbf{H}_c$ satisfies conditions 12.2.3.(1) and (2). ∎∎∎

13.2.12 REMARK. Since $\mathbf{H}_i \subseteq \mathbf{H}_c$ and $\mathbf{H}_{P-i} \subseteq \mathbf{H}_P$, we have $\mathrm{Pj}(\mathbf{H}_c) \subseteq \mathrm{Pj}(\mathbf{H}_i)$, and $\mathrm{Pj}(\mathbf{H}_P) \subseteq \mathrm{Pj}(\mathbf{H}_{P-i})$ for any set P of terms. In particular the category theoretical definition of initialness implies that $\mathrm{Pj}(\mathbf{H}_i) = \mathbf{Alg}(\underline{S})$, and therefore $\mathbf{H}_i$ satisfies conditions 12.2.3.(1) and (2). Moreover, $Pj(H_i) = Pj(I)$, and therefore we have for the closure operators connected with 12.2.3 that $H_i SF = ISF$, where $S \in \{S_i, S_w, M_i\}$ and $F \in \{e, P^{r,k}, P_+^{r,k}, P_u, P_u e\}$ (see Table 12.1). ∎

We leave a discussion of H_{P-i} and similar operators as a problem to interested readers.

13.3 Examples in Alg($\underline{S}$) for the Meta Birkhoff Theorem

We now combine the results from 12.3, 13.1 and 13.2 and get the following

13.3.1 THEOREM. Let $k \geq d^*$ be a regular cardinal, and let P and P' be any sets of terms. Table 13.1 shows, which operators **H** and **S** are now known to satisfy the conditions of Lemma 12.2.3 (the Meta Birkhoff Theorem). An entry "*" means a positive answer, whenever $P' \subseteq P$, while "+" means a positive answer in general; we use the notation from Table 12.1. Missing entries only mean that these pairs have not yet been investigated.

H \ S	S_i	S_w	M_i	S_c	M_c	$M_{P'}$	$S_{P'}$	$S_{iP'}$	M_n
I	+	+	+	+	+	+	+	+	+
H_c	+	+	+	+	+	+	+	+	+
H_f	+	+	+	+	+	+	+	+	+
H_w	+	+	+	+	+	+	+	+	+
H_P	+	+	+			*	*	*	
H_{fP}	+	+	+			*	*	*	
H_i	+	+	+						
H_b	+	+	+						

Table 13.1. Pairs of operators **H, S** satisfying the assumptions of the Meta Birkhoff Theorem (see 13.3.1) ∎∎∎

13.3.2 REMARKS. **(i)**: We are quite certain that for all the positive entries in Table 12.1 the corresponding operators **H** and **S** also satisfy conditions 12.2.3.(1) and (2) and (6), and therefore lead to examples of Theorem 12.2.5 in **Alg**($\underline{S}$).

It is quite likely that the proofs of the facts that the assumptions of 12.2.3 are satisfied for a given pair **(H,S)** above could in most instances be derived from some more general principles or results which are not known to us at this moment. Such general principles and results could make Theorem 12.2.5 and corresponding results of H.ANDRÉKA and I.NÉMETI even more valuable. The preceding investigations and results are only meant to give a feeling for its possibilities and for some methods to treat the corresponding problems.

(ii): We did not yet look for counterexamples for the cases where there is no entry in Table 13.1. Thus it might be a good exercise to look for positive or negative answers in those cases.

(iii): Table 13.1 contains 54 positive entries -- but since the **H**-operators I and H_i, respectively H_w and H_b yield the same axioms in connection with their positive entries for S-operators, we actually have to subtract 6 possibilities. 6 of these entries yield families with two (almost independent) parameters, while 18 of them yield families with one parameter, and if the type is rich enough, these families are infinite. Omitting the row corresponding to I and observing 12.3.3 we get from Table 13.1 624 different model theoretic closure operators, 336 of which correspond to infinite families. This illustrates how sensitive an instrument Theorem 12.2.5 provides for partial algebras.

(iv): However this richness of the theory makes it difficult for the user to find the appropriate axioms and operators for his applications. We do not want to hide that the situation may be even more complex, since there are model theoretic approaches to parts of implicational partial algebra, which do not fit into the "picture drawn so far": In computer science (see WIRSING et al.) so called "strong equality" is often used -- see 8.1.3.(b). In connection with other investigations, so called "weak equality" is used -- see 8.1.3.(a) --, e.g. in logics of quantum mechanics (cf. S.KOCHEN, E.P.SPECKER [KoSp 68]). For these concepts the above instruments are not sufficient. Here R.JOHN (see [J 75], [J 75a], [J 78]) has provided another approach for the description of model theoretic closure operators. But the application of his approach to partial algebras looks less elegant than the presented one, and we do not have the space to discuss it here in more detail.

(a): Restrictions on the premise $\Lambda^*\{t_i \stackrel{e}{=} t'_i \mid i \in I\}$ in case of **H** are:	
I	no restrictions
$\mathbf{H}_w$	$t_i = t'_i$ and it is a variable from M $(i \in I)$ (actually the implication is an E-equation);
$\mathbf{H}_f$	$t_i = t'_i$ and it has the form $(x_{ki} \mid k \in K_{\phi_i})\phi_i$ $(i \in I,\ \phi_i \in \Omega)$ such that $(k,i) \neq (k',i')$ implies $x_{ki} \neq x_{k'i'}$;
$\mathbf{H}_c$	$t_i = t'_i$ is arbitrary $(i \in I)$;
$\mathbf{H}_p$	$t_i = t'_i \in C_{<\underline{F}(M,\mathbf{TAlg}(\underline{S})),P>}M$, i.e. each t_i is generated from M by iterative application of conditioned terms derived from P;
$\mathbf{H}_{fp}$	$I = J \dot{\cup} K$, and for $j \in J$ the conditions for $\mathbf{H}_f$ hold, while, for $k \in K$, $t_k = t'_k \in C_{<\underline{F}(M,\mathbf{TAlg}(\underline{S})),P>}(\downarrow\{t_l \mid l \in J\} \cup M)$.

(b): Restrictions on the conclusion $t \stackrel{e}{=} t'$ with respect to $\Lambda^*\{t_i \stackrel{e}{=} t'_i \mid i \in I\}$ as premise:	
$\mathbf{S}_w$	$t,t' \in \downarrow\{t_i,t'_i \mid i \in I\} \cup M$;
$\mathbf{S}_i$	t arbitrary, $t' \in \downarrow\{t_i,t'_i \mid i \in I\} \cup M$;
$\mathbf{S}_c$	t,t' arbitrary;
$\mathbf{M}_i$	t arbitrary, $t' \in \downarrow\{t_i,t'_i \mid i \in I\} \cup M$ and not both of t,t' are variables;
$\mathbf{M}_c$	$t = t'$ arbitrary;
$\mathbf{S}_{p'}$	$t,t' \in C_{<\underline{F}(X,\mathbf{TAlg}(\underline{S})),P>}(\downarrow\{t_i,t'_i \mid i \in I\} \cup M)$;
$\mathbf{M}_{p'}$	$t = t' \in C_{<\underline{F}(X,\mathbf{TAlg}(\underline{S})),P>}(\downarrow\{t_i,t'_i \mid i \in I\} \cup M)$;
$\mathbf{S}_{ip'}$	t arbitrary, $t' \in C_{<\underline{F}(X,\mathbf{TAlg}(\underline{S})),P>}(\downarrow\{t_i,t'_i \mid i \in I\} \cup M)$.

Table 13.2 Shape of the implications (see 13.3.3)

13.3.3 SHAPE of the IMPLICATIONS. The Meta Birkhoff Theorem does not only tell us which model theoretic closure operators are around, in connection with 12.1.2 it also tells us the shape of the corresponding elementary implications, once we know Pj(**H**) and Λ^{op}(**S**).

The starting point is 12.1.2, and we only consider $\mathbf{HSP}^{r,k}$-cones ($\mathbf{k} \geq \mathbf{d^*}$ a regular cardinal), i.e. epimorphisms $e:\underline{A} \to \underline{B}$, where $\underline{A} \in Pj(\mathbf{H})$ is strongly $\mathbf{k}$-small and $e \in \Lambda^{op}(\mathbf{S})$, which have all the essential features of other kinds of axioms. For the corresponding elementary implication we can restrict our considerations to one-element conclusions, i.e. we may only consider implications of the form $(M:\Lambda^*\{t_i \overset{e}{=} t'_i | i \in I\} \Rightarrow t \overset{e}{=} t')$, $\#M, \#I < \mathbf{k}$ -- $\Lambda^*\{t_i \overset{e}{=} t'_i | i \in I\}$ corresponds to the positive diagram of $\underline{A}$ above, while $t \overset{e}{=} t'$ is taken from the positive diagram of $\underline{B}$. The characterizations of $Pj(\mathbf{H})$ and $\Lambda^{op}(\mathbf{S})$, which we have got earlier for some operators $\mathbf{H}$ and $\mathbf{S}$ then yield the restrictions collected in Table 13.2.(a) and (b) (observe that we do not know nice conditions for $\mathbf{M}_n$; "=" means formal identity of terms).

13.3.4 PROOF of THEOREM 8.3.5: All statements except for 8.3.5.(vi) follow from 12.2.3 and Table 13.2. It remains to show 8.3.5.(vi), where we have to assume that we use semantics in which each assignment interprets all of X, i.e. we can omit the reference to the set of formulas:
If $\mathbf{NTE}(\mathbf{K}) = \emptyset$, i.e. if there is no term t such that $\mathbf{K} \models \neg(t \overset{e}{=} t)$, then $\mathbf{ModSECH}^{\mathbf{k}}(\mathbf{K})$ is an ECE-variety and therefore equal to $\mathbf{H}_C\mathbf{S}_C\mathbf{P}^{r,\mathbf{k}}(\mathbf{K})$. Therefore, assume that $\mathbf{NTE}(\mathbf{K}) \neq \emptyset$, say $\neg(q \overset{e}{=} q) \in \mathbf{NTE}(\mathbf{K})$. We claim that in $\mathbf{ModSECH}^{\mathbf{k}}(\mathbf{K}) =: \mathbf{K}'$ also all formulas $\neg\Lambda^*\{t_i \overset{e}{=} t_i | i \in I\}$ are valid which are valid in $\mathbf{K}$, i.e. that $\mathbf{K}' = \mathbf{ModECH}^{\mathbf{k}}(\mathbf{K}) = \mathbf{H}_C\mathbf{S}_C\mathbf{P}_+{}^{r,\mathbf{k}}(\mathbf{K})$: Namely
$\mathbf{K} \models \neg\Lambda^*\{t_i \overset{e}{=} t_i | i \in I\}$ iff $\mathbf{K} \models \Lambda^*\{t_i \overset{e}{=} t_i | i \in I\} \Rightarrow q \overset{e}{=} q$:
validity of the implication and the fact that $q \overset{e}{=} q$ is never satisfied imply that the premise is also never satisfied; the opposite direction is trivial. ■■■

13.3.5 REMARK and COUNTEREXAMPLE. The restriction in 8.3.5.(vi) is in particular satisfied, if we restrict considerations to the homogeneous case and forbid the empty model. The restriction is necessary, since we can consider a type $\tau := (1,1,1,1)$ with unary operation symbols ϕ, ϕ_1, ϕ_2 and ϕ_3 such that for $S := \{s, s', s''\}$: $graph(\phi^{\underline{S}}) := \{((s),s)\}$, $graph(\phi_i^{\underline{S}}) := \{((s'),s'')\}$, $i \in \{1,2,3\}$. Let x be a variable of sort s, y a variable of sort s', a an element of sort s, b an element of sort s' and c an element of sort s'', and consider for all (i,j,k) with $\{i,j,k\} = \{1,2,3\}$ the partial algebras $\underline{A}_{ij} := (\{a,b,c\}; \phi^{\underline{A}_{ij}}, \phi_1^{\underline{A}_{ij}}, \phi_2^{\underline{A}_{ij}}, \phi_3^{\underline{A}_{ij}})$ such that $graph(\phi_k^{\underline{A}_{ij}}) := graph(\phi^{\underline{A}_{ij}}) := \emptyset$ and $graph(\phi_i^{\underline{A}_{ij}}) := graph(\phi_j^{\underline{A}_{ij}}) := \{((b),c)\}$. Finally, set $\mathbf{K} := \{\underline{A}_{ij} | i \neq j \text{ in } \{1,2,3\}\}$. Then in $\mathbf{K}$ the axioms $(\{x\}:\neg(x\phi \overset{e}{=} x\phi))$ and $(\{y\}:\neg(y\phi_1 \overset{e}{=} y\phi_1 \wedge y\phi_2 \overset{e}{=} y\phi_2 \wedge y\phi_3 \overset{e}{=} y\phi_3))$ are valid, but from the corresponding elementary implication
$(\{x,y\}:y\phi_1 \overset{e}{=} y\phi_1 \wedge y\phi_2 \overset{e}{=} y\phi_2 \wedge y\phi_3 \overset{e}{=} y\phi_3 \Rightarrow x\phi \overset{e}{=} x\phi)$ we can only derive
$(\{x,y\}:\neg(y\phi_1 \overset{e}{=} y\phi_1 \wedge y\phi_2 \overset{e}{=} y\phi_2 \wedge y\phi_3 \overset{e}{=} y\phi_3))$. Since we do not see any other negation of an E-equation valid in $\mathbf{K}$ and not needing a variable of sort s, $\mathbf{ModSECH}^{\mathbf{k}}(\mathbf{K})$ contains as model a partial algebra $\underline{B}$ on $B := \{b,c\}$, where $\phi_i^{\underline{B}}$ is total for $i \in \{1,2,3\}$; however, $\underline{B}$ is not contained in $\mathbf{H}_C\mathbf{S}_C\mathbf{P}_+{}^{r,\mathbf{k}}(\mathbf{K})$, while $(\{x\}:\neg(x\phi \overset{e}{=} x\phi))$ excludes the product with empty index set, of which $\underline{B}$ is a subalgebra. ■
We do not have the feeling that in the general heterogeneous case a result similar to 8.3.5.(vi) can be proved.

13.3.6 REMARK. Because of its category theoretical nature the Meta Birkhoff Theorem also applies to other categories, e.g. to E-varieties in $\mathbf{Alg}(\tau)$.

13.4 Special examples

13.4.1 Total ($\underline{S}$-) algebras. The class **TAlg**($\underline{S}$) of all total ($\underline{S}$) algebras is a TE-variety of Alg($\underline{S}$) defined by the axioms

(T_ϕ) $(var(\underline{x}):\underline{x}\phi \overset{e}{=} \underline{x}\phi)$, $\{\underline{x}v_X\}$ = dom $\phi^{\underline{S}}$, $\underline{x}$ without repetitions,

for all $\phi \in \Omega$. Thus **TAlg**($\underline{S}$) is closed among others with respect to the operators $\mathbf{H}_w$, $\mathbf{M}_c$ and $\mathbf{P}^{r,d*}$. ∎

13.4.2 Relational systems. As already mentioned earlier, relational systems of type τ may be considered as partial algebras of type τ, since with each K_R-ary relational symbol $R \in \Omega$ we can associate a K_R-ary operation symbol ϕ_R and choose some fixed $k_R \in K_R$ -- in the finitary case this will usually be the first component --, and we then replace the relational system $\underline{M}:= (M,(R^{\underline{M}})_{R\in\Omega})$ by the partial algebra $\underline{A}(\underline{M}):= (M,(\phi_R^{\underline{A}(\underline{M})})_{R\in\Omega})$ satisfying

(*) $\underline{a} \in$ dom $\phi_R^{\underline{A}(\underline{M})}$ iff $\underline{a} \in R^{\underline{M}}$; and if $\underline{a} \in R^{\underline{M}}$, then $\underline{a}\phi_R^{\underline{A}(\underline{M})} := (k_R)\underline{a}$.

The homomorphisms between relational systems are then the same as the homomorphisms between the corresponding partial algebras. It is easily seen that the class of all partial algebras corresponding in this way to relational systems of a given type forms an ECE-variety **R** in **Alg**($\underline{S}$) defined by the set $\{(Rel_R) | R \in \Omega\}$ of axioms, where for $R \in \Omega$:

(Rel_R) $(var(\underline{x}):\underline{x}\phi_R \overset{e}{=} \underline{x}\phi_R \Rightarrow \underline{x}\phi_R \overset{e}{=} (k_R)\underline{x})$, $\underline{x} \in X^{K_R}$ injective.

The shape of the axioms shows that they are $\mathbf{H}_f\mathbf{S}_w\mathbf{P}^{r,d*}$-axioms, i.e. that $\mathbf{H}_f\mathbf{S}_w\mathbf{P}^{r,d*}(\mathbf{R}) = \mathbf{R}$. -- On the other hand the special form of the partial operations in **R** implies that **R** is not $\mathbf{H}_w$-closed, since $\mathbf{H}_w$ does not impose restrictions on the inner extensions of the factor algebras.

13.4.3 Partially ordered sets (posets). According to 13.4.2 posets can be considered as partial algebras of type $\tau := (2)$, and they form an ECE-variety defined by the following axioms (we use the infix notation "xy" for "xyϕ"):

(POS1)	$xy \overset{e}{=} xy \Rightarrow xy \overset{e}{=} x$,	(relation)
(POS2)	$xx \overset{e}{=} xx$,	(reflexivity)
(POS3)	$xy \overset{e}{=} xy \wedge yx \overset{e}{=} yx \Rightarrow x \overset{e}{=} y$,	(antisymmetry)
(POS4)	$xy \overset{e}{=} xy \wedge yz \overset{e}{=} yz \Rightarrow xz \overset{e}{=} xz$,	(transitivity).

(POS3) and (POS4) are no $\mathbf{H}_f\mathbf{S}_w\mathbf{P}^{r,d*}$-axioms, since both premises violate the form for $\mathbf{H}_f$, and the conclusion of (POS4) is not of the form for $\mathbf{S}_i$, but only of the form for $\mathbf{S}_c$. Observe, however, that in **R** (see 13.4.2) the concepts of "subalgebra" and "relative subalgebra" coincide. The behaviour of posets shows that the class **POS** of all posets is indeed closed with respect to the operators $\mathbf{H}_c$, $\mathbf{S}_c$ (here equal to $\mathbf{S}_r$), $\mathbf{P}^{r,\omega_0}$ and $\mathbf{P}_u$, while it is neither closed with respect to full homomorphic images nor with respect to weak subalgebras. ∎

13.4.4 ω-continuous posets (the class POSω). It is shown in [LePa 80] that the class **POSω** of all ω-continuous posets ($\omega = \omega_0$) -- which are an important tool in D.SCOTT's fixed point approach to the description of programs in computer science -- is an infinitary homogeneous ECE-variety: $\tau := (2,\omega)$, multiplication being the binary, ϕ the ω-ary operation; axioms:

(POSωi):= (POSi) for $1 \le i \le 4$;

(POSω5) $\Lambda^*\{x_ix_{i+1} \overset{e}{=} x_ix_{i+1} | i \in \omega\} \Rightarrow (x_i | i \in \omega)\phi \overset{e}{=} (x_i | i \in \omega)\phi$,

(POSω6) $(x_i | i \in \omega)\phi \overset{e}{=} (x_i | i \in \omega)\phi \Rightarrow \Lambda^*\{x_ix_{i+1} \overset{e}{=} x_ix_{i+1} | i \in \omega\}$.

If there has to be a smallest element, usually denoted by "$\bot$" (and called "bottom"), then the following axioms have to be added:

(POS⊥7) $\bot \overset{e}{=} \bot$,

(POS⊥8) $\bot x \overset{e}{=} \bot x$.

Because of (POSω5) and (POS⊥7) these axioms only allow $\mathbf{H}_c$, $\mathbf{S}_c$ and $\mathbf{P}^{r,\#\omega_1}$ as compatible operators, and we do not see how to do better. ∎

13.4.5 Partially ordered and ω-continuously ordered algebras. Let $(\cdot,\bot,(\phi)_{c\in\Omega'})$ be any family of operation symbols corresponding to a similarity type $\tau = (2,0,(K_\phi)_{\phi\in\Omega'})$ such that (POS1), (POS2), (POS3), (POS4), (POS⊥7), (POS⊥8) as well as for each $\phi \in \Omega'$ the axioms (T_ϕ) (see 13.4.1) and (OA_ϕ) are valid, where for $\underline{x} := (x_k | k \in K_\phi)$, $\underline{y} := (y_k | k \in K_\phi)$

(OA_ϕ) $\Lambda^*\{x_k \cdot y_k \overset{e}{=} x_k \cdot y_k | k \in K_\phi\} \Rightarrow (\underline{x}\phi)\cdot(\underline{y}\phi) \overset{e}{=} (\underline{x}\phi)\cdot(\underline{y}\phi)$,

meaning that ϕ has to be monotone. Once more we deal with a homogeneous ECE-variety. Assume that we have an additional ω-ary partial operation ϕ satisfying (POSω5), (POSω6) and that we have for each $\phi \in \Omega'$ and for each sequence $\underline{x} := (\underline{x}_i | i \in \omega)$ with $\underline{x}_i = (x_{ik} | k \in K_\phi)$ $(i \in \omega)$ the axiom

(ωCA_ϕ) $\Lambda^*\{(x_{ik} | i \in \omega)\phi \overset{e}{=} (x_{ik} | i \in \omega)\phi | k \in K_\phi\} \Rightarrow (\underline{x}_i\phi | i \in \omega)\phi \overset{e}{=} ((x_{ik} | i \in \omega)\phi | k \in K_\phi)\phi$.

Thus we get the ECE-variety of ω-continuous algebras. -- This could also be done for S-algebras and partial S-algebras, although in the latter case the axioms (OA_ϕ) and (ωCA_ϕ) become a little bit more complicated, since some applicabilities of ϕ have to be assumed in the premises. -- Thus everything that has been said so far or will be said later on ECE-varieties is applicable to ω-continuously ordered algebras. In particular free objects exist, but -- as we shall see in sections 16ff -- the partial operations ϕ and $\cdot$ are "trivial" on them, i.e. only defined on constant sequences. But since also all universal solutions exist, this can be overcome by starting with a structure in which the partial order $\cdot$, and therefore the supremum ϕ, are less trivial. -- The class of ω-continuous algebras plays a fundamental rôle in the semantics oriented parts of computer science. For more details see e.g. the papers of the "ADJ-group" (see 22.2.3) or of A.PASZTOR [Pa 83] and [Pa84] and consult [AN 79f] with its references.

13.4.6 hep-varieties. The preceding examples already provide an idea about the relevance of the results of sections 12 and 13. Observe that the hep-varieties considered in PART II are QE-varieties, since their basic idea is the following one: for each partial operation ϕ its domain is the set of all solutions of some set G_ϕ of E-equations, i.e.

(*) $(M_\phi : \underline{x}\phi \overset{e}{=} \underline{x}\phi \Leftrightarrow \Lambda^*\{\underline{x}t_i \overset{e}{=} \underline{x}t'_i | i \in I_\phi\})$.

In order to avoid loops it is additionally required that there exists a mapping $h:\Omega \to \omega_0$ (hierarchy mapping) such that each operation in $(0)h^{-1}$ has to be total, and if $\phi h = n$, then in G_ϕ there occur only fundamental operation symbols ϕ' satisfying $\phi' h < n$. This "hierarchy condition" implies that each bimorphism is an isomorphism, and therefore the objects and morphisms of hep-varieties can be considered to be still relatively close to total algebras and their homomorphisms.

As it can be easily seen from the axioms for small categories and their consequences (see 8.1.1), the class **Cat** of all small categories is such a hep-variety which is even an ECE-variety (for more details see PART II).

13.4.7 Commutative monoids with partial inverse. Next, let us consider the class **CMpI** of commutative monoids with partial inverse -- commutativity is only chosen for simplicity of further observations. I.e. we consider a type $\tau = (2,0,1)$ for $\Omega = (\cdot, e, ^{-1})$ -- we use the usual infix notation for the multiplication even omitting "$\cdot$" -- satisfying the following axioms:

(CMpI1) $x(yz) \overset{e}{=} (xy)z$,

(CMpI2) $xy \overset{e}{=} yx$,

(CMpI3) $xe \overset{e}{=} x$,

(CMpI4) $x^{-1} \overset{e}{=} x^{-1} \Rightarrow xx^{-1} \overset{e}{=} e$.

Observe that each of (CMpI1) and (CMpI2) implies totality of multiplication, (CMpI3) implies totality of the nullary constant e, while formation of the inverse may remain partial. We deal here with $H_fS_cP^{r,\omega_0}$-axioms. That we cannot define this class only by E-equations is easily seen. ∎

13.4.8. If we add in 13.4.7 the axiom of the uniqueness of the inverse

(CMpI5) $xy \overset{e}{=} e \Rightarrow x^{-1} \overset{e}{=} y$,

then this is only a QE-equation; but the class defined by the axioms (CMpI1) through (CMpI5) is easily seen to be closed with respect to closed homomorphic images. Thus it has to be an ECE-variety, and (CMpI5) may be replaced by the two ECE-equations

(CMpI5') $e^{-1} \overset{e}{=} e$ and

(CMpI5") $(xy)^{-1} \overset{e}{=} (xy)^{-1} \Rightarrow x^{-1} \overset{e}{=} x^{-1}$

(the proof is left as an exercise). ∎

13.4.9. A proper QE-variety which is not an ECE-variety is given by the class **CSgI** of all commutative semigroups with generalized inverse, where = (2,1) , operations $(\cdot, ^{-1})$, and where we have besides (CMpI1), (CMpI2) and (CMpI3) the axioms

(CSgI4) $x^{-1} \overset{e}{=} x^{-1} \Rightarrow (xx^{-1})(xx^{-1}) \overset{e}{=} xx^{-1}$ and

(CSgI5) $xy \overset{e}{=} z \wedge zz \overset{e}{=} z \wedge xz \overset{e}{=} x \wedge yz \overset{e}{=} y \Rightarrow x^{-1} \overset{e}{=} y$.

I.e. we do not have a "central unit", but we may have "local units" -- such as z in (CSgI5) -- and each existing generalized inverse refers to one such local unit.

It is easy to check that **CSgI** is not closed with respect to closed homomorphic images (see [B 82], Example 5.1), i.e. (CSgI5) cannot be replaced by ECE-equations. ∎

13.4.10 Weak or relative embeddability into total algebras.

When we consider Table 13.2, then we also get a theoretical answer to the question, when a partial algebra is weakly or fully embeddable into some total algebra of some given variety **V** of total algebras. Namely, we have

$$\mathbf{IS_wP^{r,d*}(V)} = \mathbf{Mod}K_{\mathbf{IS_wP^{r,d*}(V)}} = \mathbf{IS_w(V)} \ , \quad \text{and}$$

$$\mathbf{IS_iP^{r,d*}(V)} = \mathbf{Mod}K_{\mathbf{IS_iP^{r,d*}(V)}} = \mathbf{IS_i(V)} \ ,$$

since each E-variety is closed with respect to $\mathbf{P^{r,d*}}$, i.e. to reduced products.

Therefore, in order to belong to $\mathbf{IS_w(V)}$ a partial algebra $\underline{A}$ has to satisfy all axioms of the form $(M: \wedge^*\{t_i \overset{e}{=} t'_i \mid i \in I\} \Rightarrow t \overset{e}{=} t')$, $t, t' \in \downarrow\{t_i, t'_i \mid i \in I\} \cup X$, $\#I < \mathbf{d^*}$, which are valid in **V**. If $\underline{A}$ shall belong to $\mathbf{IS_i}$, it has to satisfy all axioms of the form $(M: \wedge^*\{t_i \overset{e}{=} t'_i \mid i \in I\} \Rightarrow t \overset{e}{=} t')$, $t' \in \downarrow\{t_i, t'_i \mid i \in I\} \cup X$, t arbitrary, (where

$\#I < \mathbf{d}^*$) which are valid in **V**.

It is a great **problem** to investigate for which varieties **V** there are recursive or even finite subsets of these axioms which only have to be tested for embeddability -- or even to decide whether there are any such varieties besides **TAlg**($\underline{S}$).

We shall return to questions of embeddability in Chapter IV, but we do not have enough space to discuss results of T.EVANS [Ev 51], I.FLEISCHER [Fl 81] or I.NÉMETI [N 75].

13.4.11 Fields. The well known fact that the class F of all fields is not closed with respect to direct products (also for non-empty index sets) shows that F is not definable by universal Horn formulas, although the fact that E-equations hide some existential quantifiers allows us to define fields by a collection of universal formulas. In order to do this, let us consider the type $\tau = (2,0,1,2,0,1)$ for $\Omega = (+,0,-,\cdot,1,^{-1})$; then we have the axioms:

(F1) $(x+y)+z \overset{e}{=} x+(y+z)$,

(F2) $(x\cdot y)\cdot z \overset{e}{=} x\cdot(y\cdot z)$,

(F3) $x+y \overset{e}{=} y+x$,

(F4) $x\cdot y \overset{e}{=} y\cdot x$,

(F5) $x\cdot(y+z) \overset{e}{=} x\cdot y+x\cdot z$,

(F6) $x+0 \overset{e}{=} x$,

(F7) $x\cdot 1 \overset{e}{=} x$,

(F8) $x+(-x) \overset{e}{=} 0$,

(F9) $\neg(0 \overset{e}{=} 1)$,

(F10) $\neg(0^{-1} \overset{e}{=} 0^{-1})$,

(F11) $\neg(x \overset{e}{=} 0) \Rightarrow x\cdot x^{-1} \overset{e}{=} 1$.

Axiom (F11) is the only one which is not a universal Horn formula, but it cannot be replaced by universal Horn formulas as observed above. ∎

13.4.12. Further similar examples are elaborated in [N 75].

14 Closed classes of epimorphisms (elementary implications)

14.1 Preparatory results

14.1.1. In the sense of abstract model theory we have investigated in sections 12 and 13 different (implicational) abstract languages, and we have investigated the corresponding semantical closure operators $\mathrm{Inj}K_{\mathbf{HSF}}$ (see 12.1.2.(iii)). In this section we are concerned with a characterization of the syntactical closure operators $K_{\mathbf{HSF}}\mathrm{Inj}$ for some types of abstract languages discussed earlier, namely for the case $\mathbf{F} \in \{\mathbf{P},\mathbf{P}_+,\mathbf{P}^{r,\mathbf{k}},\mathbf{P}_+{}^{r,\mathbf{k}}\}$, and some **H**, **S** as related in Table 13.1. The closure operators for other cases of **H**,**S** and **F** are left as open problems (or exercises, since we cannot exclude that the answer might already exist somewhere).

14.1.2 GENERAL ASSUMPTIONS and NOTATION. Let $\underline{\mathbb{C}}$ be any category, $\mathbf{H},\mathbf{S} \subseteq \mathrm{Mor}(\underline{\mathbb{C}})$, **k** an infinite regular cardinal or ∞, such that $\underline{\mathbb{C}}$ is strongly **k**-algebroidal, possesses direct products and direct limits of **k**-directed systems, and let **H** and **S** satisfy assumptions

12.2.3.(1) through (7). Recall that Pj(H) designates the class of all H-projective $\underline{C}$-objects, and $Pj_k(H)$ designates the class of all strongly k-small H-projective objects. $E_S^H := \{e \in \Lambda^{op}(S) | Dom(e) \in Pj(H)\}$, and
$E_S^{H,k} := \{e \in \Lambda^{op}(S) | Dom(e) \in Pj_k(H)\}$.
designate the classes of all $\Lambda^{op}(S)$-morphisms with H-projective (strongly k-small) domain. These are HSF-morphisms.
For $P \in Ob(\underline{C})$, $K \subseteq Ob(\underline{C})$ we introduce $Mor(P,K) := \bigcup\{Mor(P,K) | K \in K\}$.
For $Ob(\underline{C}) = Alg(\underline{S})$ and $e \in E_S^H \subseteq Epi(Alg(\underline{S}))$ let $(N:P(e) \Rightarrow C(e))$ be the corresponding implication according to 8.2.2 such that e is the encoding epimorphism of this implication, i.e. if $e:\underline{A} \to \underline{B}$, then N is a generating subset of $\underline{A}$, P(e) and C(e) are sets of E-equations with variables in N such that $\underline{+}P(e)/Cong_{\underline{+}P(e)}(P(e)) \cong \underline{A}$ "over" id_N (i.e. id_N induces the isomorphism), and $\underline{+}(P(e) \cup C(e))/Cong_{\underline{+}(P(e)\cup C(e))}(P(e)\cup C(e)) \cong \underline{B}$ over $id_N \cdot e$.
We shall always assume that $F \in \{P, P_+, P^{r,k}, P_+^{r,k}\}$.

14.1.3 LEMMA. With the notation and assumptions of 14.1.2 let $\underline{C}$ be a category, let $K \subseteq Ob(\underline{C})$ satisfy $K = HSF(K)$, and let $P, P' \in Ob(\underline{C})$ be arbitrary.

(i) If $Mor(P,K) = \emptyset$, then $Mor(P,P') \neq \emptyset$ implies $Mor(P',K) = \emptyset$.

(ii) If $Mor(P,K) \neq \emptyset$, then P has a K-universal K-solution $(r_{P,K}, F(P,K)) =: (r_P, F_P)$ such that $r_P \in \Lambda^{op}(S)$ and $F_P \in K$, and if $q:P' \to F_P$, then also $Mor(P',K) \neq \emptyset$ and there exists a unique morphism $f_q:F_{P'} \to F_P$ such that $q = r_{P'} f_q$.

14.1.4. In the language of category theory 14.1.3.(ii) means: if $Mor(P,K) \neq \emptyset$ for each $P \in Ob(\underline{C})$, then K is $\Lambda^{op}(S)$-reflective in $\underline{C}$ and $(P \mapsto F_P, (e:P' \to P) \mapsto (f_{er_P}:F_{P'} \to F_P))$ is the description of the left adjoint functor with respect to the inclusion functor into $\underline{C}$ of the full subcategory of $\underline{C}$ with object class K.

PROOF of 14.1.3. (i) is obvious.
Ad (ii): Since $\underline{C}$ is $\Lambda^{op}(S)$-co-(well-powered), there is a **set** $E(P,K) \subseteq \Lambda^{op}(S)$ such that for each $e \in E(P,K)$ one has:

(1) $Dom(e) = P$.

(2) There are $f \in Mor(P,K)$ and $g \in S$ such that $eg = f$.

(3) If $f \in Mor(P,K)$, and if (h',s') is a $(\Lambda^{op}(S),S)$-factorization of f, then there exists an isomorphism i such that $h'i \in E(P,K)$.

Thus E(K) is a representative system for all $\Lambda^{op}(S)$-morphisms leaving P and occurring in a $(\Lambda^{op}(S),S)$-factorization of some $f \in Mor(P,K)$. -- For each $e \in E(P,K)$ let $Cod(e) =: C_e$. Then $C_e \in K$ by (2) and $S(K) \subseteq K$. Let $B := X(C_e | e \in E(P,K))$, then $B \in K$, since K is P_+-closed and $E(P,K) \neq \emptyset$. Let $e^*:P \to B$ be the product morphism induced by $E(P,K)$, and let (r_P, s_P) be its $(\Lambda^{op}(S),S)$-factorization. Then $F_P := Cod(r_P) \in K$, and it is routine to realize that (r_P, F_P) is the K-universal solution of P. (cf. sections 5.11 and 12.2). The rest of the statement then follows immediately from the fact that also $(r_{P'}, F_{P'})$ is the K-universal solution of P'. ■■■

14.1.5 THEOREM. Let the same assumptions be satisfied as in 14.1.2 and 14.1.3 except that we assume $F = P_+^{r,k}$, and let
$N_k(K) := \{P | P \in Pj_k(H), Mor(P,K) = \emptyset\}$,

$\mathbf{E}_k(\mathbf{K}) := \{r_P \mid (r_P, F_P)$ is a $\mathbf{K}$-universal solution of $P \in Pj_k(\mathbf{H})$ with $Mor(P,\mathbf{K}) \neq \emptyset\}$.
Then obviously $\mathbf{E}_k(\mathbf{K}) \subseteq \mathbf{E}_S^{H,k}$ and moreover:

(i) $\mathbf{K} = Inj(\mathbf{N}_k(\mathbf{K}) \cup \mathbf{E}_k(\mathbf{K}))$,

(ii) $K_{HSF}(\mathbf{K}) = \mathbf{N}_k(\mathbf{K}) \cup \{e \in \mathbf{E}_S^{H,k} \mid Dom(e) \in \mathbf{N}_k(\mathbf{K})\} \cup$
$\cup \{e \in \mathbf{E}_S^{H,k} \mid em = r_{Dom(e)}$ for some $m \in Mor(\underline{\mathbf{C}})$, $Mor(Dom(e),\mathbf{K}) \neq \emptyset\}$.

14.1.6. Observe that we get $F = \mathbf{P}_+$ for $k = \infty$, and if $F = \mathbf{P}^{r,k}$, then we have to omit in 14.1.5 the consideration of $\mathbf{N}_k$. Notice that above we have identified P with (P,) and e with (Dom(e),e), and we shall do so, too, below.

PROOF of 14.1.5. From 12.2.3 we know -- using our assumptions -- that $\mathbf{K} = InjK_{HSF}(\mathbf{K})$, and $\mathbf{N}_k(\mathbf{K}) \cup \mathbf{E}_k(\mathbf{K}) \subseteq K_{HSF}(\mathbf{K})$ by definition of these sets -- observe that each $r_P \in \mathbf{E}_k(\mathbf{K})$ is $\mathbf{K}$-extendable. Therefore $InjK_{HSF}(\mathbf{K}) \subseteq Inj(\mathbf{N}_k(\mathbf{K}) \cup \mathbf{E}_k(\mathbf{K}))$.
Ad (ii): Let $e \in \mathbf{E}_S^{H,k}$, $Dom(e) =: P$, such that $em = r_P$ for some $m \in Mor(\underline{\mathbf{C}})$, and let $f: P \to K$, $K \in \mathbf{K}$; then $P \in Pj_k(\mathbf{H})$, and there is $g: F_P \to K$ such that $f = r_P g = emg$, showing that $e \in K_{HSF}(\mathbf{K})$.
Conversely, assume that $c \in K_{HSF}(\mathbf{K})$. Then either c is a cone consisting of a single object, say $c = (P,)$, $P \in Pj_k(\mathbf{H})$ such that $Mor(P,\mathbf{K}) = \emptyset$, i.e. $c \in \mathbf{N}_k(\mathbf{K})$, or $c = (e: P \to C)$ is a $\Lambda^{op}(S)$-morphism with $P \in Pj_k(\mathbf{H})$. In the former case c is of the desired kind. In the latter case we have either $F_P \in \mathbf{K}$ for the $\mathbf{K}$-universal solution (r_P, F_P) of P, and $c \in K_{HSF}$ implies the existence of $m: C \to F_P$ such that $cm = r_P$, or $Mor(P,\mathbf{K}) = \emptyset$ and there are no further restrictions on e (in each case $c \in \mathbf{E}_S^{H,k}$) . ∎
Ad (i): Assume $A \in Inj(\mathbf{N}_k(\mathbf{K}) \cup \mathbf{E}_k(\mathbf{K}))$. Then, for each $P \in \mathbf{N}_k(\mathbf{K})$, we have $Mor(P,\mathbf{K}) = \emptyset$. Now assume that $(e: P \to D) \in K_{HSF}(\mathbf{K})$ and $f: P \to A$. If $Mor(P,\mathbf{K}) \neq \emptyset$, then -- because of (ii) -- there is an m such that $em = r_P$. Since $r_P \in \mathbf{E}_k(\mathbf{K})$, there is $g: F_P \to A$ such that $f = r_P g = e(mg)$; otherwise $e \in \mathbf{N}_k(\mathbf{K})$, contradicting the existence of f. This shows that $A \in InjK_{HSF}(\mathbf{K}) = \mathbf{K}$ and implies equality in (i). ∎∎∎

14.2 The general theorem

14.2.1. The results of 14.1 provide an idea of what closed sets of **HSF**-cones have to look like (cf. also R.JOHN [J 75a], Proposition 2). The following theorem presents an abstract description of the corresponding closure operator. As in 14.1 we restrict our considerations to the case where negations of conjunctions of equations are allowed, i.e. $F = \mathbf{P}_+^{r,k}$. The other cases for **F** can be obtained by setting $k = \infty$, respectively by neglecting the cones corresponding to negations, in particular -- with the notation of 14.2.2 below -- we then have $\mathbf{P}_2 = \emptyset$. Observe that sometimes we identify a cone with the corresponding object or mapping without mentioning it explicitly.

14.2.2 THEOREM. Let $\underline{\mathbf{C}}$ be a category, $\mathbf{H},\mathbf{S} \subseteq Mor(\underline{\mathbf{C}})$, k a regular infinite cardinal and $F = \mathbf{P}_+^{r,k}$ such that the assumptions of Lemma 12.2.3 are satisfied; moreover, let us use the notation of 14.1.2 and 14.1.3. Let $\mathbf{E} \subseteq K_{HSF}$, define $\mathbf{P}_2 := \{P \in Pj_k(\mathbf{H}) \mid (P,) \in \mathbf{E}\}$, $\mathbf{P}_1 := Pj_k(\mathbf{H}) \setminus \mathbf{P}_2$, and for $P \in \mathbf{P}_1$, $e,e' \in \mathbf{E}_S^{H,k}$, $e: P \to D$ and $e': P \to D'$, for which there is $q: D \to D'$ satisfying $eq = e'$, set $q =: q_{P,ee'}$ and $e \leq_P e'$ (observe that q is unique, if it exists, since e is an epimorphism).

(a) The following statements are equivalent:

(i) $E = K_{HSF}Inj(E)$.

(ii) The following conditions are satisfied:

(1) For each $P \in \mathbf{P_1}$ the class $\{e|e:P \to Cod(e) \text{ in } E\}$ is non-empty, quasi-ordered by $\leq_P$, and has a largest element, say $r_P:P \to F_P$.

(2) For each $P \in \mathbf{P_1}$, $P' \in \mathbf{P_2}$ one has $Mor(P',F_P) = \emptyset$.

(3) If $P,P' \in Pj_k(H)$, then

(*) $P \in \mathbf{P_2}$ and $Mor(P,P') \neq \emptyset$ imply $P' \in \mathbf{P_2}$,

(**) $P \in \mathbf{P_1}$ and $Mor(P',P) \neq \emptyset$ imply $P' \in \mathbf{P_1}$.

(4) If $P,P' \in \mathbf{P_1}$, $p \in Mor(P,F_{P'})$, then there is a unique $d_p:F_P \to F_{P'}$ such that $p = r_P d_p$.

(5) $\{r_P|P \in \mathbf{P_1}\} \subseteq \Lambda^{op}(S)$.

(6) $E = \mathbf{P_2} \cup \{e|e:P \to D \text{ in } E_S^{H,k}, P \in \mathbf{P_1} , em = r_P \text{ for some } m \in Mor(D,F_P)\} \cup$
$\cup \{e|e:P \to D \text{ in } E_S^{H,k} , P \in \mathbf{P_2}\}$.

(b) If (ii) above is satisfied, then
$Inj(E) = Inj(\mathbf{P_2} \cup \{r_P|P \in \mathbf{P_1}\}) = HSF\{F_P|P \in \mathbf{P_1}\}$.

PROOF. Ad (a): "(i) implies (ii)" is an easily verified consequence of 14.1.3 and 14.1.5 -- in particular (1) and (6) follow from 14.1.3.(ii) and 14.1.5.(ii), where here $(r_P,F_P) = (r_{P,Inj(E)},F(P,Inj(E)))$ is the Inj(E)-universal solution of P.
"(ii) implies (i)": Define $K_0:= \{F_P|P \in \mathbf{P_1}\}$, and for use in (b):
$E':= \{(P,)|P \in \mathbf{P_2}\} \cup \{(P,r_P:P \to F_P)|P \in \mathbf{P_1}\}$ (E' is usually identified with $\mathbf{P_2} \cup \{r_P|P \in \mathbf{P_1}\}$.) And let E be defined as in (ii).(6), where we now have to think of the corresponding cones. -- It is obvious that indeed $E \subseteq K_{HSF}$ and $E' \subseteq E$. We want to show that $E = K_{HSF}(K_0)$, since then $K_{HSF}(K_0) = K_{HSF}Inj(E)$, proving (a):
" $E \subseteq K_{HSF}(K_0)$ ": Let $c \in E$. If $c = (P,)$ or $c = (P,e:P\to Cod(e))$ with $P \in \mathbf{P_2}$, then $c \in K_{HSF}(K_0)$ by (ii).(2). Therefore assume that $c = (P,e:P \to D)$ with $P \in \mathbf{P_1}$, $e \in \Lambda^{op}(S)$ and some m_e satisfying $em_e = r_P$. With (4) we see that for $f:P \to F_{P'}$ $(P' \in \mathbf{P_1})$ there is $d_f:F_P \to F_{P'}$ such that $f = r_P d_f = e(m_e d_f)$, showing that $c \in K_{HSF}(K_0)$. ∎
" $K_{HSF}(K_0) \subseteq E$ ": Let $c \in K_{HSF}(K_0)$, then we have to distinguish two cases:
Case 1: $c = (P,)$, $P \in Pj_k(H)$: Then $Mor(P,K_0) = \emptyset$, i.e. P has to belong to $\mathbf{P_2}$, because for $P \in \mathbf{P_1}$ we have $r_P:P \to F_P \in K_0$.
Case 2: $c = (P,e:P \to D)$: Since $c \in K_{HSF}(K_0)$, we have $P \in Pj_k(H)$ and $e \in \Lambda^{op}(S)$, i.e $e \in E_S^{H,k}$. If $P \in \mathbf{P_2}$, then $c \in E$ by (6). Thus assume $P \in \mathbf{P_1}$. Then $c \in K_{HSF}(K_0)$ implies that for $r_P:P \to F_P$ there has to exist $g:D \to F_P$ such that $r_P = eg$, showing with (6) that $c \in E$. ∎
Ad (b): Let E be closed. From the definition of E' above, we have $Inj(E) \subseteq Inj(E')$, but the description of E in (ii).(6) shows that $Inj(E') \subseteq Inj(E)$. $E' \subseteq E \subseteq K_{HSF}$ and $E = K_{HSF}(K_0)$ show in connection with 12.2.3 that $Inj(E') = Inj(E) = HSF(K_0)$. ∎∎∎

14.2.3 REMARK. The above Theorem 14.2.2 shows what closed sets of implications have to look like, when we translate them into epimorphisms. Families of morphisms satisfying condition 14.2.2.(ii).(4) have been called "strongly natural families" in R.JOHN [J 75a], where it has been observed that in the case when $\mathbf{P_1} \subseteq Pj(\Lambda^{op}(S))$ we only have to require "natural families" $(r_P:P \to F_P|P \in \mathbf{P_1})$, i.e. we only need that for every $p:P \to P'$ $(P,P' \in \mathbf{P_1})$ there exists a unique $d_p:F_P \to F_{P'}$ such that $r_P d_p = pr_{P'}$.

We repeat that in the case when F allows products with empty index sets, then $P_1 = Pj_k(H)$.

14.3 Closed sets of E-, ECE- and QE-equations and of $H_f S_C P^{r,k}$-implications

14.3.1. Theorem 14.2.2 now allows us to describe without great additional effort closed sets of E-, ECE^k- and QE^k-equations as well as closed sets of $H_f S_C P^{r,k}$-implications ($k \geq d^*$, regular). In 14.4 we shall treat the case when we also allow negations of conjunctions of E-equations. The descriptions of the kinds of axioms listed above are easier than other ones, since we do not have any restrictions on the scope of the conclusions (having $S = S_C$), while this is different for other specifications of the operator **S**. Within the scope of this book we shall not treat the other cases which -- as far as we know -- have not yet been treated elsewhere in the literature. Theorem 14.2.2 and the following results provide an idea of how such descriptions in more general cases could be approached, but we leave it as an **open problem** to go into the details there.

14.3.2 NOTATION. In order to translate Theorem 14.2.2 for the categories **Alg**($\underline{S}$) and **TAlg**($\underline{S}$) and for some special kinds of QE-equations into a set theoretical and algebraical language we use the following representation of elementary implications $\Sigma = (M:\wedge^*\{t_i \overset{e}{=} t'_i | i \in I\} \Rightarrow t \overset{e}{=} t')$:
$\Sigma \cong (P,C)$, where $P:= \{(t_i,t'_i)|i \in I\} \cup \{(x,x)|x \in M\}$, $C:= (t,t')$.
For a given regular cardinal **k** and any set M let $\mathbf{Subs_k}(M):= \{N \subseteq M | \#N < \mathbf{k}\}$ be the set of all <u>**k**-small</u> subsets of M.
Let -- as before -- be X a set of variables, $v_X: X \to S$ such that $\#((s)v_X^{-1}) = \mathbf{k} \geq d^{\&}$ ($s \in S$), let, for $M \subseteq X$, $T_M:= F(M,\mathbf{TAlg}(\underline{S}))$ be the set of all terms with variables in M, $\Delta_T:= \{(t,t)|t \in T_X\}$, $\Delta_M:= \{(y,y)|y \in M\}$, $Eeq_X:= \{(t,t')|t,t' \in T_X , t\tilde{v}_X = t'\tilde{v}_X\}$ be the set theoretical equivalent of the set of all E-equations with variables in X; let
$\Delta_f:=\{(y,y),(\underline{x}\phi,\underline{x}\phi)|\underline{x} \in X^{K_\phi}$ injective, sort respecting, $y \in X$, $\phi \in \Omega\}$
be the set of all TE-statements with respect to the 0-th and first Baire classes of $\underline{F}(X,\mathbf{TAlg}(\underline{S}))$ with injective argument sequences, and
$\mathbf{Subs_k}^*(\Delta_f):= \{K \in \mathbf{Subs_k}(\Delta_f) | (\underline{x}\phi,\underline{x}\phi),(\underline{y}\phi',\underline{y}\phi') \in K , \phi,\phi' \in \Omega$ imply $[K_\phi]\underline{x} \cap [K_{\phi'}]\underline{y} = \emptyset\}$
be the set of all **k**-small subsets of Δ_f, which describe H_f-projective objects.
Let Prem be a variable for one of the sets $\mathbf{Subs_k}(Eeq_X)$, $\mathbf{Subs_k}(\Delta_T)$, $\mathbf{Subs_k}(\Delta_X)$ or $\mathbf{Subs_k}^*(\Delta_f)$, i.e. a variable for the set of admissible premises concerning the cases of QE-, ECE- and E-equations, respectively of $H_f S_C P^{r,k}$-implications. Thus, for a given value of Prem the set of elementary implications under consideration is described as
$Imp_{Prem}:= \{(\Phi,(t,t'))|\Phi = \{(t_i,t'_i)|i \in I\} \in Prem , var(t \overset{e}{=} t') \subseteq var(\Phi)\}$.
If $Q \subseteq Imp_{Prem}$ and $\Phi \in Prem$ with set M of variables, then
$(\Phi)Q:= \{(t,t')|(\Phi,(t,t')) \in Q\}$, $\{t,t'\} \subseteq X$ implies $\{t,t'\} \subseteq var(\Phi)$ is the set of all conclusions of Q-implications with premise Φ. Observe that in general $(\Phi)Q$ should contain (x,x) for each $x \in X$ satisfying $xv_X \in [\downarrow\Phi]\tilde{v}_X$, but we omit "new" variables here with respect to later applications.
Closed sets of implications of premise type Prem are then described as follows:

14.3.3 THEOREM. With the notation introduced in 14.3.2 let $Q \subseteq Imp_{Prem}$ be any set of elementary implications of length less than **k** and with premises restricted by Prem. Then the following statements are equivalent:

(i) $Q = \mathbf{Imp}_{Prem}\mathbf{Mod}(Q)$.

(ii) Q has the following properties (I1) through (I4) for any $\Phi,\Phi' \in$ Prem with $M := X \cap \downarrow\Phi$ respectively $M' := X \cap \downarrow\Phi'$:

(I1) $\underline{\downarrow}(\Phi)Q$ is an M-generated relative subalgebra of $\underline{F}(M,\mathbf{TAlg}(\underline{S}))$.

(I2) $(\Phi)Q$ is a closed congruence relation on $\underline{\downarrow}(\Phi)Q$.

(I3) $\Phi \subseteq (\Phi)Q$.

(I4) For every homomorphism $f:\underline{\downarrow\Phi} \to \underline{\downarrow}(\Phi')Q$ satisfying $[\Phi](f\times f) \subseteq (\Phi')Q$ there exists a homomorphic extension $f_{\Phi\Phi'}:\underline{\downarrow}(\Phi)Q \to \underline{\downarrow}(\Phi')Q$ which satisfies $[(\Phi)Q]f_{\Phi\Phi'}{}^{(2)} \subseteq (\Phi')Q$.

14.3.4 COROLLARY 1. With the assumptions and notation of Theorem 14.3.3 let $Q \subseteq \mathrm{Imp}_{Prem}$ be a closed set, i.e. $Q = \mathbf{Imp}_{Prem}\mathbf{Mod}(Q)$.

(i) If $\Phi \in$ Prem with set M of variables, then
$(\Phi)Q = \bigcap\{\ker \tilde{f} \mid f:\underline{\downarrow\Phi} \to \underline{A} \in \mathbf{Mod}(Q)\ ,\ \Phi \subseteq \ker \tilde{f}\}$.

(ii) Let $\Phi,\Phi' \in$ Prem for sets M respectively M' of variables such that $M \subseteq M'$ $[\downarrow\Phi]\tilde{v}_X = [\downarrow\Phi']\tilde{v}_X$ and $\Phi \setminus \Delta_X = \Phi' \setminus \Delta_X$, i.e. Φ and Φ' represent the same conjunction of E-equations, but for different reference sets of variables. Then there exists a homomorphism $g:\underline{\downarrow\Phi'} \to \underline{\downarrow}(\Phi)Q$ satisfying $g|M = id_M$, moreover $\downarrow(\Phi)Q = \mathbf{C}_{\underline{\downarrow}(\Phi')Q}\, M$ and $(\Phi)Q = (\Phi')Q \cap (\downarrow(\Phi)Q)^2$.

14.3.5 REMARKS. (i): (I1) through (I3) just say that, for each $\Phi \in$ Prem with set M of variables, $(\Phi)Q$ is a closed congruence -- containing Φ -- on an M-generated relative subalgebra of $\underline{F}(M,\mathbf{TAlg}(\underline{S}))$. (I4) is the crucial axiom corresponding to full invariance in connection with E-equations. Thus this result is an immediate generalization of the Second Birkhoff Theorem for equations (see 6.4.2). (I4) encodes "strong naturalness" (see 14.2.2.(ii).(4)) of the family $(r_P \mid P \in \mathbf{P_1})$ in 14.2.2.

(ii): For the different meanings of the choice of Prem see 14.3.2.
For E-equations (I3) becomes trivial, while in the case of E- and ECE-equations as well as of $\mathbf{H_fS_cP}^{r,k}$-implications (I4) reduces to

(I4)' For every homomorphism $f:\underline{\downarrow\Phi} \to \underline{\downarrow}(\Phi')Q$ there exists a homomorphic extension $f_{\Phi\Phi'}:\underline{\downarrow}(\Phi)Q \to \underline{\downarrow}(\Phi')Q$ satisfying $[(\Phi)Q]f_{\Phi\Phi'}{}^{(2)} \subseteq (\Phi')Q$.

(iii): In the case of non-empty phyla one can restrict considerations in 14.3.3.(ii) to X-generated relative subalgebras of $\underline{F}(X,\mathbf{TAlg}(\underline{S}))$ (see [ABN 80], Theorem 3.4.5 for the homogeneous case): In this case we can assume that, for every $\Phi \in$ Prem , $(\Phi)Q$ contains (x,x) for every $x \in X$, and if we include all these elements, then all phyla of $\downarrow(\Phi)Q$ are non-empty and $\downarrow(\Phi)Q$ allows assignments from X (respectively assignments starting from subsets of X can be extended to all of X without violating e.g. the assumptions in (I4)).

(iv): In the case of total algebras references to relative subalgebras have to be replaced by such to subalgebras. And in the case of non-empty phyla we can always refer to the total term algebra on X. See in particular the following Corollary 2.

14.3.6 COROLLARY 2. With the notation introduced in 14.3.2 let us consider the case of total algebras: Let $Q \subseteq \mathrm{Imp}_{Prem}$ be any set of elementary implications of length less than **k**, let $\mathbf{Mod}^t(Q)$ designate the class of all **total** algebras in which Q is valid; and let $\mathbf{Mod}^t_e(Q)$ be the class of all models of Q which are total and have non-empty phyla.

(a) The following two statements are equivalent:

$(i)^t$ $Q = \mathbf{QEqMod}^t(Q)$.

$(ii)^t$ Q has the following properties for $\Phi,\Phi' \in$ Prem with sets M respectively M' of variables:

$(I1)^t$ $(\Phi)Q$ is a congruence on $\underline{F}(M,\mathbf{TAlg}(\underline{S}))$ containing Φ.

$(I2)^t$ For every homomorphism $f:\underline{F}(M,\mathbf{TAlg}(\underline{S})) \to \underline{F}(M',\mathbf{TAlg}(\underline{S}))$ $[\Phi]f^{(2)} \subseteq (\Phi')Q$ implies $[(\Phi)Q]f^{(2)} \subseteq (\Phi')Q$.

(b) The following two statements are equivalent:

$(i)^t_\theta$ $Q = \mathbf{QEqMod}^t_\theta(Q)$.

$(ii)^t_\theta$ Q has the following properties for $\Phi,\Phi' \in$ Prem with sets X of variables:

$(I1)^t_\theta$ $(\Phi)Q$ is a congruence on $\underline{F}(X,\mathbf{TAlg}(\underline{S}))$ containing Φ.

$(I2)^t_\theta$ For every endomorphism $f:\underline{F}(X,\mathbf{TAlg}(\underline{S})) \to \underline{F}(X,\mathbf{TAlg}(\underline{S}))$ $[\Phi]f^{(2)} \subseteq (\Phi')Q$ implies $[(\Phi)Q]f^{(2)} \subseteq (\Phi')Q$. ■■■

PROOF of 14.3.3. Theorem 14.3.3 is just a translation to partial algebras of Theorem 14.2.2, and we only recall a few items:
Let $(r_P:P \to F_P,F_P)$ be the **Mod**(Q)-universal **Mod**(Q)-solution of the premise algebra $\underline{P}:= (\underline{\downarrow\Phi})/\mathrm{Cong}_{\downarrow\Phi}(\Phi)$, and let $nat_\Phi:\underline{\downarrow\Phi} \to \underline{P}$ be the corresponding natural homomorphism. Then the Diagram Completion Lemmas 2.7.1/2 and 5.8.2 tell us that $(\Phi)Q = \ker(nat_\Phi \cdot r_P)^\sim$ is a closed congruence, and that -- especially because of 14.2.2.(ii).(4) -- all the other properties are satisfied and conversely imply the corresponding properties in 14.2.2.(ii). ■■■

PROOF of 14.3.4. Ad (i): From the definitions and results in 5.11 we see that, with respect to $M:= X \cap \downarrow\Phi$ and $\underline{P}:= (\underline{\downarrow\Phi})/\mathrm{Cong}_{\downarrow\Phi}(\Phi)$, $(\Phi)Q$ is the preimage of $\mathrm{char}(\underline{P};;\mathbf{Mod}(Q))$, with respect to the natural homomorphism from $\underline{F}(M,\mathbf{TAlg}(\underline{S}))$ onto $\underline{F}(\underline{P},\mathbf{TAlg}(\underline{S}))$ induced by the natural mapping from M into $\underline{P}$, and (i) is just the algebraic translation of this fact. ■
Ad (ii): In view of (i) this can now be proved in a similar way as 6.2.3.(ii). We leave the details as an exercise to the reader. ■■■

14.3.7 REMARK. In [Hoe 82] H.-J.HOEHNKE has shown for the total homogeneous finitary case that for $Q = \mathbf{QEqMod}^t_\theta(Q)$ the set $\{(\Phi)Q \mid \Phi \in \mathbf{Subs}_{\#\omega_0}(F(X,\mathbf{TAlg}(\tau))^2)\}$ is a "fully invariant" algebraic closure system (see 14.3.8 for the definition) on $\mathrm{Cong}(\underline{F}(X,\mathbf{TAlg}(\tau)))$, X being a countably infinite set of variables. Moreover, he shows that $\{\{(\Phi)Q \mid \Phi \in \mathbf{Subs}_{\#\omega_0}(F(X,\mathbf{TAlg}(\tau))^2)\} \mid Q \subseteq \mathrm{Imp}_{Prem} , Q = \mathbf{QEqMod}^t_\theta(Q)\}$ -- $\mathrm{Prem} = \mathbf{Subs}_k(\mathrm{Eeq}_X)$ -- is the set of all fully invariant algebraic closure systems on $\mathrm{Cong}(\underline{F}(X,\mathbf{TAlg}(\tau)))$. At the end he conjectures that a similar result is true for partial algebras. We show below that he is right.
Following H.-J.HOEHNKE in [Hoe 82] we define fully invariant closure systems as follows in 14.3.8. Our definition becomes a bit more complicated than his, since besides partiality we also have to take care of the different reference sets of variables. We only give this definition here for the closure system of M-initial closed congruences ($M \subseteq Y$) of a total Peano algebra $\underline{F}(Y,\mathbf{TAlg}(\underline{S}))$ on some set Y -- usually our "standard set X of variables".

14.3.8 DEFINITION. Let Y be any ($\underline{S}$-) set.

(i): In analogy to Definition 5.9.2 we define

$\mathbf{Rsuba}(\subseteq Y, \underline{F}(Y,\mathbf{TAlg}(\underline{S}))) := \bigcup\{\mathbf{Rsuba}(M,\underline{F}(M,\mathbf{TAlg}(\underline{S}))) \mid M \subseteq Y\}$

to be the set of all M-generated relative subalgebras of $\underline{F}(Y,\mathbf{TAlg}(\underline{S}))$ for some subset M of Y. Similarly we define

$\mathrm{Cong}_C(\subseteq Y,\underline{F}(Y,\mathbf{TAlg}(\underline{S}))) := \bigcup\{\mathrm{Cong}_C(\underline{A}) \mid \underline{A} \in \mathbf{Rsuba}(\subseteq Y,\underline{F}(Y,\mathbf{TAlg}(\underline{S})))\}$

to be the set of all "M-initial closed congruences" of $\underline{F}(Y,\mathbf{TAlg}(\underline{S}))$ for subsets M of Y, thus obtaining a **k**-algebraic closure system -- $\mathbf{k} \geq \mathbf{d}^{\&}$, regular -- (with respect to set theoretical inclusion) on $\mathbf{Subs}(F(Y,\mathbf{TAlg}(\underline{S}))^2)$.

(ii): A closure system $\mathcal{H}$ on $\mathbf{Subs}(F(Y,\mathbf{TAlg}(\underline{S}))^2)$ is called <u>fully invariant</u> iff for any closed set $C \in \mathcal{H}$ and for any closed and surjective homomorphism $f:\underline{A} \to \underline{\downarrow}C$ for some $\underline{A} \in \mathbf{Rsuba}(\subseteq Y,\underline{F}(Y,\mathbf{TAlg}(\underline{S})))$ -- $\underline{\downarrow}C$ being as usual the initial relative subalgebra of $\underline{F}(Y,\mathbf{TAlg}(\underline{S}))$ generated by the domain of C -- one has for $B := [C](f^{(2)})^{-1}$ that $\downarrow B = A$ and $B \in \mathcal{H}$.

14.3.9 OBSERVATION. Let $\mathcal{H}$ be a closure system on a closure system $\mathfrak{G}$ on a complete lattice $\underline{L}$; then $\mathcal{H}$ is a closure system on $\underline{L}$ itself, since the meet on $\mathfrak{G}$ and $\underline{L}$ is the same. ∎

The next lemma characterizes some kind of fully invariant closure systems, which we shall need later:

14.3.10 LEMMA. Let $\mathcal{H}$ be a closure system on $\mathrm{Cong}_C(\subseteq Y,\underline{F}(Y,\mathbf{TAlg}(\underline{S})))$ for some ($\underline{S}$-) set Y with closure operator $\mathbf{C}_{\mathcal{H}}$ such that, for all $D \subseteq F(Y,\mathbf{TAlg}(\underline{S}))^2$, $\underline{\downarrow}\mathbf{C}_{\mathcal{H}}D$ is generated by $Y \cap \downarrow D$. Then the following statements are equivalent:

(i) $\mathcal{H}$ is fully invariant.

(ii) For every $H \in \mathcal{H}$, for every $M \subseteq Y$ and for every homomorphism $g:\underline{F}(M,\mathbf{TAlg}(\underline{S})) \to \underline{F}(Y \cap \downarrow H,\mathbf{TAlg}(\underline{S}))$ one has $[H](g^{(2)})^{-1} \in \mathcal{H}$.

(iii) For every $D \in \mathbf{Subs}(F(Y,\mathbf{TAlg}(\underline{S}))^2)$, for every $H \in \mathcal{H}$ and for every homomorphism $f:\underline{\downarrow}D \to \underline{\downarrow}H$ which satisfies $[D]f^{(2)} \subseteq H$ **there exists a** homomorphic extension $f':\underline{\downarrow}\mathbf{C}_{\mathcal{H}}D \to \underline{\downarrow}H$ such that $[\mathbf{C}_{\mathcal{H}}D]f'^{(2)} \subseteq H$.

14.3.11. (i): Observe that the condition on the closure system $\mathcal{H}$ in 14.3.10 that " $\underline{\downarrow}\mathbf{C}_{\mathcal{H}}D$ is generated by $Y \cap \downarrow D$ for all $D \subseteq F(Y,\mathbf{TAlg}(\underline{S}))^2$ " is equivalent to the fact that for each subset $M \subseteq Y$ the set $F(M,\mathbf{TAlg}(\underline{S}))^2$ is closed -- in particular, since to each subset W of $F(Y,\mathbf{TAlg}(\underline{S}))$ the set $Y \cap \downarrow W$ of "variables occuring in W" is uniquely determined. ∎

(ii): The fact that we deal in 14.3.10 with a closure system on $\mathrm{Cong}_C(\subseteq Y,\underline{F}(Y,\mathbf{TAlg}(\underline{S})))$ trivially implies that in 14.3.10.(ii) $[\downarrow D]f^{-1} = \downarrow([D](f^{(2)})^{-1})$; therefore this does not have to be required there separately. ∎

PROOF of 14.3.10. Observe that by 2.8.2.(iii) the preimage of a closed congruence with respect to a closed and surjective homomorphism is closed.

" (i) ⇒ (iii) ": Let $\mathcal{H}$ be a fully invariant closure system on $\mathrm{Cong}_C(\subseteq Y,\underline{F}(Y,\mathbf{TAlg}(\underline{S})))$, and let the assumptions of (iii) be satisfied. Let $\overline{f}:\underline{F}(\downarrow D \cap X,\mathbf{TAlg}(\underline{S})) \to \underline{F}(N,\mathbf{TAlg}(\underline{S}))$, where $N := \downarrow[D]f \cap X$, be the homomorphic extension of f. $\overline{f}$ is closed and surjective, and $H' := H \cap F(N,\mathbf{TAlg}(\underline{S}))^2$ is closed (see 14.3.11.(i)).

Thus $D \subseteq [H'](\overline{f}^{(2)})^{-1} = [H](\overline{f}^{(2)})^{-1} =: D'$, and D' belongs to $\mathcal{H}$. Therefore monotony of $C_{\mathcal{H}}$ implies $C_{\mathcal{H}}D \subseteq D'$, and $f' := \overline{f}|{\downarrow}C_{\mathcal{H}}D$ is a homomorphic extension of f: $f': \underline{{\downarrow}C_{\mathcal{H}}D} \to \underline{{\downarrow}H}$; moreover $[C_{\mathcal{H}}D]f'^{(2)} \subseteq H$. ∎

" (iii) ⇒ (ii) ": Let H, g be given as in (ii), let $D := [H](g^{(2)})^{-1}$. Then D, H and $f := g|{\downarrow}D: \underline{{\downarrow}D} \to \underline{{\downarrow}H}$ satisfy the assumptions of (iii). However, by the definition of D and f there has to hold: $f' = f$ and $C_{\mathcal{H}}D = D$, when we apply (iii). ∎

" (ii) ⇒ (i) ": Let $C \in \mathcal{H}$, $\underline{A} \in$ **Rsuba**$(\subseteq Y, \underline{F}(Y,$**TAlg**$(\underline{S})))$, and let $f: \underline{A} \to \underline{{\downarrow}C}$ be a closed and surjective homomorphism. Let M be the Peano basis of $\underline{A}$, N the one of $\underline{{\downarrow}C}$, and let $\overline{f}: \underline{F}(M,$**TAlg**$(S)) \to \underline{F}(N,$**TAlg**$(\underline{S}))$ be the homomorphic extension of f. Defining $D := [C](\overline{f}^{(2)})^{-1}$ as in " (iii) ⇒ (ii) ", closedness and surjectivity of f and 4.1.13.(iii) imply that ${\downarrow}D = A$ and $D \in \mathcal{H}$. ∎∎∎

14.3.12 THEOREM. Let $\mathbf{k} \geq \mathbf{d^*}$ be a regular cardinal, X the canonical set of variables (i.e. $\#((s)v_X^{-1}) = \mathbf{k}$ for each $s \in S$). There is a one-to-one correspondence between the closed sets Q of **k**-small QE-equations and the fully invariant **k**-algebraic closure systems $\mathcal{H}$ (respectively closure operators $C_{\mathcal{H}}$) on $\mathrm{Cong}_C(\subseteq X, \underline{F}(X,$**TAlg**$(\underline{S})))$, which satisfy that $\underline{{\downarrow}C_{\mathcal{H}}H}$ is generated by $X \cap {\downarrow}H$ for all $H \subseteq F(X,$**TAlg**$(\underline{S}))^2$.

This correspondence is given by the assignments

$Q \mapsto C_Q$ such that, for $D \subseteq F(X,$**TAlg**$(\underline{S}))^2$, $C_QD = \bigcup\{(D')Q | D' \in$ **Subs**$_{\mathbf{k}}(D)\}$.

$\mathcal{H} \mapsto Q_{\mathcal{H}}$ such that $Q_{\mathcal{H}} := \{(D,(t,t')) | D \subseteq$ **Subs**$_{\mathbf{k}}(F(X,$**TAlg**$(\underline{S}))^2)$, $(t,t') \in C_{\mathcal{H}}D\}$.

PROOF. First assume that Q = **QEeqMod**(Q) . Since $\mathbf{k} \geq \mathbf{d^*}$, 14.3.3 implies that, for each $D \subseteq F(X,$**TAlg**$(\underline{S}))^2$, C_QD is a closed congruence on a relative subalgebra $\underline{{\downarrow}C_QD}$ of $\underline{F}(X,$**TAlg**$(\underline{S}))$ generated by $X \cap {\downarrow}C_QD$ because of the **k**-inductivity of these concepts. Let $D \subseteq F(X,$**TAlg**$(\underline{S}))^2$ be arbitrary. It is obvious that always $D \subseteq C_QD$, and that $D \subseteq D'$ implies $C_QD \subseteq C_QD'$ (see (I3) and (I4)). By definition C_Q is **k**-algebraic. In order to show that $C_QC_QD \subseteq C_QD$ assume $E \subseteq C_QD$, $\#E < \mathbf{k}$. For each $\Sigma \in E$ there is $E_\Sigma \subseteq D$ such that $\#E_\Sigma < \mathbf{k}$ and $\Sigma \in C_QE_\Sigma$. Set $E' := \bigcup\{E_\Sigma | \Sigma \in E\}$. Since **k** is regular, $\#E' < \mathbf{k}$, and $E \subseteq C_QE'$. Thus $C_QE \subseteq C_QE' \subseteq C_QD$ showing that C_Q is idempotent. Finally assume $D, D' \subseteq \underline{F}(X,$**TAlg**$(\underline{S}))^2$, $f: \underline{{\downarrow}D} \to \underline{{\downarrow}C_QD'}$ such that $[D]f^{(2)} \subseteq C_QD'$. We know by 14.3.3.(ii).(I4) that for each $D'' \subseteq D$ with $\#D'' < \mathbf{k}$ we have $[C_QD'']f^{(2)} \subseteq C_QD'$. Thus $[C_QD]f^{(2)} = [\bigcup\{C_QD'' | D'' \in$ **Subs**$_{\mathbf{k}}(D)\}]f^{(2)} = \bigcup\{[C_QD'']f^{(2)} | D'' \in$ **Subs**$_{\mathbf{k}}(D)\} \subseteq C_QD'$
showing that C_Q is the closure operator of a fully invariant closure system, since the assumption of 14.3.10 is trivially satisfied. ∎

Conversely, let $\mathcal{H}$ be a fully invariant closure system on $\mathrm{Cong}_C(\subseteq X, \underline{F}(X,$**TAlg**$(\underline{S})))$, which satisfies that $\underline{{\downarrow}C_{\mathcal{H}}D}$ is generated by $X \cap {\downarrow}D$ for each $D \subseteq F(X,$**TAlg**$(\underline{S}))^2$. Then it is easily seen that $Q_{\mathcal{H}}$ satisfies properties 14.3.3.(ii).(I1) through (I4). Thus Q is a closed set of implications. ∎∎∎

The following result might be of some interest:

14.3.13 PROPOSITION. With the notation of Theorem 14.3.12 let Q be a closed set of **k**-small QE-equations. Then one has:

For all $M \subseteq X$, for all M-generated relative subalgebras $\underline{A}$ of $\underline{F}(X,$**TAlg**$(\underline{S}))$ and for all closed congruences R on $\underline{A}$:

$\underline{A}/R \in$ **Mod**(Q) iff $R \in \mathcal{H}_Q$.

In particular $(R)Q := C_Q R = \cap\{\ker \tilde{f} | f: \underline{A} \to \underline{B} \in \mathbf{Mod}(Q)\ ,\ R \subseteq \ker \tilde{f}\}$.

PROOF. It easily follows from 14.2.2.(a) and the Epimorphism Theorem that, for **k**-small M and $R \in \Pi_Q$, $\underline{A}/R$ belongs to **Mod**(Q), where $\underline{A} = \downarrow R$; namely for $R = (D)Q$ we have that $(nat_R|\downarrow D, \underline{A}/R)$ is the **Mod**(Q)-universal solution of $\downarrow D$. In the general case we get $\underline{A}/R$ as the **k**-direct limit of such **Mod**(Q)-objects, and therefore it also belongs to **Mod**(Q) (see 11.6.7).
The converse follows from similar considerations as those in connection with 14.3.4, when we start with the relation R of $\underline{F}(A \cap X, \mathbf{TAlg}(\underline{S}))$ and obtain that
$(R)Q = C_Q R = \cap\{\ker \tilde{f} | f: \underline{A} \to \underline{B} \in \mathbf{Mod}(Q)\ ,\ R \subseteq \ker \tilde{f}\}$. ■■■

14.3.14 REMARKS. (i): We have no idea whether there is a description of a similar nature as in 14.3.12 for closed sets of ECE-equations. Here the closed sets which one starts with are generated by **k**-small subsets of Δ_T.
(ii): For more practical applications one often looks for complete sets of rules of derivability, as we have already presented them for E-equations in section 6. Such a list which can easily be derived from Theorem 14.3.3 has already been presented in [AN 76] for the kinds of axioms under consideration (see also [B 82], but observe that in both papers the axiom corresponding to 14.3.15.(ii).(IR8) is incorrectly formulated, since the essential assumption " $\{(M':\Phi' \Rightarrow q_m \stackrel{e}{=} q_m) | m \in M\}$ " is missing). In what follows we use the implicational notation rather than the set theoretical one:

14.3.15 PROPOSITION. With the notation and assumptions of 14.3.2 let $Q \subseteq Imp_{Prem}$ -- i.e. we treat simultaneously E-, ECE- and QE-equations as well as $\mathbf{H_f S_c P^{r,k}}$-implications. Then the following statements are equivalent:

(i) Q is closed, i.e. $Q = \mathbf{Imp}_{Prem}\mathbf{Mod}(Q)$.

(ii) Q is closed with respect to the following derivation rules, where the premises Φ and Φ' are in each case taken from Prem, with sets $M, M' \subseteq X$ of variables, respectively, and $\#I < \mathbf{k}$:

- (IR1) $\vdash (M:\Phi \Rightarrow x \stackrel{e}{=} x)$ for all $x \in M$.
- (IR2) $\vdash (M:\Phi \Rightarrow t_i \stackrel{e}{=} t'_i)$, $i \in I$, when $\Phi = \wedge^*\{t_i \stackrel{e}{=} t'_i | i \in I\}$.
- (IR3) $(M:\Phi \Rightarrow t \stackrel{e}{=} t') \vdash (M:\Phi \Rightarrow t \stackrel{e}{=} t)$.
- (IR4) $(M:\Phi \Rightarrow t \stackrel{e}{=} t') \vdash (M:\Phi \Rightarrow t' \stackrel{e}{=} t)$.
- (IR5) $(M:\Phi \Rightarrow t \stackrel{e}{=} t'), (M:\Phi \Rightarrow t' \stackrel{e}{=} t'') \vdash (M:\Phi \Rightarrow t \stackrel{e}{=} t'')$.
- (IR6) $(M:\Phi \Rightarrow (t_k|k \in K_\phi)\phi \stackrel{e}{=} (t_k|k \in K_\phi)\phi) \vdash (M:\Phi \Rightarrow t_k \stackrel{e}{=} t_k)$, $k \in K_\phi$, $\phi \in \Omega$.
- (IR7) $\{(M:\Phi \Rightarrow t_k \stackrel{e}{=} t'_k) | k \in K_\phi\}, (M:\Phi \Rightarrow (t_k|k \in K_\phi)\phi \stackrel{e}{=} (t_k|k \in K_\phi)\phi) \vdash$
  $\vdash (M:\Phi \Rightarrow (t_k|k \in K_\phi)\phi \stackrel{e}{=} (t'_k|k \in K_\phi)\phi)$, $\phi \in \Omega$.
- (IR8) Let $\Phi = \wedge^*\{t_i \stackrel{e}{=} t'_i | i \in I\}$:
  $(M:\Phi \Rightarrow t \stackrel{e}{=} t'), \{(M':\Phi' \Rightarrow q_m \stackrel{e}{=} q_m) | m \in M\}$,
  $\{(M':\Phi' \Rightarrow (q_m|m \in M)t_i \stackrel{e}{=} (q_m|m \in M)t'_i) | i \in I\} \vdash$
  $\vdash (M':\Phi' \Rightarrow (q_m|m \in M)t \stackrel{e}{=} (q_m|m \in M)t')$,
  t, t', t_i, t'_i, q_m arbitrary terms ($i \in I$, $m \in M$) on the corresponding sets of variables.

PROOF. Assume $\Phi \in$ Prem with set M of variables, let $Q \subseteq$ Prem , and define
$D(Q;(M:\Phi)) := \{(t,t') \in T_X{}^2 | (M:\Phi \Rightarrow t \stackrel{e}{=} t')$ is derivable from Q by the rules in 14.3.14$\}$.
Then (IR1), (IR3), (IR4), (IR5), (IR6) and (IR7) are equivalent to the fact that

$D(Q;(M:\Phi))$ is a closed congruence on its domain which is an M-generated relative subalgebra of $\underline{F}(X,\mathbf{TAlg}(\underline{S}))$ (for the latter see (IR1), (IR3) and (IR6)). (IR2) takes care of the fact that $\{(t_i,t'_i)|i \in I\} \subseteq D(Q;(M:\Phi))$, when $\Phi = \Lambda^*\{t_i \overset{e}{=} t'_i|i \in I\}$. (IR8) is equivalent to (I4) in 14.3.3, it is a general substitution rule. Thus the equivalence of (i) and (ii) is established by Theorem 14.3.3. ■■■

14.3.16 REMARK. Observe that the above rules can be simplified in the case of total algebras (or of non-empty phyla). We leave a formulation of the corresponding simplifications to the reader as an exercise.

14.4 Closed sets of ECH- and Horn-formulas

14.4.1. In this subsection we want to discuss briefly the cases of closed sets of ECH-formulas respectively of universal Horn formulas, i.e. we now allow negations of the form $(M:\neg\Lambda^*\{t_i \overset{e}{=} t_i|i \in I\})$ respectively $(M:\neg\Lambda^*\{t_i \overset{e}{=} t_i'|i \in I\})$ ($\#I < \mathbf{k}$, $\mathbf{k} \geq \mathbf{d^*}$, regular) in addition to ECE-equations or QE-equations, respectively. As before, we let Prem be the set of admissible premises, i.e. in this case Prem = $\mathbf{Subs_k}(\Delta_T)$ or Prem = $\mathbf{Subs_k}(T_X{}^2)$ (see 14.3.2), where $T_X := F(X,\mathbf{TAlg}(\underline{S}))$ for the usual set X of variables. Moreover, the conjunctions of the elements of elements of Prem now yield those formulas which can be negated. As before let Q designate a closed set of $\mathbf{HS_C P^{r,k}}$-axioms for $\mathbf{H} \in \{\mathbf{H_C},\mathbf{I}\}$, and let Neg(Q') be a set of premises in Prem corresponding to a closed set of negations of conjunctions of the elements of admissible premises such that $Q' = Q \cup \{(X \cap \downarrow\Phi:\neg\Lambda^*\{\Sigma|\Sigma \in \Phi\})|\Phi \in Neg(Q')\}$ is a closed set of ECH- respectively of Horn-formulas. Then we call Q the <u>positive part</u> of Q', and we have

14.4.2 LEMMA. With the notation and assumptions of 14.4.1 and $\Phi,\Phi' \in$ Prem we have:

(i) $\Phi \in \mathbf{Neg(Q')}$ implies that $(\Phi')Q = \nabla_{\underline{F}(var(\Phi'),\mathbf{TAlg}(\underline{S}))}$, whenever there is a homomorphism (necessarily compatible with the sort homomorphisms!) from $\underline{\downarrow\Phi}/Cong_{\downarrow\Phi}(\Phi)$ into $\underline{\downarrow}(\Phi')Q/(\Phi')Q$ (in particular for $\Phi = \Phi'$).

(ii) If $\Phi \,\overline{\in}\, Neg(Q')$, $[\downarrow\Phi]\widetilde{v}_X \subseteq [\downarrow\Phi']\widetilde{v}_X$, and $(\Phi')Q = \nabla_{\underline{F}(var(\Phi'),\mathbf{TAlg}(\underline{S}))}$, then $\Phi' \in Neg(Q')$ . .

(iii) If $\Phi \in \mathbf{Neg(Q')}$ and $\Phi \subseteq \Phi'$, then $\Phi' \in \mathbf{Neg(Q')}$.

(iv) The positive part Q of a closed set Q' satisfies 14.3.3.(ii).

PROOF. When we recall that $Pj(H_C)$ consists of all partial Peano algebras and that Pj(I) consists of all partial algebras, then all these statements are immediate consequences of Theorem 14.2.2 and the fact that for a closed set Q' of ECH- respectively of Horn-formulas with positive part Q we have:
$\Phi \in Neg(Q')$ implies $(\Phi,(t,t')) \in Q$ for every admissible $(t,t') \in F(var(\Phi),\mathbf{TAlg}(\underline{S}))^2$ (since in no model of Q' Φ can be satisfied). And therefore, if the assumptions of (ii) are satisfied, but $\Phi' \notin Neg(Q')$, then $\underline{F}(var(\Phi'),\mathbf{TAlg}(\underline{S}))/\ _{\underline{F}(var(\Phi'),\mathbf{TAlg}(\underline{S}))}$ would be a model of Q' allowing a homomorphism from $\underline{\downarrow\Phi}/Cong_{\downarrow\Phi}(\Phi)$ into it, thus contradicting $\Phi \in Neg(Q')$. ■■■

14.4.3 REMARK. Observe that Neg(Q') above is in general not uniquely determined by the positive part Q of Q', since for some sets of phyla there need not exist a negation of a conjunction of (T)E-equations depending only on variables of these sorts, but there may be a premise depending only on these sorts such that all corresponding models only have one-element phyla. However, we always have

$Neg(Q') \subseteq \{\Phi \in Prem \mid (\Phi)Q = \nabla_{\underline{F}(var(\Phi),\mathbf{TAlg}(\underline{S}))}\} =: Q^*$.

And in the homogeneous case we only have the possibilities $Neg(Q') = \emptyset$ or $Neg(Q') = Q^*$.

In the general case 14.4.2.(ii) shows that every set $V \subseteq \{[\downarrow\Phi]\tilde{v}_X \mid \Phi \in Q^*\} =: \mathbf{Subs}(S,Q^*)$ which is an upper end with respect to set theoretical inclusion within $(\mathbf{Subs}(S,Q^*);\subseteq)$ is in one-to-one correspondence with a possible set $Neg(Q') := \{\Phi \in Q^* \mid [\downarrow\Phi]\tilde{v}_X \in V\}$. -- This is easily seen from 14.2.2 and the above remarks. ∎

We would also like to mention that the cases of closed sets of ECH- or Horn-formulas (and even other cases) can be treated like the corresponding ones for closed sets of ECE- or QE-equations, if we add to Δ_T respectively $T_X \times T_X$ a new element " FALSE " (not depending on any variables or sorts) such that

$\underline{A} \models (M:\wedge^*\{t_i \stackrel{e}{=} t'_i \mid i \in I\} \Rightarrow \text{FALSE})$ iff $\underline{A} \models (M:\neg\wedge^*\{t_i \stackrel{e}{=} t'_i \mid i \in I\}$.

Moreover, $\text{FALSE} \in \Phi$ implies $(\Phi)Q = \nabla_{F(X\cap\downarrow\Phi,\mathbf{TAlg}(\underline{S}))}$.

Thus we have proved the following

14.4.4 PROPOSITION. Let $\mathbf{k} \geq \mathbf{d}^*$ be a regular cardinal. Then a set $Q' = Q \cup Neg(Q')$ of encoded ECH$^{\mathbf{k}}$- respectively Horn$^{\mathbf{k}}$-formulas is closed iff Q is a closed set of ECE$^{\mathbf{k}}$- respectively QE$^{\mathbf{k}}$-equations and $Neg(Q') \subseteq Prem$ such that there is an upper end V of $(\mathbf{Subs}(S,Q^*);\subseteq)$ (see 14.4.3) such that

$$Neg(Q') = \{\Phi \in Q^* \mid [\downarrow\Phi]\tilde{v}_X \in V\} .$$ ∎∎∎

14.4.5 REMARK. The reader might have realized in particular while dealing with the last sections that a lot of -- sometimes relatively easy -- work can still be done in connection with special realizations of applications of Theorems 12.2.5 and 14.2.2. But we hope that the results discussed so far form a good basis, and that everybody who has to deal with similar problems not treated here has at least found some help in this chapter for an easier treatment of them.

We did not yet touch a related topic, which could be of some interest, namely the axiomatization by so called "notions of validity of equations in partial algebras" (see 13.3.2.(iii)). We do not have the space here and refer e.g. to R.JOHN [J 75].

CHAPTER IV

SPECIAL TOPICS

In this chapter we mainly treat or extend some fundamental concepts which we either touched so far not at all or only very briefly. Moreover, in most sections of this chapter we now investigate the intrinsic set theoretical respectively algebraical structure of some concepts rather than treating them from a category theoretical or model theoretical point of view which we mostly used in the previous chapters. In this connection the fact that we also include heterogeneous partial algebras into our considerations (but not the fact that we treat them in the form of partial $\underline{S}$-algebras) gains much more importance than in the previous sections and causes on several occasions complications which do not occur in the homogeneous case.

In **section 15** we treat two concepts of superposition of operations and derived concepts of "algebraic partial operations" of a given partial algebra. At some instances there is a close connection to the concept of term operations introduced in section 5.

In **section 16** we briefly touch some general concepts connected with possible (minimal) completions of partial algebras, but we do not go into details concerning the existence of special kinds of completions within some kind of variety.

In **section 17** some additional properties of independence and the relationship between independence classes and E-varieties are considered. The existence -- with respect to special properties -- of relatively free completions within an E-variety, say **K**, for the members of **K** and in particular for the **K**-freely generated **K**-algebras is investigated and (relatively) free partial algebras are characterized. Finally, the so-called representation of "initial objects" of (QE-) varieties by canonical terms is investigated.

In **section 18** we return to the investigation of algebraic (respectively term) operations from the point of view of "dependence on variables". Moreover, the existence of a minimal generating subset (element basis) of a free generating subset for any given element of a free partial algebra is investigated.

In **section 19** we characterize the process of generation of closed congruences of finitary partial algebras, and we characterize the congruences belonging to P-reflecting surjective homomorphisms for some set P of terms closed with respect to the formation of subterms.

In **section 20** several concepts of subdirect representability of partial algebras are investigated, the corresponding subdirectly irreducible objects are characterized, and representation theorems are proved.

In **section 21** we show, how some of the tools used in particular for model theoretic investigations within the whole species of partial algebras can be relativized to E-varieties satisfying some special conditions which are to a great extent satisfied in the homogeneous case.

15 Algebraic operations

15.1 Superposition

15.1.1. In subsection 5.7 we have already briefly treated term operations and polynomial operations on a given partial algebra. In this section we return to this concept from a different starting point, namely from the one of algebraic partial operations, which are derived directly from the fundamental partial operations and total projection operations of a given partial $\underline{S}$-algebra.
There are several ways to define "partial operations composed from the fundamental operations (and total projections) of a given partial algebra." While some of them yield the same results, others result in sets of partial operations which differ from each other. In each case one includes some collection of "identity operations" as starting points, namely either the total identity operation of each phylum, or a collection of total projections. A comprehensive study of superposition of partial functions is contained in [Hoe 72] by H.-J.HOEHNKE (see also A.I.MAL'CEV [Mal 74], §2).
Observe that we always more or less tacitly consider partial $\underline{S}$-algebras and S-sets -- if not stated otherwise -- for some "sort algebra" $\underline{S}$, i.e. each set, say M, is connected with a "sort mapping" $v_M:M \to S$ -- which is to be a homomorphism from $\underline{M}$ into $\underline{S}$, whenever M carries some given partial algebraic structure.

15.1.2 DEFINITION of basic concepts. Let A, M be any S-sets, i.e. they are connected with sort mappings $v_A:A \to S$ and $v_M:M \to S$.

(i): An M-ary partial operation with output sort s on A is some mapping $f:D \to A_s$ (for some $s \in S$, $A_s = (s)v_A^{-1}$), where D is a subset of the set A^M of all sort-respecting mappings from M into A, i.e. of all mappings $q:M \to A$ satisfying $qv_A = v_M$.
-- Observe that for some partial operation we always assume that all its values are of the same sort. Moreover, observe that an M-ary partial operation, say m, does not only consist of its graph, but it is actually a triplet $(A^M, \text{graph}(m), A_s)$ consisting of the "global" domain which contains the actual domain (i.e. in the above case: D), its graph and its codomain. Thus partial operations with empty "actual" domain can well be distinguished by their arities and the sort of their target sets.

(ii): By $PO^M(A;S)$ we denote the set of all M-ary partial operations on A, i.e.

$$PO^M(A;S) := \bigcup\{A_s{}^D \mid s \in S \text{ , } D \subseteq A^M\} \text{ , and by}$$

$$O^M(A;S) := \bigcup\{A_s{}^{(A^M)} \mid s \in S\}$$

its subset of all total M-ary (sort compatible) operations. If S has only one element, then we also simply write $PO^M(A)$, respectively $O^M(A)$.

(iii): For each $m \in M$ we denote by $e_m{}^M \in O^M(A;S)$ the m-th total M-ary projection $e_m{}^M:A^M \to A_s$ $(s:= (m)v_A)$: $\underline{a}e_m{}^M := (m)\underline{a}$ for each $\underline{a} \in A^M$.

$E^M(A;S) := \{e_m{}^M \mid m \in M\}$ designates the set of all M-ary total projections of A.

(iv): Let $\underline{A}$ be any partial $\underline{S}$-algebra of type τ with cardinal dimension $\mathbf{d}$, and let X be any (S-) set, such that each phylum of X has cardinality $\mathbf{d}^*$. Then we define for $\mathbf{d}^\& = \max(\mathbf{d}^*, (\#S)^+)$:

$$PO(\underline{A};X;\underline{S}) := \bigcup\{PO^M(A;S) \mid M \subseteq X \text{ , } \#M < \mathbf{d}^\&\} \text{ ,}$$

$$O(\underline{A};X;\underline{S}) := \bigcup\{O^M(A;S) \mid M \subseteq X \text{ , } \#M < \mathbf{d}^\&\} \text{ ,}$$

$$E(\underline{A};X;\underline{S}) := \bigcup\{E^M(A;S) \mid M \subseteq X \text{ , } \#M < \mathbf{d}^\&\} \text{ ,}$$

the sets of all partial, respectively total projection operations on A with $\mathbf{d}^{\&}$-small arity in X.

On some occasions we shall need ordinals as arities. Therefore, let $\omega(\mathbf{d}^{\&})$ be the least ordinal β such that $\#\beta = \mathbf{d}^{\&}$. Then we define

$$PO(\underline{A};;\underline{S}) := PO(\underline{A};\omega(\mathbf{d}^{\&});\underline{S}) := \mathbf{U}\{PO^{n,v}(A;S) \mid n < \omega(\mathbf{d}^{\&}),\ v: n \to S\},$$
$$O(\underline{A};;\underline{S}) := O(\underline{A};\omega(\mathbf{d}^{\&});\underline{S}) := \mathbf{U}\{O^{n,v}(A;S) \mid n < \omega(\mathbf{d}^{\&}),\ v: n \to S\},$$
$$E(\underline{A};;\underline{S}) := E(\underline{A};\omega(\mathbf{d}^{\&});\underline{S}) := \mathbf{U}\{E^{n,v}(A;S) \mid n < \omega(\mathbf{d}^{\&}),\ v: n \to S\}.$$

Fundamental for the definition of algebraic structure on $PO(\underline{A};;\underline{S})$ and $PO(\underline{A};X;\underline{S})$ are the following two definitions of superposition (for a better understanding see the examples in 15.1.5):

15.1.3 DEFINITION.

(i) Superposition of the first kind: Let $h \in PO^{M}(A;S)$, let $g_m \in PO^{L_m}(A;S)$ have output sort mv_M for each $m \in M$, and let $N := \dot{\mathbf{U}}_M L_m := \mathbf{U}\{\{m\} \times L_m \mid m \in M\}$, -- or, if M and L_m $(m \in M)$ are each ordinals, then let $\nu := \Sigma_M L_m := \Sigma(L_m \mid m \in M)$ in the sense of ordinal arithmetic (see section 23) -- be the disjoint union -- respectively the ordinal sum -- of the arities, and let v_N be the disjoint union of the sort mappings of the sets L_m $(m \in M)$ (respectively of those connected with the given ordinals). Then we define a partial operation $\underline{g}*h := (g_m \mid m \in M)*h \in PO^{N}(A;S)$ (or $\underline{g}^\circ h := (g_m \mid m \in M)^\circ h \in PO^{n,v}(\underline{A};;\underline{S})$ in the ordinal case) with the same output sort, say s, as h as follows:

For every $\underline{a} \in A^{N}$ and some $a \in A_s$ we have

$\underline{a}[(g_m \mid m \in M)*h] = a$ iff $(((m,l)\underline{a} \mid l \in L_m)g_m \mid m \in M)h = a$;

i.e. $\underline{g}*h$ is defined on an N-ary (sort-respecting!) sequence $\underline{a}$ and has value "a", iff for each $m \in M$ g_m is defined on the sequence $((m,l)\underline{a} \mid l \in L_m)$ and takes a value, say $b_m := ((m,l)\underline{a} \mid l \in L_m)g_m$, and h is defined on $(b_m \mid m \in M)$ and takes the value a: $(b_m \mid m \in M)h = a$.

In the case of "ordinal superposition of the first kind" we have for $M = \zeta$ and $L_\xi = \lambda_\xi$ $(\xi < \zeta)$

$\underline{a}[(g_\xi \mid \xi < \zeta)^\circ h] = a$ iff $(((n)\underline{a} \mid \Sigma(\lambda_\sigma \mid \sigma < \xi) \le n < \Sigma(\lambda_\sigma \mid \sigma \le \xi))g_\xi \mid \xi < \zeta)h = a$.

(ii) Superposition of the second kind: Let $h \in PO^{M}(A;S)$, $g_m \in PO^{L}(A;S)$ of output sort mv_M for each $m \in M$; then we define a partial algebraic operation $(g_m \mid m \in M)\cdot h \in PO^{L}(A;S)$ as follows:

For every $\underline{a} \in A^{L}$ and for every $a \in A$ one has

$\underline{a}[(g_m \mid m \in M)\cdot h] = a$ iff $(\underline{a}g_m \mid m \in M)h = a$,

i.e. iff for every $m \in M$ there exists some $b_m \in A$ such that $\underline{a}g_m = b_m$ and $(b_m \mid m \in M)h = a$.

15.1.4 REMARKS and further DEFINITIONS. **(i):** Observe that we can state equalities as in 15.1.3 iff both sides of the equation are defined at the same time and equal, i.e. if we use in the metalanguage the same conventions for "=" as we use in the object language for "$\stackrel{e}{=}$" .

(ii): For superposition of the first kind the ordinal version is often more convenient than the "set theoretical one"; namely, if we have a fixed set, say X, from which we choose the arities under consideration as subsets, then it is usually not guaranteed that for $L_m \subseteq X$ $(m \in M)$ and $M \subseteq X$ we will also have $\dot{\mathbf{U}}_M L_m \subseteq X$. On the other hand one is usually interested in having -- for some regular bound on the arities -- a representative system for all partial operations on a set. In this connection the switch to ordinals can provide such a possibility: One may choose a well-ordering for the given set X; then this well-ordering will also induce a well-ordering on each of its subsets, and this can be used for the superposition -- usually it is of no great importance what specific set we have used as arity: we may define two partial operations, say $f \in PO^M(A;S)$ and $g \in PO^N(A;S)$ to be **equivalent**, iff there is a sort respecting bijection, say $i:N \to M$, such that for each $\underline{a} \in A^M$ and $a \in A$ one has $\underline{a}f = a$ iff $(i\underline{a})g = a$ (i.e. i and id_A induce a sort preserving bijection between graph(f) and graph(g): $(\underline{a},a) \in$ graph(f) iff $(i\underline{a},a) \in$ graph(g)).

(iii): If $\underline{A}$ is any partial algebra, then the ordinal superposition of the first kind induces a total algebraic structure, say $(\phi^\circ)_{\phi\in\Omega}$, of type τ on $PO(\underline{A};;\underline{S})$ (respectively on $PO(\underline{A};X;\underline{S})$, when we well-order X and relate to each subset of X its induced well-order type) in the following way, when we well-order each K_ϕ ($\phi \in \Omega$), say by an ordinal κ_ϕ:
Let $f_\sigma \in PO(\underline{A};;\underline{S})$ be a μ_σ-ary partial operation on A for each $\sigma < \kappa_\phi$; then define the $\Sigma(\mu_\sigma|\sigma < \kappa_\phi)$-ary partial operation

$$(f_\sigma|\sigma < \kappa_\phi)\phi^\circ := (f_\sigma|\sigma < \kappa_\phi)\circ\phi^{\underline{A}} .$$

-- Similarly we get from the superposition of the second kind a partial algebraic structure, say $(\overline{\phi})_{\phi\in\Omega}$, on $PO(\underline{A};X;\underline{S})$, defined as follows:

Let $(f_k|k \in K_\phi) \in PO^M(A;S)^{K_\phi}$ for some $\mathbf{d}^{\&}$-small subset M of X be sort compatible with K_ϕ, then:

$$(f_k|k \in K_\phi)\overline{\phi} := (f_k|k \in K_\phi)\cdot\phi^{\underline{A}}$$

-- observe the remark on empty mappings in 15.1.3.(i).

The corresponding total and partial algebras are denoted by $\underline{PO}(\underline{A};;\underline{S})$ ($=:\underline{PO}(\underline{A};\omega(\mathbf{d}^{\&});\underline{S})$) respectively by $\underline{PO}(\underline{A};X;\underline{S})$.

One should observe that, for each $M \subseteq X$ (with $\#M < \mathbf{d}^{\&}$) and for each non-nullary $\phi \in \Omega$, $\phi^{\underline{PO}^M(\underline{A};\underline{S})} := \overline{\phi}||PO^M(\underline{A})$ (" $||$ " meaning that the domain is restricted to $(PO^M(A;S))^{K_\phi}$, and the target is restricted to $PO^M(A;S)$), is a total operation on $PO^M(A;S)$. However, if $K_\phi = \emptyset$, then $\emptyset_M\cdot\phi^{\underline{PO}^M(\underline{A};\underline{S})}$ ($\emptyset_M$ here denoting the empty sequence of M-ary partial operations) is either a total M-ary operation with value $\phi^{\underline{A}}$, if this constant is defined in $\underline{A}$, otherwise it is the empty M-ary operation with the same output sort as ϕ.

Thus $\underline{PO}^M(\underline{A};\underline{S}) := (PO^M(A;S),(\phi^{\underline{PO}^M(\underline{A};\underline{S})})_{\phi\in\Omega})$ is a total algebra of type τ, while the corresponding restriction of $(\phi^\circ)_{\phi\in\Omega}$ to $PO^{n,v}(\underline{A};\underline{S})$ ($n < \omega(\mathbf{d}^{\&})$, $v:n \to S$) is usually partial, in most cases even empty. ∎

15.1.5 REMARKS and EXAMPLES. (i): Observe that by the above definition $\underline{PO}^M(\underline{A};\underline{S})$ is **not** just a relative subalgebra of $\underline{PO}(\underline{A};X;\underline{S})$ (this will only be the case if we change the definition for the nullary constants, the non-nullary partial operations are just the restrictions of the corresponding global ones).

(ii): For the following examples let us assume that each arity is finite, say $n := \{0,1,\dots,n-1\}$, and for simplicity let us consider the homogeneous case, where we do not mention "$\underline{S}$" in the notations:

Each $h \in PO^n(A)$ can then be considered as a "black box" with n "entrances" and one "outlet" (see Figure 15.1) such that for an "admissible" sequence $\underline{a} = (a_0,\dots a_{n-1})$ -- i.e. $\underline{a} \in$ dom h, each component given on the corresponding entrance -- yields $(\underline{a})h$ at the outlet; else there

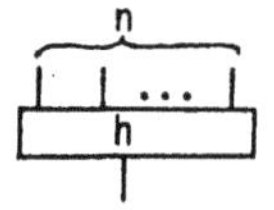

Figure 15.1

is no reaction at the outlet.

In the case of superposition of the first kind, say for $(g_0,g_1,g_2)°h$, where, say, $g_0 \in PO^3(A)$, $g_1 \in PO^2(A)$ and $g_2 \in PO^4(A)$, we connect the corresponding "black boxes" as shown in Figure 15.2, i.e. all the entries of the operations g_0, g_1 respectively g_2 are independent. It is easily seen in this connection that it does not matter whether or not the arity of h is empty.

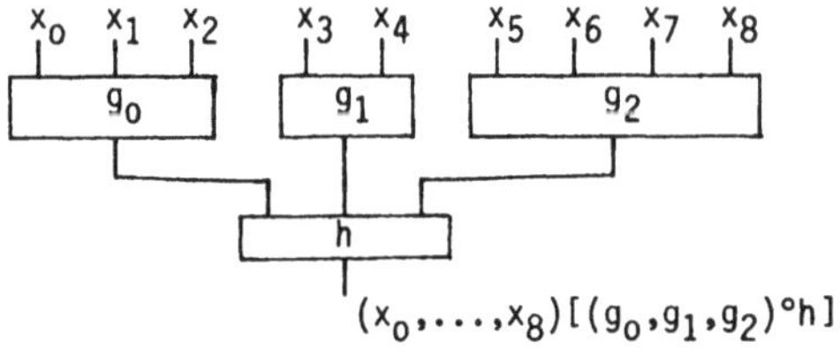

Figure 15.2

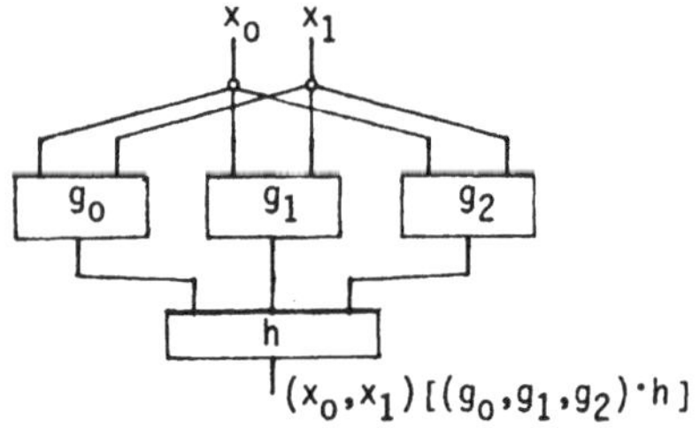

Figure 15.3

In the case of superposition of the second kind the corresponding picture for the "black boxes" looks as in Figure 15.3, when we consider $h \in PO^3(A)$ and $g_0,g_1,g_2 \in PO^2(A)$; i.e. first the variables are "copied" as often as needed (here 3 times) into the corresponding entrances of the partial operations, and the outlets of the g_i are led into the appropriate entrances of h.

Obviously these are two rather extremal possibilities for the combination of partial operations. Using the unary total identity operation $e_0^1 = id_A$ (for which in the diagrams the corresponding "box" is usually omitted) we can write e.g.

$((x_0,x_1,x_2)g,x_3,x_4)h$ as $(x_0,\dots,x_4)[(g,e_0^1,e_0^1)*h]$,

but we have difficulties in representing e.g. $((x_0,x_1,x_0)g,x_2,x_1)h$ by superposition of the first kind.

In both cases superposition of the second kind is applicable and yields

$(x_0,x_1,x_2,x_3,x_4)[((e_0^5,e_1^5,e_2^5)\cdot g,e_3^5,e_4^5)\cdot h]$, respectively

$(x_0,x_1,x_2)[((e_0^3,e_1^3,e_0^3)\cdot g,e_2^3,e_1^3)\cdot h]$ (the reader should draw the corresponding "black box schemes").

This shows that the concept of superposition of the second kind is more flexible when we include for each arity the total projection operations among the basic tools which usually consist of the fundamental operations on a partial algebra $\underline{A}$.

(iii): It easily follows from Definition 15.1.3 that for any set M one has for the corresponding total projections:

(*) If $h \in PO^M(A;S)$, then $h = (e_m^M \mid m \in M)\cdot h$, but

(**) If $g_m \in PO^M(A;S)$ $(m \in M)$, then $(g_n \mid n \in M)\cdot e_m^M = g_m|\cap(\mathrm{dom}\; g_n \mid n \in M)$.

Thus the total projections behave "left neutral," but they are "right neutral" only in a restricted sense.

We have already observed similar features in connection with conditioned term operations which we had to introduce in 10.3 (cf. 10.3.2) in order to be able to describe **Mod**(F_P)-extendable epimorphisms. We shall see below that there is indeed a close connection.

(iv): Observe that the total projection e_m^M behaves on a partial algebra $\underline{A}$ in the same way as the term operation $(M:m)^{\underline{A}}$, when M is a subset of the set of variables. Since $\underline{PO}^M(\underline{A};\underline{S})$ is a total algebra, the mapping $e:M \to E^M(\underline{A};\underline{S})$, $m \mapsto e_m^M$, has a homomorphic extension $eval_M:\underline{F}(M,\mathbf{TAlg}(\underline{S})) \to \underline{PO}^M(\underline{A};\underline{S})$ (if necessary, we write $eval_M^{\underline{A}}$).

15.2 Algebraic partial operations of the first and second kind

15.2.1 LEMMA. Let $\underline{A}$ be any partial algebra, M a $\mathbf{d}^{\&}$-small set of variables. Then the homomorphism $eval_M:\underline{F}(M,\mathbf{TAlg}(\underline{S})) \to \underline{PO}^M(\underline{A};\underline{S})$, which has been defined in 15.1.5.(iv), satisfies for each $t \in F(M,\mathbf{TAlg}(\underline{S}))$: $(t)eval_M = (M\!:\!t)^{\underline{A}}$;
i.e. $eval_M$ provides a link for the evaluation of terms as term functions on a more abstract level.

The straightforward proof by algebraic induction starting with 15.1.5.(iv) is left as an exercise to the reader. ■■■

This observation gives rise to the following definition, where we also treat substitution of the first kind.

15.2.2 DEFINITION. Let $\underline{A}$ be any partial ($\underline{S}$-) algebra.
(i): The elements of the subalgebra of $\underline{PO}(\underline{A};\omega(\mathbf{d}^{\&});\underline{S})$ (with respect to superposition of the first kind) generated by the set $\{id_{A_s} \mid s \in S\}$ of total identity operations for the phyla of $\underline{A}$ are called <u>algebraic (partial) operations of the first kind</u> on $\underline{A}$, and the whole set is denoted by $APO1(\underline{A};\omega(\mathbf{d}^{\&});\underline{S})$ $(= \mathbf{C}_{\underline{PO}(\underline{A};;\underline{S})}\{id_{A_s} \mid s \in S\})$.
(ii): Let M be any set. Then

$$[F(M,\mathbf{TAlg}(\underline{S}))]eval_M = \mathbf{C}_{\underline{PO}^M(\underline{A};\underline{S})}E^M(A;S) =: APO^M(\underline{A};\underline{S})$$

is called the set of all <u>M-ary algebraic (partial) operations (of the second kind)</u> on $\underline{A}$ -- the additions in brackets are often suppressed.
For any set X with $\#X_s \geq \mathbf{d}^*$ for each $s \in S$ we denote by $APO2(\underline{A};X;\underline{S}) := \bigcup(APO^M(\underline{A}) \mid M \subseteq X\ ,\ \#M < \mathbf{d}^{\&})$ the set of all <u>algebraic partial operations of the second kind on $\underline{A}$ with arities in</u> X (restricted in cardinality by $\mathbf{d}^{\&}$).
By $\underline{APO1}(\underline{A};;\underline{S})$, $\underline{APO2}(\underline{A};X;\underline{S})$ respectively $\underline{APO}^M(\underline{A};\underline{S})$ we denote the relative subalgebras of the corresponding partial algebras of operations.

15.2.3 REMARK. Observe that $APO2(\underline{A};X;\underline{S})$ is generated as a subalgebra of $\underline{PO}(\underline{A};X;\underline{S})$ by $G := \bigcup(E^M(A;S) \cup C^M(\underline{A};\underline{S}) \mid M \subseteq X\ ,\ \#M < \mathbf{d}^{\&})$, where

$$C^M(\underline{A};\underline{S}) := \{c_\phi^{M,\underline{A}} \in O^M(\underline{A};\underline{S}) \mid \phi \in \Omega\ ,\ K_\phi = \emptyset\ ,\ \phi^{\underline{A}} \text{ is total}\}\ ,$$

such that $\underline{a}c_\phi^{M,\underline{A}} = \phi^{\underline{A}}$ for each $\underline{a} \in A^M$.
-- Observe that $C^{\emptyset}(\underline{A};\underline{S})$ could be omitted from G, since it will be provided by the algebraic structure of $\underline{PO}(\underline{A};X;\underline{S})$. Within each $\underline{PO}^M(\underline{A};\underline{S})$ the subset $APO^M(\underline{A};\underline{S})$ is generated by the M-ary total projections only, but for the "global" set $APO(\underline{A};X;\underline{S})$ we also need the "extended fundamental constants" (i.e. extended to each $M \subseteq X$, $\#M < \mathbf{d}^{\&}$) in order to get everything that is induced by terms with $\mathbf{d}^{\&}$-small subsets of X for variables.
The relationship between algebraic operations of the first and of the second kind will follow from the subsequent results.

15.2.4 LEMMA. Superposition of the second kind is "associative." I.e. let $\underline{A} \in \mathbf{Alg}(\underline{S})$, $f \in PO^M(A;S)$, $g_m \in PO^N(A;S)$ for each $m \in M$ and $h_n \in PO^L(A;S)$ for each $n \in N$. Then $(h_n \mid n \in N)\cdot[(g_m \mid m \in M)\cdot f] = ((h_n \mid n \in N)\cdot g_m \mid m \in M)\cdot f$.

PROOF. The statement is easily derived from Definition 15.1.3 . ■■■

15.2.5 DEFINITION. Let A, M and N be any sets, $p:M \to N$ any mapping, and let $h \in PO^M(A;S)$. Then

$$h_{p,N} := (e_{(m)p}{}^N \mid m \in M)\cdot h$$

is an N-ary operation on A obtained from h by a <u>transformation of variables along</u> p.

15.2.6 PROPOSITION. Let $\underline{A} \in \mathbf{Alg}(\underline{S})$, M, N sets, $p:M \to N$ a mapping. Then p induces a homomorphism $p^\circ:\underline{PO}^M(\underline{A};\underline{S}) \to \underline{PO}^N(\underline{A};\underline{S})$, $h \mapsto h_{p,N}$ $(h \in PO^M(A;S))$, whose restriction to $APO^M(\underline{A};\underline{S})$ is a homomorphism $p^*:\underline{APO}^M(\underline{A};\underline{S}) \to \underline{APO}^N(\underline{A};\underline{S})$ between the corresponding total algebras of algebraic operations on $\underline{A}$.

PROOF. From 15.2.5 it immediately follows that $p^\circ:PO^M(A;S) \to PO^N(A;S)$ is a mapping. Let $\phi \in \Omega$, $\{h_k \mid k \in K_\phi\} \subseteq PO^M(A;S)$, then:

$((h_k \mid k \in K_\phi)\phi^{\underline{PO}^M(\underline{A};\underline{S})})p^\circ = ((h_k \mid k \in K_\phi)\cdot\phi^{\underline{A}})p^\circ = (e_{(m)p}{}^N \mid m \in M)\cdot[(h_k \mid k \in K_\phi)\cdot\phi^{\underline{A}}] =$

$= ((e_{(m)p}{}^N \mid m \in M)\cdot h_k \mid k \in K_\phi)\cdot\phi^{\underline{A}}$ (using 15.2.4)

$=(h_k p^\circ \mid k \in K_\phi)\cdot\phi^{\underline{A}} = (h_k p^\circ \mid k \in K_\phi)\phi^{\underline{PO}^N(\underline{A};\underline{S})}$.

Thus $p^\circ:\underline{PO}^M(\underline{A};\underline{S}) \to \underline{PO}^N(\underline{A};\underline{S})$ is a homomorphism.

Now assume that $h = e_m{}^M$; then $h_{p,N} = (e_{(m')p}{}^N \mid m' \in M)\cdot e_m{}^M$, and this implies that $(e_m{}^M)_{p,N} = (e_m{}^M)p^\circ = e_{(m)p}{}^N$. The compatibility of homomorphisms with the algebraic closure operator now implies $[APO^M(\underline{A};\underline{S})]p^\circ \subseteq APO^N(\underline{A};\underline{S})$, and the last statement follows. ■■■

15.2.7 PROPOSITION. For every partial algebra $\underline{A}$ each algebraic partial operation on $\underline{A}$ of the first kind is equivalent to some algebraic partial operation of the second kind, but in general not vice versa.

PROOF (by algebraic induction). For simplicity we assume that X and each K_ϕ are ordinals -- see 15.1.4.(ii): Obviously $id_{A_s} = e_0{}^1 \in APO2(\underline{A};X;\underline{S})$, when the only element 0 of 1 is mapped to s $(s \in S)$. Therefore assume that $h = (h_k \mid k \in K_\phi)\circ\phi^{\underline{A}}$, and that each h_k is equivalent to $\bar{h}_k \in APO^{L_k,v_k}(\underline{A};\underline{S})$ $(k \in K_\phi)$. Let $N:= \Sigma(L_k \mid k \in K_\phi)$ with the induced sort mapping v_N. Because of 15.2.6

$g:= ((e_\beta{}^N \mid \Sigma(L_\gamma \mid \gamma < k) \le \beta < \Sigma(L_\gamma \mid \gamma \le k))\cdot\bar{h}_k \mid k \in K_\phi)\cdot\phi^{\underline{A}} \in APO^{N,v_N}(\underline{A};\underline{S})$; and it is easy to check that for each $\underline{a} \in A^N$ and $b \in A$:

$\underline{a}[(h_k \mid k \in K_\phi)\circ\phi^{\underline{A}}] = b$ iff $\underline{a}g = b$, i.e. again $(h_k \mid k \in K_\phi)\circ\phi^{\underline{A}}$ is equivalent to some algebraic partial operation of the second kind. ■

In order to see that the converse implication does not hold, one may consider a binary fundamental operation symbol, say ϕ $(\tau = (2))$, and $\underline{A}:= \underline{F}(\{x,y\},\mathbf{TAlg}(\tau))$ -- homogeneous case! --, $h:= (e_0{}^1,e_0{}^1)\cdot\phi^{\underline{A}} \in APO^1(\underline{A})$, i.e. $xh:= (x,x)\phi^{\underline{A}}$ $(x \in A)$. But when we consider the algebraic (partial) operations of the first kind on $\underline{A}$, then the only unary one is id_A, the only binary one is $\phi^{\underline{A}}$, etc.; however we do not get h, which is unary, but not the identity. ■■■

The following consequence corresponds to 15.2.4 and yields another generalization of the associativity of the composition of unary functions (observe that in the case of arbitrary sets one has to include a bijection between the arities):

15.2.8 LEMMA. Let M, N_m, L_n ($m \in M$, $n \in \Sigma(N_m|m \in M)$) be arbitrary ordinals. Let $h \in PO^M(A;S)$, $g_m \in PO^{M_m}(A;S)$ ($m \in M$), $f_n \in PO^{L_n}(A;S)$ ($n \in \Sigma(N_m|m \in M)$). Then $(f_n|n \in \Sigma(N_m|m \in M))^\circ[(g_m|m \in M)^\circ h] = ((f_n|\Sigma(N_1|1 < m) \leq n < \Sigma(N_1|1 \leq m))^\circ g_m|m \in M)^\circ h$.
■■■

The connection between algebraic partial operations of the first and second kind is described by the following two results:

15.2.9 LEMMA. Let $\underline{A}$ be any partial algebra, M any set. Then every element of $APO^M(\underline{A};\underline{S})$ can be derived from some element of $APO1(\underline{A};;\underline{S})$ by transformation of variables.

PROOF (by algebraic induction). (a): We assume a fixed well-ordering for each K_ϕ, $\phi \in \Omega$. For each $m \in M$, $e_m{}^M = (e_m{}^M)\cdot id_{A_s}$ where $s := (m)v_M$ -- $p:1 \to M$, $0 \mapsto m$.
(b): Let $\phi^{\underline{A}}$ be a nullary fundamental operation of $\underline{A}$, which is defined in $\underline{A}$. Then the empty mapping $p:\emptyset \to M$ yields the total M-ary constant operation $\phi^{\underline{A}}{}_{p,M}$ with value $\phi^{\underline{A}}$ of $APO^M(\underline{A};\underline{S})$.
(c): Assume $h \in APO^M(\underline{A};\underline{S})$, $h = (h_k|k \in K_\phi)\cdot\phi^{\underline{A}}$, such that for each h_k there exists $g_k \in APO1(\underline{A};;\underline{S})$ of arity ζ_k (and sort mapping v_k) with $p_k:\zeta_k \to M$ such that $h_k = (g_k)p_k{}^\circ$. We then define for $\nu := \Sigma(\zeta_k|k \in K_\phi)$ a mapping $p:\nu \to M$ by $(\xi)p := \xi' p_k$, where $\nu_k := \Sigma(\zeta_\sigma|\sigma < k) \leq \xi =: \nu_k + \xi' < \nu_{k+1}$, i.e. ξ' is uniquely determined by ordinal arithmetic. And we obtain for every $\underline{a} \in A^M$ and $b \in A$ that

$$\underline{a}h = \underline{a}[(h_k|k \in K_\phi)\cdot\phi^{\underline{A}}] = ((\underline{a})(g_k)_{p_k,M}|k \in K_\phi)\phi^{\underline{A}} = \underline{a}((g_k|k \in K_\phi)^\circ\phi^{\underline{A}})_{p,M}$$

(the reader should fill in missing details as an exercise). ■■■

15.2.10 PROPOSITION. Let $X = \omega(\mathbf{d}^{\&})$ and each K_ϕ ($\phi \in \Omega$) be ordinals together with their sort mappings such that each phylum of X has cardinality at least $\mathbf{d}^*$. Then $APO2(\underline{A};X;\underline{S})$ is the closure of $APO1(\underline{A};;\underline{S})$ with respect to transformation of variables.

15.2.11 REMARK. In this connection one has to distinguish between the fact that X has to be well-ordered in order that each subset carries a well-ordering and that we may have for a given family of subsets a unique well-ordering on the sum of their order types. Since the sort mapping of this sum is given by the disjoint union of the sort mappings of the summands, we cannot just take such a sum as the initial segment of X, but we have to take the subset with the smallest elements of X having the corresponding well-ordering type and the same sort mapping as the ordinal obtained as sum. This can always be obtained by the properties of well-orderings, and we shall not make this more explicit here.

PROOF of 15.2.10. From 15.2.6 we infer that $APO2(\underline{A};X;\underline{S})$ is always closed with respect to transformation of variables, and from 15.2.7 we then obtain that all partial operations with $\mathbf{d}^{\&}$-small arity in X obtained from elements of $APO1(\underline{A};;\underline{S})$ by transformation of variables belong to $APO2(\underline{A};X;\underline{S})$. The converse inclusion follows from 15.2.9. ■■■

15.2.12. In the finitary case one can say more about the basic principles (i.e. operations on $PO(A;\omega(\mathbf{d}^{\&});\underline{S})$), by which one can generate the algebraic partial operations of the second kind from those of the first kind -- see e.g. W.M.GLUSCHKOW, G.J.ZEITLIN, J.L.JUSTSCHENKO [GlZJu 80], p. 73f, and also H.-J.HOEHNKE [Hoe 72].

15.2.13 LEMMA. $APO1(\underline{A};;\underline{S})$ and $APO2(\underline{A};X;\underline{S})$ are closed with respect to superposition of the first kind, where only algebraic partial operations of the first kind are allowed "on the outer part": Let $h \in APO1(\underline{A};;\underline{S})$ with arity, say , and let $(g_\xi|\xi < \zeta)$ be a sequence of algebraic partial operations of $\underline{A}$ of the first kind (respectively of the second kind). Then $(g_\xi|\xi < \zeta)^\circ h$ is an algebraic partial operation of the first (second) kind. -- And only if $\underline{A}$ is total, $APO2(\underline{A};X;\underline{S}) = AO2(\underline{A};X;\underline{S})$ is closed with respect to superposition of the second kind with respect to arbitrary elements of $AO2(\underline{A};X;\underline{S})$.

PROOF (by algebraic induction on $APO1(\underline{A};;\underline{S})$). If $h = id_{A_s}$, then the statement is trivial. Let $h = (f_k|k \subset K_\phi)^\circ\phi^{\underline{A}}$ for some $\phi \in \Omega$ and $\{f_k|k \in K_\phi\} \subseteq APO1(\underline{A};X;\underline{S})$ of arity ζ (assuming that X and all arities are ordinals or at least well-ordered; moreover, we only treat the second kind, the other case is similar). Assume that, for each $k \in K_\phi$, $h_k := (g_\xi|\Sigma(\nu_\sigma|\sigma < k) \leq \xi < \Sigma(\nu_\sigma|\sigma \leq k))^\circ f_k \in APO2(\underline{A};X;\underline{S})$, and let ν_k be the corresponding arity; then the mappings $p_k:\nu_k \to \zeta$, $\xi \mapsto \Sigma(\nu_\sigma|\sigma < k) + \xi$ $(\xi < \nu_k, k \in K_\phi)$ yield algebraic partial operations $(h_k)_{p_k,\zeta}$, and we get $(g_\xi|\xi < \zeta)^\circ h = (h_k|k \in K_\phi)^\circ\phi^{\underline{A}} = ((h_k)_{p_k,\zeta}|k \in K_\phi)\cdot\phi^{\underline{A}}$, showing that we obtain indeed an algebraic partial operation of the second kind. ■
The statement about total algebras follows from the fact that in this case the projection operations are also right neutral -- 15.1.5.(iii).(**) does no longer yield any restriction on the domain which is now all of A^M. ■■■

15.2.14 REMARKS. (i): It is easily seen that $APO1(\underline{A};;\underline{S})$ is not closed with respect to superposition of the second kind: Let $\underline{A}$ be the free semigroup (type (2)) on an at least two-element free generating set M. Then we get for the binary operation ϕ: $(x,y) \mapsto ((x,y)\phi,(x,y)\phi)\phi = (x,y)[(\phi,\phi)\cdot\phi]$ is an algebraic operation of the second, but not of the first kind. ■
(ii): $APO2(\underline{A};X;\underline{S})$ is not closed with respect to superposition of the second kind: Let $\tau := (2,1,0)$, $\Omega := (\phi,h,n)$, and let $\underline{F} := \underline{F}(\{x,y\},\mathbf{K})$ be the **K**-free partial algebra on $\{x,y\}$ in the E-variety **K** defined by the E-equations $xy\phi \overset{e}{=} xy\phi$ and $nh \overset{e}{=} n$, i.e. $h^{\underline{F}}$ has a one-element domain, while $n^{\underline{F}}$ and $\phi^{\underline{F}}$ are total. Consider $g := (e_0{}^2, e_1{}^2\cdot h^{\underline{F}})\cdot e_0{}^2$; then $\mathrm{dom}\, g = F \times \{n^{\underline{F}}\}$. We claim that g is not an algebraic partial operation of the second kind: g is a binary partial projection onto the first component. We scan the first Baire classes of $APO^2(\underline{F})$ with respect to $E^2(\underline{F})$:
0th class: $e_0{}^2$, $e_1{}^2$ are total projections.
1st class: $e_0{}^2\cdot h^{\underline{F}}$, $e_1{}^2\cdot h^{\underline{F}}$, $\emptyset_2\cdot n^{\underline{F}}$ ($\emptyset_2$ the empty sequence of binary algebraic partial operations on $\underline{F}$), $(e_0{}^2,e_1{}^2)\cdot\phi^{\underline{F}} = \phi^{\underline{F}}$, $(e_1{}^2,e_0{}^2)\cdot\phi^{\underline{F}}$, $(e_0{}^2,e_0{}^2)\cdot\phi^{\underline{F}}$, $(e_1{}^2,e_1{}^2)\cdot\phi^{\underline{F}}$.
2nd class: $((e_0{}^2,e_1{}^2)\cdot\phi^{\underline{F}})\cdot h^{\underline{F}}$, $((e_1{}^2,e_0{}^2)\cdot\phi^{\underline{F}})\cdot h^{\underline{F}}$, $((e_0{}^2,e_0{}^2)\cdot\phi^{\underline{F}})\cdot h^{\underline{F}}$, $((e_1{}^2,e_1{}^2)\cdot\phi^{\underline{F}})\cdot h^{\underline{F}}$, and operations "ending with $\phi^{\underline{F}}$ at their right end."
g does not coincide with any of the above operations, and it can be seen that the Baire classes of higher rank cannot contain g either. Therefore, although the projection operations are algebraic, it is in general not allowed within the scope of algebraic partial operations of the second kind to substitute algebraic partial operations for their variables. ■
(iii): Let us just mention without proof that for $\underline{A} \in \mathbf{Alg}(\underline{S})$, $h \in APO^1(\underline{A};\underline{S})$ and $g \in APO(\underline{A};X;\underline{S})$ one always has $g\cdot h \in APO(\underline{A};X;\underline{S})$. ■

15.3 Algebraic partial operations of the third kind

15.3.1. In the theory of homogeneous total algebras so-called "clones" of (total!) operations play an important rôle, i.e. for a given set A and a set X of variables one considers subsets C of OM(A;X) which are <u>closed with respect to superposition</u> (of the second kind): if $\{h_l \mid l \in L\} \subseteq C_M := \{h \in C \mid h \text{ has arity } M\}$ and $f \in C_L$, then $(h_l \mid l \in L)\cdot f \in C$. And it is well known that for any total algebra $\underline{A}$ the set $APO2(\underline{A};X)$ of all algebraic operations of $\underline{A}$ is a clone (see 15.2.13), and if a clone for A contains all M-ary projections for all **d***-small subsets of X, then it is the clone of all algebraic operations (of the second kind) on a suitable total (homogeneous) algebraic structure on A.

The observations at the end of subsection 15.2 show that the same is not true for partial algebras. We have already realized in other situations that the term operations (which are by definition the same as the algebraic operations of the second kind) are not sufficient to describe everything in connection with partial algebras, and we had to introduce conditioned term operations. We shall see that these conditioned term operations -- which also arise by category theoretical investigations, see A.OBTUŁOWICZ [O 81] -- form clones, namely those generated by the algebraic partial operations of the second kind with respect to superposition (of the second kind).

In order to get a better feeling for this third kind of algebraic partial operations we start with a seemingly weaker definition and show that we indeed get clones.

15.3.2 DEFINITION. Let $\underline{A}$ be any partial algebra and M any set. Then we define the set $APO3^M(\underline{A};\underline{S})$ of all <u>M-ary algebraic partial operations of the third kind</u> to be the subset of $PO^M(A;S)$ which is generated by $APO^M(\underline{A};\underline{S})$ with respect to superposition of the second kind, where we allow "at the top" only algebraic partial operations of the second kind, i.e. we take $APO2(\underline{A};X;\underline{S})$ (for the usual set X of variables) as set of fundamental operations of a partial algebra $\underline{A}^*$ (and define $\underline{S}^*$ in a similar way) and define $APO3^M(\underline{A};\underline{S})$ as $APO2^M(\underline{A}^*;\underline{S}^*)$.

$APO3(\underline{A};X;\underline{S}) := \bigcup(APO3^M(\underline{A};\underline{S}) \mid M \subseteq X,\ \#M < \mathbf{d}^{\&})$ designates the set of all algebraic partial operations of the third kind with $\mathbf{d}^{\&}$-small sets of variables in X.

By $\underline{APO3}^M(\underline{A};\underline{S})$ respectively $\underline{APO3}(\underline{A};X;\underline{S})$ we denote the relative subalgebras of $\underline{PO}^M(\underline{A};\underline{S})$ respectively $\underline{PO}(\underline{A};X;\underline{S})$.

Since all fundamental operations $\phi^{\underline{A}}$ are algebraic partial operations of the second kind ($\phi^{\underline{A}} = (e_k{}^{K_\phi} \mid k \in K_\phi)\cdot\phi^{\underline{A}}$), we get

15.3.3 LEMMA. For each $\underline{A} \in \mathbf{Alg}(\underline{S})$ and for each set M, $\underline{APO3}^M(\underline{A};\underline{S})$ is a (total) subalgebra of (the total algebra) $\underline{PO}^M(\underline{A};\underline{S})$. ■■■

The following characterization of algebraic partial operations of the third kind provides more insight into their nature and relates them to conditioned terms:

15.3.4 PROPOSITION. Let $\underline{A}$ be any partial algebra, M any set. Then $h \in APO3^M(\underline{A};\underline{S})$ iff there exist a $\mathbf{d}^{\&}$-small set L, an $l' \in L$ and a sequence $\underline{f} := (f_l \mid l \in L) \in (APO^M(\underline{A};\underline{S}))^L$ such that $h = \underline{f}\cdot e_{l'}{}^L$.

PROOF (by twofold algebraic induction). If already $h \in APO^M(\underline{A};\underline{S})$, then we may choose $L = \{1'\} = 1$, $\underline{f} = (h)$: $h = (h)e_0{}^1$. Thus assume that $h = (g_n|n \in N)\cdot h^*$, where $h^* \in APO^N(\underline{A};\underline{S})$ ($N \subseteq X$, $d^{\&}$-small), and that for each g_n we already have a representation $g_n =: (g_{nl}|l \in L_n)\cdot e_{l'_n}{}^{L_n}$ of the desired kind. If $h^* = e_{n'}{}^N$, then we define $L := \bigcup(\{n\} \times L_n|n \in N)$ and get

$h = ((g_{nl}|l \in L_n)\cdot e_{l'_n}{}^{L_n}|n \in N)\cdot e_{n'}{}^N = (g_{nl}|(n,l) \in L)\cdot e_{(n',l'_{n'})}{}^L$,

i.e. h has the desired representation.

Next assume that $h^* = (f_k|k \in K_\phi)\cdot\phi^{\underline{A}}$ and that $(g_n|n \in N)\cdot f_k =: (f_{kl}|l \in L_k)\cdot e_{l'_k}{}^{L_k} =: f_k{}^*$ is of the desired form for each $k \in K_\phi$. Then we have for each $\underline{a} \in A^M$ and for each $b \in A$ that

$\underline{a}h = b$ iff $\underline{a} \in \bigcap(\text{dom } f_{kl}|l \in L_k , k \in K_\phi)$, $\underline{b} := (\underline{a}f_k{}^*|k \in K_\phi) \in \text{dom } \phi^{\underline{A}}$ and $b = \underline{b}\phi^{\underline{A}}$,

iff $\underline{a} \in \bigcap(\text{dom } f_{kl}|l \in L_k , k \in K_\phi)$, $\underline{b}' := (\underline{a}f_{kl'_k}|k \in K_\phi) \in \text{dom } \phi^{\underline{A}}$ and $b = \underline{b}\phi^{\underline{A}}$;

however, $h' := (f_{kl'_k}|k \in K_\phi)\cdot\phi^{\underline{A}} \in APO^M(\underline{A};\underline{S})$ because of all the intermediate assumptions. Therefore we may define for some new index l': $L := \{l'\} \cup \bigcup(\{k\} \times L_k|k \in K_\phi)$ and $f_{l'} := h'$, $f_{(k,l)} := f_{kl}$ ($k \in K_\phi$, $l \in L_k$). Then $h = (f_l|l \in L)\cdot e_{l'}{}^L$ is of the desired form. ■

Conversely, if h is of the above form, then it follows from the definition of $APO3(\underline{A};X;\underline{S})$ that h belongs to this set. ■■■

15.3.5 REMARK. 15.3.4 shows that the graph of each algebraic partial operation of the third kind is a subset of the graph of an algebraic partial operation of the second kind (i.e. term operation), while its domain is just the intersection of the domains of algebraic operations of the second kind (i.e. term operations). Since each algebraic partial operation of the second kind is a term operation, we get for algebraic partial operations of the third kind the following representation:

Let $h \in APO3^M(\underline{A};\underline{S})$, i.e. $h = (f_l|l \in L)\cdot e_{l'}{}^L$; since each f_l is of the second kind, we have $f_l = (M:t_l)^{\underline{A}}$ for some $t_l \in F(M,\mathbf{TAlg}(\underline{S}))$, and therefore $h = (M:t_{l'}|\{t_l|l \in L\})^{\underline{A}}$ is a conditioned term operation. Conversely it is easily seen that each conditioned term operation, where the condition is $d^{\&}$-small, yields an algebraic partial operation of the third kind. Thus we get:

15.3.6 COROLLARY. For each partial algebra $\underline{A}$ and the usual set X of variables the sets $APO3(\underline{A};X;\underline{S})$ of all algebraic partial operations of the third kind and the set of all conditioned term operations on $\underline{A}$, for which the restricting set of terms is $d^{\&}$-small, are identical. ■■■

15.3.6 PROPOSITION: For each partial algebra $\underline{A}$ and each set M the sets $APO3^M(\underline{A};\underline{S})$ and $APO3(\underline{A};X;\underline{S})$ are closed with respect to superposition of the second kind.

PROOF. Using 15.3.4, let $g = (f_n|n \in N)\cdot e_{n'}{}^N \in APO3^L(\underline{A};\underline{S})$, and, for each $l \in L$, $g_l = (f_{li}|i \in I_l)\cdot e_{i'_l}{}^{I_l} \in APO3^M(\underline{A};\underline{S})$. Then for each $\underline{a} \in A^M$ and $b \in A$ we have

$\underline{a}\,(g_l|l \in L)\cdot g = b$ iff $\underline{a} \in \bigcap(\text{dom } f_{l''i}, \text{dom}(f_{li'_l}|l \in L)\cdot f_n|l'' \in L , i \in I_{l''} , n \in N)$ and $(\underline{a}f_{li'_l}|l \in L)\cdot f_{n'} = b$; and an argument similar to the one at the end of the proof of 15.3.4 shows that $(g_l | l \in L) \cdot g \in APO^M(\underline{A};\underline{S})$. ∎∎∎

15.3.7 REMARK. This shows that indeed the algebraic partial operations of the third kind (and this means: the $d^{\&}$-small conditioned term operations) form a **clone** of partial operations. We cannot go into more details here, however we refer in particular to the quoted papers of A.OBTUŁOWICZ.

15.3.8 REMARK. We have seen that the set of algebraic partial operations of the second kind is not closed with respect to superposition; and if we want to apply arbitrary superposition (of the second kind), we are led to algebraic partial operations of the third kind which can be considered as special "domain restricted" algebraic partial operations of the second kind. However, when we remember that each algebraic partial operation of the second kind stems from some term, we may avoid this extension of $APO2(\underline{A};X;\underline{S})$ to $APO3(\underline{A};X;\underline{S})$ in connection with superposition: if we have to superpose algebraic partial operations of the second kind -- as term operations! --, we superpose the associated terms and then consider the resulting term operation on $\underline{A}$ -- the graph of the resulting algebraic partial operation (of the second kind) then contains the graph of the algebraic partial operation of the third kind which we would get by the usual superposition. One should only observe that this requires the choice of a specific term for each involved algebraic partial operation, but -- as subsection 15.5 will show -- different choices will lead to the same results.

We mention this effect, since it has an analogue in some programming languages which allow to pass the names of functions as parameters to another function or subroutine (procedure). We have detected this possibility e.g. in some implementations of the programming language PASCAL, for instance in PASCAL MT+ (from DIGITAL RESEARCH for personal computers). The short programs TestVal and TestParm below show this effect:

```
PROGRAM TestVal (INPUT,OUTPUT);
VAR Argument : REAL;
FUNCTION Square (Arg1 : REAL) : REAL;
  BEGIN                                   (* Square *)
    Square := Arg1 * Arg1
  END;                                    (* Square *)
FUNCTION Inverse (Arg2 : REAL) : REAL;
  BEGIN                                   (* Inverse *)
    Inverse := 1 / Arg2
  END;                                    (* Inverse *)
FUNCTION Test1 (V1, V2 : REAL) : REAL;
  BEGIN                                   (* Test1 *)
    Test1 := V1
  END;                                    (* Test1 *)
BEGIN                                     (* TestVal *)
  WRITELN('Enter any real number: ');
  READLN(Argument);
  WRITELN;
  WRITELN('TestVal(',Argument,') = ',Test1(Square(Argument),Inverse(Argument)))
END.                                      (* TestVal *)

PROGRAM TestPar(INPUT,OUTPUT);
VAR Argument:REAL;
FUNCTION Square(Arg1:REAL):REAL;
  BEGIN                                   (* Square *)
    Square:= Arg1 * Arg1;
```

```
  END;                                   (* Square *)
FUNCTION Inverse(Arg2:REAL):REAL;
  BEGIN                                  (* Inverse *)
    Inverse:= 1 / Arg2;
  END;                                   (* Inverse *)
FUNCTION Test2(FUNCTION F1(X:REAL):REAL;FUNCTION F2(U:REAL):REAL):REAL;
  BEGIN                                  (* Test2 *)
    Test2:= F1(Argument);
  END;                                   (* Test2 *)
BEGIN                                    (* TestPar *)
  WRITELN('Enter any real number: ');
  READLN(Argument);
  WRITELN;
  WRITELN('TestPar(',Argument,') = ',Test2(Square,Inverse));
END.                                     (* TestPar *)
```

TestVal as well as TestPar are unary algebraic partial operations on $\underline{R}$:= (REALS;Square,Inverse) , where REALS is the set of real numbers accepted by the PASCAL compiler under consideration. The algebraic partial operation of the third kind under consideration is
(Square,Inverse)$\cdot e_0^2$, i.e. substitution into a projection.
It is evaluated by TestVal (i.e. by Test1) in the usual way of superposition of the second kind. Thus, entering 0 (zero) in TestVal yields an error message, since division by zero is detected in evaluating Inverse(0). TestPar corresponds to the formal evaluation using the terms, therefore TestPar yields for the argument 0 the value Square(0), i.e. 0. For all arguments different from 0 TestVal and TestPar yield the same value.
Usually the implementations of the two kinds of passage of variables as discussed above have so-called side-effects which we do not discuss here.
We would like to mention that similar effects as those described in connection with TestPar occur in programs, which define some function by using "IF...THEN...ELSE..." in connection with the values of partial functions.

15.4 Algebraic total operations

Before we study in some more detail the algebras of algebraic total M-ary operations of a given partial algebra $\underline{A}$ let us recall from 5.1.4 that a subset M of a partial algebra $\underline{A}$ is called $\underline{B}$-independent (= $\{\underline{B}\}$-independent), if each mapping from M into B can be extended to a homomorphism from $\underline{C}_A M$ into $\underline{B}$. Closely related to results in section 6.2 is the following one which we shall need below:

15.4.1 LEMMA. Let $\underline{A}$, $\underline{B}$ be similar partial algebras, M a $\underline{B}$-independent subset of $\underline{A}$ and $N \subseteq M$. If one of the following three conditions is satisfied, then N is $\underline{B}$-independent in $\underline{A}$, too:
(i) There exists an ($\underline{S}$-) mapping from M into B.
(ii) There is no ($\underline{S}$-) mapping from N into B.
(iii) $[N]v_A = [M]v_A$.

PROOF. If (ii) is satisfied, then the statement is trivial. If (i) or (iii) is satisfied, and if $f:N \to B$ is any mapping, then f can be extended to a mapping $g:M \to B$, $g|N = f$. If $\overline{g}:\underline{C}_A M \to \underline{B}$ extends g, then its restriction to $\underline{C}_A N$ is the desired homomorphic extension of f. ■■■

15.4.2. It is easy to realize that the conditions in Lemma 15.4.1 are necessary as long as empty phyla are allowed; in particular, in the homogeneous case the conditions of 15.4.1 can only be violated when $N = B = \emptyset$, but $M \neq \emptyset$.

Another lemma which we shall need below is the following one:

15.4.3 LEMMA. Let $\underline{A}$ be any partial algebra, $\underline{B}$ a (relatively) normal (see 5.3.2 and 9.2.7ff) relative subalgebra of $\underline{A}$, and let $M \subseteq B$ be a generating subset of $\underline{A}$. Then M also generates $\underline{B}$.

PROOF. Let $C := C_{\underline{B}}M$, then $C \subseteq B$. We prove $B \subseteq C$ by algebraic induction on A: Obviously $M \subseteq B \cap C$. Assume that $b = \underline{a}\phi^{\underline{A}} \in B$, and that $(k)\underline{a} \in B$ implies $(k)\underline{a} \in C$ for each $k \in K_\phi$. Since $\underline{B}$ is relatively normal in $\underline{A}$, and since $b \in B$, we have $\underline{a} \in B^{K_\phi}$; thus $\underline{a} \in C^{K_\phi}$, and finally $b = \underline{a}\phi^{\underline{A}} = \underline{a}\phi^{\underline{B}} \in C$ by the definition of C. ■

We now return to operations on $\underline{A}$:

15.4.4 LEMMA. For any partial algebra $\underline{A}$ and for any set M the relative subalgebra of $\underline{PO}^M(\underline{A};\underline{S})$ of all total M-ary operations on $\underline{A}$, i.e. $\underline{O}^M(\underline{A};\underline{S})$, carries the induced structure of the power $\underline{D} := \underline{A}^{(A^M)}$; and it is a normal relative subalgebra of $\underline{PO}^M(\underline{A};\underline{S})$.

PROOF. Let $\phi \in \Omega$, $\underline{f} := (f_k \mid k \in K_\phi) \in D^{K_\phi}$, $g \in D$; then $\underline{f}\phi^{\underline{D}} = g$ iff for every $\underline{a} \in A^M$: $\underline{a}[\underline{f}\phi^{\underline{D}}] = \underline{a}g$, iff $(\underline{a}f_k \mid k \in K_\phi)\phi^{\underline{A}} = \underline{a}g$, i.e. iff $g = \underline{f}\cdot\phi^{\underline{A}}$. -- If $\{g_k \mid k \in K_\phi\} \subseteq PO^M(\underline{A};\underline{S})$, then $\mathrm{dom}[(g_k \mid k \in K_\phi)\cdot\phi^{\underline{A}}] \subseteq \bigcap(\mathrm{dom}\ g_k \mid k \in K_\phi)$. I.e. if at least one g_k is not total, then $(g_k \mid k \in K_\phi)\cdot\phi^{\underline{A}}$ cannot be total. This shows that $\underline{O}^M(\underline{A};\underline{S})$ is normal in $\underline{PO}^M(\underline{A};\underline{S})$. ■■■

15.4.5 DEFINITION. For every set M and for every partial algebra $\underline{A}$ we denote by

$$F^M(\underline{A};\underline{S}) := APO^M(\underline{A};\underline{S}) \cap O^M(\underline{A};\underline{S})$$

the set of all <u>algebraic total M-ary operations</u> on $\underline{A}$ (sometimes also called in the literature the set of "full" M-ary algebraic operations), and $\underline{F}^M(\underline{A};\underline{S})$ designates the relative subalgebra of $\underline{APO}^M(\underline{A};\underline{S})$ on $F^M(\underline{A};\underline{S})$.

15.4.6 PROPOSITION. Let $\underline{A}$ be any partial algebra, M any set. Then $\underline{F}^M(\underline{A};\underline{S})$ is a normal relative subalgebra of $\underline{APO}^M(\underline{A};\underline{S})$ freely and $\underline{A}$-freely generated by $E^M(A;S)$. Moreover, $\underline{F}^M(\underline{A};\underline{S})$ is a subalgebra of $\underline{O}^M(\underline{A};\underline{S})$.

PROOF. Obviously $E^M(A;S) \subseteq F^M(\underline{A};\underline{S})$. That $\underline{F}^M(\underline{A};\underline{S})$ is normal in $\underline{APO}^M(\underline{A};\underline{S})$ is shown in the same way as the corresponding result in 15.4.4. -- Now, 15.4.3 implies that $E^M(A;S)$ generates $\underline{F}^M(\underline{A};\underline{S})$, and from 15.4.4 and 15.4.5 we infer that $F^M(\underline{A};\underline{S})$ has to be a closed subset of $\underline{O}^M(\underline{A};\underline{S})$. Thus $\underline{F}^M(\underline{A};\underline{S}) \in \mathbf{H}_w\mathbf{S}_c\mathbf{P}(\{\underline{A}\})$, and therefore 5.1.6.(ii) implies that it suffices to show that $E^M(A;S)$ is $\underline{A}$-independent in $\underline{F}^M(\underline{A};\underline{S})$: Let $f: E^M(A;S) \to A$ be any mapping; then let $f': M \to A$, $m \mapsto (e_m{}^M)f$ $(m \in M)$, consider the projection homomorphism $pr_{f'}: \underline{A}^{(A^M)} \to \underline{A}$, and let $\overline{f} := pr_{f'} \mid F^M(\underline{A};\underline{S})$. We claim that $\overline{f} \mid E^M(A;S) = f$:
If $g \in A^{(A^M)}$ then $(g)pr_{f'} = (f')g$, and $(e_m{}^M)pr_{f'} = (f')e_m{}^M = (m)f' = (e_m{}^M)f$. ■■■

15.4.7 REMARK. Let $\underline{A}$ be any partial algebra, and let M be any set which does not allow any mapping into A -- this is equivalent to the fact that $[M]v_M \not\subseteq [A]v_A$. Then all M-ary projections on A -- and therefore also all algebraic partial or total operations -- will have empty domain and can only be distinguished by their output sort. In particular all elements of $APO^M(\underline{A};\underline{S})$ are (many-sorted) total. All phyla of $E^M(A;S)$ and of $APO^M(\underline{A};\underline{S})$ have at most one element.

It should also be observed that for those $s \in S$ for which A has a one-element phylum and M_s is non-empty (but no further restriction on M), the corresponding phylum of $E^M(A;S)$ (respectively, of $F^M(\underline{A};\underline{S})$) also consists of one element only, since we only have constant mappings onto this phylum of A.

In both situations the mapping $m \mapsto e_m{}^M$ $(m \in M)$ is injective only in the case when each phylum of M has at most one element.

15.4.8 PROPOSITION. For every partial algebra $\underline{A}$ and for any sets M and N one has:

(i) $\underline{F}^M(\underline{A};\underline{S}) \in \mathbf{S_C P}(\{\underline{A}\})$, $\underline{F}^M(\underline{A};\underline{S}) \cong \underline{F}(M,\mathbf{S_C P}(\{\underline{A}\}))$.

(ii) If one of the following conditions is satisfied, then $\underline{F}^M(\underline{F}^N(\underline{A};\underline{S});\underline{S}) \cong \underline{F}^M(\underline{A};\underline{S})$:

- (a) $M \subseteq N$ and $[M]v_M = [N]v_N$.
- (b) $M \subseteq N$ and $[N]v_N \subseteq [A]v_A$.
- (c) $\mathbf{H_w S_C P}(\{\underline{F}^N(\underline{A};\underline{S})\}) = \mathbf{H_w S_C P}(\{\underline{A}\})$.

PROOF. The proof of (i) is contained in the one of 15.4.6. Therefore it remains only to prove (ii):

Let us assume (a) or (b): If there is no mapping from M into A, then there is also no mapping from N into A, and the statement easily follows from 15.4.7. Thus let us assume that we have at least one mapping from N into A -- in case (b) this is always guaranteed --, and then we also have at least one mapping from M into A as well as into $F^N(\underline{A};\underline{S})$.

First let us observe that $E^M(F^N(\underline{A};\underline{S});S)$ is $\mathbf{H_w S_C P}(\{\underline{F}^N(\underline{A};\underline{S})\})$-independent in $\underline{F}^M(\underline{F}^N(\underline{A};\underline{S});\underline{S})$. Therefore the mapping $f:E^M(F^N(\underline{A};\underline{S});S) \to \underline{F}^N(\underline{A};\underline{S})$, $e_m{}^M(F^N(\underline{A};\underline{S})) \mapsto e_m{}^N(\underline{A})$ $(m \in M)$ -- observe that we have to distinguish the projections on different sets and that our assumptions (by 15.4.7) indeed yield a mapping -- extends to a homomorphism $\overline{f}:\underline{F}^M(\underline{F}^N(\underline{A};\underline{S});\underline{S}) \to \underline{F}^N(\underline{A};\underline{S})$.

Since $\mathbf{H_w S_C P}(\{\underline{F}^N(\underline{A};\underline{S})\}) \subseteq \mathbf{H_w S_C P}(\{\underline{A}\})$, the mapping $g:E^N(A;S) \to E^M(F^N(\underline{A};\underline{S});S)$, defined by $(e_n{}^N(A))g := e_n{}^M(F^N(\underline{A};\underline{S});S)$, if $n \in M$, and $(e_n{}^N(A))g := e_{m'}{}^M(F^N(\underline{A};\underline{S});S)$ for some fixed $m' \in M$, if $n \notin M$, extends to a homomorphism $\overline{g}:\underline{F}^N(\underline{A};\underline{S}) \to \underline{F}^M(\underline{F}^N(\underline{A};\underline{S});\underline{S})$. Obviously fg is the identity mapping on $E^M(F^N(\underline{A};\underline{S});S)$, and therefore $\overline{fg}$ is the identical isomorphism of $\underline{F}^M(\underline{F}^N(\underline{A};\underline{S});\underline{S})$. This shows (see 11.4) that $\overline{f}$ is a section -- and therefore closed and injective. Since we have a mapping from N -- and therefore from $E^N(A;S)$ -- into A, the assumptions of 15.4.1 are satisfied for $\{e_m{}^N(A;S) \mid m \in M\} \subseteq E^N(A;S)$, i.e. $\{e_m{}^N(A;S) \mid m \in M\}$ is $\underline{A}$-independent in $\underline{F}^N(\underline{A};\underline{S})$, and therefore $\underline{F}^M(\underline{F}^N(\underline{A};\underline{S});\underline{S}) \cong \underline{C}_{\underline{F}^N(\underline{A};\underline{S})}\{e_m{}^N(A;S) \mid m \in M\} \cong \underline{F}(M,\mathbf{S_C P}(\{\underline{A}\})) \cong \underline{F}^M(\underline{A};\underline{S})$.

Finally assume that (c) is true: Then obviously $\mathrm{ind}_{\underline{F}^M(\underline{F}^N(\underline{A};\underline{S});\underline{S})} E^M(F^N(\underline{A};\underline{S});S) \supseteq$ $\supseteq \mathbf{H_w S_C P}(\{\underline{A}\})$. Since $\underline{F}^M(\underline{F}^N(\underline{A};\underline{S});\underline{S})$ belongs to the E-variety $\mathbf{H_w S_C P}(\{\underline{A}\})$, we again get $\underline{F}^M(\underline{F}^N(\underline{A};\underline{S});\underline{S}) \cong \underline{F}(M,\mathbf{H_w S_C P}(\{\underline{A}\})) \cong \underline{F}^M(\underline{A};\underline{S})$. ■■■

The following result will be useful in the next subsection:

15.4.9 LEMMA. Let $\underline{A}$ be any partial algebra, M, N any sets.

(i) If $\underline{g}:=(g_n|n \in N) \in (F^M(\underline{A};\underline{S}))^N$ and $f \in APO^N(\underline{A};\underline{S})$, then $\underline{g}\cdot f \in APO^M(\underline{A};\underline{S})$.

(ii) Each endomorphism of $\underline{F}^M(\underline{A};\underline{S})$ is extendable to an endomorphism of $\underline{APO}^M(\underline{A};\underline{S})$.

PROOF. Ad (i): We prove (i) by algebraic induction on f: If $f = e_{n'}{}^N$, then $\underline{g}\cdot f = g_{n'}$ (since totality of all the g_n does not impose any domain restriction). Thus, assume that $f = (f_k|k \in K_\phi)\cdot\phi^{\underline{A}}$ and that $\underline{g}\cdot f_k \in APO^N(\underline{A};\underline{S})$ for all $k \in K_\phi$. Then it is obvious that also $\underline{g}\cdot f = (\underline{g}\cdot f_k|k \in K_\phi)\cdot\phi^{\underline{A}}$ belongs to $APO^M(\underline{A};\underline{S})$. ■

Ad (ii): We define for $f \in End(\underline{F}^M(\underline{A};\underline{S}))$ and $h \in APO^M(\underline{A};\underline{S})$: $(h)\overline{f} := ((e_m{}^M)f|m \in M)\cdot h$. By (i) it is easily realized that $\overline{f}$ is indeed a mapping $\overline{f}:APO^M(\underline{A};\underline{S}) \to APO^M(\underline{A};\underline{S})$. We leave the details of a proof that $\overline{f}$ is an endomorphism of $\underline{APO}^M(\underline{A};\underline{S})$ as an exercise. ■■■

For further results on free partial algebras, e.g. on $\underline{F}^M(\underline{A};\underline{S})$, see section 17.

15.5 More about algebraic partial operations and terms; algebraic partial functions

15.5.1. We already know by Definition 15.2.2 that each M-ary algebraic partial operation, say h, of the second kind of some partial algebra $\underline{A}$ is the homomorphic value with respect to $eval_M:\underline{F}(M,\mathbf{TAlg}(\underline{S})) \to \underline{PO}^M(\underline{A};\underline{S})$ -- induced by mapping $m \in M$ onto the m-th M-ary projection $e_m{}^M(\underline{A})$ -- of some term t_h with variables in M: $h = (M:t_h)^{\underline{A}}$. And in 15.3.6 we have seen that each M-ary algebraic partial operation of the third kind corresponds in a similar way to a conditioned term operation. Now, in general, $eval_M$ is not injective, and here we briefly want to investigate its kernel -- it should be noticed that $eval_M$ heavily depends on the chosen partial algebra $\underline{A}$, although we do not always indicate it in the notation.

15.5.2 PROPOSITION. Let $\underline{A}$ be any partial algebra, M a set, $t,t' \in F(M,\mathbf{TAlg}(\underline{S}))$. Then $(M:t)^{\underline{A}} = (M:t')^{\underline{A}}$ (i.e. $(t,t') \in ker(eval_M)$) iff $\underline{A} \models ((M:t \overset{e}{=} t \vee t' \overset{e}{=} t' \Rightarrow t \overset{e}{=} t')$ [iff $\underline{A} \models ((M:(t \overset{e}{=} t \Rightarrow t \overset{e}{=} t') \wedge (t' \overset{e}{=} t' \Rightarrow t \overset{e}{=} t'))]$.

15.5.3. Observe that the axioms in 15.5.2 correspond to the concept of "strong validity" of equations in partial algebras as mentioned in 8.1.3.(ii).(b).

PROOF. If $(M:t)^{\underline{A}} = (M:t')^{\underline{A}}$, then, for all $\underline{a} \in A^M$ and $b \in A$, $\underline{a}(M:t)^{\underline{A}} = b$ iff $\underline{a}(M:t')^{\underline{A}} = b$, but this just means that in $\underline{A}$ the above formula is valid. Conversely it is as easily seen that for $t,t' \in F(M,\mathbf{TAlg}(\underline{S}))$ validity in $\underline{A}$ of the above formula implies $(M:t)^{\underline{A}} = (M:t')^{\underline{A}}$. ■■■

15.5.4 PROBLEMS. We do not know which congruences of $\underline{F}(M,\mathbf{TAlg}(\underline{S}))$ can occur as kernels of canonical homomorphisms $eval_M{}^{\underline{A}}$ onto algebras of algebraic partial operations. The interested reader will find some information in this direction in J.SŁOMIŃSKI [SŁ 68] in connection with the concept of "validity." Our Lemma 15.4.9 provides an idea that not every congruence R on $\underline{F}(M,\mathbf{TAlg}(\underline{S}))$ can occur as kernel of some $eval_M{}^{\underline{A}}$, since every mapping $f:M \to M$ induces an endomorphism $\overline{f}:\underline{F}(M,\mathbf{TAlg}(\underline{S})) \to \underline{F}(M,\mathbf{TAlg}(\underline{S}))$ which has to satisfy $[R]f^{(2)} \subseteq R$, but it need not be fully invariant either (cf. section 6.3).

We also do not know whether it has already been investigated which other properties the set of all kernels of evaluation morphisms $eval_M^{\underline{A}}$ for fixed M and arbitrary $\underline{A}$ has. Is it closed with respect to intersections? Is it **d***-algebraic? etc.
Some of the results in the next subsection will also provide some more information about these questions.

15.5.5 PROPOSITION. Let $\underline{A}$ be any partial algebra, $M \subseteq A$, $a \in \mathbf{C}_{\underline{A}}M$. Then there exists $h \in APO^M(\underline{A};\underline{S})$ such that $a = (id_M)h$ -- i.e. every $a \in \mathbf{C}_{\underline{A}}M$ is the value of a suitable M-ary algebraic partial operation applied to the identity sequence of M.

PROOF. $i := id_M : M \to M$ extends to a closed homomorphism $\tilde{i}:\underline{dom}\ \tilde{i} \to \underline{A}$ (out of $\underline{F}(M,\mathbf{TAlg}(\underline{S}))$) -- onto $\mathbf{C}_{\underline{A}}M$. Therefore there is $t \in dom\ \tilde{i}$ such that $(t)\tilde{i} = (i)(M:t)^{\underline{A}} =$
$= (id_M)((t)eval_M^{\underline{A}})$, $(t)eval_M^{\underline{A}} \in APO^M(\underline{A};\underline{S})$. ■■■

In the same way one proves

15.5.6 COROLLARY 1. For each sequence $\underline{a} \in A^M$ and for each element b of the closed subset $\mathbf{C}_{\underline{A}}([M]\underline{a})$ generated by the image of $\underline{a}$ there is an M-ary algebraic partial operation $h \in APO^M(\underline{A};\underline{S})$ with value b at $\underline{a}$: $b = \underline{a}h$. ■■■

15.5.7 COROLLARY 2. Each mapping $f:E^M(A;S) \to \underline{A}$ has a closed homomorphic extension out of $APO^M(\underline{A};\underline{S})$: $f^*:\underline{dom}\ f^* \to \underline{A}$. In particular, if $g:M \to \underline{A}$ is defined by $mg := (e_m{}^M)f$, then $dom\ f^* = [dom\ \tilde{g}]eval_M^{\underline{A}}$, and $\underline{dom}\ f^*$ is an $E^M(A;S)$-generated relative subalgebra of $\underline{APO}^M(\underline{A};\underline{S})$. Moreover we have, for each $h \in APO^M(\underline{A};\underline{S})$: $h \in dom\ f^*$ iff $g \in dom\ h$.

PROOF. We have $g \in dom(M:t)^{\underline{A}}$ iff $t \in dom\ \tilde{g}$, iff $g \in dom((t)eval_M^{\underline{A}})$. From these equivalences all the statements are easily derived. ■■■

15.5.8. In the notation of POYTHRESS [Po 73] 15.5.6 says that $E^M(A;S)$ is p-free for $\underline{A}$ -- in the notation of KERKHOFF [Ke 70] $E^M(A;S)$ is called **conform free** for $\underline{A}$ in $\underline{APO}^M(\underline{A};\underline{S})$.

Finally we would like to mention another concept closely related to the one of algebraic partial operations: the one of **algebraic partial functions** or **polynomial functions**:

15.5.9 DEFINITION. Let M be any set, $\underline{A}$ any partial algebra. An M-ary partial operation $p \in PO^M(A;S)$ is called an <u>M-ary algebraic (partial) function</u> (also <u>M-ary polynomial function</u>) on $\underline{A}$, if there exist a set N containing M and an algebraic partial operation h, such that p is obtained from h by replacing each argument belonging to some element of $N \setminus M$ by some fixed element of A, i.e. iff there exists a mapping $\underline{b}:(N \setminus M) \to A$ such that for each $\underline{a} \in A^M$ -- and for $\underline{a}\hat{\ }\underline{b}$ denoting the sequence whose graph is obtained as the disjoint union of the graphs of $\underline{a}$ and $\underline{b}$ -- $\underline{a}p := (\underline{a}\hat{\ }\underline{b})h$.

15.5.10. The set of all algebraic partial functions on a partial algebra $\underline{A}$ derivable from algebraic partial operations with $\mathbf{d}^{\&}$-small arities in X can be considered as the closure of the set of all total projections with $\mathbf{d}^{\&}$-small arities in X with respect to the structure on $\underline{PO}(\underline{A}_A;X;\underline{S}_S)$, where $\underline{A}_A := (A,(\phi^{\underline{A}})_{\phi\in\Omega} \cup (\phi_a)_{a\in A})$, and each ϕ_a is a (non-empty) nullary constant with value a.

This shows how all the results on algebraic partial operations can be extended to algebraic partial functions. This concept is useful in particular in connection with the investigation of congruence relations (see section 19.1). They also arise in connection with the evaluation of global partial polynomials in $\underline{A}$, as can easily be realized.

15.6 Algebraic partial operations and homomorphisms

15.6.1. From 5.7.5 we know that for each M-ary term (M:t) and each homomorphism $f:\underline{A}\to\underline{B}$ one has that $\underline{a}(M:t)^{\underline{A}} = b$ implies $\underline{a}f(M:t)^{\underline{B}} = bf$ for each $\underline{a}\in A^M$ and $b\in A$, i.e. that homomorphisms are compatible with term operations. Thus, if $(M:t)^{\underline{A}} = h \in APO^M(\underline{A};\underline{S})$, $(M:t)^{\underline{B}} = h' \in APO^M(\underline{B};\underline{S})$, then $\underline{a}h = b$ implies $\underline{a}fh' = bf$ for each $\underline{a}\in A^M$, $b\in A$. But this does not mean that f induces in general a homomorphism, say $f^{\#}$, from $\underline{APO^M}(\underline{A};\underline{S})$ into $\underline{APO^M}(\underline{B};\underline{S})$, since this would be equivalent with the fact:

(*) ker $eval_M^{\underline{A}} \subseteq$ ker $eval_M^{\underline{B}}$, i.e. for any terms $t,t'\in F(M,\mathbf{TAlg}(\underline{S}))$:
$(M:t)^{\underline{A}} = (M:t')^{\underline{A}}$ implies $(M:t)^{\underline{B}} = (M:t')^{\underline{B}}$.

But in general this cannot be expected when f is not surjective and not full. However, even when f is full and surjective this need not be the case as the homomorphism f in Figure 11.1 in 11.6.1 shows: f is full and surjective, but $APO^1(\underline{A}) = \{e_0^{\,1},\phi^{\underline{A}},\emptyset\}$, $APO^1(\underline{B}) = \{e_0^{\,1},\phi^{\underline{B}},\phi^{\underline{B}}\cdot\phi^{\underline{B}},\emptyset\}$, and f cannot induce a homomorphism from $\underline{APO^1}(\underline{A})$ into $APO^1(\underline{B})$, since the identity would go onto the identity, and $f^{\#}$ would have to be surjective as a homomorphism between total algebras.
However we have the following

15.6.2 PROPOSITION. Let $f:\underline{A}\to\underline{B}$ be a **closed** and surjective homomorphism. Then, for each set M, f induces a homomorphism $f^{\#}:\underline{APO^M}(\underline{A};\underline{S})\to\underline{APO^M}(\underline{B};\underline{S})$ such that, for every $t\in F(M,\mathbf{TAlg}(\underline{S}))$, $(M:t)^{\underline{B}} = (M:t)^{\underline{A}}f^{\#}$.

PROOF. We prove 15.6.1.(*): Assume $t,t'\in F(M,\mathbf{TAlg}(\underline{S}))$ such that $(M:t)^{\underline{A}} = (M:t')^{\underline{A}}$. Let $\underline{b}\in dom(M:t)^{\underline{B}}$ be arbitrary. Since f is surjective and closed, there is $\underline{a}\in dom(M:t)^{\underline{A}}$ such that $\underline{a}f = \underline{b}$ (Axiom of Choice). $(M:t)^{\underline{A}} = (M:t')^{\underline{A}}$ now implies that $\underline{a}(M:t)^{\underline{A}} = \underline{a}(M:t')^{\underline{A}}$, and therefore
$\underline{b}(M:t)^{\underline{B}} = (\underline{a}f)(M:t)^{\underline{B}} = (\underline{a}(M:t)^{\underline{A}})f = (\underline{a}(M:t')^{\underline{A}})f = (\underline{a}f)(M:t')^{\underline{B}} = \underline{b}(M:t')^{\underline{B}}$.
Symmetry of the assumptions implies $(M:t)^{\underline{B}} = (M:t')^{\underline{B}}$. Thus 15.6.1.(*) is satisfied, and the Diagram Completion Lemma implies the existence of $f^{\#}:\underline{APO^M}(\underline{A};\underline{S})\to\underline{APO^M}(\underline{B};\underline{S})$ having the required properties. ■■■

In a similar way one proves a corresponding result for algebraic **total** operations, which needs weaker conditions on f, since we do not have to bother with definedness:

15.6.3 PROPOSITION. Let M be any set and $f:\underline{A}\to\underline{B}$ a surjective homomorphism. Then f induces a homomorphism $f^{\&}:\underline{F^M}(\underline{A};\underline{S})\to\underline{F^M}(\underline{B};\underline{S})$ such that, for every $t\in[F^M(\underline{A};\underline{S})](eval_M^{\underline{A}})^{-1}$, $(M:t)^{\underline{B}} = (M:t)^{\underline{A}}f^{\&}$. ■■■

15.6.4. Notice that surjectivity of f is necessary in 15.6.2 as well as in 15.6.3, even in the case of total algebras, since algebraic partial operations which coincide on [A]f may differ outside of this subset of B (example?).

16 Semilattices of (weak) completions

Although we have claimed in the introduction that we want to treat partial algebras in their own right, it cannot be denied that the topic of completions of partial algebras or questions about their weak, full or even normal embeddability into a total algebra of some prespecified kind are very important, in particular in computer science, but also in mathematics in general. If, for instance, in computer science a program is specified which uses partial functions -- i.e. functions which are not everywhere defined or which are "badly defined" e.g. because of the possibility of ending in a loop which they cannot leave -- then anyway the program should stop after some time in one way or another, either yielding a so-called error value or by some implemented or "enforced" time limit (meaning that the program is interrupted from the outside after some time). Each such "ending" of the program could be connected with some new value, thus completing the operation defined by the program, if this is done for all possibilities. In this way the implementation yields a total operation, although it was originally partial with respect to its basic meaning. A neat treatment of the program should distinguish these different kinds of values. These different kinds should not be mistaken for different sorts as occurring in many-sorted algebras - these different sorts just hide some kind of partiality, and a good compiler will tell by an error message, when one has used at some place a wrong sort (often called "data type", e.g. in the programming language PASCAL) of variables, thus completing this many-sorted operation to a one-sorted one. In connection with a many-sorted partial operation (given by some program, or often specified in connection with some data type) also the error-values will have a specific sort (if there are no "error-messages" in connection with the use of wrong sorts (data types)). For the treatment of error-values in connection with the specification of abstract data types by means of many-sorted (total) algebras see the papers of the ADJ-group concerning this topic, in particular [GoTWa 78] or [Go 78].
In pure mathematics we refer to the famous paper of T. EVANS [Ev 51] relating in connection with a given variety, say **K**, of total algebras the solvability of the (full) embeddability problem for a special class of partial algebras to the solution of the word problem for **K**. We do not want to go into details about this here. Another area where the completion of partial algebras plays an important rôle is the question of whether the universal solution of a so-called partial lattice within the variety of all lattices is finite or infinite, and related topics in lattice theory. Another instance is given by the constructions of direct sums (coproducts, often called "free products") within given equational classes of total algebras. Here we want to mention e.g. lattice theory (see e.g. GRÄTZER [G 78]). Also the questions in strong amalgamation theory are actually of the kind concerning whether special partial algebras constructed from the amalgams can be fully embedded into total algebras of the given variety (but we cannot treat this within the scope of this book, either). This shows that also within the theory of total algebras the completion of partial algebras plays an important rôle.
In section 17 we shall also mainly study completions, but this time free completions within E-varieties.

16.1 The semilattice of (weak) minimal completions

The material of this subsection is basically contained in [BSch 67].

16.1.1 DEFINITION. (i): Let $\underline{A}$ be any partial algebra; a total ($\underline{S}$-) algebra $\underline{B}$ is called a weak completion of $\underline{A}$, if $\underline{A}$ is a weak subalgebra of $\underline{B}$, a completion, if $\underline{A}$ is a relative subalgebra; and a normal completion, if $\underline{A}$ is normal in $\underline{B}$ (see 5.3.2). If A generates its (weak) completion $\underline{B}$, then $\underline{B}$ is called a minimal (weak or normal) completion of $\underline{A}$.
(ii): Let $\underline{B}$ and $\underline{C}$ be minimal weak completions of $\underline{A}$ such that id_A induces a (necessarily surjective) homomorphism from $\underline{B}$ onto $\underline{C}$; then we say that $\underline{B}$ is greater than $\underline{C}$ and denote this by

$$\underline{B} >_A \underline{C} \quad \text{or} \quad \underline{B} > \underline{C} \text{ ,}$$

assuming that the reference to $\underline{A}$ is obvious.

16.1.2 REMARKS. (i): Notice that we want this definition to be understood in such a way that the ... completion of any partial $\underline{S}$-algebra is again an $\underline{S}$-algebra and therefore in general not homogeneous. But what we shall say in the following will also apply to "homogeneous completions" of partial $\underline{S}$-algebras, when we just forget about $\underline{S}$.
(ii): A minimal weak completion $\underline{B}$ of $\underline{A}$ is normal iff $id_A : \underline{A} \to \underline{B}$ is relatively normal in the sense of section 9 (see 9.2.7).
(iii): It is easy to realize that ">" is a quasi-order on the class of all minimal weak completions of any given partial algebra $\underline{A}$, and if there holds $\underline{B} > \underline{C}$ as well as $\underline{C} > \underline{B}$, then obviously $\underline{B}$ and $\underline{C}$ are isomorphic over id_A, and we say that they are equivalent and denote this by $\underline{B} \sim \underline{C}$.
It is well known that any quasi-order factored with respect to the induced equivalence relation yields a partial order. This also holds for classes, when one chooses a representative system, and such a representative system is easily provided in our case:
Let $\underline{A}\hat{} := \underline{F}(\underline{A}, \mathbf{TAlg}(\underline{S}))$ be the free completion of $\underline{A}$. Then we know from section 5 that $\underline{A}\hat{}$ is a normal minimal completion of $\underline{A}$, and that every homomorphism from $\underline{A}$ into some total algebra has a homomorphic extension to $\underline{A}\hat{}$. Thus every minimal weak completion, say $\underline{B}$, of $\underline{A}$ corresponds to a unique congruence, say $R_{\underline{B}}$ on $\underline{A}\hat{}$.

As an immediate consequence of the Homomorphism Theorem we get the

16.1.3 PROPOSITION. For any two weak completions $\underline{B}$ and $\underline{C}$ of a partial algebra $\underline{A}$ one has:

(i) $\underline{B} > \underline{C}$ iff $R_{\underline{B}} \subseteq R_{\underline{C}}$.

(ii) $\underline{B} \sim \underline{C}$ iff $R_{\underline{B}} = R_{\underline{C}}$. ∎∎∎

Notice that the quasi-order between weak minimal completions of $\underline{A}$ is just dual to the inclusion between the corresponding congruences.

16.1.4 THEOREM. Let R be an arbitrary congruence on $\underline{A}\hat{}$. Then:

(i) $R = R_{\underline{B}}$ for some weak minimal completion $\underline{B}$ of $\underline{A}$ iff $R \cap (A \times A) = id_A$.

(ii) $R = R_{\underline{B}}$ for some minimal completion $\underline{B}$ of $\underline{A}$ iff

$$R \cap (A \times (A \cup \mathbf{U}\{[A^{K_\phi}]\phi^{\underline{A}\hat{}} \mid \phi \in \Omega\})) = id_A \text{ .}$$

16.1.5. 16.1.4.(ii) can be interpreted as follows: If $ag = xg$, $a \in A$, and

$x \in A \cup \bigcup\{[A^{K_\phi}]\phi^{\underline{A}^\wedge} \mid \phi \in \Omega\} = D_{\underline{A}^\wedge}A$ (where a and x are both of the same sort), then $x = a$, where g is a homomorphism with kernel R.

PROOF of Theorem 16.1.4. Ad (i): $\underline{B}$ is a minimal weak completion of $\underline{A}$ iff it contains $\underline{A}$ as a generating weak subalgebra, i.e. iff the induced restriction to A of the corresponding homomorphism, say g, from $\underline{A}^\wedge$ onto $\underline{B}$ is injective, and this is equivalent to $R_g \cap (A \times A) = id_A$. ∎

Ad (ii): First, consider $R = R_{\underline{B}}$, where $\underline{B}$ is a minimal completion of $\underline{A}$. Assume $a \in A$, $x \in A \cup \bigcup\{[A^{K_\phi}]\phi^{\underline{A}^\wedge} \mid \phi \in \Omega\}$, and $ag = xg$, where $g := g_{\underline{B}}$ is the induced surjection from $\underline{A}^\wedge$ onto $\underline{B}$. If $x \in A$, then $a = x$ by (i), since $\underline{B}$ is also a weak minimal completion of $\underline{A}$. If $x = \underline{a}\phi^{\underline{A}^\wedge}$, where $(k)\underline{a} \in A$ for all $k \in K_\phi$, then we have

$$a = ag = xg = (\underline{a}g)\phi^{\underline{B}} = \underline{a}\phi^{\underline{B}},$$

hence $a = \underline{a}\phi^{\underline{A}}$, since $\underline{A}$ is a relative subalgebra of $\underline{B}$. Therefore $a = \underline{a}\phi^{\underline{A}^\wedge} = x$, since $\underline{A}$ is a relative subalgebra of $\underline{A}^\wedge$.

Conversely, let the congruence R satisfy the condition of 16.1.4.(ii). Then there is a total algebra $\underline{B}'$ with a homomorphism g' from $\underline{A}^\wedge$ onto $\underline{B}'$, which induces R (as its kernel). By hypothesis, the restriction h of g' to A is injective, and we can construct a total algebra $\underline{B}$ which contains $\underline{A}$ at least as a weak relative subalgebra, and which is isomorphic to $\underline{B}'$, say via h', induced by h: $hh' = id_A$ (cf. Appendix, section 23.2).

Then id_A is a homomorphism from $\underline{A}$ into $\underline{B}$, and we claim that $\underline{A}$ is even a relative subalgebra of $\underline{B}$: In fact, assume that $a = \underline{a}\phi^{\underline{B}}$ for some $\underline{a} \in A^{K_\phi}$ and some $\phi \in \Omega$, and consider $x := \underline{a}\phi^{\underline{A}^\wedge} \in A^\wedge$. Then

$$(a)g'h' = a = \underline{a}\phi^{\underline{B}} = (\underline{a}g'h')\phi^{\underline{B}} = xg'h'.$$

Since $\ker g'h' = R$, this implies by hypothesis $a = x$ or $\underline{a}\phi^{\underline{A}^\wedge} = a \in A$. By (FC1) (cf. 5.3.1) we obtain $a = \underline{a}\phi^{\underline{A}}$. Thus $\underline{B}$ is a completion of $\underline{A}$, and even a minimal one, since A generates $\underline{A}^\wedge$ and the homomorphism $g'h'$ is surjective, carrying A onto A. Moreover, $R = \ker g'h' = R_{\underline{B}}$, completing the proof. ∎∎∎

16.1.6 REMARKS. (i): Theorem 16.1.4 provides a comparatively concrete survey on all weak minimal respectively on all minimal completions of a given partial algebra $\underline{A}$. Notice that the set of all inner completions is characterized with respect to 16.1.4.(i) by the additional property that A has a non-empty (and hence a one-element) intersection with each congruence class of the corresponding congruence relation R of $\underline{A}^\wedge$. (This is obvious, since the corresponding surjective homomorphism is onto A.)

(ii): By replacing each (weak) minimal completion $\underline{B}$ of $\underline{A}$ by its corresponding factor structure $\underline{A}^\wedge/R_{\underline{B}}$ we do not get a weak completion but a structure which is canonically isomorphic to it. Thus a representative system for all weak minimal completions of $\underline{A}$ forms a set, which bijectively corresponds to the set of all congruence relations R on $\underline{A}^\wedge$ satisfying $R \cap (A \times A) = id_A$. And this set also contains a representative from each isomorphism class of minimal completions of $\underline{A}$.

Moreover, let $(\underline{B}_j)_{j \in J}$ be any non-empty family of representatives of isomorphism classes of weak minimal completions of $\underline{A}$. Then this family has a supremum in the poset introduced above, namely the isomorphism class of minimal completions corresponding to $R := \bigcap\{R_{\underline{B}_j} \mid j \in J\}$, which obviously also satisfies the condition in 16.1.4.(i). -- Notice that the corresponding weak completion is canonically isomorphic to the subalgebra generated by the image of A in the direct product of the family $(\underline{B}_j | j \in J)$.

(iii): Instead of dealing with a poset of isomorphism types of weak minimal completions of $\underline{A}$ we can switch, because of the preceding results, to the poset of congruence relations of $\underline{A}^{\wedge}$ satisfying the condition in 16.1.4.(i) -- or in 16.1.4.(ii) in the case of minimal completions. Since the intersection of a non-empty family of such congruence relations still satisfies these conditions, they form in each case a conditionally complete meet-semilattice, which has in general - in each case - no greatest element but many maximal elements: In the case of weak minimal completions each inner completion obviously corresponds to a maximal congruence relation, i.e. inner completions are minimal elements in the set of isomorphism types of weak minimal completions. In the case of minimal completions (with full embeddings) one has to add to $\underline{A}$ at least one new element to each phylum which belongs to the output sort of a really partial operation in order to get a full embedding of $\underline{A}$ into such a minimal completion, and all sequences in $A^{K_\phi} \setminus \text{dom}\,\phi^{\underline{A}}$ $(\phi \in \Omega)$ have then to be mapped onto the corresponding new element. Let B be the corresponding new carrier set. There is now no condition for the values of fundamental operations on B for sequences which contain at least one of the new elements except that they have to belong to the appropriate phylum, and it is easily realized that any choice will then yield in the end a minimal element in the poset of isomorphism types of minimal completions of $\underline{A}$, since the corresponding congruence relation cannot be extended without violating condition 16.1.4.(ii) (proof as an exercise!). Thus in general we also have in this subsemilattice of minimal completions more than one minimal element (see [BSch 67] for an example that a minimal element of this subsemilattice of minimal completions of even a homogeneous partial algebra $\underline{A}$ can have more than one additional element).

We thus get the following

16.1.7 PROPOSITION. The representative weak minimal completions respectively minimal completions of a partial algebra $\underline{A}$ constitute in each case a conditionally complete join semilattice with greatest element $\underline{A}^{\wedge}$. In general this is not a lattice. ■■■

16.2 The complete lattice of normal minimal completions

16.2.1 The (heterogeneous) normal one-point completion: The situation becomes much nicer, when we consider normal minimal completions (see 5.3.2). Let us first observe that in this case the ordered set of representatives of minimal normal completions of a given partial algebra, say $\underline{A}$, has (up to isomorphism) exactly one minimal element, namely the so-called <u>(heterogeneous) normal one-point completion</u>, which we shall denote by $\underline{A}^{\circ}$:

For each $s \in S$ we add a new element, say $*_s$, to $A_s := (s)v_A^{-1}$ (such that $*_s \notin A_s$ and $(*_s)v_{A*} := s$ for each $s \in S$). Moreover we define for each sort respecting $\phi \in \Omega$ and for each $\underline{a} \in A^{*K_\phi}$:

$$\underline{a}\phi^{\underline{A}^*} := \begin{cases} \underline{a}\phi^{\underline{A}}, & \text{if } \underline{a} \in \text{dom}\,\phi^{\underline{A}}, \\ *_s, & \text{else, where } s \text{ is the output sort of } \phi. \end{cases}$$

$\underline{A}°$ is then the subalgebra of $\underline{A}*$ generated by A (i.e. new elements are added to some phylum only when they are absolutely necessary for the completion).
It is easy to realize that $\underline{A}°$ is in fact a normal minimal completion of $\underline{A}$, and that it corresponds to the congruence relation

$$R:= R_{\underline{A}°}:= id_A \cup \bigcup\{((A^{\wedge})_s \setminus A_s)^2 | s \in S\}$$

(the normality of $\underline{A}$ within $\underline{A}^{\wedge}$ guarantees that R is in fact a congruence relation). Any of the two descriptions of $\underline{A}°$ above shows that $\underline{A}°$ is minimal (a homomorphic image over id_A) with respect to all other normal minimal completions of $\underline{A}$.

Next we consider a criterion similar to those in 16.1.4:

16.2.2 THEOREM. A minimal completion $\underline{B}$ of $\underline{A}$ is **normal** iff the corresponding congruence relation $R_{\underline{B}}$ of $\underline{A}^{\wedge}$ satisfies

$$R_{\underline{B}} \cap (A \times A^{\wedge}) = id_A .$$

16.2.3. Since id_A is the restriction of $g:\underline{A}^{\wedge} \to \underline{B}$ to A, this condition (strengthening the one of Theorem 16.1.4) means that $xg \in A$ implies (and is implied by) $x \in A$, for all $x \in A^{\wedge}$ (and implies $xg = x$). This observation is helpful for the

PROOF of Theorem 16.2.2. Let $\underline{B}$ be a normal minimal completion of $\underline{A}$. We prove the implication of 16.2.3 by algebraic induction on x: The inductive beginning for $x \in A$ is trivial. Induction hypothesis: $(x_k)g \in A$ implies $x_k \in A$, for all $k \in K_\phi$, for some fundamental operation symbol ϕ, and for some family $\underline{x}:= (x_k|k \in K_\phi)$ of elements $x_k \in A^{\wedge}$. We have to prove our implication for $x:= \underline{x}\phi^{\underline{A}^{\wedge}}$. Thus assume $xg \in A$, i.e. $((x_k)g|k \in K_\phi)\phi^{\underline{B}} \in A$. Since $\underline{B}$ is a normal completion of $\underline{A}$, this implies $(x_k)g \in A$. Thus even $\underline{xg}\phi^{\underline{B}} = \underline{xg}\phi^{\underline{A}}$, and by induction hypothesis: $x_k \in A$, i.e. $(x_k)g = x_k$ for all $k \in K_\phi$. Thus $\underline{x}\phi^{\underline{A}}$ is defined, and since $\underline{A}$ is a relative subalgebra of $\underline{A}^{\wedge}$, $x = \underline{x}\phi^{\underline{A}^{\wedge}} = \underline{x}\phi^{\underline{A}} \in A$.
Conversely, let a minimal completion $\underline{B}$ of $\underline{A}$ fulfill the condition of Theorem 16.2.2, and assume $\underline{b}\phi^{\underline{B}} \in A$ $(\underline{b} = (b_k|k \in K_\phi))$. By the Axiom of Choice there are elements $x_k \in A^{\wedge}$ such that $b_k = (x_k)g$, for all $k \in K_\phi$, and we have $((x_k|k \in K_\phi)\phi^{\underline{A}^{\wedge}})g = \underline{b}\phi^{\underline{B}} \in A$. By assumption $(x_k|k \in K_\phi)\phi^{\underline{A}^{\wedge}} \in A$; by (FC1) (normality of $\underline{A}$ in $\underline{A}^{\wedge}$), we get $b_k = (x_k)g = x_k \in A$, for all $k \in K_\phi$, completing the proof. ■■■

16.2.4. Since the relation $R_{\underline{B}} \subseteq A^{\wedge} \times A^{\wedge}$ above is reflexive and symmetric, the condition of Theorem 16.2.2 is equivalent with

$$R_{\underline{B}} \subseteq id_A \cup ((A^{\wedge} \setminus A) \times (A^{\wedge} \setminus A)) ;$$

or, as

$$id_A \cup ((A^{\wedge} \setminus A) \times (A^{\wedge} \setminus A)) = R_{\underline{A}°} ,$$

our condition in Theorem 16.2.2 means that $R_{\underline{B}} \subseteq R_{\underline{A}°}$. This again makes evident (as we already noted above) that the (heterogeneous) normal one-point completion $\underline{A}°$ of $\underline{A}$ is a normal extension of $\underline{A}$ and minimal with respect to all other such completions.
Moreover we have the

16.2.5 COROLLARY 1. A minimal completion $\underline{B}$ of $\underline{A}$ is normal iff $\underline{B} > \underline{A}°$. ■■■

It is even more important that now, when we have a minimal element, we have the following counterpart to 16.1.7:

16.2.6 COROLLARY 2. The representative normal minimal completions of any partial algebra $\underline{A}$ constitute a complete lattice with greatest element $\underline{A}^\wedge$ and least element $\underline{A}^\circ$. ■■■

16.2.7 REMARKS. (i): The heterogeneous one-point completion plays an important rôle in computer science, when the algebraic specification of an abstract data type yields a partial algebra which has to be (heterogeneously) completed in a most simple way. Here the heterogeneous one-point completion is the easiest way and produces the minimal possible number of so called "error values", as the new elements are then called. In any way one usually wants a normal minimal completion of the data type specified. A detailed discussion of the completion of heterogeneous partial algebras by "error-values" can be found in [GoTWa 78] and in [Go 78] (notice that Proposition 7 in [GoTWa 78] is more or less vacuous, since it only applies in the case when the algebra, which is to be extended, is already total). It should however be mentioned that the concept of "error algebras" as introduced in [Go 78] by GOGUEN is not so easily subsumed by the concept of a free partial algebra in our sense, which is then normally completed by some error values, since one of the main conditions in [Go 78] is that a homomorphism between error algebras has to carry error elements into error elements.
(ii): We would like to mention as a sort of natural completion of a partial algebra the one using an algebra which is often called the "global" of a given partial algebra (this construction is also used for total algebras):
Recall that **Subs**(A) designates the disjoint union of all power sets of phyla of $\underline{A}$; then we equip this set with a total (heterogeneous) structure as follows:
For every $\phi \in \Omega$ and for every sequence $\underline{M} := (M_k | k \in K_\phi) \in (\mathbf{Subs}(A))^{K_\phi}$ we set:

$$\underline{M}\phi^{\underline{\mathbf{Subs}(A)}} := \{(m_k | k \in K_\phi)\phi^{\underline{A}} | m_k \in M_k \text{ for each } k \in K_\phi\} .$$

It is easy to realize that the set of all one-element subsets in **Subs**(A) forms a relative subalgebra of $\underline{\mathbf{Subs}(A)}$ which is canonically isomorphic to $\underline{A}$, and the subalgebra generated by this set is canonically isomorphic to the normal heterogeneous one-point completion $\underline{A}^\circ$ of $\underline{A}$.

16.2.8 DEFINITION. Let $(M: t \stackrel{e}{=} t')$ be any E-equation. It is called <u>regular</u>, if $var(t) = var(t')$, i.e. if both terms contain the same variables.

16.2.9 PROPOSITION. Let G be any set of regular E-equations, and $\underline{A} \in \mathbf{Mod}(G)$ be any model of G. Then the heterogeneous one-point completion $\underline{A}^\circ$ of $\underline{A}$ is also a model of G.

PROOF. If $\underline{A}$ is total, then $\underline{A}^\circ = \underline{A}$, and the statement is trivial. Let $e := (M: t \stackrel{e}{=} t')$ be any E-equation in G, and let $u: M \to A^\circ$ be any assignment. Let $N := var(t) = var(t')$. Then we have to distinguish two cases:
(i) $[N]u \subseteq A$: Then e is already satisfied in $\underline{A}$ with respect to u.
(ii) $[N]u \not\subseteq A$: Then there is at least one new element contained in [N]u. Therefore the normality of the completion implies that both $(u)(M:t)^{\underline{A}^\circ}$ and $(u)(M:t')^{\underline{A}^\circ}$ are new elements not contained in A (proof by algebraic induction), and they have to be the same new element (since t and t' are of the same sort). Thus e is also satisfied in $\underline{A}^\circ$ with respect to u in this case. ■■■

17 Independence, free partial algebras and primitive classes

In earlier sections (e.g. in sections 5, 6 and 15) independence and partial algebras freely generated by some set -- or even by some other partial algebra -- have been discussed to some extent as well as their relationship to primitive classes. In this section we want to continue this discussion and add further results. At the beginning we want to discuss some inheritance properties of independence, followed by an investigation of the free completions within primitive classes as well as by a characterization of free partial algebras freely generated by some set -- a similar characterization for free extensions of partial algebras within some kind of variety (in particular of universal solutions) is missing in general and might be an interesting problem (however see 17.6.4). Free partial algebras -- or more generally maximal free extensions of partial algebras -- play an important rôle in computer science in connection with the initial algebraic specification of abstract data types. Thus this will also be discussed in PART II in some more specific context. As in most sections before we treat arbitrary partial $\underline{S}$-algebras for some given nonempty homogeneous partial algebra $\underline{S}$ of the given similarity type without usually mentioning it explicitly.

17.1 Some properties of independence and independence classes

17.1.1. We recall from 5.1.4 that a subset M of some partial algebra $\underline{A}$ is $\underline{B}$-independent in $\underline{A}$ for some partial algebra $\underline{B}$, iff every assignment $w:M \to \underline{B}$ has a homomorphic extension $w':\underline{E}:= \underline{C}_A M \to \underline{B}$ -- satisfying $w'v_B = v_E$. Recall, too, that $ind_{\underline{A}}(M)$ designates the class of all partial algebras $\underline{D}$ for which M is $\underline{D}$-independent in $\underline{A}$, and that it is called the independence class of M in $\underline{A}$. In addition we know (cf. 5.1.6) that every independence class is primitive, i.e. it is closed with respect to (direct) products, (closed) subalgebras and (weak) homomorphic images.

17.1.2 LEMMA. Let M be any subset of some partial algebra $\underline{A}$, and let $\underline{B}$ be any partial algebra. Then the following statements are equivalent:

(i) M is $\underline{B}$-independent in $\underline{A}$.

(ii) Every subset N of M of cardinality less than $d^{\&}$, and which satisfies $[N]v_A = [M]v_A$, is $\underline{B}$-independent in $\underline{A}$.

If $[M]v_A \subseteq [B]v_B$, then the condition "$[N]v_A = [M]v_A$" is not necessary in (ii). This applies in particular, when $\underline{B} = \underline{A}$, i.e. when we deal with a so-called algebraically independent subset M of $\underline{A}$.

PROOF. The implication from (i) to (ii) is trivial, since the assumptions of (ii) imply that every assignment for N into $\underline{B}$ can be extended to an assignment of M into $\underline{B}$. The converse is an immediate consequence of 4.4.9. If $[M]v_A \subseteq [B]v_B$, then there exists at least one assignment from M into $\underline{B}$, and even in the case that $[N]v_A = [M]v_A$ is violated, any assignment from N into $\underline{B}$ can be extended to one starting from M; hence the equivalence of (i) and (ii) can also be proved. ■

17.1.3 REMARK. Notice that in general the condition "$[N]v_A = [M]v_A$" is essential even when we are in the homogeneous case, since otherwise it might happen that there does not exist any assignment from M into $\underline{B}$ but that it exists for its subset N -- in the homogeneous case this might only happen when B and N are void and M is nonvoid.

Let us recall from 1.4 that -- in the homogeneous case -- the rank **r** of a given similarity type is defined as $\mathbf{r} := \sup(n|n < \mathbf{d})$. Moreover recall from sections 5, 6, and 15 that, for every quasi-primitive class **K**, $\underline{F}(M,\mathbf{K})$ designates the free partial **K**-algebra, **K**-freely generated by $[M]r_{M,\mathbf{K}}$, where $r_{M,\mathbf{K}}: M \to \underline{F}(M,\mathbf{K})$ is the **K**-universal homomorphism for M, and that in general $r_{M,\mathbf{K}}$ is injective iff there exists at least one injective assignment $w: M \to \underline{B}$ from M into some **K**-algebra $\underline{B}$.

17.1.4 PROPOSITION. Let **K** be any primitive class of partial algebras. Then the following statements are true:

(i) In the case of **homogeneous** partial algebras it is true for every set M of cardinality $\#M \geq \mathbf{r} + 1$ that
$\mathrm{ind}_{\underline{F}(M,\mathbf{K})}(M) = \mathbf{K} \cup \{\underline{\emptyset}\}$, or even more precisely $\mathbf{K} = \mathrm{ind}_{\underline{F}(M,\mathbf{K})}(M) \cap \mathrm{ind}_{\underline{F}(\emptyset,\mathbf{K})}(\emptyset)$.

(ii) In the case of arbitrary partial $\underline{S}$-algebras let M be such that, for every $s \in S$, $\#(s)v_M^{-1} \geq \mathbf{r} + 1$, and for every subset $Z \subseteq S$ of S let $M_Z := [Z]v_M^{-1}$ be the preimage of Z under v_M. Then we have $\mathbf{K} = \bigcap\{\mathrm{ind}_{\underline{F}(M_Z,\mathbf{K})}(M_Z) | Z \subseteq S\}$.

PROOF. Since M is always nonempty by assumption, the empty partial algebra always trivially belongs to the independence class of M, and whether or not the empty partial algebra $\underline{\emptyset}$ belongs to K is decided by $\mathrm{ind}_{\underline{F}(\emptyset,\mathbf{K})}(\emptyset)$, which always contains **K**, but which contains $\underline{\emptyset}$ iff $\underline{F}(\emptyset,\mathbf{K})$ is itself empty. This observation easily shows that (i) is an immediate consequence of (ii), which - as a matter of fact - is a consequence of Theorem 6.4.2.(iii) in combination with the fact that the **K**-universal homomorphism $r_{M,\mathbf{K}}: M \to \underline{F}(M,\mathbf{K})$ represents the set of all E-equations with variables in M which are valid in **K**. Since each E-equation needs from each phylum of M less than **d**+1 and hence at most **r**+1 elements, the statement can easily be derived from 6.4.2.(iii). The details are left as an exercise. ∎∎∎

17.2 Relatively free completions of partial algebras

17.2.1 REMARK. In this subsection we continue the study of completions of partial algebras started in section 16; but this time we restrict considerations to completions within a given E-variety. Although we are especially interested in the relation between **K**-freely generated **K**-algebras for some given E-variety **K** and their "free completions" within **K**, part of this can be discussed in a more general context, and we shall do this in this subsection. In the next one we shall then consider the other problem, and in a later subsection we shall even realize that there is in some respect no great difference between the free completion -- if it exists -- of a **K**-free partial algebra and the one of an arbitrary partial algebra.

Actually, we shall see that the situation is relatively simple in the homogeneous case, which has already been represented in [B 73], but in the heterogeneous case, things become relatively complicated, since, in some way, E-varieties of heterogeneous partial algebras are -- from the "homogeneous model theoretic point of view" -- already sort of implicational classes (one introduces a fundamental relation for each sort, and the heterogeneous E-equations $(Y: t \overset{e}{=} t')$ can then be considered as an implication with a conjunction in the premise saying that for each non-empty sort of Y there exists an element satisfying the corresponding relation, and with $t \overset{e}{=} t'$ as conclusion). Actually we shall observe already for E-varieties of heterogeneous partial algebras phenomena which appear in the homogeneous case at the earliest in connection with ECE-equations. But since for ECE-equations the description of cases of full embeddability into the "relatively free completion" is even more complicated, we do not treat this situation within the scope of this book. Nevertheless the results of this subsection can provide a slight idea of what can happen in this more general case.

As a preparation of our results we need some notation:

17.2.2 DEFINITION. **(i)**: Let **K** be any class of partial algebras, then we designate by $\mathbf{K}_t$ the class of all total **K**-algebras.
(ii): Let $\underline{A}$ be any partial algebra; then we designate by $U_{\underline{A}}$ the disjoint union of the set of all complements of domains of fundamental operations of $\underline{A}$ (i.e. the set of all "undefined sequences" of $\underline{A}$ -- observe many-sortedness!):

$$U_{\underline{A}} := \bigcup\{\{\phi\} \times (A^{K\phi} \setminus \operatorname{dom} \phi^{\underline{A}}) \mid \phi \in \Omega\} .$$

(iii): Let $\underline{A}$ and $\underline{B}$ be similar partial algebras. Then $\underline{B}$ is called an inner extension of $\underline{A}$, iff $A = B$ and $\underline{A}$ is a weak subalgebra of $\underline{B}$. If, in particular, $\underline{B}$ is total, then $\underline{B}$ is called an inner completion of $\underline{A}$.
It is easy to realize for the case, when $\underline{B}$ is an inner extension of $\underline{A}$, that there exists a mapping $z: \operatorname{dom} z \to A$ such that $\operatorname{dom} z \subseteq U_{\underline{A}}$, and for every $\phi \in \Omega$ one has

$$\operatorname{graph}(\phi^{\underline{B}}) = \operatorname{graph}(\phi^{\underline{A}}) \cup \{(\underline{a},(\phi,\underline{a})z) \mid (\phi,\underline{a}) \in \operatorname{dom} z\} .$$

In connection with this mapping we also call $\underline{B}$ a z-extension of $\underline{A}$ (and we denote $\underline{B}$ by $\underline{A}^z$), and if $\underline{B}$ is total, then we speak of a z-completion.

As a preparation for our embeddability results we prove

17.2.3 LEMMA. Let **K** be any class of similar partial algebras closed with respect to subalgebras and homomorphic images, let $\underline{A}$ be any partial algebra of the same type, and let $r_{\underline{A},\mathbf{K}}$ be the **K**-universal epimorphism for $\underline{A}$ (which exists according to section 5.11), then we have

(i) If $\underline{A}$ allows an injective homomorphism into some **K**-algebra, then $r_{\underline{A},\mathbf{K}}$ is injective, too.

(ii) If $\underline{A}$ allows a full and injective homomorphism into some **K**-algebra, then $r_{\underline{A},\mathbf{K}}$ is full and injective, too.

PROOF. Since injectivity as well as the property to be full and injective (i.e. initial and injective), can be described by the reflection of formulas, which are preserved by homomorphisms (cf. section 9, e.g. 9.2.15), this is an immediate consequence of Lemma 9.1.7.(ii). ∎∎∎

17.2.4 PROPOSITION. Let **K** be any E-variety, let $\underline{A}$ be any **K**-algebra, and consider the following conditions:

(*) A is empty.

(**) To $\underline{A}$ and $\underline{F}(\underline{A},\mathbf{K}_t)$ there belong the same subsets of S of sorts for which these partial algebras have nonempty phyla, i.e. $[A]v_A = [F(\underline{A},\mathbf{K}_t)]v_{F(\underline{A},\mathbf{K}_t)}$.

(***) $\mathbf{K}$ has a set, say G, of defining E-equations such that for any $(Y: t \overset{e}{=} t') \in G$ there exists an assignment from Y into A.

(****) For those elements of S for which $\underline{A}$ has a non-empty phylum, $\underline{F}(\underline{A},\mathbf{K}_t)$ has an at least two-element phylum.

Then we have the following statements:

(i) $\mathbf{K}_t$ is again an E-variety, each inner extension of $\underline{A}$ belongs to $\mathbf{K}$, and if (**) holds, then $\underline{A}$ allows inner completions, and these belong to $\mathbf{K}_t$.

(ii) (**) holds iff $[A]v_A$ is a closed subset of $\underline{S}$.

(iii) If $\underline{A}$ satisfies one of the conditions (*), (**), or (***), then $r_{\underline{A},\mathbf{K}_t}$ is injective.

(iv) If $\underline{A}$ satisfies one of the conditions (*), ((**) and (****)) or ((***) and (****)), then $r_{\underline{A},\mathbf{K}_t}$ is injective and full.

(v) If $\mathbf{K}$ is definable by so-called regular E-equations (cf. 16.2.8), then $\underline{F}(\underline{A},\mathbf{K}_t)$ is always a normal minimal completion of $\underline{A}$, i.e. $r_{\underline{A},\mathbf{K}_t}$ is injective and relatively normal (and in particular full).

PROOF. Ad (i): Since for a given E-variety $\mathbf{K}$ homomorphic images, subalgebras and direct products of total $\mathbf{K}$-algebras are again total, we immediately get the first part of the statement. If $\Sigma := (Y: t \overset{e}{=} t') \in \mathbf{Eeq}(\mathbf{K})$, and if $\underline{B}$ is an inner extension of $\underline{A}$, then Y allows an assignment into $\underline{B}$ iff it allows one into $\underline{A}$, and since Σ is valid in $\underline{A}$ it is valid in $\underline{B}$. -- If (**) is satisfied, then all fundamental operations whose input sorts are non-empty in $\underline{A}$ have also output sorts which are non-empty in $\underline{A}$, and therefore $\underline{A}$ allows inner completions, and they belong to $\mathbf{K}$ and therefore to $\mathbf{K}_t$. ∎

Ad (ii): Since $v_{F(\underline{A},\mathbf{K}_t)}$ is closed, $[F(\underline{A},\mathbf{K}_t)]\,v_{F(\underline{A},\mathbf{K}_t)}$ is always a closed subset of $\underline{S}$, and the statement easily follows. ∎

Ad (iii): If A is empty, then the statement is trivial. -- Thus assume that A is non-empty. If (**) is satisfied, then the statement is implied by (i) and 17.2.3.(i). -- Finally assume that (***) holds, but that (**) is violated: Then we construct a total $\mathbf{K}$-algebra $\underline{B}$ containing $\underline{A}$ as a weak subalgebra: Let

$\Omega_A := \{\phi \in \Omega \mid$ the elements of graph$(\phi^{\underline{S}})$ are contained in $[A]v_A\}$,

and let $B := A \mathbin{\dot{\cup}} \{b_s \mid s \in S \setminus [A]v_A\}$ such that

$$bv_B := \begin{cases} bv_A, & \text{if } b \in A, \\ s, & \text{if } b = b_s. \end{cases}$$

Define $\underline{B}$ to contain $\underline{A}$ as relative subalgebra and having no additional structure ($\phi^{\underline{B}} := \phi^{\underline{A}}$ for every $\phi \in \Omega$). Then (***) implies that $\underline{B}$ and each inner completion of $\underline{B}$ belong to $\mathbf{K}$. Therefore the statement follows from 17.2.3.(i). ∎

Ad (iv): If $A = \emptyset$, then the statement is trivial. Therefore we may assume that $A \neq \emptyset$. Let us abbreviate $\underline{F} := \underline{F}(\underline{A},\mathbf{K}_t)$. Consider $g: \underline{F}(F,\mathbf{K}) \to \underline{F}$ to be the homomorphism induced by id_F, and let $(\phi^*)_{\phi\in\Omega}$ be the quotient structure on F with respect to g joined with the structure of $\underline{A}$ (observe that A can be considered as a subset of F, because of the injectivity of $r_{\underline{A},\mathbf{K}_t}$ which follows from (i) or (iii) and our assumptions). Obviously $\underline{F}^* := (F,(\phi^*)_{\phi\in\Omega})$ belongs to $\mathbf{K}$ as an inner extension of some $\mathbf{K}$-algebra. Let A_s and F_s be the phyla of sort s of $\underline{A}$ and $\underline{F}$, respectively, and let S_A and S_F be the subsets of S of sorts, for which the corresponding phyla in $\underline{A}$ and $\underline{F}$, respectively, are non-empty. For each $s \in S_F$ we now choose $a_s, b_s \in F_s$ such that $a_s \neq b_s$ whenever

$s \in S_A$, and we define on F the following two structures:

$$(\underline{c})\phi_a := \begin{cases} (\underline{c})\phi^*, & \text{if } \underline{c} \in \operatorname{dom} \phi^*, \\ a_s & \text{, otherwise, when } s \text{ is the output sort of } \phi; \end{cases}$$

and

$$(\underline{c})\phi_b := \begin{cases} (\underline{c})\phi^*, & \text{if } \underline{c} \in \operatorname{dom} \phi^*, \\ b_s & \text{, otherwise, when } s \text{ is the output sort of } \phi; \end{cases}$$

for every $\phi \in \Omega$ and for every $\underline{c} \in F^{K_\phi}$. Define $\underline{F}_c := (F,(\phi_c)_{\phi\in\Omega})$ for $c \in \{a,b\}$, let $\underline{G} := \underline{F}_a \times \underline{F}_b$, and let $i:\underline{A} \to \underline{G}$ be defined by $ai := (a,a)$ for each $a \in A$. Then it is easy to realize from the above construction of $\underline{G}$ that i is a full and injective homomorphism, and that $\underline{G}$ is a $\mathbf{K}_t$-algebra. This proves (iv). ∎

Ad (v): This statement is a consequence of 16.2.9, since $\underline{A}°$ will belong to $\mathbf{K}_t$. ∎∎∎

Why we cannot state much more in the heterogeneous case is shown in the following

17.2.5 EXAMPLE. Since in the many-sorted case E-equations are also provided with a set of variables, their influence also depends on the evaluability of this set, and the situation of completions within E-varieties can become rather complicated for heterogeneous partial algebras. Consider:

$\Omega := (\phi,\phi',\phi'')$, $\tau := (1,1,1)$, $S := \{s,s'\}$, $\phi^{\underline{S}},\phi'^{\underline{S}}: s \mapsto s$, $\phi''^{\underline{S}}: s \mapsto s'$.

Moreover, let x,y be distinct variables of sort s, x',y' be distinct variables of sort s', respectively, and let

$g := (\{x\}: x\phi \overset{e}{=} x\phi)$,

$g' := (\{x,x'\}: x\phi \overset{e}{=} x\phi')$,

$g'' := (\{x,y,x'\}: x \overset{e}{=} y)$,

$g''' := (\{x,x',y'\}: x' \overset{e}{=} y')$,

$g'''' := (\{x\}: x\phi' \overset{e}{=} x\phi')$.

Let $\underline{A}$ be the following partial $\underline{S}$-algebra: $A_s := \omega_0 = \{0,1,2,3,...\}$, $A_{s'} := \emptyset$, $\operatorname{graph}(\phi^{\underline{A}}) := \{(i,i+1) \mid i \in \omega_0\}$, $\operatorname{graph}(\phi'^{\underline{A}}) := \operatorname{graph}(\phi''^{\underline{A}}) := \emptyset$.

Then it is easy to realize that $\underline{A}$ belongs to each of the following E-varieties:

$\mathbf{K}_1 := \mathbf{Mod}(\{g\})$,

$\mathbf{K}_2 := \mathbf{Mod}(\{g,g'\})$,

$\mathbf{K}_3 := \mathbf{Mod}(\{g,g''\})$, (omit ϕ' from the type!)

$\mathbf{K}_4 := \mathbf{Mod}(\{g,g',g''\})$,

$\mathbf{K}_5 := \mathbf{Mod}(\{g,g'',g'''\})$, (omit ϕ' from the type!)

$\mathbf{K}_6 := \mathbf{Mod}(\{g'\})$,

$\mathbf{K}_7 := \mathbf{Mod}(\{g,g',g''''\})$,

since even $\underline{A} = \underline{F}(\{0\},\mathbf{K}_i)$ for $1 \le i \le 5$ (0 having sort s), as only g has some influence on $\underline{A}$, but the other three E-equations cannot be evaluated in $\underline{A}$.

On the other hand the phylum corresponding to s' has to be non-empty in each $\underline{F}(\underline{A},(\mathbf{K}_i)_t)$, and therefore now all the E-equations under consideration are really evaluable, yielding the following properties of $r_{\underline{A},(\mathbf{K}_i)_t}$:

$r_{\underline{A},(\mathbf{K}_1)_t}$ is full and injective,

$r_{\underline{A},(\mathbf{K}_2)_t}$ is injective, but not full,

$r_{\underline{A},(\mathbf{K}_3)_t}$ and $r_{\underline{A},(\mathbf{K}_5)_t}$ are full, but not injective, (ϕ' being omitted!)

$r_{\underline{A},(\mathbf{K}_4)_t}$ is neither full, nor injective.

We have also considered $\mathbf{K}_5$, since in this case each phylum of each total $\mathbf{K}_5$-algebra has

at most one element, i.e. $(\mathbf{K}_5)_t$ is a trivial E-variety, although $\mathbf{K}_5$ is non-trivial. Finally, let $\underline{B} := \underline{F}(\{0\},\mathbf{K}_6)$ $(= \underline{\{0\}}_d)$, and $\underline{C} := \underline{F}(\{0\},\mathbf{K}_7)$ (0 having sort s). Then $\underline{C}$ is a partial Peano algebra in which $\phi^{\underline{C}}$ and $\phi'^{\underline{C}}$ are total and $\phi''^{\underline{C}}$ is empty, and $\underline{C}$ also belongs to $\mathbf{K}_6$, $\underline{B}$ is a relative subalgebra of $\underline{C}$ as well as of $\underline{F}(\underline{B},(\mathbf{K}_6)_t)$, and $\underline{F}(\underline{B},(\mathbf{K}_6)_t)$ and $\underline{F}(\underline{C},(\mathbf{K}_6)_t)$ are isomorphic -- the isomorphism being induced by id_B --, but $r_{\underline{C},(\mathbf{K}_6)_t}$ is not injective (in this case it is full, but a similar example, only with different E-varieties, would show that it need not even be full under the circumstances above).
-- For ECE-varieties $\mathbf{K}$, where $\mathbf{K}_t$ is also always an E-variety (!), one gets similar effects even in the homogeneous case by taking as axioms besides g ECE-equations with $x\phi'' \overset{e}{=} x\phi''$ as premise and one of the E-equations g' or g" (or g'") as conclusion.

This example shows that we need in the heterogeneous case relatively strong restrictions on the E-varieties $\mathbf{K}$ (or, equivalently, on their defining E-equations) in order to get injective and/or full homomorphisms $r_{\underline{A},\mathbf{K}}$. 17.2.4 is the best we can do at the moment in general. But in the homogeneous case we have the

17.2.6 COROLLARY. Let $\mathbf{K}$ be any E-variety of **homogeneous** partial algebras, and let $\underline{A}$ be any $\mathbf{K}$-algebra. Then $r_{\underline{A},\mathbf{K}_t}$ is always injective, and it will only happen that $r_{\underline{A},\mathbf{K}_t}$ is not full, when both A and $\underline{F}(\underline{A},\mathbf{K}_t)$ contain exactly one element and $\underline{A}$ is not total. ■■■

17.2.7 REMARK. Notice that the last part of the example above shows that for E-varieties $\mathbf{K}$ of heterogeneous partial algebras a $\mathbf{K}$-algebra need not have a completion within $\mathbf{K}$, not even a weak completion in the sense of section 16 on completions in general. In the homogeneous case there exists at least a weak completion, and if $\underline{F}(\underline{A},\mathbf{K}_t)$ has at least two elements, then $\underline{A}$ also has a completion within $\mathbf{K}$. As we have indicated in 17.2.1, "heterogeneous E-varieties" already have features which one observes in the homogeneous case only for ECE-varieties. That the results above are not so far from the study of $\mathbf{K}$-free partial $\mathbf{K}$-algebras, $\mathbf{K}$-freely generated by some set, follows from the results in the subsection on initial partial algebras below. In this subsection here we now draw the special conclusions from the above results for $\mathbf{K}$-free partial $\mathbf{K}$-algebras on sets. We restrict the results to criteria for $r_{\underline{F}(M,\mathbf{K}),\mathbf{K}_t}$ to be full and injective -- the case of injectivity alone is handled similarly.

17.2.8 COROLLARY. Let $\mathbf{K}$ be any E-variety, and let M be any set such that $\underline{F}(M,\mathbf{K})$ satisfies one of the following conditions:

(+) If a phylum of $\underline{F}(M,\mathbf{K}_t)$ is non-empty, then it contains at least two elements, and the corresponding phylum of $\underline{F}(M,\mathbf{K})$ is non-empty, too.

(++) $\mathbf{K} = ind_{\underline{F}(M,\mathbf{K})}(M)$, and $\underline{F}(M,\mathbf{K}_t)$ has an at least two-element phylum for each $s \in [F(M,\mathbf{K})]\, v_{\underline{F}(M,\mathbf{K})}$.

Then $r_{\underline{F}(M,\mathbf{K}),\mathbf{K}_t}$ is full and injective.

PROOF. Notice that (+) implies (**) and (****) of 17.2.4; thus the statement follows from 17.2.4. -- If $\mathbf{K}$ is the independence class of M with respect to $\underline{F}(M,\mathbf{K})$, then it is easily seen that 17.2.4.(***) has to be satisfied (cf. section 6), while the second part of (++) is just 17.2.4.(****); hence $r_{\underline{F}(M,\mathbf{K}),\mathbf{K}_t}$ is full and injective. ■■■

17.3 Relatively free completions of relatively free partial algebras

In order to consider the structure of $\underline{F}(M,\mathbf{K}_t)$ more intensively we present Definition 17.3.1. For the homogeneous case the following material can be found in [B 73].

17.3.1 DEFINITION. (i): Let $\underline{S}$ be any sort-algebra, and let S' be any subset of S. Then the relative structure of $\underline{S}$ on S' defines a similarity type of many-sorted partial algebras, which is called the S'-reduction of the given one. It satisfies
$\Omega(S') := \{\phi \mid \phi \in \Omega$, and each input and output sort of ϕ belongs to $S'\}$;
$\tau(S') := (K_\phi)_{\phi\in\Omega(S')}$; $\phi^{\underline{S}'} := \phi^{\underline{S}}$ for each $\phi \in \Omega(S')$.
If $\underline{A}$ is any partial algebra, and if we forget about all phyla of $\underline{A}$ of sorts not in S' and about all fundamental operations of $\underline{A}$ not in $\Omega(S')$, then we call this partial algebra the S'-reduct of $\underline{A}$ and denote it by $\underline{A}|_{S'}$. -- Notice that this can also be defined for arbitrary weak subalgebras $\underline{S}'$ of $\underline{S}$, only $\Omega(S')$ and $\tau(S')$ have then to be modified into the corresponding $\Omega(\underline{S}')$ and $\tau(\underline{S}')$, where now only those operations are considered, which are non-empty in $\underline{S}'$.
If **K** is any class of partial $\underline{S}$-algebras and $\underline{S}'$ any weak subalgebra of $\underline{S}$, then we denote by $\mathbf{K}(\underline{S}'\text{-t})$ the class of all partial **K**-algebras for which each fundamental operation corresponding to some $\phi \in \Omega(\underline{S}')$ is total, and we call it the class of all $\underline{S}'$-total **K**-algebras. If $\underline{S}'$ is the relative subalgebra of $\underline{S}$ on S', then we also use the notation $\mathbf{K}(S'\text{-t})$ instead of $\mathbf{K}(\underline{S}'\text{-t})$.
(ii): Let $\underline{B} := \underline{A}^z$ be a z-extension of a given partial algebra $\underline{A}$ (cf. 17.2.2.(iii)). Then we define on $\underline{B}$ a mixed step operator $\mathbf{G}_{\underline{A},z}$ as follows:

$$\mathbf{G}_{\underline{A},z}\, M := \mathbf{C}_{\underline{A}}\mathbf{D}_{\underline{A}^z}\, M$$

for all $M \subseteq A = B$. Moreover, we define as usual

$\mathbf{G}_{\underline{A},z}{}^{0}M := M$, and $\mathbf{G}_{\underline{A},z}{}^{\alpha}M := \mathbf{G}_{\underline{A},z}(\bigcup\{\mathbf{G}_{\underline{A},z}{}^{\beta}M \mid \beta < \alpha\})$ for all ordinal numbers $\alpha > 0$.

Let $\delta = \delta(\tau)$ be the ordinal dimension of the given similarity type τ (see 1.3.1), then

17.3.2 LEMMA.

(i) Let $\underline{A} \in \mathbf{Alg}(\underline{S})$ and let $z : \mathrm{dom}\, z \subseteq U_{\underline{A}} \to A$ be any mapping such that the sort of $(\phi,\underline{A})z$ is the output sort of ϕ ($(\phi,\underline{a}) \in U_{\underline{A}}$). Then we have for all $M \subseteq A$:

$$\mathbf{C}_{\underline{A}^z}\, M = \mathbf{G}_{\underline{A},z}{}^{\delta}M ,$$

and in the case of $\delta > 1$ we even get $\mathbf{C}_{\underline{A}^z}\, M = \bigcup\{\mathbf{G}_{\underline{A},z}{}^{\alpha}M \mid \alpha < \delta\}$.

(ii) Let $\mathbf{K} \subseteq \mathbf{Alg}(\underline{S})$ be any E-variety, $\underline{A} \in \mathbf{K}$ and z as in (i). Then we have for every ordinal number $\alpha > 0$ and for every subset M of A that $(\mathbf{G}_{\underline{A},z}{}^{\alpha}M,(\phi^{\underline{A}})_{\phi\in\Omega})$ and $(\mathbf{G}_{\underline{A},z}{}^{\alpha}M,(\phi^{\underline{A}^z})_{\phi\in\Omega})$ are relative subalgebras of $\underline{A}$, respectively of $\underline{A}^z$, and they are both **K**-algebras. Moreover $(\mathbf{G}_{\underline{A},z}{}^{\alpha}M,(\phi^{\underline{A}^z})_{\phi\in\Omega})$ is generated by M.

PROOF. (i) can easily be proved in the same way as for the step operator **D** (cf. subsection 3.4). -- (ii) is also easily checked, since, for each $\alpha > 0$, $G_\alpha := \mathbf{G}_{\underline{A},z}{}^{\alpha}M$ is a closed subset of the **K**-algebra $\underline{A}$ (by the definition of $\mathbf{G}_{\underline{A},z}$); hence the relative subalgebras $(\mathbf{G}_{\underline{A},z}{}^{\alpha}M,(\phi^{\underline{A}})_{\phi\in\Omega})$ and $(\mathbf{G}_{\underline{A},z}{}^{\alpha}M,(\phi^{\underline{A}^z})_{\phi\in\Omega})$ of $\underline{A}$ and $\underline{A}^z$, respectively, are **K**-algebras. That M generates $(G_{\underline{A},z}{}^{\alpha}M,(\phi^{\underline{A}^z})_{\phi\in\Omega})$ can be proved by transfinite induction. ■■■

The remaining part of this subsection is basically dedicated to the proof and to the discussion of consequences of the following

17.3.3 THEOREM. Let K be any primitive class of similar partial algebras, such that K contains a K-algebra with at least two elements in each phylum, and let M be any set such that $[F(M,K)]v_{\underline{F}(M,K)}$ is a closed subset of $\underline{S}$ (or $K = ind_{\underline{F}(M,K)}(M)$).

Then there exist a set N containing M and a bijective mapping $z:U_{\underline{F}(N,K)} \to (N \setminus M)$ -- satisfying $(\phi,\underline{a})z \notin [K_\phi]\underline{a}$ for all $(\phi,\underline{a}) \in U_{\underline{F}(N,K)}$ -- such that the identity mapping id_M induces an isomorphism between $\underline{F}(M,K_t)$ and $\underline{F}(N,K)^z$.

Before we prove this result, we want first to discuss its immediate consequences, and, moreover, to prove some useful preparatory facts. In particular we immediately get a result similar to the inner axiomatic characterization of absolutely free completions of partial algebras.

17.3.4 COROLLARY Let K be a primitive class of partial algebras which contains a K-algebra with at least two elements in each phylum for each sort s in S. Let $\underline{A} \in K_t$ be a total K-algebra and M any subset of A such that either $[F(M,K)]v_{\underline{F}(M,K)}$ is a closed subset of $\underline{S}$ or $K = ind_{\underline{F}(M,K)}(M)$. Then id_M induces an isomorphism between $\underline{F}(M,K_t)$ and $\underline{A}$, iff $\underline{A}$ has the following properties:

(RFC 1) M generates $\underline{A}$.

(RFC 2) There exists a weak subalgebra $\underline{B}$ of $\underline{A}$ with carrier set $B = A$ such that the assignment $z:U_{\underline{B}} \to A$, defined by $(\phi,\underline{a})z := \underline{a}\phi^{\underline{A}}$ for every $(\phi,\underline{a}) \in dom\ z$, and the set $N := M \cup [U_{\underline{B}}]z$ fulfill the following conditions:
- (a) $M \cap [U_{\underline{B}}]z = \emptyset$,
- (b) z is injective, and $(\phi,\underline{a})z \notin [K_\phi]\underline{a}$ for all $(\phi,\underline{a}) \in U_{\underline{B}}$,
- (c) id_N induces an isomorphism between $\underline{B}$ and $\underline{F}(N,K)$. ■■

17.3.5 ASSUMPTIONS and NOTATION. For the remainder of this subsection we shall use the following assumptions and abbreviations:

$K \subseteq Alg(\underline{S})$ shall designate an E-variety which contains a K-algebra with at least two elements in each phylum, and as before K_t designates the subclass of all total K-algebras. We assume that $M \subseteq N \neq \emptyset$ are any two sets and $z:dom\ z \subseteq U_{\underline{F}(N,K)} \to (N \setminus M)$ shall be an injective mapping satisfying $(\phi,\underline{a})z \notin [K_\phi]\underline{a}$ (the sort of $(\phi,\underline{a})z$ being the output sort of ϕ) for all $(\phi,\underline{a}) \in dom\ z$. Finally assume that either $[F(M,K)]v_{\underline{F}(M,K)}$ is a closed subset of $\underline{S}$ or that $K = ind_{\underline{F}(M,K)}(M)$, and in the latter case additionally $\underline{F}(\emptyset,K) \cong C_{\underline{F}(Q,K)}\,\emptyset$ for each non-empty set Q.

Moreover, we introduce the abbreviations

$\underline{F} := \underline{F}(N,K)$, $\quad G := G_{\underline{F},z}$, $\quad (\phi^*) := (\phi^{\underline{F}})_{\phi\in\Omega}$, $\quad (\phi^z) := (\phi^{\underline{F}^z})_{\phi\in\Omega}$,

and we shall also denote by (ϕ^*) and (ϕ^z) the structures of relative subalgebras of $\underline{F}$ and $\underline{F}^z$, respectively (for single operations we just omit the brackets).

17.3.6 LEMMA. In addition to the assumptions in 17.3.5 let $\underline{A}$ be any M-generated relative subalgebra of $\underline{F}^z$. Then

(i) $A = C_{\underline{F}}(N \cap A)$.

(ii) $\underline{F}(M,K) \cong \underline{C_F}M \subseteq (A,(\phi^*)) \cong \underline{F}(N \cap A,K)$.

(iii) The restriction mapping $z' := z|([N \cap A]z^{-1})$ is a bijection from dom z' onto $(N \cap A) \setminus M$, and $\text{dom } z' \subseteq U_{(A,(\phi^*))}$.

(iv) $N_\alpha := N \cap G^\alpha M = N \cap D_{\underline{F}z}(\mathbf{U}\{G^\beta M | \beta < \alpha\}) =$

$= M \cup [\text{dom } z \cap (\mathbf{U}\{\{\phi\} \times ((\mathbf{U}\{G^\beta M|\beta < \alpha\})^{K_\phi} \setminus \text{dom } \phi^*)|\phi \in \Omega\})]z$, $(\alpha > 0)$

(v) $G^\alpha M = \underline{C_F}N_\alpha$, $(\alpha > 0)$.

(vi) $\underline{C}_{\underline{F}z}$ M (and each $G^\alpha M$ $(\alpha > 0)$) is K_t-freely generated by M.

(vii) If additionally $\text{dom } z = U_{\underline{F}}$, then id_M induces an isomorphism between $\underline{C_F}M$ and $\underline{F}(M,K_t)$.

PROOF. Ad (i): Since $\underline{A} \in K$, $id_{N \cap A}$ can be extended to a homomorphism, say $g:C_{\underline{F}}(N \cap A) =: \underline{B} \to \underline{A}$, such that $g = id_B$ -- observe that, because of the fact that M generates $\underline{A}$ and because of the assumptions on $\underline{F}(M,K)$ or **K**, we may apply 17.1.2 in order to realize that $\underline{B}$ is isomorphic to $\underline{F}(N \cap A,K)$. Hence we have $M \subseteq (N \cap A) \subseteq B \subseteq A$. Since $[\text{dom } z']z' \subseteq (N \cap A) \subseteq B$ (for z' as defined in (iii)), $\underline{B}^{z'}$ is a subalgebra of $\underline{A}$ containing M -- observe that B cannot contain any sequence of $U_{\underline{A}}$, which z maps outside of B --, therefore $B = C_{\underline{A}} M = A$, and we even get equality between $\underline{A}$ and $\underline{B}$. ∎

Ad (ii): Again the assumptions on **K** and M imply the first "equality" and the inclusion, while the last "equality" has already been shown in connection with the proof of (i). ∎

Ad (iii): This is an immediate consequence of the proof of (i). ∎

Ad (iv): These statements are an immediate consequence of the definitions of $G_{\underline{A},z}$ and (ϕ^z), and of the fact that $N \cap C_{\underline{F}} Q = Q$ for each subset Q of N because of the $\underline{A}$-freeness of N within $\underline{F}$ (cf. (v)). ∎

Ad (v): This follows from (i) and 17.3.2.(ii). ∎

Ad (vi): This statement is proved by transfinite induction: Assume $\underline{D} \in K_t$, and let $g:M \to D$ be any mapping. Since M is **K**-free in $\underline{F}(M,K)$, g has a homomorphic extension $g':\underline{F}(M,K) \to \underline{D}$, whose restriction to M is a homomorphism $g_0:(M,(\phi^z)) \to \underline{D}$ extending g. Now let $\alpha > 0$ be any ordinal number such that for every ordinal number $\beta < \alpha$ there exists a homomorphism $g_\beta:(G^\beta M,(\phi^z)) \to \underline{D}$ satisfying $g_\beta|M = g$. Since $(G^\beta M,(\phi^z))$ is generated by M, there holds, for any two ordinal numbers γ and ε which satisfy $\gamma < \varepsilon < \alpha$, that $\text{graph}(g_\gamma) \subseteq \text{graph}(g_\varepsilon)$. Hence g'_α with $\text{graph}(g'_\alpha) := \mathbf{U}\{\text{graph}(g_\beta)|\beta < \alpha\}$ maps $\underline{A_\alpha} := (\mathbf{U}\{G^\beta M|\beta < \alpha\},(\phi^z))$ homomorphically into $\underline{D}$. The other statements of the Lemma imply that $G^\alpha M = C_{\underline{F}} N_\alpha$, where N_α is defined as in (iv) and satisfies $N_\alpha = \mathbf{U}\{N_\beta|\beta < \alpha\} \cup [U_{(A_\alpha,(\phi^*))} \cap \text{dom } z]z$.

Let g" be the restriction of g'_α to $N \cap A_\alpha$, and let $g°:N_\alpha \to D$ be defined as follows: $\text{graph}(g°) := \text{graph}(g") \cup \{((\phi,\underline{a})z,(\underline{a}g')\phi^{\underline{D}})|(\phi,\underline{a}) \in U_{(A_\alpha,(\phi^*))} \cap \text{dom } z\}$.

Because of the assumptions on z, g° is well defined. Therefore we can assume g_α to be the homomorphic extension of g° to $C_{\underline{F}} N_\alpha = (G^\alpha M,(\phi^z))$ (into $\underline{D}$). Let δ be the ordinal dimension of the given similarity type, then $g^* := g_\delta:C_{\underline{F}z} M \to \underline{D}$ is the homomorphic extension of g, whose existence has been claimed. ∎

Ad (vii): This follows from (vi) and the uniqueness up to isomorphism of the K_t-universal K_t-solution. ∎∎∎

The preceding results now make it relatively easy to prove Theorem 17.3.3:

17.3.7 PROOF of THEOREM 17.3.3. We first show that there exists a set N' containing M such that each phylum of $N' \setminus M$ has a cardinality not less than the cardinality of $U_{\underline{F}(N',K)}$, since this will then allow an injective mapping $z':U_{\underline{F}(N',K)} \to N'$. Thus take

$$n := 2^{(\#M + \#S + \#\Omega + \#\delta + \#\omega_0)},$$

where δ is the ordinal dimension of the given type. Consider any set N' which contains M and whose phyla, say $N_s' := (s)v_{N'}^{-1}$, for each $s \in S$ satisfy $\#(N_s' \setminus M_s) = n$. Then N' is a set of the kind we have looked for (cf. the arguments in subsection 5.10): For each $\phi \in \Omega$ we have $\#(F(N',K)^{K\phi}) \le (n\cdot\#S)^{\#K\phi} \le n$.

This suffices to allow an injective mapping z' as described above, since it now easily follows that also $\#U_{\underline{F}(N',K)} \le n$.

Define $\underline{A} := \underline{C}_{\underline{F}(N',K)}z^M$, $N := N' \cap A$ and $\underline{B} := \underline{F}(N,K)$ considered as a weak subalgebra of $\underline{A}$; then (RFC 1) and (RFC 2) (cf. 17.3.4) are easily seen to be satisfied (z is defined to be the restriction of z' to $U_{\underline{A}}$). ■■■

17.3.8 REMARK. We can get another kind of proof of Theorem 17.3.3, since (RFC 2)(b) can be achieved recursively by using a direct limit construction and observing facts in the proof of 17.3.6.(vi). Thus we get a version where we do not need the above cardinality argument. But notice that then we have also to find sets disjoint from the given ones, whose elements may become the distinct values for the operations undefined in the previous step (see 23.2).

17.4 Characterization of heterogeneous relatively free partial algebras

As before we assume throughout this section that $\underline{S}$ is a heterogeneous sort algebra. The results in the previous sections and subsections now enable us to formulate and prove the following Characterization Theorem of Free Partial Algebras for the nontrivial case, and we shall show in the subsequent remarks that the theorem remains true - except for the equivalence between (v) and the other items - even for the cases when $\underline{A}$ has one-element phyla.

17.4.1 THEOREM (Characterization Theorem of Free Partial Algebras). Let $\underline{A}$ be any partial $\underline{S}$-algebra such that each nonempty phylum of $\underline{A}$ has at least two elements, and let M be any **generating** subset of $\underline{A}$. Then the following statements are equivalent:

(i) $\underline{A}$ is a free partial algebra with M as a basis (i.e. M is $\underline{A}$-independent in $\underline{A}$).

(ii) $\underline{A} = \underline{F}(M,K)$ for some nontrivial primitive class K of partial $\underline{S}$-algebras.

(iii) $\underline{A}$ is isomorphic to the partial algebra $\underline{F}^M(\underline{A};\underline{S})$ of all total algebraic operations of type M on $\underline{A}$. This isomorphism is induced by the mapping $e_m^M \mapsto m$ $(m \in M)$ from $E^M(A;S)$ onto M.

(iv) For every element $a \in A$ there exists exactly one algebraic partial operation $g_a \in APO^M(\underline{A};\underline{S})$ such that $a = (id_M)g_a$. And for every $g \in APO^M(\underline{A};\underline{S})$ $id_M \in \mathrm{dom}\, g$ implies that g is total, i.e. $g \in F^M(\underline{A};\underline{S})$.

(v) There exists a free **total** algebra $\underline{B}$ with basis M such that $\underline{A}$ is a relative subalgebra of $\underline{B}$, and such that for every algebraic operation $h \in F^M(\underline{B};\underline{S})$ of $\underline{B}$

the following is true: If $(id_M)h \in A$, then the restriction $h|A^M:A^M \to B$ is an element of $F^M(\underline{A})$, i.e. an algebraic total operation on $\underline{A}$.

(vi) Let $\underline{F}:= \underline{F}(M,\mathbf{TAlg}(\underline{S}))$ be the ($\underline{S}$-) total Peano (or term-) algebra on M of type τ, and let i be the identity mapping id_M on M considered as a mapping out of F into A. Then $i:\underline{F} \supseteq \underline{\mathrm{dom}\ \tilde{i}} \to \underline{A}$ is a closed, fully invariant and surjective homomorphism, and $\underline{\mathrm{dom}}\ \tilde{i}$ is freely generated by M.

PROOF. The equivalence of (i) and (ii) is trivial, since we can choose $K:= \overline{\mathrm{ind}_{\underline{A}}(M)}$, and the equivalence of (i) and (iii) easily follows from 15.4.6 and 15.4.8. The equivalence of (i) (or (ii)) with (vi) has been proved in section 6 (cf. e.g. 6.3.9). ∎

(i) ⇒ (iv): We already know that $e_m^M \mapsto m$ induces an isomorphism, say $f:\underline{F}^M(\underline{A};\underline{S}) \to \underline{A}$ such that $(h)f = (id_M)h$ for every $h \in F^M(\underline{A};\underline{S})$. Thus there corresponds to every element of A exactly one from $F^M(\underline{A};\underline{S})$. Now assume $h \in APO^M(\underline{A};\underline{S})$ such that $id_M \in \mathrm{dom}\ h$, and let $\underline{a} \in A^M$ be arbitrary. Since M is a basis of $\underline{A}$, $\underline{a}$ extends to an endomorphism, say $\underline{a}^*$ of $\underline{A}$. From 5.7.5 and 15.2.1 we may infer that algebraic partial operations commute with homomorphisms; therefore $((id_M)h)\underline{a}^* = (id_M \cdot \underline{a}^*)h = (\underline{a})h$. This shows that $\underline{a} \in \mathrm{dom}\ h$, and that h is therefore total on $\underline{A}$. ∎

(iv) ⇒ (i): By 15.4.6 $e_m^M \mapsto m$ induces a homomorphism, say $f:\underline{F}^M(\underline{A};\underline{S}) \to \underline{A}$. The assumptions of (iv) imply that f is injective. From 15.5.7 we know that f has a closed homomorphic extension, say f*, and from (iv) there follows that f* is injective and that dom f = dom f*, i.e. f = f* showing that f is closed and therefore also surjective, hence an isomorphism. Thus $\underline{A}$ is freely generated by M. ∎

The equivalence of (i) and (iv) is therefore proved. ∎∎

(i) ⇒ (v): Because of the assumptions on $\underline{A}$ we may apply 17.2.8 in connection with $K:= \overline{\mathrm{ind}_{\underline{A}}(M)}$, and $\underline{A}$ is fully embeddable over M into $\underline{F}(M,K_t) =: \underline{B}$. Without loss of generality we may even assume that $\underline{A}$ is a relative subalgebra of $\underline{B}$. Consider $h \in F^M(\underline{B};\underline{S})$ such that $(id_M)h \in A$, and let $\underline{a} \in A^M$ be any assignment -- which is obviously also an assignment from M into B. Since M freely generates $\underline{A}$ and $\underline{B}$, $\underline{a}$ can be extended to endomorphisms, say $\underline{a}'$ and $\underline{a}''$, of $\underline{A}$ and $\underline{B}$, respectively, where $\underline{a}'$ is the restriction of $\underline{a}''$ to $\underline{A}$. Therefore $A \ni (id_M)h\underline{a}' = (id_M)h\underline{a}'' = (id_M \cdot \underline{a}'')h = (\underline{a})h$. This shows that the restriction of h to $\underline{A}$ is a total operation on $\underline{A}$. By algebraic induction it can also be shown that the restriction of h to $\underline{A}$ is even an algebraic operation on $\underline{A}$; therefore it belongs to $F^M(\underline{A};\underline{S})$. ∎

(v) ⇒ (i): Assume that $\underline{A}$ is a relative subalgebra of the free total algebra $\underline{B}$ with basis M, such that (v) holds. Let $f:M \to \underline{A}$ be any assignment. Then f has a homomorphic extension $f':\underline{B} \to \underline{B}$. Since $\underline{B}$ is isomorphic to $\underline{F}^M(\underline{B},\underline{S})$, there exists for every element b of B exactly one algebraic operation $h_b \in F^M(\underline{B};\underline{S})$ such that $b = (id_M)h_b$. Now assume that b belongs to A; then we have $bf' = (id_M)h_bf' = (id_M \cdot f')h_b = (f)h_b$. Since $A \ni b = (id_M)h_b$ and $f \in A^M$ we can infer with (v) that $bf' = (f)h_b \in A$. This shows that the restriction $f'|A$ of f' is an endomorphism of $\underline{A}$; therefore M is free in $\underline{A}$. Since M generates $\underline{A}$ by general assumption, M is a basis of $\underline{A}$. ∎

Thus the equivalence of (i) and (v) has been proved. ∎∎

Hence all the equivalences stated in 17.4.1 have been proved. ∎∎∎

17.4.2 REMARKS. (i): Before we discuss the nontriviality assumptions of 17.4.1 let us observe that 17.4.1.(iv) and (v) generalize E.MARCZEWSKI's concept of freeness of total algebras (cf. [Mar 58]), and that freeness with respect to sets is closely related to algebraic total operations, as we have already seen in section 6.

(ii): From our observations in section 15 and on other occasions it is easy to realize that items (i), (ii), (iii) and (vi) of 17.4.1 are equivalent for every partial algebra $\underline{A}$ generated by M. Since M is assumed to be a subset of A, the proof of the equivalence between (i) and (iv) carries over to the arbitrary case.

Only in connection with (v) do we need the assumption of non-triviality of $\underline{A}$: As a simple example consider the case when we have only two sorts, say s and s', and two unary operations, say $\phi: s \mapsto s'$ and $\phi': s' \mapsto s'$, and $A_s := M_s := \{a_1, a_2\}$, and $A_{s'} := M_{s'} := \{a\}$, such that $\phi^{\underline{A}}$ maps a_1 and a_2 on a and $\phi'^{\underline{A}}$ is empty (see Figure 17.1)

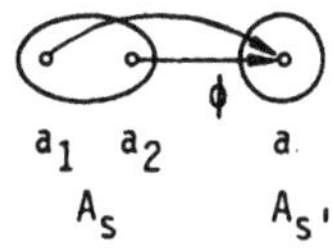

Figure 17.1.

Then it is easy to see that $\underline{A}$ is freely generated by M, but because of $a \in M$ each completion $\underline{B}$ of $\underline{A}$ which has M as basis can only have a one-element phylum of sort s' and therefore $\underline{A}$ cannot be a relative subalgebra of $\underline{B}$. Because of this (v) cannot be equivalent to (i). -- Since a is generated by $\{a_1\}$ or $\{a_2\}$, M is not independent in a closure theoretical sense, while in the homogeneous case an algebraically independent set M is also "closure independent": $m \notin \mathbf{C}_{\underline{A}}(M \setminus \{m\})$ for each $m \in M$.

Notice that we cannot find an example similar to the above one, when S has only one element: Let $A := \{a\}$; if $M = \emptyset$, then a has to be a fundamental constant of $\underline{A}$, and if $\underline{A}$ is partial, then it is easily seen that $\underline{F} := \underline{F}(\underline{A}, \mathbf{TAlg}(\tau))$ is freely generated by M and contains $\underline{A}$ as a relative subalgebra; if $M = \{a\}$ and $\underline{A}$ is partial, then define on F a new structure $(\phi^*)_{\phi \in \Omega}$ as follows:

$$\underline{a}\phi^* := \begin{cases} b & \text{, if } \underline{a} \text{ is a constant sequence with value b, and } \operatorname{graph}(\phi^{\underline{A}}) \neq \emptyset \text{ ,} \\ \underline{a}\phi^{\underline{F}} & \text{, else.} \end{cases}$$

Then $\underline{\mathbf{C}}_{(F,(\phi^*)_{\phi \in \Omega})}M$ is freely generated by M and contains $\underline{A}$ as a relative subalgebra. Since then also the rest of the proof of the equivalence between (i) and (v) in 17.4.1 remains true, we can state the

17.4.3 COROLLARY In the homogeneous case 17.4.1.(i) through 17.4.1.(vi) are equivalent for each partial algebra $\underline{A}$ with respect to some generating subset M of $\underline{A}$. ■■■

17.5 Generators and relations

17.5.1. In particular in group theory it is customary to specify groups by presenting a set, say M, of generators and a set, say G, of equations among terms in these generators -- usually called "relations" -- and to say that the group under consideration is the "free group generated by the set M satisfying the relations in G". This procedure can easily be extended to arbitrary varieties of total algebras, but it seems to be not so well known that it also applies to partial algebras. We shall treat -- for lack of space -- in this subsection only the mere existence and the general principles. More detailed information in connection with hep-varieties can be obtained in PART II.

The main difference between the partial and the total case lies in the fact that in the total case one need not care about which terms are defined and which are not; but in the partial case this is an essential item. The concept of E-equations takes care of this necessity and provides a suitable tool to handle definedness, while the principle of "freeness" takes care of undefinedness in the specification.

Let us make this more precise by an example:

We first want to specify for $\tau := (2)$ $(\Omega := \{\phi\})$ within the whole species **Alg**(τ) the free partial algebra, say $\underline{F}$, generated by $M := \{a,b,c\}$ and satisfying the set (we use the reverse Polish notation) $G := \{(aa\phi, bb\phi), (cc\phi, cc\phi), (ab\phi c\phi, ac\phi)\}$ of relations - actually these relations are formulated in a language for the extended type $\tau' := (2,0,0,0)$, extended by three constants for a, b and c (which we have also written as a,b and c, as is often done in this connection):

First a partial Peano algebra, say $\underline{P}$, is constructed as a relative subalgebra of $\underline{F}(M,\mathbf{TAlg}(\underline{S}))$ consisting of all terms and subterms contained in G, i.e.

$P := \{a, b, c, aa\phi, bb\phi, cc\phi, ab\phi, ac\phi, ab\phi c\phi\}$.

$\underline{P}$ describes the scope of the partial algebra to be specified; and in order to take care of the relations we only have to factor $\underline{P}$ with respect to the congruence relation generated by G in $\underline{P}$, i.e.

$\underline{F} := \underline{P}/\mathrm{Cong}_P G$.

Thus $\underline{F}$ is described by Figure 17.2.

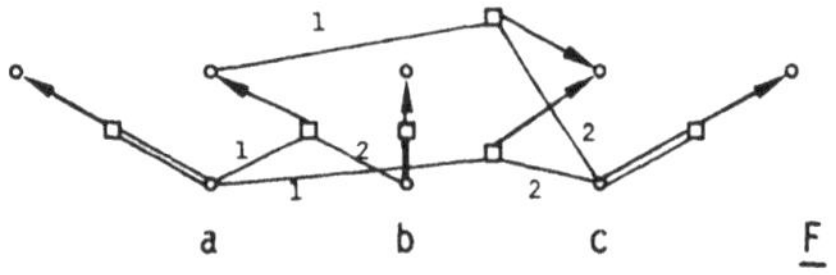

Figure 17.2.

If it is desired that for instance $xx\phi$ should be a total algebraic operation, then one has to specify the free partial algebra "presented by M and G in the E-variety **K** " characterized by the set $Q := \{(xx\phi, xx\phi)\}$ of E-equations, i.e. one has specified $\underline{F}' := \underline{F}(\underline{F},\mathbf{K})$ (which would be described in connection with Figure 17.2 by letting start from each element of the upper row an infinite path, each arrow representing the formation of $(d,d)\phi^{\underline{F}'}$, where d is the corresponding argument).

What has been sketched above is made more precise in 17.5.2 and 17.5.3:

17.5.2 DEFINITION. Let M be any set, let $G \subseteq F(M,\mathbf{TAlg}(\underline{S})) \times F(M,\mathbf{TAlg}(\underline{S}))$ be a set of E-equations, let Q be any set of E-equations with respect to any set X of variables, and let $\mathbf{K} := \mathbf{ModEeq}(Q)$ be the E-variety defined by Q. Then we say that a partial algebra $\underline{A}$ together with a mapping $r_{\underline{A}}: M \to A$ is a <u>free</u> **K**-<u>algebra presented by</u> M and G, if $\underline{A}$ is a **K**-algebra generated by $[M]r_{\underline{A}}$ such that $G \subseteq \ker \tilde{r}_{\underline{A}}$, and for every mapping g from M into any **K**-algebra $\underline{B}$ which satifies $G \subseteq \ker \tilde{g}$, there exists a homomorphism $\bar{g}: \underline{A} \to \underline{B}$ satisfying $g = r_{\underline{A}}\bar{g}$ (see Figure 17.3).

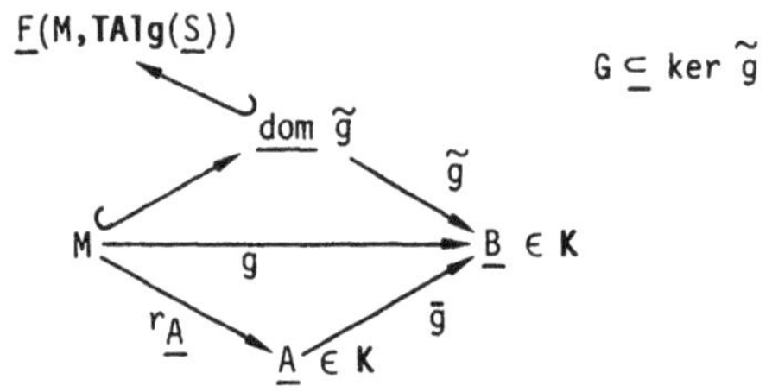

Figure 17.3

17.5.3 PROPOSITION. Let M be any set, let G be a set of E-equations with respect to $\underline{F}(M,\mathbf{TAlg}(\underline{S}))$, let Q be any set of QE-equations with respect to some set X of variables, and let $\mathbf{K} := \mathbf{ModQEeq}(Q)$ be the QE-variety defined by Q. Then the free partial **K**-algebra presented by M and G exists and is determined up to unique isomorphism, and it is denoted by $\underline{F}((M,G),Q)$ (or by $\underline{F}((M,G),\mathbf{K})$, if **K** is the QE-variety defined by Q), and the canonical mapping from M into $\underline{F}((M,G),Q)$ is denoted by $r_{M,G}$.

In particular, $\underline{F}((M,G),Q)$ can be obtained as follows:
Let $\underline{P} := \downarrow(M \cup G)$ be the relative subalgebra of $\underline{F}(M,\mathbf{TAlg}(\underline{S}))$ on the set M joined with the set of all subterms of terms occurring in G; then $(r_{M,G}, \underline{F}((M,G),Q))$ is the **K**-universal **K**-solution of $\underline{P}/Cong_{\underline{P}}(G)$.

PROOF. The description of $\underline{F}((M,G),Q)$ at the end of 17.5.3 together with 17.5.2 easily verifies the statement in 17.5.3. ■■■

17.5.4. Notice that we have already used similar constructions in section 6 and Chapter II, when we translated validity of sets of E- or QE-equations into injectivity with respect to special epimorphisms.

17.6 Initial objects and representation by canonical terms

17.6.1. In connection with the algebraic specification of abstract data types in computer science so-called initial objects within varieties specified by equations or elementary implications play a fundamental rôle. We recall that in category theory an object A of a category $\underline{\mathbf{C}}$ is called an <u>initial object</u> of $\underline{\mathbf{C}}$ iff for every $\underline{\mathbf{C}}$-object B there exists exactly one $\underline{\mathbf{C}}$-morphism from A into B.
Roughly speaking the initial algebra approach to algebraic specification of abstract data types then consists in finding a set of axioms such that the algebraic structure under consideration becomes the initial object in the subcategory of (partial) algebras in which the axioms are valid.
Lemma 17.6.2 shows the connection between initial objects and free (partial) algebras:

17.6.2 LEMMA. Let **K** be any non-empty Horn-variety (cf. 8.1.2). Then one has:
(i) For every partial algebra $\underline{A}$ which allows at least one homomorphism into some **K**-algebra, the **K**-universal solution $\underline{F}(\underline{A},\mathbf{K})$ belongs to **K**.
(ii) **K** has an initial object, and this is naturally isomorphic with $\underline{F}(\emptyset,\mathbf{K})$, the **K**-free **K**-algebra **K**-freely generated by the empty set.

PROOF. Ad (i): **K** is assumed to be a universal Horn-variety, therefore the statement follows from 5.11.9 and 8.3.5.(vii). ■
Ad (ii): $\underline{F} := \underline{F}(\emptyset,\mathbf{K})$ exists in **K**, since the empty set allows a mapping (homomorphism) into every partial algebra, and **K** is assumed to be non-empty. The same reason and the **K**-freeness of $\underline{F}$ allows at least one homomorphism from $\underline{F}$ into each **K**-algebra. And since a homomorphism is uniquely determined by its restriction to a generating subset, and since the empty set allows exactly one mapping into each set, this homomorphism is unique, therefore $\underline{F}$ is an initial object in **K**. It is well known in category theory (exercise!) that there can be up to isomorphism at most one initial object in any category. Thus (ii) is proved. ■■■

Since each E-, ECE-, or QE-variety is always a non-empty HORN-variety, we have the

17.6.3 COROLLARY. Each E-, ECE- or QE-variety **K** of partial algebras has an initial object, and this is represented by $\underline{F}(\emptyset,\mathbf{K})$. ■■■

The "limitations" of the above results and of the initial algebra approach to abstract data types in computer science are shown by the following result:

17.6.4 PROPOSITION. Let $\underline{A}$ be any partial algebra of type τ (with respect to some sort algebra $\underline{S}$), and let M be a generating subset of $\underline{A}$. Moreover, let τ_M be the similarity type obtained from τ by adding a nullary operation symbol, say c_m, to the set Ω of operation symbols for each $m \in M$ such that c_m is of the same sort as m (the set of operation symbols thus obtained will be designated by Ω_M). We then convert $\underline{S}$ into a partial τ_M-algebra $\underline{S}_M$ by defining $c_m^{\underline{S}_M} := (m)\dot{v}_A$, and similarly we convert $\underline{A}$ into $\underline{A}_M$ by defining $c_m^{\underline{A}_M} := m$ (in each case for each $m \in M$). Then $\underline{A}_M$ is an initial **K**-algebra for some suitable E-variety **K** of partial $\underline{S}_M$-algebras.

Notice that the above procedure is well known in model theory as the extension of the underlying language by a suitable set of constants corresponding to some subset M of A.

PROOF Since M generates $\underline{A}$, there exists for each a A a term, say $(M:t_a)$ on M (considering M also as a set of variables) such that $a = (id_M)(M:t_a)^{\underline{A}}$. Let $i: M \to \{c_m | m \in M\}$, $(m)i := c_m$ $(m \in M)$ be a bijective mapping. Define for $\underline{F}_M := \underline{F}(\emptyset, \mathbf{TAlg}(\underline{S}_M))$

$G := \{(\emptyset : (i)(M:t)^{\underline{F}_M} \overset{e}{=} (i)(M:t')^{\underline{F}_M}) \mid t,t' \in F(M,\mathbf{TAlg}(\underline{S}))$ and $(id_M)(M:t)^{\underline{A}} = (id_M)(M:t')^{\underline{A}}\}$

(notice that equality in the meta-language is used with the same semantics as existence equality in the object language).

Then it is easy to check (exercise!) that $\underline{A}_M$ is an initial object in the E-variety $\mathbf{K} := \mathbf{Mod}_{\underline{S}_M}\mathbf{Eeq}_{\underline{S}_M}(G)$. ■■■

17.6.5 REMARKS. (i): The above result shows that theoretically each partial algebra can be algebraically specified as an abstract data type in connection with the initial algebra approach. We "have only forgotten to mention" that this specification has to be an explicit, i.e. a recursive one. Namely, the real problem only starts when one tries to find a **recursive** algebraic specification. We cannot go into more details here on this problem, we only wanted to show the limitations for the praxis of a too general formulation of the initial algebra approach.

(ii): A similar effect is met in connection with the following method of representing partial algebras, which also has a - non-constructive - general solution. Nevertheless it is a very useful tool - like the initial algebra approach -, whenever there is an explicit way to effectively determine such a representation as we want to discuss.

The following definition and result are slight generalizations of Definition 10 and Theorem 8, respectively, in [GoTWa 78].

17.6.6 DEFINITION. A partial algebra $\underline{A}$ is said to be a <u>canonical term algebra</u> iff A is a subset of $F(M,\mathbf{TAlg}(\underline{S}))$ for some term algebra $\underline{F}(M,\mathbf{TAlg}(\underline{S}))$ on a suitable set M of variables, such that

(i) A is closed with respect to the formation of subterms, and

(ii) the relative subalgebra on A of $\underline{F}(M,\mathbf{TAlg}(\underline{S}))$ is a weak subalgebra of $\underline{A}$.

This can also be formulated by saying that a weak subalgebra of $\underline{A}$ with carrier set A is an initial relative subalgebra of some suitable term algebra. While in [GoTWa 78] this concept is restricted to total heterogeneous algebras generated by the empty set, we allow here a more general situation!

17.6.7 THEOREM. Each partial algebra $\underline{A}$ is isomorphic to some canonical term algebra.

PROOF. Let M be any generating subset of $\underline{A}$, and let $f := (id_M)^{\sim} : \underline{dom}\ (id_M)^{\sim} \to \underline{A}$ be the closed initial extension of id_M with respect to the term algebra $\underline{F}(M,\mathbf{TAlg}(\underline{S}))$. Let

$I := \{I \mid I$ is an initial segment of dom f, and the restriction of f to I is injective$\}$.

It is easy to see that the union of any chain of elements of I ordered by set theoretical inclusion also belongs to I; therefore I is inductive and has maximal elements according to Zorn's Lemma (notice that M belongs to I, therefore I is non-empty). Let $C \in I$ be such a maximal element; we then claim that $f|C$ is surjective and therefore bijective (namely we show by algebraic induction that each element of dom f is congruent modulo ker f to some element of C, and since f is surjective, this also implies surjectivity of $f|C$):

First we show that each element of $\mathbf{B}_{\underline{dom}\ f}{}^1C$ is congruent to some element of C: Assume that $t := (t_k \mid k \in K_\phi)\phi \in \mathbf{B}_{\underline{dom}\ f}{}^1C$. $\overline{\text{If}}$ $tf \notin [C]f$, then $f|(C \cup \{t\})$ would be injective, and $C \cup \{t\}$ would also be an initial segment of $\underline{F}(M,\mathbf{TAlg}(\underline{S}))$, contradicting the maximality of C. Now assume that for each ordinal α satisfying $0 \leq \alpha < \beta$ each element of $\mathbf{B}_{\underline{dom}\ f}{}^\alpha C$ is congruent modulo ker f to some element of C, and assume $t := (t_k \mid k \in K_\phi)\phi \in \mathbf{B}_{\underline{dom}\ f}{}^\beta C$, then there is by induction hypothesis for each $k \in K_\phi$ some term $t'_k \in C$ such that $(t_k, t'_k) \in \ker f$. Since ker f is a closed congruence relation and $(t_k \mid k \in K_\phi) \in \text{dom}\ \phi^{\underline{dom}\ f}$, we get $t' := (t'_k \mid k \in K_\phi)\phi \in \mathbf{B}_{\underline{dom}\ f}{}^1C$ and $(t,t') \in \ker f$. By assumption there is $t'' \in C$ such that $(t',t'') \in \ker \overline{f}$; therefore $(t,t'') \in \ker f$, which is what we wanted to show.

$f|C$ is then a bijective homomorphism from the relative subalgebra of $\underline{F}(M,\mathbf{TAlg}(\underline{S}))$ on C onto $\underline{A}$, and the structure on C can be enriched in such a way that $f|C$ becomes an isomorphism from $\underline{C}$ onto $\underline{A}$, while $\underline{C}$ is a canonical term algebra. ∎∎∎

17.6.8 REMARKS and EXAMPLES. (i): Following more the idea of proof in [GoTWa 78] one can proceed in a different way in order to prove 17.6.7: Again let M be any generating subset of $\underline{A}$, and let $f := (id_M)^{\sim} : \underline{dom}\ (id_M)^{\sim} \to \underline{A}$ be the closed initial extension of id_M with respect to the term algebra $\underline{F}(M,\mathbf{TAlg}(\underline{S}))$. For brevity we set $\underline{B} := \underline{dom}\ f$ and denote the congruence class of $t \in B$ with respect to ker f by $[t]$. Let $\delta = \delta(\tau)$ be the ordinal dimension of the type τ, and for each $t \in B$ let $\sigma(t)$ designate the step number of t with respect to M (cf. subsection 3.4). This time we proceed by transfinite induction on the step number of terms to produce a family $(C_n \mid n \leq \delta)$ such that:

(1) If $t \in C_n$, then $\sigma(t) \leq n$.

(2) If $t \in B$ such that $[t]$ has a representative t' with $\sigma(t') \leq n$, then there is a unique representative of t in C_n.

(3) If $(t_k \mid k \in K_\phi)\phi^{\underline{B}} \in C_n$, then $t_k \in C_n$ for each $k \in K_\phi$.

One can now fill in the details of the proof (exercise!) in such a way that one can infer that the set C of canonical terms which one ends up with in the above proof can even be chosen in such a way that for each $(t,t') \in \ker f$ with $t \in C$ the step number of t with respect to M is always less or equal to the one of t' (exercise!), i.e. that the representative in C is always of "minimal complexity".

(ii): The above result is sort of a "normal form theorem" for arbitrary partial algebras with only the drawback that neither C nor the equivalence of any element of dom f to its representative in C is in general computable. Nevertheless the above theorem can be a useful tool in proving freeness (or correct specification) of some partial algebra under consideration, whenever a canonical term algebra representing it can be explicitly constructed. Actually such a construction was a first item for the author showing the usefulness of applying partial algebras (cf. the proof of Satz 2.2 in [B 68]).

(iii): Several applications of the above principle of representation of partial algebras can be found in [GoTWa 78]. Another example is given by representing a free semigroup freely generated by a set, say {a,b,c,d}, by introducing additional constants a, b, c, d to the given type $\tau = (2)$ and considering in $\underline{F}(\emptyset,\mathbf{TAlg}(2,0,0,0,0))$ the subset generated from the empty set by the nullary operations corresponding to the introduced four constants and the unary operations corresponding to the terms xa*, xb*, xc* and xd* ("*" being the operation symbol for the binary operation). The resulting subset of the Peano algebra $\underline{F}(\emptyset,\mathbf{TAlg}(2,0,0,0,0))$ may be endowed in a natural way with a binary multiplication (composition of the corresponding term operations) in order to yield a canonical term algebra for the free semigroup freely generated by a,b,c,d . In this example we have applied another principle often used in connection with the algebraic specification of abstract data types in computer science, namely the **principle of hidden functions** (here the hidden functions are the four auxiliary nullary constants) , which is even more useful in connection with problems of finite axiomatization (cf. [Go 78]).

(iv): In order to show that for the study of initial objects in computer science the aspects taken from universal algebra yield only the trivial part of the problem, while the main difficulties lie in questions of computability, we present the following lemma, whose last statement shows us that each partial algebra can be described as the initial object of a suitable E-variety just by introducing enough additional constants and formulating a suitable set of E-equations.

17.6.9 LEMMA. Let **K** be any E-variety of partial $\underline{S}$-algebras, $\underline{A} \in \mathbf{K}$ be any **K**-algebra, and τ_A be the extension of the underlying similarity type τ by nullary fundamental constants, say c_a, for each element a of A according to 17.6.4. Let $\mathbf{K}_{\underline{A}}$ be the E-variety of all partial $\underline{S}_A$-algebras defined by the set $\mathbf{Eq}(\mathbf{K}) \cup D^+(\underline{A})$ of E-equations, where $D^+(\underline{A}) := \{(\emptyset: t \overset{e}{=} t') \mid t,t' \in F(\emptyset,\mathbf{TAlg}(\underline{S}_A))$, and $\underline{A} \models (\emptyset: t \overset{e}{=} t')\} \cup \{(\emptyset: c_a \overset{e}{=} c_a) \mid a \in A\}$ is equivalent to the model theoretic positive diagram of $\underline{A}$ (cf. CHANG, KEISLER [ChKs 73]). Then one has:

(i) There is a natural bijection β_A between $\mathbf{K}_{\underline{A}}$ and the class $\mathrm{Hom}^*(\underline{A},\mathbf{K}) := \{(f,\underline{B}) \mid f:\underline{A} \to \underline{B}$ and $\underline{B} \in \mathbf{K}\}$, where for any $\underline{C} \in \mathbf{K}_{\underline{A}}$ we define $(\underline{C})\beta_A := (f_{\underline{C}},(C,(\phi^{\underline{C}})_{\phi\in\Omega}) =: \underline{C}_{*A})$ ($\underline{C}_{*A}$ is called the Ω-reduct of $\underline{C}$; $f_{\underline{C}}$ is induced by $a \mapsto c_a^{\underline{C}}$ $(a \in A)$).

(ii) This bijection β_A satisfies $(r_{\underline{A},\mathbf{K}_t}, \underline{F}(\underline{A},\mathbf{K}_t)) = \underline{F}(\underline{A}_A,(\mathbf{K}_{\underline{A}})_t)\beta_A$, i.e. the Ω-reduct of the $(\mathbf{K}_{\underline{A}})_t$-universal $(\mathbf{K}_{\underline{A}})_t$-solution of $\underline{A}_A$ is naturally isomorphic to the $\mathbf{K}_t$-universal $\mathbf{K}_t$-solution of $\underline{A}$.

(iii) $\underline{A}_A$ is an initial object of $\mathbf{K}_{\underline{A}}$.

PROOF. Ad (i): Assume $\underline{C} \in \mathbf{K}_A$, and as indicated let $\underline{C}_{*A}$ designate the Ω-reduct of $\underline{C}$. Since $\underline{C} \models D^+(\underline{A})$, it is easy to realize that $f_{\underline{C}}:\underline{A}_A \to \underline{C}$, $a \mapsto c_a^{\underline{C}}$ $(a \in A)$ is a homomorphism, which remains a homomorphism for the reducts, therefore we can define $(\underline{C})\beta_A := (f_{\underline{C}}, \underline{C}_{*A})$ as indicated. Since $f_{\underline{C}}$ is the only possible homomorphism from $\underline{A}_A$ into $\underline{C}$, β_A is a mapping, and it is obviously injective; that it is also surjective holds, since $f:\underline{A} \to \underline{B}$ implies that

$\underline{B}_A := (B, (\phi^{\underline{B}})_{\phi\in\Omega} \cup (c_a^{\underline{B}_A} := af)_{a\in A}) \in \mathbf{K}_A$, and $(\underline{B}_A)\beta_A = (f, \underline{B})$. ∎

Ad (ii): This is an immediate consequence of (i), since $\underline{C}_A \in (\mathbf{K}_A)_t$ iff $\underline{C}_{*A} \in \mathbf{K}_t$. ∎

Ad (iii): This now follows from 17.6.4. ∎∎∎

18 Dependence of operations on arguments and element basis property

18.1 General concepts

Closely related to the concept of a basis is -- at least in the homogeneous case -- the one of an element basis, which concept is connected with the one of dependence of operations on their arguments. In this subsection we try a generalization to many-sorted partial algebras of material contained in J. SCHMIDT [Sch 64a].

Let us note that as before we denote for every set N and for every sort s (of S) by N_s the subset of N of all elements of sort s: $N_s := (s)v_N^{-1}$.

18.1.1 DEFINITION. Let $\underline{A}$ be any partial algebra, M any subset and x any element of A. A subset M' of M is called an <u>element basis</u> of x in $\underline{A}$ with respect to M, iff the following two conditions hold:

(EB 1) $x \in C_{\underline{A}}M'$;

(EB 2) for every subset M" of M one has $x \in C_{\underline{A}}M"$ iff $M' \subseteq M"$.

If an element basis of some element x of A exists with respect to M, then it is uniquely determined and we shall denote it by xM.

We say that the subset M of A has the <u>element basis property</u> in $\underline{A}$, iff each element x of $C_{\underline{A}}M$ has an element basis in $\underline{A}$ with respect to M.

18.1.2 REMARKS. **(i)**: Let $\underline{A}$ be any partial algebra; let M be any subset of A and $x \in A$ such that the element basis xM of x in $\underline{A}$ for M exists. Then xM is obviously uniquely determined by the other data and is given by

$$xM = \bigcap\{M' \mid M' \subseteq M \text{ and } x \in C_{\underline{A}}M'\} .$$

(ii): So far nothing at all has been said about the existence of element bases in general. It is obvious (cf. section 5) that the Peano basis of any partial Peano algebra (especially of any term algebra) has the element basis property. And if a subset M of a partial algebra $\underline{A}$ has the element basis property, then so has each subset of M.

We shall see below that it is not easy in general -- at least in the really many-sorted or in the infinitary case -- to give criteria for when the element basis exists. In the finitary homogeneous case we can show that except for some restrictions on the values of constant functions all algebraically independent sets have the element basis property.

(iii): One instance when the concept of an element basis may become important, is given in connection with "relativized logic", i.e. when one tries to formulate equations or more general formulas already within some special theory, as is e.g. often done in group theory, where one formulates equations -- or so-called "relations" -- for groups within a given free group rather than in the abstract term algebra thus being able to use associativity and other features in order to simplify the presentation of the axioms.

Before we discuss in more detail the existence of element bases, let us introduce the concepts of "dependence of a partial operation on an argument" and of "constant algebraic operations" together with some facts about them.

18.1.3 DEFINITION. (i): Let M be any set and L any subset of M. Moreover let $\underline{A}$ be any partial algebra and h an M-ary partial operation on $\underline{A}$. We say that h <u>does not depend on (is independent of)</u> L iff for all elements c of A and for all M-ary sequences $\underline{a}, \underline{b} \in A^M$ satisfying $\underline{a}|(M \setminus L) = \underline{b}|(M \setminus L)$ one has $\underline{a}h = c$ iff $\underline{b}h = c$.
Thus h <u>does depend</u> on L iff there are two M-ary sequences $\underline{a}$ and $\underline{b}$ of A which coincide on all but the elements of L such that either h is defined on one of them but not defined on the other, or h is defined on both of them but $\underline{a}h \neq \underline{b}h$, i.e. the values are different.
In case that L = {1} we also say that h does not depend (or depends) on its 1-th argument -- or simply that h does not depend (or depends, respectively) on (the variable) 1.
(ii): Let us define a <u>constant</u> partial operation to be such a partial operation which does not depend on any of its arguments.
(iii): Finally let $C(\underline{A})$ designate the set of all values of constant algebraic operations of $\underline{A}$.

18.1.4. It is easy to realize that the above definition automatically implies that a constant operation is either total or empty. Namely when for one argument there is no element in the partial algebra for which the operation takes a value, then it obviously has an empty domain, and otherwise it has to be total.

18.1.5 LEMMA. For any partial algebra $\underline{A}$, $C(\underline{A})$ is a closed subset of $\underline{A}$; and if $\underline{A}$ is **homogeneous** and if there is at least one non-empty nullary fundamental constant, then $C(\underline{A}) = \mathbf{C}_{\underline{A}}\emptyset$.

PROOF. If $C(\underline{A})$ is empty, then there are no constant algebraic operations in $\underline{A}$ (in particular if there are no nullary constants in the type under consideration), and $\emptyset$ is closed in $\underline{A}$. If $C(\underline{A})$ is non-empty, then assume $\phi \in \Omega$, $\underline{a} \in \operatorname{dom} \phi^{\underline{A}} \cap C(A)^{K_\phi}$ and $b := \underline{a}\phi^{\underline{A}}$. For each $k \in K_\phi$ there then exist some constant, say N_k-ary, algebraic operation h_k and some N_k-ary sequence $\underline{c}_k$ such that $(k)\underline{a} = (\underline{c}_k)h_k$. Obviously $h := (h_k | k \in K_\phi)*\phi^{\underline{A}}$ (superposition of the first kind !) is a constant algebraic operation of $\underline{A}$ with b among its values, i.e. $b \in C(\underline{A})$. --
Let finally $\underline{A}$ be homogeneous and $\mathbf{C}_{\underline{A}}\emptyset$ be non-empty, i.e. it contains at least one element which then may be used to generate within $\mathbf{C}_{\underline{A}}\emptyset$ all elements of $C(\underline{A})$. The converse inclusion is trivial. ■■■

In the case of many-sorted partial algebras $\underline{A}$ with at least two sorts $C(\underline{A})$ may contain several other closed subsets even when $\mathbf{C}_{\underline{A}}\emptyset$ is non-empty. The following example may provide an idea of what the situation might then look like:

18.1.6 EXAMPLE. Let us assume that $S := \{s,s'\}$, $\Omega := (\phi,\phi',\phi'')$, $\tau := (0,1,2)$, and $\phi^{\underline{S}} := \{s\}$, $\phi'^{\underline{S}} : s' \mapsto s'$, $\phi''^{\underline{S}} : (s,s') \mapsto s$. The partial algebra $\underline{A}$ which we want to consider is then described in Figure 18.1.

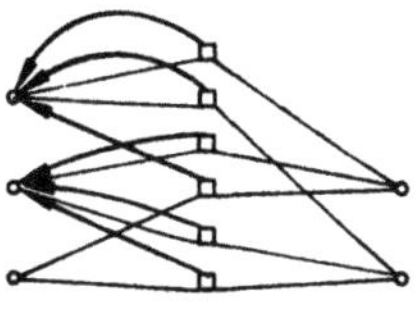

Figure 18.1.

It is then easy to realize that $C(\underline{A}) = \{a,b,c\}$, that $\mathbf{C}_{\underline{A}}\emptyset = \{a\}$, and that $\{b\}$ is another closed subset of $\underline{A}$ properly contained in $C(\underline{A})$ and even disjoint from $\mathbf{C}_{\underline{A}}\emptyset$. We have added two additional elements in A in order to make $\phi''^{\underline{A}}$ dependent on each of its arguments.
It should be easy to realize that -- by using more sorts -- the lattice of closed subsets of $\underline{A}$ contained in $C(\underline{A})$ can be made arbitrarily large.
The situation is once more different when we only allow closed subsets with non-empty phyla, since then it may happen that $C(\underline{A})$ is not of this kind. We do not discuss this further here.

18.1.7 OBSERVATION. Since we are just discussing $C(\underline{A})$, we would like to mention that -- by introducing also $D(\underline{A})$ as the intersection of all non-empty closed subsets of $\underline{A}$ -- J.SCHMIDT has shown in [Sch 64a] (Corollary to Theorem 4) that there are exactly the following four possible cases for **homogeneous** partial algebras $\underline{A}$:

I. $\emptyset \subset \mathbf{C}_{\underline{A}}\emptyset = C(\underline{A}) = D(\underline{A})$;

II. $\emptyset = \mathbf{C}_{\underline{A}}\emptyset \subset C(\underline{A}) = D(\underline{A})$;

III. $\emptyset = \mathbf{C}_{\underline{A}}\emptyset = C(\underline{A}) \subset D(\underline{A})$;

IV. $\emptyset = \mathbf{C}_{\underline{A}}\emptyset = C(\underline{A}) = D(\underline{A})$.

That even for partial homogeneous algebras these are the only possibilities ([Sch 64a] only deals with total algebras) can easily be derived from Lemma 18.1.5. In order to show that they are all possible, let us quote the following examples from [Sch 64a]:
EXAMPLE I. Remember that $\mathbf{C}_{\underline{A}} \neq \emptyset$ iff $\underline{A}$ has at least one non-empty nullary constant, and this can easily be realized. The equalities then follow from 18.1.5.
EXAMPLE II. Consider a group $\underline{A}$ as an algebra with multiplication and inversion as the only fundamental operations, both non-nullary. Then $x \mapsto x \cdot x^{-1}$ represents a constant algebraic operation with the unit element e as constant value. And since {e} is a closed subset, $C(\underline{A}) = \{e\}$. Nevertheless $\mathbf{C}_{\underline{A}}\emptyset = \emptyset$, since one has "forgotten" to introduce the nullary operation corresponding to the unit element e as a fundamental operation.
EXAMPLE III. Let A consist of two different elements a, b; let $\phi^{\underline{A}}$ be the only non-trivial permutation of A considered as the only fundamental operation. Then there is no constant algebraic operation at all, and therefore $C(\underline{A}) = \emptyset$, whereas $D(\underline{A}) = A$, since A is the only non-empty closed subset of $\underline{A}$.
EXAMPLE IV. In order to realize IV we can choose any algebra with at least two elements such that any one-element subset is a subalgebra (e.g. let $\underline{A}$ be a lattice, an affine space, etc.). ■■■

18.2 Dependence on variables and algebraic independence

A first connection between the concepts of dependence on variables and the element basis property on one side and algebraic independence on the other is indicated by the following

18.2.1 LEMMA. Let M be some algebraically independent subset of some partial algebra $\underline{A}$, let M', M" be subsets of M such that for, some $s \in S$, M'_s and M''_s are disjoint, and assume that $g \in APO^{M'}(\underline{A};\underline{S})$ and $h \in APO^{M''}(\underline{A};\underline{S})$ are algebraic partial operations such that $id_{M'}g = id_{M''}h$. Then g (and similarly h) does not depend on any argument of sort s. In particular, if M' and M" are disjoint, then g is constant.

PROOF. Clearly we assume that M'_s is non-empty. Let $c := id_{M'}g = id_{M''}h$; since M' and M" are independent, this implies that g and h are total algebraic operations on $\underline{A}$, i.e. we do not have to worry about the domains. Assume that $\underline{a},\underline{b} \in A^{M'}$ are any sequences which differ at most on arguments from M'_s, and let $f,f':M \to A$ be any mappings such that $f|M' = \underline{a}$, $f'|M' = \underline{b}$, and $f|M'' = f'|M''$ (this is possible, since $M'_s \cap M''_s = \emptyset$). Since M is independent in $\underline{A}$, f and f' have homomorphic extensions, say $f^*,f'^*:\underline{C_A}M \to \underline{A}$, respectively. As f* and f'* permute with algebraic partial operations, we get

$$\underline{a}g = (id_{M'}f^*)g = (id_{M'}g)f^* = (id_{M''}h)f^* = (id_{M''}f^*)h =$$
$$= (id_{M''}f'^*)h = (id_{M''}h)f'^* = (id_{M'}g)f'^* = (id_{M'}f'^*)g = \underline{b}g ,$$

showing that in fact g takes the same value on $\underline{a}$ and $\underline{b}$. ■■■

18.2.2 LEMMA. Let M be some algebraically independent subset of some partial algebra $\underline{A}$ with algebraic closure operator $\mathbf{C} := \mathbf{C}_{\underline{A}}$. Let M',M" be subsets of M. Then

$$\mathbf{C}M' \cap \mathbf{C}M'' \begin{cases} \subseteq C(\underline{A}) & , \text{ if } M' \cap M'' = \emptyset , \\ = \mathbf{C}(M' \cap M'') & , \text{ if } M'_s \cap M''_s \neq \emptyset \text{ for all } s \in S . \end{cases}$$

For the case when for some phyla the intersections are empty while they are nonempty for others, see the remarks below.

PROOF. As we have shown earlier each element of, say, **C**M' is the value of an M'-ary algebraic partial operation applied to $id_{M'}$. Since M' is independent in $\underline{A}$ as a subset of the independent set M, each such algebraic operation is total on $\underline{A}$. If M' and M" are disjoint, then each element of $\mathbf{C}M' \cap \mathbf{C}M''$ is the value of a total algebraic constant operation and therefore it belongs to $C(\underline{A})$ (cf. 18.2.1).
Now assume that $M'_s \cap M''_s \neq \emptyset$ for each sort s. Then there is a homomorphism f from $\underline{\mathbf{C}M}$ into $\underline{A}$ such that $af \in M' \cap M''$ for all $a \in M'$ and $bf = b$ for all $b \in M''$. A similar calculation as in the proof of 18.2.1 yields for every $c \in \mathbf{C}M' \cap \mathbf{C}M''$ that $c = (id_{M'})g = (id_{M''})h$ for appropriate algebraic partial operations g and h, i.e.

$$c = (id_{M''})h = (id_{M''}\cdot f)h = ((id_{M''})h)f = ((id_{M'})g)f = (id_{M'}\cdot f)g \in \mathbf{C}(M' \cap M'') .$$

Thus $\mathbf{C}M' \cap \mathbf{C}M'' \subseteq \mathbf{C}(M' \cap M'')$; since the converse inequality is trivial, even equality holds. ■■■

18.2.3 COROLLARY. Let $\underline{A}$ be any partial algebra such that $\mathbf{C}_{\underline{A}}\emptyset = C(\underline{A})$, and let M be an independent subset of $\underline{A}$ such that all elements of M are of the same sort. Then

$$\mathbf{C}_{\underline{A}}M' \cap \mathbf{C}_{\underline{A}}M'' = \mathbf{C}_{\underline{A}}(M' \cap M'')$$

for all subsets $M',M'' \subseteq M$.

Notice that the assumption on M is in particular satisfied when $\underline{A}$ is homogeneous. ■■■

18.2.4 REMARK and EXAMPLE. (i): The hypothesis " $C_{\underline{A}}\emptyset = C(\underline{A})$ " in 18.2.3 means that the sets of values of constant algebraic operations of a certain arity L and of arity $\emptyset$, respectively, are the same; or as $C_{\underline{A}}\emptyset = C(\underline{A})$ is equivalent with $C(\underline{A}) \subseteq C_{\underline{A}}\emptyset$, we may simply say that any value of a constant algebraic operation is the value of a nullary algebraic operation. As a matter of fact, this condition is -- at least in the homogeneous case -- a very weak one; for in the only case when it does not hold originally for homogeneous partial algebras (see case 18.1.7.II), one may slightly alter the algebraic structure of $\underline{A}$ by introducing one single new fundamental operation, namely the nullary operation corresponding to some certain element of $C(\underline{A})$, thus obtaining a partial algebra, say $\underline{A}^*$, "with constants", whose closure operator, say **C***, fulfills condition $C^*\emptyset = C(\underline{A}^*) = C(\underline{A})$.
In the inhomogeneous case $C_{\underline{A}}\emptyset = C(\underline{A})$ can also be obtained by adding "forgotten nullary constants" to the type, but in general more than one might be necessary.
(ii): While in the homogeneous case only the values of constant algebraic total operations may cause some trouble in connection with the element basis property for independent subsets, the situation is much more complicated in the inhomogeneous case. Let us illustrate the situation in the **many-sorted** case by an **example**:
$S := \{s,s'\}$, $\Omega := \{\phi\}$, $\tau := (2)$, $\mathrm{graph}(\phi^{\underline{S}}) := \{((s,s'),s')\}$.
The (total) $\underline{S}$-algebra $\underline{A}$ is indicated in Figure 18.2.
Let $M := \{a,b,a'\}$, $M' := \{a,a'\}$, $M'' := \{b,a'\}$.
It is easy to see that M is independent, that M generates $\underline{A}$, that $M'_s \cap M''_s = \emptyset$ and $M'_{s'} \cap M''_{s'} = \{a'\} \neq \emptyset$, and that $CM' \cap CM''$ ($= \{a',b',c'\}$) is neither contained in $C(\underline{A})$ ($= \emptyset$), nor is it equal to $C(M' \cap M'')$ ($= \{a'\}$).
Notice that $\phi^{\underline{A}}$ does not depend on its 0-th argument but on its 1st one.

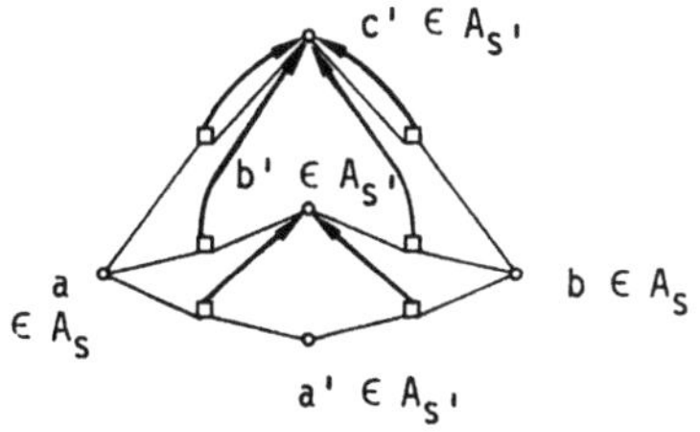

Figure 18.2.

(iii): This simple example may give an idea how complicated the situation can become in the many-sorted case. We leave a more detailed discussion as a problem if there is any need for it. Let us only finally observe that in the example of $\underline{A}$ above the element b' has no element basis with respect to M.

In connection with the following results one should observe the different assumptions concerning arities (finitary or infinitary) and number of sorts (homogeneous or arbitrary).

18.2.5 PROPOSITION. Let $\underline{A}$ be any partial algebra such that -- for $C := C_{\underline{A}}$ -- $C(\underline{A}) = C\emptyset$, let h be an algebraic partial operation of type L on $\underline{A}$ such that either all elements of L are of the same sort or each non-empty phylum of L has at least two elements, and let $\underline{a} \in \mathrm{dom}\, h$ be a sequence without repetitions (i.e. injective) of type L in $\underline{A}$. Let the range M* of $\underline{a}$ be minimal for the generation of $x := (\underline{a})h$, i.e. $x \notin CM'$ for all proper subsets M' of M*. Then h depends on each of its variables.

PROOF. Let us assume the existence of an index $l_0 \in L$ such that h does not depend on l_0 (i.e. on its l_0-th argument). In case $L = \{l_0\}$ h would be constant; therefore $x \in C(\underline{A})$, i.e. $x \in \mathbf{C}\emptyset$ by hypothesis, contradicting the minimality of $M^* = \{(l_0)\underline{a}\} \neq \emptyset$. In case $L \neq \{l_0\}$ we might define a transformation σ of L onto $K := (L \setminus \{l_0\})$ which carries l_0 into $l_1 \neq l_0$, leaving all other elements of L fixed (this is possible, since the phylum of L containing l_0 has to contain by our assumptions at least one further element). As the sequences $\underline{a}$ and $(((l)\sigma)\underline{a} | l \in L)$ only differ on the argument $l = l_0$, and as h does not depend on its l_0-th argument, we obtain

$$x = \underline{a}h = (((l)\sigma)\underline{a} | l \in L)h ;$$

therefore $x \in \mathbf{C}M'$, where $M' := \{((l)\sigma)\underline{a} | l \in L\}$. But since $\underline{a}$ is a sequence without repetitions, we get $(l_0)\underline{a} \notin M'$, thus M' is a proper subset of M*, again contradicting the minimality of M*. ∎

18.2.6 REMARK. Our example in 18.2.4.(ii) shows that restrictive assumptions on M are necessary in the above proposition. Namely in this example we can generate c from M only by using $\phi^{\underline{A}}$ which is neither constant nor does it depend on its 0-th argument, nor can it be transformed into an algebraic operation depending on every argument, since its arity does not allow a transformation (over $\underline{S}$!) into a strictly smaller arity.

18.2.7 PROPOSITION. Let $\underline{A}$ be a partial algebra with **finitary** fundamental partial operations such that $C(\underline{A}) = \mathbf{C}\emptyset$ -- again we set $\mathbf{C} := C_{\underline{A}}$ --, and let M be a subset of A such that all elements of M are of the same sort. Then each element $x \in \mathbf{C}M$ can be represented in the form $x = \underline{a}h$, where h is an algebraic partial operation of some (finite!) type L depending on each of its arguments, and $\underline{a}$ is an L-ary sequence without repetition and arguments in M.
We may choose $L = M^*$, where M* is a finite subset of M, and $(l)\underline{a} = l$ for all $l \in L = M^*$, i.e. $\underline{a} = id_{M^*}$.

PROOF. Since the fundamental operations are finitary, **C** is an ($\#\omega_0$-) algebraic closure operator; therefore there exists for each $x \in \mathbf{C}M$ a finite subset $F \subseteq M$ such that $x \in \mathbf{C}F$. Thus (by Tarski's finiteness criterion) there exists a minimal subset, say $M^* \subseteq F$, such that $x \in \mathbf{C}M^*$, and as a consequence x can be represented in the form $x = (id_{M^*})h$, where h is an M*-ary algebraic operation. Since id_{M^*} obviously is a sequence without repetitions and all of its arguments are of the same sort, Proposition 18.2.5 implies that h depends on each of its variables. ∎

We shall show in 18.3.2 that this result cannot be generalized to partial algebras with infinitary operations.
While the last results did not need independence, this concept becomes relevant in the following one.

18.2.8 PROPOSITION. Let $\underline{A}$ be a partial algebra such that $C(\underline{A}) = \mathbf{C}\emptyset$ -- $\mathbf{C} := C_{\underline{A}}$ --, let h be an algebraic partial operation of type L on $\underline{A}$ which depends on each of its variables, let $\underline{a} \in \mathrm{dom}\, h$ be a sequence without repetitions of type L in $\underline{A}$, and let its range, say M*, be a subset of the independent subset M of A. Then M* is a minimal subset of M for the generation of $x := \underline{a}h$, and if all elements of M are of the same sort -- i.e. in particular if $\underline{A}$ is homogeneous --, then it even is **the minimum** subset with this property, i.e. $M^* \subseteq M'$ for all $M' \subseteq M$ such that $x \in CM'$.

PROOF. First, let us assume the existence of some proper subset $M' \subset M^*$ such that $x \in CM'$. Then there would be an index $l_0 \in L$ such that $(l_0)\underline{a} \notin M'$. Therefore $x \in C\{(l)\underline{a} | l \in K\}$ where $K := L \setminus \{l_0\}$, hence $x = ((k)\underline{a} | k \in K)g$ where g is a partial algebraic operation of type K on $\underline{A}$. Let i be the identity transformation from K into L, then (cf. 15.2.5) $\underline{a}h = x = ((k)\underline{a} | k \in K)g = (((k)i)\underline{a} | k \in K)g = \underline{a}g_{i,L}$.
Since $g_{i,L}$ is an algebraic partial operation as well as h, and since $\underline{a}$ is a sequence without repetitions in the independent set M, we obtain $h = g_{i,L}$ (cf. Theorem 17.4.1 -- in the total case this is Marczewski's criterion of independence in [Mar 58]). But then h does not depend on its l_0-th argument; for let $\underline{c}$ and $\underline{d}$ be sequences of type L in $\underline{A}$ which differ only on l_0, and observe that h is total -- M being independent --, then

$$(\underline{c})h = (\underline{c})g_{i,L} = ((k)\underline{c} | k \in K)g = ((k)\underline{d} | k \in K)g = (\underline{d})g_{i,L} = (\underline{d})h .$$

This contradicts our hypothesis on h. Thus, as $x \in CM^*$, M^* is a minimal subset $M' \subseteq M$ such that $x \in CM'$.
Now assume that all elements of M are of the same sort. Then M^* even is the minimum of all these subsets M'; for $\underline{A}$ satisfies $C(\underline{A}) = C\emptyset$. Therefore 18.2.3 implies that $x \in CM^* \cap CM' = C(M^* \cap M')$. Thus $M^* \cap M' = M^*$, i.e. $M^* \subseteq M'$. ■■■

18.2.9 THEOREM. Let $\underline{A}$ be a partial algebra such that $C(\underline{A}) = C\emptyset$ -- $C := C_{\underline{A}}$ -- and let M be an independent subset of $\underline{A}$ such that all elements of M are of the same sort. Then for an element $x \in CM$ the following properties are equivalent:
(i) The element basis for x in $\underline{A}$ with respect to M exists.
(ii) x can be represented in the form $x = (id_{M^*})h$, where h is an M^*-ary algebraic (even total) operation on $\underline{A}$ depending on each of its variables, and $M^* \subseteq M$.
Moreover, when one of these statements is true, then the above representation of x is unique and M^* is the element basis for x in $\underline{A}$ with respect to M ("unique irredundant coordinate representation").

PROOF. **"(i) ⇒ (ii)"**: Let M^* be the element basis for x in $\underline{A}$ with respect to M. Because of $x \in CM^*$, x can be represented in the form $x = (id_{M^*})h$, where h is an algebraic (total!) operation of type M^* in $\underline{A}$. But sequence id_{M^*} being without repetitions, its range being a minimal set M' such that $x \in CM'$, and all its elements being of the same sort, h depends on each of its variables according to Proposition 18.2.5. ■
"(ii) ⇒ (i)": The algebraic operation h depending on each of its variables, sequence id_{M^*} being without repetions, its range M^* being a subset of the independent set M, all of whose elements are of the same sort, M^* is the element basis of x in $\underline{A}$ with respect to M according to Proposition 18.2.8 This proves also the uniqueness of M^* and h in this representation of x (M^* itself being independent!). ■■■

We now turn to the element basis property; as we have already seen in 18.2.4.(ii) the element basis of an element need not always exist even with respect to independent subsets (see also 18.3.2 below). For purely closure theoretic results concerning the element basis and the element basis property see [Sch 64a]. Here we state as a consequence of the preceding results:

18.2.10 THEOREM. Let $\underline{A}$ be a partial algebra with **finitary** fundamental partial operations such that $C(\underline{A}) = \mathbf{C}_{\underline{A}}\emptyset$, and let M be an independent subset of $\underline{A}$ such that all elements of M are of the same sort. Then M has the element basis property in $\underline{A}$; moreover, the element bases for x in $\underline{A}$ with respect to M are finite for all elements of $\mathbf{C}_{\underline{A}}M$. ■■■

18.3 A counterexample for the infinitary homogeneous case

Let us now consider a counterexample for the infinitary case(see. [Sch 64a]): I.e. we want to show that even in a total homogeneous algebra $\underline{A}$ without constants, with infinitary fundamental operations, an independent subset M of $\underline{A}$ need not have the element basis property. As a preparation we prove

18.3.1 PROPOSITION. For a homogeneous partial algebra $\underline{A}$ let h be a non-constant algebraic partial operation of type L which does not depend on any of its arguments, let $\underline{a}$ be an L-ary sequence without repetitions in A, and let its range M be an independent subset of $\underline{A}$. Then the element $x := (\underline{a})h$ (which belongs to $\mathbf{C}_{\underline{A}}M$) does not possess an element basis in $\underline{A}$ with respect to M.

PROOF. Let $M^* \subseteq M$ be an arbitrary subset such that $x \in \mathbf{C}_{\underline{A}}M^*$. Let K be the set of all $l \in L$ such that $(l)\underline{a} \in M^*$; then there is an algebraic operation g of type K on $\underline{A}$ such that $x = ((k)\underline{a}|k \in K)g$. Let i be the identity transformation from K into L and $\underline{b} := ((k)\underline{a}|k \in K)$; then we get (cf. the proof of 18.2.8)

$$(\underline{a})h = x = (\underline{b})g = (((k)i)\underline{a}|k \in K)g = \underline{a}g_{i,L} .$$

Since $\underline{a}$ is a sequence without repetitions in the independent set M, and since h and $g_{i,L}$ are algebraic partial operations which coincide on $\underline{a}$, we obtain $g_{i,L} = h$, and therefore $(c_l|l \in L)h = (c_l|l \in K)g$ for each sequence $(c_l|l \in L) \in A^L$. Therefore, since h is non-constant and does not depend on any of its arguments, the same is true with g. One concludes that K is infinite (since otherwise g could not be non-constant without depending on any of its arguments), and in particular, K contains two distinct elements, say k_0 and k_1. We define a transformation p from K onto $K \setminus \{k_0\}$ which carries k_0 into k_1, leaving all other elements of K fixed. As in the proof of 18.2.5 we obtain $x = (((k)p)\underline{a}|k \in K)g$. Therefore $x \in \mathbf{C}_{\underline{A}}M'$, where $M' := \{((k)p)\underline{a}|k \in K\} = M^* \setminus \{(k_0)\underline{a}\}$ is a proper subset of M^*: M^* is therefore not the element basis of x in $\underline{A}$ with respect to M (in fact, k_0 has been an arbitrary element of K, thus the intersection of all such M' is empty!). ■■■

18.3.2 EXAMPLES. (i): While every finitary partial operation not depending on any of its arguments is constant, this need not be the case for an infinitary partial operation, as any example of a (not too trivial) Fréchet-Urysohn limit space shows (for instance take the limit-space connected with the real or the rational line): We may define a partial operation h of type ω_0 by $(x_n|n \in \omega_0)h := \lim_{n\to\infty} x_n$, if this limit exists, otherwise undefined.

It is easy to realize that h does not depend on any of its arguments, nor is it constant. If one chooses one fixed element, say a, of the carrier set and defines a as value of h whenever h is originally undefined, one even gets a total operation with these properties.

(ii): An example of infinitary partial algebras in which independent sets need not have the element basis property can now be constructed as follows:

Let **K** be the class of all (total) algebras $\underline{A}$ with precisely one fundamental operation of type ω_0 which does not depend on any of its variables. It is easily shown that **K** is an E-variety: For each argument n one has to take two ω_0-sequences $\underline{a}_n$ and $\underline{b}_n$ (of variables) without repetitions which differ exactly in the n-th argument, and one has to consider the E-equation $(\underline{a}_n)h \overset{e}{=} (\underline{b}_n)h$. The set of all these E-equations defines **K**. Example (i) above shows that **K** is non-trivial in the sense that it contains at least one algebra of more than one element. Therefore the **K**-free **K**-algebra $\underline{F}:= \underline{F}(\omega_0,\mathbf{K})$ exists (with an infinite basis). Obviously the only fundamental operation $\phi^{\underline{F}}$ is non-constant and does not depend on any of its arguments (since the examples constructed in (i) belong to **K** -- in particular the one with the rational line as carrier set; therefore we may construct a surjective homomorphism from $\underline{F}$ onto this algebra --, showing that $\phi^{\underline{F}}$ is non-constant); hence by 18.3.1 no element of F possesses an element basis with respect to ω_0 in $\underline{F}$, although it is even true that $C(\underline{F}) = C_{\underline{F}}\emptyset$, i.e. ω_0 does not have the element basis property in $\underline{F}$.

19 Congruence relations revisited

19.1 Closed congruence relations of finitary partial algebras

19.1.1 INTRODUCTORY REMARKS and DEFINITIONS. In section 2 we have already studied congruence relations of partial algebras in general and closed congruence relations in particular, and we have seen that not every pair of elements (of the same sort) in a given partial algebra will belong to a closed congruence relation, but that in the finitary case there exists a largest closed congruence relation containing all those pairs , and that any congruence relation contained in a closed one is also closed. Hence a syntactical description of this largest closed congruence relation is equivalent to a syntactical description of those pairs generating a closed congruence relation.

Moreover, we have already seen that the lattice of all closed congruence relations behaves much more like the congruence lattices of total algebras than the lattice of all congruence relations does. We shall give some more evidence for this observation in this subsection.

First of all let us recall that the set of all M-ary algebraic partial functions of a given partial algebra $\underline{A}$ is just the set $APO^M(\underline{A}_A;\underline{S}_A)$ of all M-ary algebraic operations of the partial algebra $\underline{A}_A$, where all elements of A are additional fundamental constants. In this subsection we shall need in particular the sets $APF_s(\underline{A};\underline{S}):= APO^{\{s\}}(\underline{A}_A;\underline{S}_A)$ of all unary partial algebraic functions of input sort s for each $s \in S$. Finally, we consider $APF(\underline{A};\underline{S}):= \bigcup\{APF_s(\underline{A};\underline{S}) \mid s \in S\}$, the set of all unary algebraic partial functions of $\underline{A}$. Let us observe in this connection that one can obtain any algebraic partial function of $\underline{A}$ by substituting elements of A of the appropriate sort for some set (possibly empty) of variables in an algebraic partial operation of $\underline{A}$.

Finally, for any given finitary partial algebra $\underline{A}$, let $R_c:= R_c(\underline{A})$ designate the largest closed congruence relation of $\underline{A}$.

19.1.2 LEMMA. Let $\underline{A}$ be any partial algebra. Then an equivalence relation on A is a (closed) congruence relation of $\underline{A}$ iff it is a (closed) congruence relation of $\underline{A}_A$.

PROOF. This statement is obvious, since every equivalence relation is compatible with constant operations. ■■■

Of interest for us are the following consequences:

19.1.3 COROLLARY. Each congruence relation of a given partial algebra is compatible with all algebraic partial functions. ■■■

19.1.4 PROPOSITION. Let $\underline{A}$ be a **finitary** partial algebra, and let R be any equivalence relation on A. Then the following statements are equivalent:

(i) R is a closed congruence relation of $\underline{A}$.

(ii) R is compatible with all unary algebraic partial functions of $\underline{A}$, and if f is any unary algebraic partial function on $\underline{A}$, and $(a,b) \in R$ any element of R, then f is defined on a iff f is defined on b.

PROOF. It is an immediate consequence of 19.1.3 that (i) implies (ii). Thus let us assume (ii): Let $\underline{a},\underline{b} \in A^{K_\phi}$ such that for each $l \in K_\phi$ one has $((l)\underline{a},(l)\underline{b}) \in R$, and $\underline{a} \in \mathrm{dom}\ \phi^{\underline{A}}$. For simplicity we assume that $K_\phi = \{1,\dots,k\}$ is the set of the first k natural numbers ($\neq 0$), and that $(i)\underline{c} = c_i$ for $\underline{c} \in \{\underline{a},\underline{b}\}$ and $1 \leq i \leq k$. Let x_i be a variable of the suitable sort and define for $1 \leq i \leq k$
$(x_i)p_i :\doteq (b_1,\dots,b_{i-1},x_i,a_{i+1},\dots,a_k)\phi^{\underline{A}}$, whenever this is defined and makes sense. Thus we obtain k unary algebraic partial functions on $\underline{A}$. We then have $a_1 \in \mathrm{dom}\ p_1$, and it is easy to realize because of the assumptions on R -- using the "infix notation" for relations -- that all of the following elements are defined and satisfy

$$\underline{a}\phi^{\underline{A}} = a_1p_1 \ R \ b_1p_1 = a_2p_2 \ R \ \dots \ R \ b_{k-1}p_{k-1} = a_kp_k \ R \ b_kp_k = \underline{b}\phi^{\underline{A}} .$$

Because of the transitivity of R this shows that $\underline{b} \in \mathrm{dom}\ \phi^{\underline{A}}$ and $(\underline{a}\phi^{\underline{A}},\underline{b}\phi^{\underline{A}}) \in R$, i.e. that R is a closed congruence relation on $\underline{A}$. ■■■

This suggests the following

19.1.5 THEOREM. Let $\underline{A}$ be any finitary partial algebra, and let a,b A be any two elements of the same sort, say s. Then the following statements are equivalent:

(i) $(a,b) \in R_C$.

(ii) The so-called <u>principal congruence relation</u> $R_{(a,b)} :=$ **Cong**$_{\underline{A}}(\{(a,b)\})$ is closed.

(iii) Each unary algebraic partial function of $\underline{A}$ is defined on a iff it is defined on b.

(iv) For $M := \{s\}$ each unary term $(M:t) \in \underline{F}(M,\mathbf{TAlg}(\underline{S}_A))$ satisfies $a \in \mathrm{dom}\ (M:t)^{\underline{A}_A}$ iff $b \in \mathrm{dom}\ (M:t)^{\underline{A}_A}$.

PROOF. The equivalence of (i) and (ii) is obvious from what has already been said in 19.1.1. Because of 19.1.4 (iii) is an immediate consequence of (i). Thus it remains only to show that (iii) implies (i): For this purpose we define

$R^\circ := \{(a,b) \mid a$ and b are elements of A of the same sort, and each algebraic partial function is defined on a iff it is defined on $b\}$.

From what we have shown so far, we can infer that $R_c \subseteq R^\circ$. The Theorem will be proved, when we have shown that R° is a closed congruence relation on $\underline{A}$ (since R_c is the greatest one):

It is obvious from the definition that R° is reflexive and symmetric; and transitivity of R° is not less obvious. Hence R° is an equivalence relation. Because of Proposition 19.1.4 it is therefore sufficient to show that R° is compatible with all unary algebraic partial functions of $\underline{A}$: Therefore we assume $(a,b) \in R^\circ$ and $p \in APF_s(\underline{A};\underline{S})$, where s is the sort of a and b, moreover consider $q \in APF_{s'}(\underline{A};\underline{S})$, where s' is the output sort of p. From subsections 15.2 and 15.5 we know that the composition pq is then a unary algebraic partial function, too. Thus we have by the assumptions on R:

$$ap \in \operatorname{dom} q \quad \text{iff} \quad a \in \operatorname{dom} pq \quad \text{iff} \quad b \in \operatorname{dom} pq \quad \text{iff} \quad bp \in \operatorname{dom} q \ ,$$

showing that also $(ap,bp) \in R^\circ$, completing the proof that (iii) implies (i).

The equivalence between (iii) and (iv) is an immediate consequence of the close relationship between algebraic partial operations and partial term operations (cf. subsection 15.5). ■

The important rôle that is played by unary algebraic partial functions in connection with closed congruence relations is also stressed by the following result which is the analogue for closed congruence relations to the corresponding result for total algebras (cf. GRÄTZER [G 68], §10, Theorem 3).

19.1.6 THEOREM. Let $\underline{A}$ be any finitary partial algebra, and $(a,b) \in R_c(\underline{A})$. Then for any two elements x and y of A of the same sort, say s, the following statements are equivalent:

(i) $(x,y) \in R_{(a,b)}$.

(ii) There exist a natural number n, a sequence $x = z_0, z_1, \ldots, z_n = y$ of elements of A of the same sort s, and a sequence $p_0, \ldots, p_{n-1}$ of unary algebraic partial functions of $\underline{A}$ such that

$$\{ap_i, bp_i\} = \{z_i, z_{i+1}\} \ , \quad \text{for } i = 0,1,\ldots,n-1 \ .$$

See Figure 19.1 in 19.1.8.(i) below for an illustration.

PROOF. Because of 19.1.3, (i) is trivially implied by (ii), since a (closed) congruence relation is transitive and compatible with all unary algebraic partial functions. In order to show that (i) implies (ii) it is sufficient to show that the relation

$$Q := \{(x,y) \in A^2 \mid x \text{ and } y \text{ satisfy (ii)}\}$$

is a congruence relation on $\underline{A}$, namely closedness will then follow from the facts that (a,b) is contained in $R_c(\underline{A})$ -- and that (ap,bp) belongs to $R_{(a,b)}$ for each unary algebraic partial function p of $\underline{A}$, i.e. all of Q is contained in $R_{(a,b)}$ and therefore $Q = R_{(a,b)}$.

The formulation of the definition of Q immediately implies that Q is symmetric and transitive (one just reverses or composes the sequences under consideration). Reflexivity of Q follows from the fact that each element c of A induces a total unary algebraic function for each input sort with constant value c. Hence Q is an equivalence relation.

Because of 19.1.4 and our observations so far it is now sufficient to show the substitution property (i.e. compatibility with the algebraic structure) only for unary algebraic partial functions:

Thus let p be any unary algebraic partial function of $\underline{A}$, and assume $(u,v) \in Q$, $u \in \operatorname{dom} p$. By the definition of Q there exists a sequence $u = z_0, \ldots, z_n = v$, with the associated sequence of unary partial algebraic functions $p_0, \ldots, p_{n-1}$, such that $\{z_i, z_{i+1}\} = \{ap_i, bp_i\}$, $0 \le i \le n-1$.

Notice that, for each i, $z_i \in \operatorname{dom} p$, since $u \in \operatorname{dom} p$, and $(z_i, z_{i+1}) \in Q \subseteq R_c$ $(0 \le i \le n-1)$ -- apply 19.1.4. We have to show that $(up, vp) \in Q$. But this is immediately realized because of the sequences

$$up = z_0 p, \ldots, z_n p = vp \quad \text{and} \quad p_0 p, \ldots, p_{n-1} p,$$

which then satisfy $\{z_i p, z_{i+1} p\} = \{ap_i p, bp_i p\}$, $0 \le i \le n-1$. ■■■

Since the supremum of closed congruence relations in a finitary partial algebra is the equivalence theoretical one, we immediately get the

19.1.7 COROLLARY. Let $\underline{A}$ be any finitary partial algebra, and $H \subseteq R_c$. Then the following statements are equivalent for any $c,d \in A$ (of the same sort):

(i) $(c,d) \in \mathbf{Cong}_{\underline{A}}(H)$.

(ii) There exist a natural number n, a sequence $c = z_0, z_1, \ldots, z_n = d$ of elements of A of the same sort s, pairs of elements $(a_i, b_i) \in H$, and unary algebraic partial functions p_i of $\underline{A}$, $1 \le i \le n$, such that

$$\{a_i p_i, b_i p_i\} = \{z_{i-1}, z_i\}, \quad \text{for } i = 1, \ldots, n. \quad ■■■$$

19.1.8 REMARKS. (i): The description of a principal closed congruence relation $R_{(a,b)}$ given by 19.1.6 can be visualized by Figure 19.1.

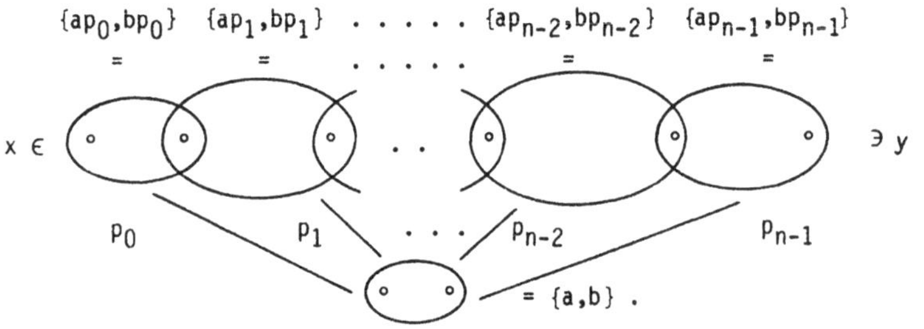

Figure 19.1.

(ii): The above results show the important rôle that is played by unary algebraic partial functions in connection with closed congruence relations, and the equivalences in the results above also imply that this will no longer be true when we drop the assumption of closedness. Actually there cannot be given in general an upper bound below the cardinal dimension for the cardinality of the arities of algebraic partial functions needed to describe compatibility of arbitrary congruence relations:

As a simple example consider the following homogeneous partial algebra $\underline{A}$ of type $\tau := (n)_{n \in \omega_0 \setminus \{0\}}$ i.e. we have $\Omega = \{\phi_n \mid n \in \omega_0 \setminus \{0\}\}$, and

$A := \dot{\bigcup}\{a_1^n, \ldots, a_n^n, b_1^n, \ldots, b_n^n, c^n, d^n \mid n \in \omega_0 \setminus \{0\}\}$; finally we assume

$\operatorname{graph}(\phi_n) := \{((a_1^n, \ldots, a_n^n), c^n), ((b_1^n, \ldots, b_n^n), d^n)\}$, for each $n \in \omega_0 \setminus \{0\}$.

When all the elements listed for A are distinct, then it is easy to see that there are equivalence relations on A, for which one needs algebraic partial functions of arbitrary high finite arity in order to show that they are congruence relations on $\underline{A}$. For instance consider the equivalence relation on A generated by
$R := \{(a_j^n, b_j^n), (c^n, d^n) \mid 1 \le j \le n$, and $n \in \omega_0 \setminus \{0\}$ even$\}$.
(iii): A characterization of $R_c(\underline{A})$ different from the one in 19.1.5 has been given by I.KUPKA in his habilitation thesis (see [Ku 77]), where the main line of the ideas and representations in this connection are correct, but the proofs and Lemma 2.3.(21) there contain some incorrect parts; a correct version of this description is contained in [Si 82] (see also the appendix of [Si 84]), while a third description has been communicated by KUPKA to the author, but we do not know whether or where it will be published. Since the version in [Ku 77] needs additional replacement operators and is somewhat more involved, we do not give here any more details, since we hope that Theorem 19.1.5 will serve most purposes.

From section 2 we already know that the congruence lattice of any finitary partial algebra is algebraic, and that the closed congruence relations form a principal ideal in it. We conclude this subsection with a general representation theorem in this connection, which has been discovered independently by J.BERMAN (cf. [Bm 71]) and the author (cf. [B 71a]) for the case of homogeneous partial algebras. As a preparation we need the following

19.1.9 LEMMA. Let $\underline{A}$ be some finitary total algebra, let R be any congruence relation of $\underline{A}$, and define $\underline{B} := (A, (\varphi^{\underline{A}})_{\varphi \in \Omega} \cup (id_C)_{C \in A/R})$. Then one has
(i) $\underline{Cong}(\underline{A}) = \underline{Cong}(\underline{B})$, and
(ii) $Cong_c(\underline{B})$ is the principal ideal of $\underline{Cong}(\underline{A})$ generated by R.

PROOF. Since unary partial operations which behave like the identity on their domain in the corresponding phylum, do not influence compatibility of an equivalence relation with the algebraic structure, the first statement is obvious. But the second statement is not less trivial: Since all the operations of $\underline{A}$ are total, closedness of a congruence relation of $\underline{B}$ only depends on the additional operations of $\underline{B}$. Let R' be any congruence relation of $\underline{B}$. If $R' \subseteq R$, and $C \in A/R$, then for any $(a,b) \in R'$ one has $a \in dom\ id_C$ iff $b \in dom\ id_C$, showing that R' is closed; but in case that $R' \not\subseteq R$ there is $(a,b) \in R' \setminus R$. Let $a \in C \in A/R$; then, by assumption, $b \notin C$, i.e. $a \in dom\ id_C$, but $b \notin dom\ id_C$, showing that R' is not closed. ■■■

Since each algebraic lattice is isomorphic to the congruence lattice of some finitary total (homogeneous) algebra (see [G 68] for references and a proof, and the second edition of [G 68] for even more references on more simple proofs and additional results), we immediately get from 19.1.9 the

19.1.10 THEOREM. Let $\underline{L}$ be any algebraic lattice, and let I be any principal ideal of $\underline{L}$. Then there exists a finitary homogeneous partial algebra, say $\underline{A}$, such that $\underline{L}$ is isomorphic to the congruence lattice of $\underline{A}$, and the restriction to I of this isomorphism is a bijection from I onto the set of all closed congruence relations of $\underline{A}$. ■■■

19.2 P-reflecting homomorphisms and P-closed congruence relations

19.2.1. In subsection 9.2 we have introduced the notion of a P-reflecting homomorphism as a concept which combines those of ordinary and closed homomorphisms in a common generalization. We have the feeling that such a concept might become of interest in connection with some questions in computer science, e.g. when one tries to define a concept of reduction for automata which are given as partial automata with states, input and output sorts, and where for reduction only such partial term operations or functions are of interest, which have as a resulting sort an output sort -- one might even restrict the consideration to such algebraic partial functions, whose variables only take states as values. As we shall see in the next section, P-reflecting homomorphisms are also important in connection with subdirect representations in E-varieties.
We recall that for a given set P of terms homomorphisms are called P-reflecting iff they reflect all terms in P -- here "reflection of a term", say of (M:t), means reflection of the TE-expression $(M: t \stackrel{e}{=} t)$. Thus the properties of homomorphisms reflecting a given set of formulas as described in subsection 9.1 especially apply in this case.
In this subsection we want to continue the study of P-reflecting homomorphisms especially by considering the congruence relations which they induce. For this purpose we introduce the

19.2.2 DEFINITION. Let P be any set of terms for some type τ, and let R be any congruence relation of some partial algebra $\underline{A}$ of the same type. Then R is called a P-closed congruence relation of $\underline{A}$, iff the natural projection $nat_R:\underline{A} \to \underline{A}/R$ is a P-reflecting homomorphism.
$Cong_P(\underline{A})$ designates the set of all P-closed congruence relations of $\underline{A}$.

19.2.3 PROPOSITION. $Cong_P(\underline{A})$ is always an initial segment of $\underline{Cong}(\underline{A})$, the congruence lattice of $\underline{A}$. And any full, P-reflecting and surjective homomorphism $g:\underline{A} \to \underline{B}$ induces a bijection between the set of all P-closed congruence relations of $\underline{A}$ which contain ker g (on one side) and $Cong_P(\underline{B})$ (on the other side).

PROOF. Let R be any P-closed congruence relation of $\underline{A}$, and assume $R' \subseteq R$. Then there is a homomorphism $g:\underline{A}/R' \to \underline{A}/R$ such that $nat_R = nat_{R'} \cdot g$ (cf. the Homomorphism Theorem). Since g is a homomorphism and therefore preserves all terms, Lemma 9.1.7.(ii) implies that $nat_{R'}$ is P-reflecting. The surjectivity of $nat_{R'}$ implies that g is P-reflecting, too; therefore ker g = R/R' is P-closed. The fact that the composition of P-reflecting homomorphism is again P-reflecting then implies the rest of the above proposition. ∎

19.2.4 PROPOSITION. Let $\underline{A}$ be any partial algebra of cardinal dimension **d**. Then $Cong_P(\underline{A})$ is **d***-inductive for $\mathbf{d}^* := \max\{\#\omega_0, \mathbf{d}\}$.

PROOF. Let $\mathbf{R} \subseteq Cong_P(\underline{A})$ be any **d***-directed subset. Then
$\mathbf{L} := (\underline{A}/R, nat_{R,R'}:\underline{A}/R \to \underline{A}/R')_{R \subseteq R' \in \mathbf{R}}$ is a **d***-directed system of P-reflecting homomorphisms. Let $f_0:\underline{A} \to \varinjlim \mathbf{L}$ be the induced homomorphism from $\underline{A} = \underline{A}/\Delta$, then sections 4.4 and 11.3 imply that f_0 is full and P-reflecting and it is easily seen that $\ker f_0 = \bigcup\{R \mid R \in \mathbf{R}\}$, i.e. $\ker f_0$ is P-closed. ∎

19.2.5 REMARK. An internal characterization of P-closed congruence relations is not as simple as one might think at the first glance, where one would hope that the following concept might work:

Let P be any set of terms, $\underline{A}$ a partial algebra of the corresponding type, and let R be a congruence relation of $\underline{A}$. Let us call R <u>weakly</u> P-<u>closed</u> iff, for every $(M:t) \in P$, and for every $\underline{a},\underline{b} \in A^M$, which satisfy $\underline{a} \in \mathrm{dom}\,(M:t)^{\underline{A}}$ and $((m)\underline{a},(m)\underline{b}) \in R$ for each $m \in M$, one has $\underline{b} \in \mathrm{dom}\,(M:t)^{\underline{A}}$.

Then it is easy to realize that every P-closed congruence relation is weakly P-closed, but the converse is not true, as the following simple example shows: Let $\tau := (1,1)$, $\Omega := \{\phi,\phi'\}$, $A := \{x, x\phi, y, \phi'\}$, and $\underline{A}$ the relative subalgebra on A of the homogeneous Peano algebra with Peano basis $\{x,y\}$. Finally let $R := \mathbf{Cong}_{\underline{A}}(\{(x\phi,y)\})$ and $P := \{z, z\phi, z\phi\phi'\}$ (for simplicity we omit the one-element arities!). Then it is easy to realize that R is weakly P-closed, but not P-closed, since $(\phi\phi')^{\underline{A}/R}$ is defined on $[x]R$, but $(\phi\phi')^{\underline{A}}$ is not defined on x.

This example and especially the corresponding Figure 19.2 show, what an internal characterization of P-closed congruence relations has to look like. Roughly speaking this can be described as follows:

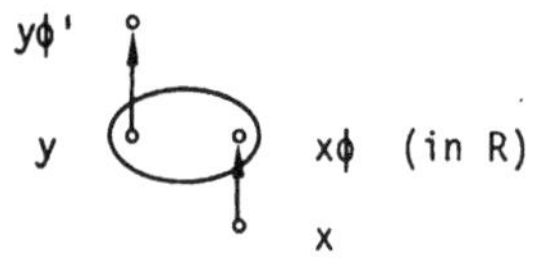

Figure 19.2

If one represents each term in P by its corresponding tree, then one has to split each node into two new ones, thus disconnecting the tree (now a forest) at this place, labeling the new leaf by a new variable of the corresponding sort, and leaving the old label at the new root. We show this by a more complex example: $S := \{s,s',s''\}$, $\mathrm{graph}(\phi^{\underline{S}}) := \{((s),s')\}$, $\mathrm{graph}(\phi'^{\underline{S}}) := \{((s',s''),s)\}$, $\mathrm{graph}(\phi''^{\underline{S}}) := \{((s',s,s''),s'')\}$, and where we want to consider for variables x: of sort s, y: of sort s', and z: of sort s'' the term $(\{x,x',y,y',z,z',z''\}: x\phi x'\phi' z\phi' yy'z'\phi' z''\phi''\phi'') =: (M,t)$. Figure 19.3 then shows first the tree representation of (M:t) on the left, and on the right the "corresponding R-split forest," whose interpretation (including the R-relationships indicated by ellipses -- a root represents a whole element) leads in the factor to the existence of the corresponding term operations (see 19.2.6 below):

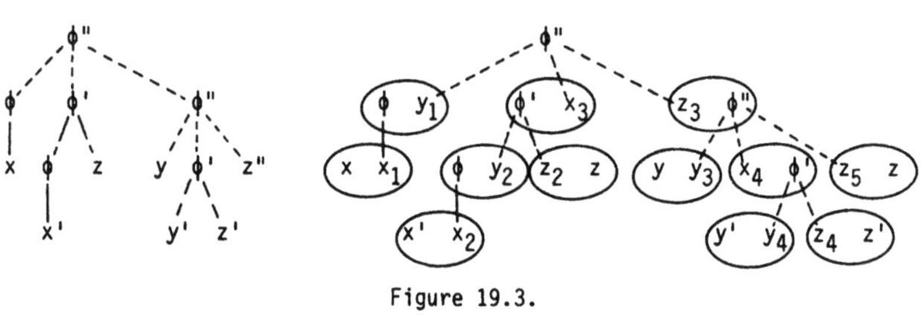

Figure 19.3.

We do not present here an explicit formulation for the general situation, since we have the feeling that this is better illustrated in connection with the above examples rather than by a clumsy formal description: R is (M:t)-closed iff: whenever the above variables are evaluated by two sequences of elements of A, say (a,a',b,b',c,c',c") and (a*,a'*,b*,b'*,c*,c'*,c"*) such that these sequences are componentwise R-congruent, and if all the elements of the R-split forest are interpreted with respect to the first sequence, then they are interpreted with respect to the second sequence, and for all

these latter interpretations the root is R-congruent with the root of the interpretation connected with the first sequence.
In the case that P is closed with respect to subterms, we can give more easily a more formal description of P-closed congruence relations. (Recall that we have already shown in 9.2.6 that the set of terms reflected by a homomorphism need not be closed with respect to the formation of subterms!)

19.2.6 PROPOSITION. Let P be a set of terms which is closed with respect to the formation of subterms, and let $\underline{A}$ be any partial algebra of the type under consideration. Then for any congruence relation R of $\underline{A}$ the following statements are equivalent:
(i) R is P-closed.
(ii) If $(M:p) = (M:(p_k|k \in K_\phi)\phi) \in P$ is any term, if $\underline{a} \in \operatorname{dom} \phi^{\underline{A}}$ is any sequence, and if $\underline{w}:M \to A$ is any assignment such that $\underline{w} \in \operatorname{dom} p_k^{\underline{A}}$ and $((k)\underline{a}, \underline{w}p_k^{\underline{A}}) \in R$ for every $k \in K_\phi$, then $(\underline{w}p_k^{\underline{A}}|k \in K_\phi) \in \operatorname{dom} \phi^{\underline{A}}$.

PROOF. From what has already been observed in 19.2.5 it is easily seen that (i) implies (ii) (since P is closed with respect to subterms!). The converse implication is proved in showing by algebraic induction that (ii) implies (*), where:
(*) If $(M:p) \in P$ and $\underline{w}\cdot nat_R \in \operatorname{dom} p^{\underline{A}/R}$, then $\underline{w} \in \operatorname{dom} p^{\underline{A}}$ (for every $\underline{w} \in A^M$).

If p is a variable, then $(M:p)^{\underline{A}}$ and $(M:p)^{\underline{A}/R}$ are at the same time either empty or total, and (*) is trivially satisfied. Now assume that p is given as in (ii), and assume $\underline{w}\cdot nat_R \in \operatorname{dom} (M:p)^{\underline{A}/R}$ for some $\underline{w} \in A^M$. Then $\underline{w}\cdot nat_R \in \operatorname{dom} (M:q)^{\underline{A}/R}$ for every subterm (M:q) of (M:p), and in particular $\underline{w}\cdot nat_R \in \operatorname{dom} (M:p_k)^{\underline{A}/R}$ for every $k \in K_\phi$. By the induction hypothesis we therefore have $\underline{w} \in \operatorname{dom} p_k^{\underline{A}}$ for every $k \in K_\phi$. Moreover, since nat_R is full, there exists some $\underline{a} \in \operatorname{dom} \phi^{\underline{A}}$ such that $\underline{a}\cdot nat_R = ((\underline{w}p_k^{\underline{A}})nat_R|k \in K_\phi)$. Because of (ii) we now get $(\underline{w}p_k^{\underline{A}}|k \in K_\phi) \in \operatorname{dom} \phi^{\underline{A}}$, and this implies (*).
Since (*) has now been proved for every $(M:p) \in P$, R is shown to be P-closed. ∎∎∎

19.2.7 PROPOSITION. Let **K** be any E-variety, let D:= ↓**Eeq**(**K**) be the set of all terms with $\mathbf{d}^{\&}$-small sets of variables in X inducing total term operations on each **K**-algebra, and let $P \subseteq D$ be any set of terms. Then each homomorphism between any two **K**-algebras is P-reflecting, and in particular we have for any **K**-algebra $\underline{A}$ that $\operatorname{Cong}_P(\underline{A}) = \operatorname{Cong}(\underline{A})$, i.e. each congruence relation of $\underline{A}$ is P-closed.

PROOF. This is an immediate consequence of the fact that the TE-statements corresponding to the terms in D are already valid in each **K**-algebra. ∎∎∎

19.2.8 CONCLUDING REMARKS. (i): From the results in section 9 quite a few conclusions can be drawn for P-reflecting homomorphisms. For this purpose let us define for a given set **H** of homomorphisms **Term**(**H**) to be the set of all terms (with arities less than $\mathbf{d}^{\&}$ contained in a suitable set X of variables) which are reflected by all homomorphisms within **H**, and for a given set P of terms let **Mor**(P) be the set of all homomorphisms reflecting all terms in P. It is easily seen that **Term** and **Mor** are the operators of a Galois connection, and that **TermMor** and **MorTerm** are closure operators on their corresponding underlying classes. As we have already observed **TermMor**(P) is usually not closed with respect to the formation of subterms, but it trivially contains all terms (M:x) for any variable $x \in M$ of any given sort, and for any subset M of the set X of

variables under consideration (notice that one starts with assignments taking their values in the domain of a homomorphism!). Moreover, the results in section 9 show that **TermMor**(P) is always closed with respect to substitution, and if it is closed with respect to subterms, then for any fixed subset $M \subseteq X$ the terms $(M:p) \in$ **TermMor**(P) form the carrier set of a relative subalgebra of $\underline{F}(M,\mathbf{TAlg}(\underline{S}))$ freely generated by M.
(ii): In the **homogeneous case** let X be a set of cardinality $\mathbf{d}^* := \max\{\mathbf{d}, \#\omega_0\}$, and consider all terms as elements of $\underline{F}(X,\mathbf{TAlg}(\tau))$. Then $P^* :=$ **TermMor**(P) -- for some subset $P \subseteq F(X,\mathbf{TAlg}(\tau))$ which is closed with respect to subterms --, iff P^* is the carrier set of a free relative subalgebra of $F(X,\mathbf{TAlg}(\tau))$ freely generated by X, and then at the same time we get that **Mor**(P^*) = **Mor**(P) is the set of all $\mathrm{ind}_{\underline{P}^*}(X)$-perfect homomorphisms in **Alg**(τ). -- A similar result can be formulated in the heterogeneous case with respect to the TE-variety defined by **TermMor**(P), as long as P is closed with respect to subterms.

20 Subdirect representations of partial algebras

20.1 The general concept and remarks

20.1.1. In the theory of total algebras the concept of subdirect representation of an algebra has been introduced in order to have a description of algebras by smaller and hopefully simpler building blocks. Despite the fact that one has the theorem of G.Birkhoff that every algebra is a subdirect product of subdirectly irreducible algebras, the theory of subdirect representations has rather more theoretical than practical importance in most instances, since for only very few varieties one definitely knows all the subdirectly irreducible members. On the other hand the fact that a variety possesses -- up to isomorphism -- only a set of subdirectly irreducible members, i.e. that it is residually small, usually allows one to draw quite a few other conclusions.
In this section we want to generalize the concept of subdirect representation to partial algebras. But there are many different properties of homomorphisms which could play a rôle (and actually will do); e.g. there is a concept of subdirect representation of partial algebras contained in the book of R.S.PIERCE [Pi 68], another one in an exercise of G.GRÄTZER's book [G 68] (cf. exercise 96 to Chapter 3); both of them will be considered later.
In order to get a more general background we introduce in this subsection what might be a general concept of subdirect representation. As we have been told, such a concept has already been investigated by I.SAIN in [Sa 74] (in Hungarian).

20.1.2 DEFINITION. Let $\underline{\mathbf{C}}$ be any category with products, and let E and **M** be classes of epimorphisms respectively monomorphisms of $\underline{\mathbf{C}}$ which are each closed with respect to composition and both contain all isomorphisms of $\underline{\mathbf{C}}$. A family $(e_i : A \to B_i \mid i \in I)$, $I \neq \emptyset$, of $\underline{\mathbf{C}}$-morphisms is called an (**E**,**M**)-<u>subdirect representation</u> of an object A of $\underline{\mathbf{C}}$, and A is called an (E,**M**)-<u>subdirect product</u> of the family $(B_i)_{i \in I}$, if each morphism e_i belongs to E $(i \in I)$, and if moreover the induced $\underline{\mathbf{C}}$-morphism $m : A \to \mathsf{X}(B_i \mid i \in I)$ from A into the product of the family $(B_i \mid i \in I)$ belongs to **M**. A is called (**E**,**M**)-<u>subdirectly</u>

irreducible, if each (**E**,**M**)-subdirect representation of A contains an isomorphism (i.e. if there is an $i \in I$ such that e_i is an isomorphism).

20.1.3 REMARKS. **(i)**: The above concept may seem to be too general, but when one considers existing or so far proposed concepts of subdirect representations in connection with relational systems or partial algebras, then it seems to be the most narrow one covering all of them. In a so-called concrete category, i.e. when there is a full and faithful functor into the category **Set** of all sets with ordinary mappings as morphisms -- as it is the case with **Alg**($\underline{S}$) or **TAlg**($\underline{S}$) --, then it is quite customary to restrict the class E of epimorphisms to be contained in the class of all morphisms whose corresponding value in **Set**, i.e. whose "underlying mapping", is surjective; and we shall do so, too, in what follows . But since we know that each epimorphism starting from some partial algebra $\underline{A}$ can be characterized by an initial closed congruence relation of $\underline{F}(\underline{A}$,**TAlg**($\underline{S}$)), it seems by no means necessary to make this restriction in order to get satisfactory results (observe that in [BSi 84] also epimorphisms in general are treated, and cf. [Wi 70]).

(ii): In the case of total algebras the most powerful tool in connection with subdirect representations and the decision for subdirect irreducibility are congruence relations -- and in the case of relational systems H.E.PICKETT shows in [Pc 64] that here equivalence relations also play a fundamental rôle. Our results and observations below will follow the usual pattern as far as possible.

(iii): For partial algebras the discussion of (**E**,**M**)-subdirect representations will usually be quite simple, as long as **M** is chosen to be the class of all monomorphisms. If then **E** is a class of epimorphisms reflecting some given set, say **F**, of formulas, then -- by the results of subsection 9.1 -- **M** can without loss of generality be also restricted to the class of all monomorphisms reflecting **F**, since the embedding morphism will inherit this reflection property. The situation will become more involved, when **M** is supposed to satisfy additional requirements -- e.g. when the monomorphisms have to be full or P-reflecting, and P-reflection is not required for the elements of **E**.

(iv): As in the case of total algebras the proof of the existence of enough (**E**,**M**)-subdirectly irreducible objects -- in order to have the result that each partial algebra has an (**E**,**M**)-subdirect representation with (**E**,**M**)-subdirectly irreducible objects -- needs the Axiom of Choice(as a matter of fact, this existence is even equivalent with the Axiom of Choice, as has been shown by G.GRÄTZER in [G 85]). Therefore we shall have to restrict our considerations to finitary similarity types.

20.2 Classes of full and surjective homomorphisms -- arbitrary monomorphisms

20.2.1 LEMMA. Let τ be a finitary similarity type, let **M** be the class of all **injective** homomorphisms in **Alg**($\underline{S}$), and let **E** be a composition closed class of **full and surjective** homomorphisms in **Alg**($\underline{S}$) such that for each partial algebra $\underline{A}$ the class of all **E**-morphisms starting from $\underline{A}$ is represented up to isomorphism by an **inductive** set $\mathrm{Cong}_E(\underline{A})$ of congruence relations of $\underline{A}$, and such that **E** is left cancellable with respect to **E**-morphisms (i.e. $fg \in E$ and $f \in E$ imply $g \in E$). Then one has for $\underline{A} \in$ **Alg**($\underline{S}$) :

(i) Let $e:\underline{A} \to \underline{B}$ be any **E**-morphism, then

(a) e induces an isomorphism, say e^*, (with respect to set theoretical inclusion) between $\{R \in \mathrm{Cong}_E(\underline{A}) \mid \ker e \subseteq R\}$ and $\mathrm{Cong}_E(\underline{B})$.

(b) $\underline{B}$ is (E,M)-subdirectly irreducible, iff

(SI) $\bigcap\{R \in \mathrm{Cong}_E(\underline{A}) \mid \ker e \subset R\} \neq \ker e$, whenever the index set for the intersection on the left side is non-empty;

-- ker e will then (when (SI) is satisfied) be called a <u>completely E-meet irreducible</u> congruence relation of $\underline{A}$.

(ii) $\underline{A}$ is (E,M)-subdirectly irreducible iff Δ_A satisfies (SI), i.e. iff Δ_A is completely E-meet irreducible.

(iii) Let $a,b \in A$ be elements of the same sort, $a \neq b$. Then there exists a maximal E-congruence relation, say $R_E(a,b)$, in $\mathrm{Cong}_E(\underline{A})$ such that $(a,b) \notin R_E(a,b)$; and $R_E(a,b)$ is completely E-meet irreducible.

20.2.2. Notice that the congruence relation $R_E(a,b)$, which we consider in 20.2.1.(iii), is in general not uniquely determined by (a,b), but that the rest of the statement of 20.2.1.(iii) applies to each such congruence relation. And if we use later on the notation $R = R_E(a,b)$, then this means that R could serve as $R_E(a,b)$.

PROOF. Ad (i): Let $e:\underline{A} \to \underline{B}$ be any E-morphism (thus $\underline{B}$ is isomorphic to $\underline{A}/\ker e$) . In order to prove (a) let us first consider $R \in \mathrm{Cong}_E(\underline{A})$ such that $\ker e \subseteq R$. Since E is assumed to be left cancellable with respect to E-morphisms, $\mathrm{nat}_{R/\ker e}$ also belongs to E, i.e. $(R)e^* := R/\ker e \in \mathrm{Cong}_E(\underline{B})$. This shows that e^* is the restriction of the isomorphism between the corresponding intervals of congruence relations as induced by the Homomorphism Theorem, and therefore e^* preserves set theoretical inclusion and is injective. Surjectivity follows from the fact that E is assumed to be closed with respect to composition; thus, for $R' \in \mathrm{Cong}_E(\underline{B})$, we get that $e \cdot \mathrm{nat}_{R'}$ belongs to E, and its kernel is the preimage of R' with respect to e^*.

-- Now let us prove (b): To begin with, let ker e be completely E-meet irreducible, and let $(e_i:\underline{B} \to \underline{C}_i \mid i \in I)$ be an (E,M)-subdirect representation of $\underline{B}$. Let $\underline{C}$ designate the direct product of the family $(\underline{C}_i)_{i\in I}$, and let $m:\underline{B} \to \underline{C}$ be the product homomorphism induced by the family $(e_i)_{i\in I}$. By assumption $\Delta_B = \ker m = \bigcap\{\ker e_i \mid i \in I\}$. It is then easy to realize that $(\Delta_B)e^{*-1} = \ker e = \bigcap\{\ker ee_i \mid i \in I\}$, and because of (SI) this can only happen when $\ker e = \ker ee_{i'}$ for some $i' \in I$. This, however, implies $\ker e_{i'} = \Delta_B$, and $e_{i'}$ has to be an isomorphism, as it is in addition full and surjective. Thus $\underline{B}$ is (E,M)-subdirectly irreducible.

Next, let us assume that ker e does not satisfy (SI). It is then easy to see that $(\mathrm{nat}_{R/\ker e}:\underline{B} \to \underline{A}/R \mid R \in \mathrm{Cong}_E(\underline{A})$, and $\ker e \subset R)$ yields an (E,M)-subdirect representation of $\underline{B}$, which contains no isomorphism (cf. (a)). Thus the equivalence of (i)(b) has been proved. ∎

(ii) is just a corollary of (i). ∎

Ad (iii): Let $\mathbf{R}_E(a,b)$ be the set of all elements of $\mathrm{Cong}_E(\underline{A})$ avoiding (i.e. not containing) (a,b). Then $\Delta_A \in \mathbf{R}_E(a,b)$, since E contains all isomorphisms. Moreover, the union of any non-empty chain (or directed set) in $\mathbf{R}_E(a,b)$ also avoids (a,b) and belongs to $\mathrm{Cong}_E(\underline{A})$, since $\mathrm{Cong}_E(\underline{A})$ is assumed to be inductive. Thus $\mathbf{R}_E(a,b)$ is itself inductive and therefore contains maximal elements, one of which will be denoted by $R_E(a,b)$. It now easily follows that $R_E(a,b)$ satisfies (SI), since each element of $\mathrm{Cong}_E(\underline{A})$, which properly contains $R_E(a,b)$ also contains (a,b), and the intersection of all such congruence relations (if there are any) will also contain (a,b) and has therefore to be different from $R_E(a,b)$. ■■■

As an immediate consequence we get the following generalization to partial algebras of G.Birkhoff's theorem on subdirect representations:

20.2.3 THEOREM. Let $\mathbf{M}$ be the class of all injective homomorphisms of a species $\mathbf{Alg}(\underline{S})$ with finitary operations, and let $\mathbf{E}$ be a composition closed class of full and surjective homomorphisms such that $\mathbf{E}$ contains all isomorphisms, and is left cancellable with respect to $\mathbf{E}$-morphisms, and, for each $\underline{A} \in \mathbf{Alg}(\underline{S})$, $\mathrm{Cong}_E(\underline{A})$ is inductive.
Then every $\mathbf{Alg}(\underline{S})$-algebra $\underline{A}$ has an $(\mathbf{E},\mathbf{M})$-subdirect representation of $(\mathbf{E},\mathbf{M})$-subdirectly irreducible partial algebras.

PROOF. According to Lemma 20.2.1.(iii) there exists for any partial algebra $\underline{A}$ and for any distinct elements $a,b \in A$ of the same sort a maximal -- completely $\mathbf{E}$-meet irreducible -- congruence relation $R_E(a,b)$ avoiding (a,b). If $R_E(a,b) = \Delta_A$, then $\underline{A}$ is itself $(\mathbf{E},\mathbf{M})$-subdirectly irreducible. Otherwise we get
$\cap\{R_E(a,b) \mid (a,b) \in A^2 \text{ , } a \neq b \text{ , } av_A = bv_A\} = \Delta_A$, and
$(nat_{R_E(a,b)}: \underline{A} \to \underline{A}/R_E(a,b) \mid (a,b) \in A^2 \text{ , } a \neq b \text{ , } av_A = bv_A)$ is an $(\mathbf{E},\mathbf{M})$-subdirect representation of $\underline{A}$ of $(\mathbf{E},\mathbf{M})$-subdirectly irreducible partial algebras. ■■■

20.2.4 COROLLARY 1. Theorem 20.2.3 applies in particular in the case of finitary partial algebras, when $\mathbf{M}$ is the class of all injective homomorphisms: $\mathbf{M} := \mathbf{S}_w$, and $\mathbf{E}$ is the class of all full and surjective

(i) homomorphisms: $\mathbf{E} := \mathbf{H}_f$; or
(ii) P-reflecting homomorphisms (for any set P of terms): $\mathbf{E} := \mathbf{H}_{fP}$ (in this case we may even take $\mathbf{M} := \mathbf{S}_P$, the class of all P-reflecting and injective homomorphisms); or
(iii) closed homomorphisms: $\mathbf{E} := \mathbf{H}_c$ (in this case we may even choose $\mathbf{M} := \mathbf{S}_c$, the class of all closed and injective homomorphisms).

PROOF. For each of the three cases above we already know that E is closed with respect to composition and contains all isomorphisms; moreover, if fg and f belong to $\mathbf{E}$, then 2.4.5.(iii).(c) implies that g is full, and that g is surjective, is trivial; since f preserves all TE-statements and is surjective, 9.1.7.(iii) implies that g reflects the same TE-statements, which are reflected by fg: Thus E is also left cancellable with respect to $\mathbf{E}$-morphisms, and therefore satisfies the assumptions of Theorem 20.2.3. That $\mathrm{Cong}_E(\underline{A})$ is inductive for each $\underline{A}$ has been proved earlier for each of the three cases. The statements about $\mathbf{M}$ follow from 20.1.3.(iii). ■■■

Thus we have in the case of partial algebras the following three "Birkhoff-Theorems" for subdirect representability:

20.2.5 COROLLARY 2. Let $\mathbf{Alg}(\underline{S})$ be a finitary species of partial algebras. Then we have for each set P of terms:

(i) Each $\mathbf{Alg}(\underline{S})$-algebra is the $(\mathbf{H}_f,\mathbf{S}_w)$-subdirect product of $(\mathbf{H}_f,\mathbf{S}_w)$-subdirectly irreducible $\mathbf{Alg}(\underline{S})$-algebras.

(ii) Each **Alg**($\underline{S}$)-algebra is the (H_{fp},S_{wp})-subdirect product of (H_{fp},S_{wp})-subdirectly irreducible **Alg**($\underline{S}$)-algebras.

(iii) Each **Alg**($\underline{S}$)-algebra is the (H_c,S_c)-subdirect product of (H_c,S_c)-subdirectly irreducible **Alg**($\underline{S}$)-algebras. ■■■

20.2.6 REMARK. It is our opinion that the three concepts of (E,M)-subdirect representation presented above are among the most important ones for partial algebras. The result on (H_f,S_w)-subdirect products is already contained in the book of R.S.PIERCE on abstract algebras (cf. [Pi 68], Chapter 2). The importance of the results in 20.2.4.(ii) and (iii) will follow from the following results about their relationships to E- and ECE-varieties, respectively. -- It is easily seen that for any E-variety **K** and for the case that E is the class of all full and surjective homomorphisms all the (H_f,S_w)-subdirectly irreducible partial algebras needed for an (H_f,S_w)-subdirect representation of some **K**-algebra belong to **K**. But it is also easy to realize that not each (H_f,S_w)-subdirect product of (H_f,S_w)-subdirectly irreducible **K**-algebras belongs to **K**. This is a point where P-reflection is needed for suitable sets P of terms, as we shall see below, while closedness is an adequate concept for ECE-varieties in general. These concepts have also been studied and applied in the thesis of M.SIEGMUND-SCHULTZE ([Si 84]), where they have been needed in order to transfer the results of W.TAYLOR on residually small varieties and equational compactness of total algebras to partial algebras.

20.2.7 DEFINITION. Let **K** be any class of partial algebras, and let E and **M** be suitable classes of epi- and monomorphisms. Then we define:

(i) $si^{(E,M)}(K)$ is the class of all (E,**M**)-subdirectly irreducible **K**-algebras.

(ii) $Ps^{(E,M)}(K)$ is the class of all (E,**M**)-subdirect products with respect to families of E-morphisms into **K**-algebras.

20.2.8 LEMMA. Let **K** be any E-variety, let $D := \deg(K) := \downarrow\mathbf{Eq}(K)$ be the completeness degree of **K**, i.e. the set of all (admissible) terms which induce total term operations in each **K**-algebra, let $P \supseteq D$ be a set of terms, and let $m:\underline{A} \to \underline{B}$ be an injective and P-reflecting homomorphism into some **K**-algebra $\underline{B}$. Then $\underline{A}$ also belongs to **K**.

PROOF. Let $(M: t \overset{e}{=} t')$ be any E-equation valid in **K**, and let $w:M \to A$ be any assignment. Since by assumption (M:t) and (M:t') belong to P, since both terms are interpreted in $\underline{B}$ with respect to wm, and since m reflects these terms, both of them are interpreted in $\underline{A}$ with respect to w; moreover they are equally interpreted, since this is the case in $\underline{B}$, and m is injective. Thus the E-equation is also valid in $\underline{A}$, i.e. $\underline{A}$ belongs to **K**. ■■■

20.2.9 THEOREM. Let **K** be any E-variety of finitary partial algebras. Then we have for any set P of terms which contains $D = \downarrow\mathbf{Eq}(K)$:

$$K = Ps^{(H_{fp},S_{wp})}(si^{(H_{fp},S_{wp})}(K)) ,$$

i.e. for $D \subseteq P$ any E-variety **K** is just the class of all (H_{fp},S_{wp})-subdirect products of (H_{fp},S_{wp})-subdirectly irreducible **K**-algebras.

PROOF. It is obvious that K is contained in $Ps^{(H_{fp},S_{wp})}(si^{(H_{fp},S_{wp})}(K))$ (because of 20.2.5.(ii), and since an E-variety is closed with respect to arbitrary homomorphic images and therefore also with respect to P-reflecting homomorphic images). Now 20.2.8 shows that each domain of a P-reflecting injective homomorphism into some **K**-algebra belongs to **K**, if $D \subseteq P$, therefore also the other inclusion follows, since **K** is closed with respect to products of **K**-algebras. ∎

20.2.10 THEOREM. Let **K** be any ECE-variety of finitary partial ($\underline{S}$-) algebras. Then

$$K = Ps^{(H_c,S_c)}(si^{(H_c,S_c)}(K)) ,$$

i.e. each ECE-variety **K** is the class of all (H_c,S_c)-subdirect products of (H_c,S_c)-subdirectly irreducible **K**-algebras.

PROOF. Since **K** is closed with respect to closed homomorphic images, all the (H_c,S_c)-subdirectly irreducible partial algebras needed for an (H_c,S_c)-subdirect representation of a **K**-algebra belong to **K**. And since **K** is closed with respect to subalgebras (and therefore with respect to the domains of closed and injective homomorphisms into **K**-algebras), all (H_c,S_c)-subdirect products of families of **K**-algebras belong to **K**. From these observations the equality stated above follows. ∎

20.2.11 REMARK and PROBLEM. In considering the last two theorems one can realize that in connection with subdirect representations ECE-varieties have the smoothest representation theorem -- when one uses (H_c,S_c)-subdirect representations. As we have already mentioned above the results of M.SIEGMUND-SCHULTZE in [Si 84] show that this is really a very suitable concept and is closely connected with a "concept of equational compactness", where one not only allows E-equations, but also negations of TE-expressions. This observation also backs our opinion that ECE-varieties are an adequate generalization to partial algebras of the concept of variety for total algebras. But we do not want to deny that the concept of hep-varieties (which are in general very special QE-varieties) as introduced by H.REICHEL (cf. Part II) is sometimes more useful, and still has to a great extent a nice theory. We did not yet consider the question of what might be the most convenient concept of **(E,M)**-subdirect representation in connection with hep-varieties -- if there is any at all --, and we leave this investigation as an open problem for the reader.

20.3 The cases of full respectively closed monomorphisms

20.3.1. The cases which we want to discuss in this subsection are analogues to a subdirect representation theorem for relational systems of H.E.PICKETT (cf. [Pc 64]), where we now have to assume that **M** is the class of all full (or even closed) and injective homomorphisms, while **E** is a class satisfying the same assumptions as in 20.2.1 or 20.2.3.

Since the monomorphisms are now supposed to be full (or even closed), we have to separate each sequence which is not contained in the graph (domain) of a fundamental operation, say $\phi^{\underline{A}}$, from the graph (domain) of this operation by a suitable **E**-congruence relation. As a preparation we define:

20.3.2 DEFINITION. Let A be any set, E an equivalence relation on A, and R a K-ary relation on A. Then we introduce

$R*E := \{\underline{b} \in A^K \mid$ there is $\underline{a} \in R$ such that $((k)\underline{a},(k)\underline{b}) \in E$ for each $k \in K\}$;
i.e. $R*E$ consists of all K-sequences of A which are "E-equivalent" with some sequence in R.

Quite useful in the sequel will also be the following

20.3.3 LEMMA. Let E and E' be classes of epimorphisms and $\mathbf{M}$ and $\mathbf{M}'$ be classes of monomorphisms closed with respect to composition and each containing all isomorphisms. Moreover assume that $\mathbf{E}' \subseteq \mathbf{E}$ and $\mathbf{M}' \subseteq \mathbf{M}$. Then each $(\mathbf{E},\mathbf{M})$-subdirectly irreducible object will also be $(\mathbf{E}',\mathbf{M}')$-subdirectly irreducible.

PROOF. Each $(\mathbf{E}',\mathbf{M}')$-subdirect representation of an object A is also an $(\mathbf{E},\mathbf{M})$-subdirect representation of A, by assumption, and if A is $(\mathbf{E},\mathbf{M})$-subdirectly irreducible, then it contains an isomorphism; thus A is also $(\mathbf{E}',\mathbf{M}')$-subdirectly irreducible. ∎∎∎

20.3.4 LEMMA. Let $\underline{S}$ be a finitary partial sort algebra, let $\mathbf{M}$ be the class of all full (respectively closed) and injective $\mathbf{Alg}(\underline{S})$-homomorphisms, and let the class $\mathbf{E}$ of epimorphisms have the same properties as in 20.2.1 or 20.2.3. Then a family $(e_i:\underline{A} \to \underline{B}_i \mid i \in I)$ of $\mathbf{E}$-morphisms is an $(\mathbf{E},\mathbf{M})$-subdirect representation of $\underline{A}$ iff (SR) and (SRf) (in the closed case (SR) and (SRc)) below are satisfied, where

(SR) $\bigcap\{\ker e_i \mid i \in I\} = \Delta_A$,

(SRf) $\bigcap\{(\mathrm{graph}(\phi^{\underline{A}}))*(\ker e_i) \mid i \in I\} = \mathrm{graph}(\phi^{\underline{A}})$ for each $\phi \in \Omega$.

(SRc) $\bigcap\{(\mathrm{dom}\ \phi^{\underline{A}})*(\ker e_i) \mid i \in I\} = \mathrm{dom}\ \phi^{\underline{A}}$ for each $\phi \in \Omega$.

PROOF. We only treat the case of full and injective homomorphisms; the other case is similar. It is clear that (SR) is eqivalent to the fact that the induced homomorphism $e:\underline{A} \to \mathsf{X}(\underline{B}_i \mid i \in I)$ is injective. Furthermore it is easy to see that the set described in (SRf) is equal to

$$\bigcap\{[\mathrm{graph}(\phi^{\underline{B}_i})](e_i^{K_\phi} \times e_i)^{-1} \mid i \in I\} ,$$

observing that all the e_i are full. And when (SR) is satisfied, then the equality of the above intersection with $\mathrm{graph}(\phi^{\underline{A}})$ ($\phi \in \Omega$) is equivalent with the fact that e is full. ∎∎∎

20.3.5 LEMMA. Let $\mathbf{E}$ and $\mathbf{M}$ satisfy the same assumptions as in Lemma 20.3.4, let $\underline{A}$ be any partial algebra, and let $R \in \mathrm{Cong}_{\mathbf{E}}(\underline{A})$ be arbitrary.

(i) Assume that for some $\phi \in \Omega$ we have $(\underline{a},a) \in (A^{K_\phi} \times A) \setminus \mathrm{graph}(\phi^{\underline{A}})$ (respectively $\underline{a} \in A^{K_\phi} \setminus \mathrm{dom}\ \phi^{\underline{A}}$). Then there exist maximal congruence relations $R_{\mathbf{E},f,\underline{a},a}$ (respectively $R_{\mathbf{E},c,\underline{a}}$) in $\mathrm{Cong}_{\mathbf{E}}(\underline{A})$ with respect to the property $(\underline{a},a) \notin (\mathrm{graph}(\phi^{\underline{A}}))*R_{\mathbf{E},f,\underline{a},a}$ (respectively $\underline{a} \notin (\mathrm{dom}\ \phi^{\underline{A}})*R_{\mathbf{E},c,\underline{a}}$).

(ii) $\underline{A}/R$ is $(\mathbf{E},S_f)$-subdirectly irreducible iff either $R = R_{\mathbf{E}}(a,b)$ for some $(a,b) \in A^2$, or $R = R_{\mathbf{E},f,\underline{a},a}$ for some $(\underline{a},a) \in (A^{K_\phi} \times A) \setminus \mathrm{graph}(\phi^{\underline{A}})$, for some $a \in A_{s_\phi}$, and for some $\phi \in \Omega$.

(iii) $\underline{A}/R$ is $(\mathbf{E},S_c)$-subdirectly irreducible iff either $R = R_{\mathbf{E}}(a,b)$ for some $(a,b) \in A^2$, or $R = R_{\mathbf{E},c,\underline{a}}$ for some $\underline{a} \in A^{K_\phi} \setminus \mathrm{dom}\ \phi^{\underline{A}}$, and for some $\phi \in \Omega$.

PROOF. We only prove the existence of $R_{\mathbf{E},f,\underline{a},a}$ and (ii); the other case is similar.

Ad (i): As in the proof of 20.2.1.(iii) we have to show that
$\mathbf{R}_{E,f,\underline{a},a} := \{R \in Cong_E(\underline{A}) | (\underline{a},a) \notin (graph(\phi^{\underline{A}}))^*R\}$ is inductive:
Obviously, $\mathbf{R}_{E,f,\underline{a},a}$ is non-empty, since it contains Δ_A. Let **D** be any directed subset of $\mathbf{R}_{E,f,\underline{a},a}$, and let R' be its set theoretical union. Since $Cong_E(\underline{A})$ is inductive, R' is an E-congruence. If we had $(\underline{a},a) \in (graph(\phi^{\underline{A}}))^*R'$, then there would exist $(\underline{b},b) \in graph(\phi^{\underline{A}})$ and $R_k \in \mathbf{D}$ for each $k \in K_\phi$ and for $k := s_\phi$ such that $((k)\underline{a},(k)\underline{b}) \in R_k$ $(k \in K_\phi)$, and $(a,b) \in R_{s_\phi}$. Since **D** is directed, there would be $R'' \in \mathbf{D}$ containing all R_k and satisfying all the equivalences contrary to the assumption that $\mathbf{D} \subseteq \mathbf{R}_{E,f,\underline{a},a}$. Thus $\mathbf{R}_{E,f,\underline{a},a}$ is inductive, and therefore it contains maximal elements, one of which we may denote by $R_{E,f,\underline{a},a}$.
Ad (ii): First let us show that in both cases for R in (ii) $\underline{A}/R$ is (E,S_f)-subdirectly irreducible: If $R = R_E(a,b)$, then this follows from 20.2.1.(iii) and 20.3.3, since $S_f \subseteq S_w$. Now assume that $R = R_{E,f,\underline{a},a}$. We already know from 20.2.1.(i) that we have a bijection e* from the set of all E-congruence relations of $\underline{A}$ which contain R, onto $Cong_E(\underline{A}/R)$. By assumption on R we have $(\underline{a}\cdot nat_R,(a)nat_R) \notin graph(\phi^{\underline{A}/R})$, but for any $R' \in Cong_E(\underline{A})$ which properly contains R, we have $(\underline{a}\cdot nat_{R'},(a)nat_{R'}) \in graph(\phi^{\underline{A}/R'})$. Therefore $(nat_{R'/R}:\underline{A}/R \to \underline{A}/R' | R \subset R' \in Cong_E(\underline{A}))$ cannot yield an (E,S_f)-subdirect representation of $\underline{A}/R$, or -- expressed in another way --: Each (E,S_f)-subdirect representation of $\underline{A}/R$ has to contain an isomorphism, showing that $\underline{A}/R$ is (E,S_f)-subdirectly irreducible.
Conversely, let $\underline{A}/R$ be (E,S_f)-subdirectly irreducible, and consider $(nat_{R'/R}:\underline{A}/R \to \underline{A}/R' | R \subset R' \in Cong_E(\underline{A}))$. This cannot be an (E,S_f)-subdirect representation of $\underline{A}/R$. Let $\underline{C}$ be the direct product of the $\underline{A}/R'$ occurring in this family of E-morphisms, and let $m:\underline{A}/R \to \underline{C}$ be the induced product homomorphism. Then -- if (SR) is violated -- $R = R_E(a,b)$ for a suitable pair $(a,b) \in A^2$, $a \neq b$, $av_A = bv_A$, for which $((a)nat_R)m = ((b)nat_R)m$; or -- if (SRf) is violated -- there exists $\phi \in \Omega$ such that $\bigcap\{(graph(\phi^{\underline{A}}))^*R' | R \subset R' \in Cong_E(\underline{A})\} \neq graph(\phi^{\underline{A}})$. Hence we have $(\underline{a},a) \in (A^{K_\phi} \times A) \setminus graph(\phi^{\underline{A}})$ such that nat_R does not identify $(\underline{a},a)$ with any element of $graph(\phi^{\underline{A}})$, but each $nat_{R'}$ (for $R \subset R' \in Cong_E(\underline{A})$) does. This shows that $R = R_{E,f,\underline{a},a}$. ∎

20.3.6 COROLLARY. Let $\underline{A}$ be any finitary partial algebra, and let E be a composition closed class of full and surjective homomorphisms, left cancellable with respect to E-morphisms and containing all isomorphisms, and for which $Cong_E(\underline{A})$ is inductive. Then:

(i) $\underline{A}$ is (E,S_w)-subdirectly irreducible, iff there is $(a,b) \in A^2$ such that each non-trivial E-congruence relation of $\underline{A}$ contains (a,b).

(ii) $\underline{A}$ is (E,S_f)-subdirectly irreducible, iff either there is $(a,b) \in A^2$ such that each non-trivial E-congruence relation of $\underline{A}$ contains (a,b), or there is $(\underline{a},a) \in (A^{K_\phi} \times A) \setminus graph(\phi^{\underline{A}})$ for some $\phi \in \Omega$ such that for every non-trivial E-congruence relation R one has $(\underline{a},a) \in (graph(\phi^{\underline{A}}))^*R$.

(iii) $\underline{A}$ is (E,S_c)-subdirectly irreducible, iff either there is $(a,b) \in A^2$ such that each non-trivial E-congruence relation of $\underline{A}$ contains (a,b), or there is $\underline{a} \in A^{K_\phi} \setminus dom\, \phi^{\underline{A}}$ for some $\phi \in \Omega$ such that for every non-trivial E-congruence relation R it is true that $\underline{a} \in (dom\, \phi^{\underline{A}})^*R$.

PROOF. (i), (ii) and (iii) follow directly from 20.3.5 and its proof. ■■■

20.3.7 THEOREM. Let $\underline{S}$ be a finitary partial sort algebra, and let E be a composition closed class of full and surjective homomorphisms, left cancellable with respect to E-morphisms and containing all isomorphisms, and for which $Cong_E(\underline{A})$ is inductive for every partial algebra $\underline{A}$. Then one has for every partial algebra $\underline{A}$:

(i) $\underline{A}$ has an (E,S_f)-subdirect representation of E-morphisms into (E,S_f)-subdirectly irreducible partial algebras.

(ii) $\underline{A}$ has an (E,S_c)-subdirect representation of E-morphisms into (E,S_c)-subdirectly irreducible partial algebras.

PROOF. The proof is similar to the one of Theorem 20.2.3, except that we now have to consider families of $R_E(a,b)$ and $R_{E,f,\underline{a},a}$ (respectively $R_{E,c,\underline{a}}$). Therefore we omit the details. ■■■

With the notation from 20.2.7 we then get the

20.3.8 COROLLARY. Let K be an E-variety of finitary partial algebras. Then

$$K = Ps^{(H_f,S_c)}(si^{(H_f,S_c)}(K)) .$$

The argument is similar to the one in 20.2.10. ■■■

20.3.9 REMARKS. **(i):** Corollary 20.3.8 shows that the concept of (H_f,S_c)-subdirect representation is very suitable for E-varieties, but it is similarly adequate for all H_f-closed ECE-varieties. Actually the results of this subsection provide suitable concepts of **(E,M)**-subdirect representation for quite a lot of special concepts of ECE-varieties as discussed in Chapter III, especially in section 13. But the explicit treatment of the concepts of **(E,M)**-subdirect representation as discussed in this subsection is much more difficult than for the cases treated in subsection 20.2.

(ii): Notice that for total algebras all the concepts of subdirect representation considered so far coincide with the usual one for total algebras.

(iii): We have restricted our consideration here to the case when E is contained in the class of all full and surjective homomorphisms of **Alg**($\underline{S}$) in order to provide an idea of possibilities of subdirect representations of partial algebras and of their treatment. In [BSi 84] one can also find a discussion of the cases when E is the class of all surjective homomorphisms (and **M** is the class of all full and injective homomorphisms) -- which was introduced by G.GRÄTZER in exercise 96 to Chapter 3 of [G 68] --, and of the case when E is the class of all epimorphisms (and **M** is the class of all closed and injective homomorphisms) -- this case was first introduced by R.WILLE in [Wi 70]. In both cases the congruences on $\underline{A}$ are no longer sufficient to characterize the **(E,M)**-subdirectly irreducible "components" of $\underline{A}$, but one has to consider the A-initial closed congruence relations of $\underline{F}(\underline{A},\mathbf{TAlg}(\underline{S}))$ -- and for surjective homomorphisms only those for which each congruence class has a non-empty intersection with A.

20.3.10 TABLE. As a résumé of the results in this section -- combined with those in [BSi 84] -- we get Table 20.1 for concrete positive cases of Birkhoff type results on the existence of **(E,M)**-subdirect representations with respect to **(E,M)**-subdirectly irreducible partial algebras (the numbers refer to the theorem where the corresponding

result is proved, "*" meaning that this is a quotation from [BSi 84]).
It is not difficult to add further examples, which are easily derivable with the previous methods and statements, but we think that the examples presented here will do for most cases.

E	M	Quotation
full and surjective hom.s	injective hom.s	20.2.5
full and surjective hom.s	full and injective hom.s	20.3.7
full and surjective hom.s	closed and injective hom.s	20.3.7
P-reflecting full and surj. hom.s	P-reflecting and inj. hom.s	20.2.5
closed and surjective hom.s	closed and injective hom.s	20.2.5
surjective hom.s	full and injective hom.s	*5.8
epimorphisms	closed and injective hom.s	*6.4

Table of positive cases of (**E**,**M**)-subdirect representation theorems:

Table 20.1 (see 20.3.10)

20.4 Subdirectly irreducible mono-unary partial algebras

20.4.1 REMARKS. **(i)**: In order to get a better feeling for the different concepts of subdirect irreducibility, we present these concepts in more detail in the most simple case where we have only one unary partial operation. This is also the only similarity type, where in the total case all subdirectly irreducible algebras of this type are known (cf. [JoSe 67] and [We 70]). In the total case all subdirectly irreducible algebras are known only for very few varieties. The characterizations of all (H_f,S_w)-respectively of all (H_c,S_c)-subdirectly irreducible mono-unary partial algebras was first given in [Si 82], while all the examples presented here can be found in [BSi 84]. Since all subdirect products of total algebras only use total subdirect factors, all subdirectly irreducible total (mono-unary) algebras will also be subdirectly irreducible for all other concepts.
(ii): Since for mono-unary partial algebras and for every non-trivial set P of terms which is closed with respect to subterms every P-reflecting homomorphism is already closed, we will not dicuss P-reflection here; the interested reader may investigate for himself cases of P-reflection where P is not closed with respect to subterms.
(iii): Observe that mono-unary partial algebras can be considered as directed graphs -- in the sense of graph theory -- without multiple edges and with outdegree ≤ 1 , namely by taking graph($\phi^{\underline{A}}$) as the corresponding relation between the elements of $\underline{A}$, which then will also be called <u>vertices</u>, while the elements of graph($\phi^{\underline{A}}$) will be called (directed) <u>edges</u>. Thus to a mono-unary partial algebra will be attributed some graph theoretical property, whenever the corresponding graph has this property; especially such a mono-unary partial algebra will be called <u>connected</u>, whenever the corresponding graph is connected (as an undirected graph, i.e. by forgetting directions of the edges), and a maximal connected part of a given graph will be called a <u>(connected) component</u>: Notice that every graph is a disjoint union of its connected components. We shall also represent a mono-unary partial algebra by its corresponding graph, whenever necessary.

20.4.2 NOTATION. Let us give names to some special mono-unary partial algebras:
(i): In what follows, "+" will always denote disjoint union of graphs, respectively of the corresponding mono-unary partial algebras. Moreover, n and m will designate natural numbers ($\in \omega_0$) or "∞" (meaning "infinity" in the sense that a "path" of the graph has no finite beginning or end point).
(ii): For every $n \in \omega_0$ and $m,m' \in \omega_0 \cup \{\infty\}$, $m,m' \notin \{0\}$, we consider the mono-unary partial algebras described by their graphs in Figure 20.1 (notice that $C_1 = I_0$).

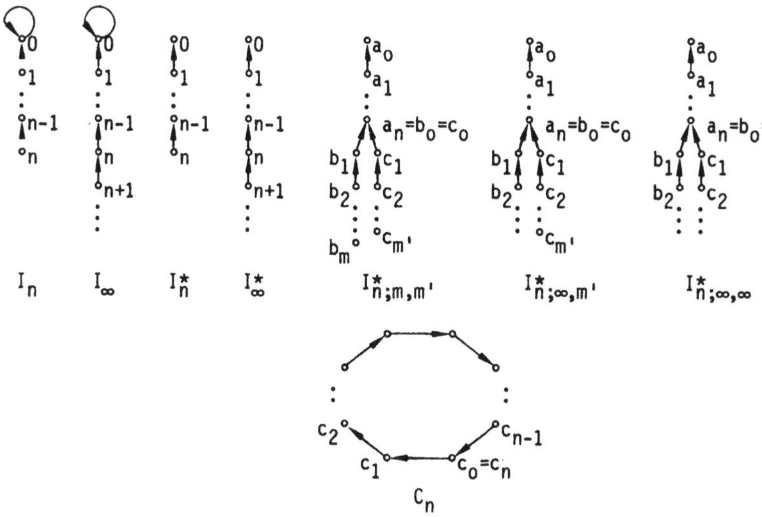

Figure 20.1 (see 20.4.2)

In the proof of the next theorem one needs the following lemma and its corollary taken from [Si 82].

20.4.3 LEMMA. Let $\underline{A}$ be a mono-unary partial algebra with "partial part" $\underline{P}$ (union of all connected components which are really partial, i.e. contain at least one "partial element") and "total part" $\underline{T}$; then one has the isomorphism

$$\underline{\mathrm{Cong}_c}(\underline{A}) = \underline{\mathrm{Cong}_c}(\underline{P}) \times \underline{\mathrm{Cong}}(\underline{T}) \text{ ,}$$

where the right side is considered as the usual direct product of lattices.

SKETCH of PROOF. The assignment

$$\mathrm{Cong}_c(\underline{P}) \times \mathrm{Cong}(\underline{T}) \to \mathrm{Cong}_c(\underline{A}) \text{ , } \quad (R,R') \mapsto R \,\dot{\cup}\, R'$$

is a lattice isomorphism. We omit the somewhat long computation, whose details are available in the preprint [Si 82] (or see the appendix of [Si 84]). ■■■

From this lemma there follows that the atoms in $\mathrm{Cong}_c(\underline{A})$ are exactly the closed congruence relations $R \cup \Delta_T$ and $\Delta_P \cup R'$ with atoms R in $\mathrm{Cong}_c(\underline{P})$ and R' in $\mathrm{Cong}(\underline{T})$, respectively, and we have the

20.4.4 COROLLARY. A mono-unary partial algebra $\underline{A} = \underline{P} \,\dot{\cup}\, \underline{T}$ is (H_c,S_c)-subdirectly irreducible, iff one of the two parts is (H_c,S_c)-subdirectly irreducible, and the other one is closed-simple. ■■■

20.4.5 THEOREM. Up to isomorphism all (E,M)-subdirectly irreducible mono-unary partial algebras are given in the cases below by the following lists:

(i) Total case:

(A) I_n , $C_{n'}$, $C_{n'} + I_0$, and $I_0 + I_0$,

where $n' = p^k$, p prime, $k \in \omega_0$ $(k \neq 0)$, $n \in \omega_0 \cup \{\infty\}$.

(ii) $E:= H_f$, $M:= S_w$:

The list (A) and

(B) I_0^* , I_1^* , $I_0 + I_0^*$, and $I_0^* + I_0^*$.

(iii) $E:= H_c$, $M:= S_c$:

The list (A) and

(C) I_n^* , $I_{n';m,m'}^*$, $I_{n';m,m'}^* + I_0$, $I_n^* + I_{n''}^*$, $I_n^* + I_{n''}$,

$I_n^* + C_{p^k}$, $I_n^* + I_{n''}^* + I_0$, $I_n^* + C_{p^k} + I_0$,

where $n,n'' \in \omega_0 \cup \{\infty\}$, $p,k,n' \in \omega_0$, p prime, and $m,m' \in (\omega_0 \setminus \{0\}) \cup \{\infty\}$.

(Notice that $I_n^* + I_0 + \dot{I}_0$ is contained in the last case.)

(iv) $E:= H_f$, $M:= S_c$:

The lists (A), (B) and

(D) I_n^* , and $I_n^* + I_0$, where $n \in \omega_0 \cup \{\infty\}$.

(v) $E:= H_f$, $M:= S_f$:

The lists (A), (B) and

(E) I_2^* .

For the proof we refer to [BSi 84] or leave it as an exercise. ∎

20.4.6 REMARK. We would like to mention a paper of W.BARTOL, Z.RAŚ and A.SKOWRON [BtRaSk 75], where they study mono-unary partial algebras (called "iterative systems") used as models of digital computers of the von Neumann type. For a mono-unary partial algebra $\underline{A}$ they explicitly describe the greatest element in $Cong_c(\underline{A})$ concluding that $\underline{Cong_c}(\underline{A})$ is a complete lattice, and they give a necessary and sufficient condition for $\underline{Cong_c}(\underline{A})$ to be a distributive lattice or a Boolean algebra.
Moreover, they show that every mono-unary partial algebra is the (H_c,S_c)-subdirect product of (H_c,S_c)-subdirectly irreducible mono-unary partial algebras, but they do not use this representation in the subsequent discussion of a "stored program computer" -- being a mono-unary partial algebra with additional structure imposed on the carrier set and the fundamental partial operation.

21 More facts about E-varieties

Throughout this book there are already scattered many facts about E-varieties. In this section we want to present some additional results especially about this kind of variety, and in particular such, which will show us that -- at least in the homogeneous case -- E-varieties and the free (partial and total) algebras behave almost in the same way as the corresponding whole species and its absolutely free (partial and total) algebras. In particular one can formulate results about equational and implicational logic in them

(almost) in the same way, as this has been done earlier within a whole species. Main restrictions in addition to homogeneity will be the assumptions that the bases of the free partial and total algebras under consideration have the element basis property, and that the variety has a partial algebra with at least two elements; we have seen in section 17 that this is not always satisfied. We conjecture that the element basis property is not necessary for the results below, but we have not succeeded in managing without this assumption.
Nevertheless the basic generalization of the Recursion Theorem holds without such an assumption, and we shall prove this first. To some extent we shall also treat arbitrary heterogeneous partial algebras, but the main conclusions will then be restricted to homogeneous ones. Most of the results presented here are generalizations of results first presented in [B 71].

21.1 Initial homomorphic extensions

21.1.1 REMARK. In 4.1.13 we have proved a quite general homomorphism extension lemma; however, for this section it is not yet general enough. Namely in 4.1.13 we have assumed -- in the notation of that lemma -- that $\underline{D}$ is a relative subalgebra of $\underline{B}$, while in this section we need a similar result for the case when $\underline{D}$ need only be a weak subalgebra of $\underline{B}$. Since we want to apply such a result in the case when $\underline{B} = \underline{F}(\underline{D},K_t)$ for some E-variety K, the results in section 17 show that in the heterogeneous case $\underline{D}$ need not be a relative subalgebra of $\underline{B}$ -- even "weak subalgebra" is not guaranteed, while in the homogeneous case this is always satisfied (see 17.2.6).

21.1.2 LEMMA. Let $\underline{A}$, $\underline{A}'$, $\underline{B}$ and $\underline{B}'$ be similar partial algebras, such that $\underline{A}$ is a weak subalgebra of $\underline{A}'$, and $\underline{B}$ is a weak subalgebra of $\underline{B}'$. Moreover, let $f:\underline{A} \to \underline{B}$ be any homomorphism which possesses a homomorphic extension $f':\underline{A}' \to \underline{B}'$. Then the following statements are true:

(i) f^* with $\text{graph}(f^*) := C_{\underline{A}'\times\underline{B}}\,\text{graph}(f)$ is a homomorphism $f^*:\underline{E} \to \underline{B}$, and A generates $\underline{E}$, where $E := \text{dom}\, f^*$, and $\underline{E}$ carries the largest weak substructure of $\underline{A}'$ which still allows f^* to be a homomorphism. -- If, in particular, $\underline{B}$ is a relative subalgebra of $\underline{B}'$, then $\underline{E}$ is a relative subalgebra of $\underline{A}'$.

(ii) If $\underline{C}$ is any weak subalgebra of $\underline{A}'$ generated by A, and if $g:\underline{C} \to \underline{B}$ is any homomorphic extension of f, then $\text{graph}(g) \subseteq \text{graph}(f^*)$, and $\underline{C}$ is a weak subalgebra of $\underline{E}$.

(iii) If $\underline{A}$ is total, then $f^*:\underline{E} \to \underline{B}$ (from (i)) is a closed homomorphism. -- And if $g:\underline{D} \to \underline{B}$ is any other closed homomorphic extension of f starting from an A-generated weak subalgebra $\underline{D}$ of $\underline{A}'$, then $g = f^*$ (including $\underline{D} = \underline{E}$).

PROOF. Ad (i): Because of 4.1.11.(i) f^* is a homomorphism from $\underline{\text{dom}}\, f^*$ -- the relative subalgebra of $\underline{A}'$ with carrier dom f^* -- into the relative subalgebra of $\underline{B}'$ with carrier B, since obviously $\text{graph}(f^*) \subseteq \text{graph}(f')$. -- From this we already get the last statement in (i). -- For each $\phi \in \Omega$ we define

$$\text{graph}(\phi^{\underline{E}}) := \{ (\underline{a},a) \in \text{graph}(\phi^{\underline{A}'}) \mid \underline{a} \in \text{dom}\, \phi^{\underline{\text{dom}}\, f^*} \text{ and } \underline{a}f^* \in \text{dom}\, \phi^{\underline{B}} \}.$$

It is easy to realize that this is indeed the largest weak substructure of $\underline{A}'$ on E allowing f^* to be a homomorphism from $\underline{E}$ into $\underline{B}$. It can now easily be shown (e.g. by algebraic induction or by induction on the step numbers) that also graph(f^*) = = $C_{\underline{E}\times\underline{B}}$ graph(f) , and 4.1.12 implies that A generates $\underline{E}$. ∎

(ii) and (iii) are proved in a similar way as in 4.1.13. ∎∎∎

21.1.3 REMARK. The results above generalize Theorem 2 (Hutter's Theorem) in J.SCHMIDT [Sch 70], where it is required that $\underline{A}'$ be a normal extension of $\underline{A}$ (only the homogeneous case is considered). But the next theorem below will show that normality is not necessary for the existence of a largest -- and then closed -- initial homomorphic extension, say f^*, within a relatively free completion $\underline{F}(\underline{A},\mathbf{K}_t)$ of a given homomorphism $f:\underline{A} \to \underline{B}$ within a primitive class $\mathbf{K}$. But if we want the domain of f^* to be a relative subalgebra of $\underline{F}(\underline{A},\mathbf{K}_t)$, then $\underline{B}$ has in general to be a relative subalgebra of $\underline{F}(\underline{B},\mathbf{K}_t)$, as follows from 21.1.2.(i) -- if there is a sort $s \in S$ for which each $\mathbf{K}$-algebra has an at most one-element phylum, then this can usually easily be violated.

In this context the main application of Lemma 21.1.2 is the following

21.1.4 THEOREM. Let $\mathbf{K}$ be any E-variety, and $\mathbf{K}_t$ its subclass of all total $\mathbf{K}$-algebras. Let $\underline{A}$ and $\underline{B}$ be $\mathbf{K}$-algebras such that $\underline{A}$ and $\underline{B}$ are weak subalgebras of $\underline{F}(\underline{A},\mathbf{K}_t)$ and $\underline{F}(\underline{B},\mathbf{K}_t)$, respectively, and let $f:\underline{A} \to \underline{B}$ be any homomorphism. Then there exists a largest initial homomorphic extension, say $f^*:\underline{E} \to \underline{B}$, of f from an A-generated weak relative subalgebra $\underline{E}$ of $\underline{F}(\underline{A},\mathbf{K}_t)$ into $\underline{B}$ -- $\underline{E}$ having as much structure as possible. In particular we have

(i) graph(f^*) = $C_{\underline{F}(\underline{A},\mathbf{K}_t)\times\underline{B}}$ graph(f) .

(ii) $f^*:\underline{E} \to \underline{B}$ is a closed homomorphism.

(iii) f^* is the right factor of a ($\mathbf{K}_t$-extendable epimorphisms, $\mathbf{K}_t$-perfect homomorphisms)-factorization of f (see 10.2.16), and the embedding of $\underline{A}$ into $\underline{E}$ is the corresponding left factor.

(iv) $\underline{E}$ is a $\mathbf{K}$-algebra (generated by A).

(v) If $\underline{B}$ is a relative subalgebra of $\underline{F}(\underline{B},\mathbf{K}_t)$, then $\underline{E}$ is a relative subalgebra of $\underline{F}(\underline{B},\mathbf{K}_t)$.

(vi) If we are in the case of **homogeneous** partial algebras, then $\underline{A}$ and $\underline{B}$ are always at least weak subalgebras of $\underline{F}(\underline{A},\mathbf{K}_t)$ and $\underline{F}(\underline{B},\mathbf{K}_t)$, respectively, and if $\mathbf{K}$ contains an at least two-element partial algebra, then $\underline{E}$ is always a relative subalgebra of $\underline{F}(\underline{A},\mathbf{K}_t)$ -- possibly apart from the case that both $\underline{F}(\underline{A},\mathbf{K}_t)$ and $\underline{F}(\underline{C}_{\underline{B}}([A]f),\mathbf{K}_t)$ contain only one element.

PROOF. Since f has a homomorphic extension $f':\underline{F}(\underline{A},\mathbf{K}_t) \to \underline{F}(\underline{B},\mathbf{K}_t)$ -- f can also be considered as a homomorphism from $\underline{A}$ into $\underline{F}(\underline{B},\mathbf{K}_t)$ -- Lemma 21.1.2 is applicable, where $\underline{A}' := \underline{F}(\underline{A},\mathbf{K}_t)$ and $\underline{B}' := \underline{F}(\underline{B},\mathbf{K}_t)$. Thus the main statement of 21.1.4 as well as (i), (ii) and (v) immediately follow. Since $\underline{E}$ is a homomorphic image with respect to the first projection applied to the subalgebra on graph(f^*) of the direct product of the $\mathbf{K}$-algebras $\underline{F}(\underline{A},\mathbf{K}_t)$ and $\underline{B}$, (iv) is also true. ∎

Ad (iii): Let $i:\underline{A} \to \underline{E}$ be the canonical embedding; we first show that i is a $\mathbf{K}_t$-extendable epimorphism: That i is an epimorphism follows from the fact that A generates $\underline{E}$; and since each homomorphism from $\underline{A}$ into some $\mathbf{K}_t$-algebra even extends to all of $\underline{F}(\underline{A},\mathbf{K}_t)$, it especially extends to $\underline{E}$, showing that i is $\mathbf{K}_t$-extendable. -- It remains to show that f^* is $\mathbf{K}_t$-perfect: Consider the commutative diagram in Figure 21.1, where h is assumed to be K_t-extendable, and where we have already added the bottom rectangle for the K_t-universal solutions.

Since f' belongs to $\Lambda(\mathrm{Ext}(K_t))$ (i.e. is K_t-perfect), there is a homomorphism $d':\underline{V} \to \underline{F}(\underline{A},K_t)$ such that $hd' = pi_1$ and $d'f' = qi_2$. Since f* is just the restriction of f' to E, and since f* is closed, we only have to show that [V]d' is contained in E (the rest of the proof then being straightforward), and we prove this by algebraic induction on the elements of V, using that [U]h generates $\underline{V}$, and that [U]p is contained in E:

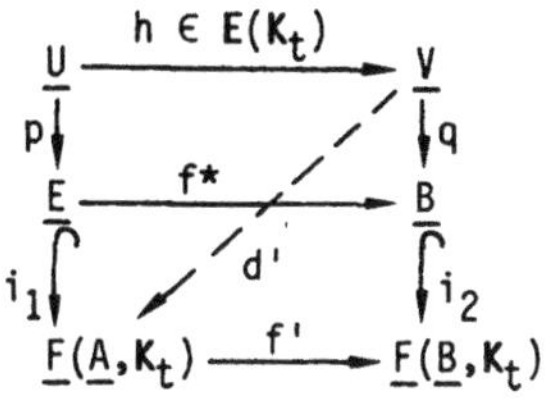

Figure 21.1

Thus $uhd' = upi_1 = up$ $(u \in U)$ provides the initial condition; assume that $\underline{a} \in \mathrm{dom}\, \phi^{\underline{V}}$, and that d' maps all elements of $\underline{a}$ into E; then $\underline{a}q \in \mathrm{dom}\, \phi^{\underline{B}}$, and therefore $\underline{a}d' \in \mathrm{dom}\, \phi^{\underline{E}}$ (by the definition of f* and $\underline{E}$). Thus $(\underline{a}d')\phi^{\underline{E}} = (\underline{a}\phi^{\underline{V}})d'$ belongs to E, completing the proof. ■

Ad (vi): If $\underline{F}(\underline{B},K_t)$ contains at least two elements, or if B is empty, then $\underline{B}$ is a relative subalgebra of $\underline{F}(\underline{B},K_t)$, and the statement follows from (v). Thus it remains only the case that -- without loss of generality -- B and $F(\underline{B},K_t)$ both have exactly one element; but if A and $F(\underline{A},K_t)$ are both empty, then the statement is trivial, and if $F(\underline{A},K_t)$ has only one element, we are in the possibly exceptional case; thus we may assume that $F(\underline{A},K_t)$ has at least two elements. Since this remaining situation probably will not occur that often in practice, we only briefly sketch the proof: One may use the fact that each operation in $\underline{B}$ is either total or empty, and one will realize that the relative subalgebra of $\underline{F}(\underline{A},K_t)$ generated by A with respect to the non-empty operations of $\underline{B}$ is identical with $\underline{E}$. ■■■

Because of its importance we introduce the

21.1.5 DEFINITION. Let **K** be any E-variety, and let $f:\underline{A} \to \underline{B}$ be any homomorphism between **K**-algebras. Then the largest closed initial homomorphic extension, if it exists according to Theorem 21.1.4, will always be designated by $f^*:\underline{d}\mathrm{om}\, f^* \to \underline{B}$; except for the case when K_t is the class $\mathbf{TAlg}(\underline{S})$ of all total algebras, when we shall mostly use the old notation $\tilde{f}$ instead of f*. If we know that $\underline{d}\mathrm{om}\, f^*$ is indeed a relative subalgebra of $\underline{F}(\underline{A},K_t)$, then we shall use as before the notation $\underline{\mathrm{dom}}\, f^*$ for the domain of f*, otherwise the fact that we only underline "d" of "dom" shall indicate that it might only be a weak subalgebra of $\underline{F}(\underline{A},K_t)$.

Since each mapping $f:M \to \underline{B}$ into some homogeneous **K**-algebra $\underline{B}$ has a homomorphic extension to $\underline{F}(M,\mathbf{K})$, the above theorem immediately yields the

21.1.6 COROLLARY. With the same assumptions as in Theorem 21.1.4, and the additional one that we are in the **homogeneous case**, each mapping $f:M \to \underline{B}$ into some **K**-algebra $\underline{B}$ has a largest closed initial homomorphic extension $f^*:\underline{d}\mathrm{om}\, f^* \to \underline{B}$, to which all the facts in Theorem 21.1.4 apply, and for which there always holds that $\underline{F}(M,\mathbf{K})$ is at least a weak subalgebra of $\underline{d}\mathrm{om}\, f^*$. ■■■

21.1.7 COROLLARY. In every non-trivial E-variety **K** of **homogeneous** partial algebras each **K**-algebra, say $\underline{A}$, is a closed homomorphic image of a weak **K**-subalgebra of a suitable K_t-

freely generated K_t-algebra. More precisely: If M is any generating subset of $\underline{A}$, then $(id_M)^*:\underline{dom}(id_M)^* \to \underline{A}$ is a closed and surjective homomorphism from an M-generated weak **K**-subalgebra of $\underline{F}(\underline{A},K_t)$ onto $\underline{A}$ -- and if $\underline{F}(M,K_t)$ has at least two elements, it is even a relative subalgebra -- and the pair $(\underline{dom}(id_M)^*, \ker(id_M)^*)$ is a presentation of $\underline{A}$ within $\underline{F}(\underline{A},K_t)$ which contains all the information about $\underline{A}$. ■■■

21.1.8 REMARKS. (i): In some papers on partial algebras, like in [Ke 70] by R.KERKHOFF and in [Po 73] by V.S.POYTHRESS a concept of structure compatible mappings is used which is based on partial mappings. In [Ke 70] these mappings are called "Konformismen", and in [Po 73] p-morphisms, and if we use them, we shall call them conformisms: Let $\underline{A}$ and $\underline{B}$ be any partial algebras; a conformism between $\underline{A}$ and $\underline{B}$ is a closed homomorphism from a weak subalgebra of $\underline{A}$ into $\underline{B}$ (notice that in this concept the structure in the preimage is in some way stronger than the one in the image, while this is just the other way round for the homomorphisms which we use in this presentation). Then -- in the terminology of the authors quoted above -- we have just proved that for every E-variety **K** of homogeneous partial algebras and for every set M the K_t-free K_t-algebra $\underline{F}(M,K_t)$ K_t-freely generated by M is also conformism-freely generated by M with respect to **K** in the sense of Corollary 21.1.7.

(ii): The results above also show that to some extent the K_t-universal solutions $\underline{F}(\underline{A},K_t)$ of **K**-algebras behave at least in the homogeneous case in a similar way as we have already observed it with respect to the general **TAlg**($\underline{S}$)-universal solutions $\underline{F}(\underline{A},$**TAlg**$(\underline{S}))$ -- i.e. the absolutely free completions -- in the whole species do.

A connection between the results above and closed initial homomorphic extensions within absolutely free completions is given by the following

21.1.9 PROPOSITION. Let **K** be any E-variety, and $f:\underline{A} \to \underline{B}$ any homomorphism between **K**-algebras. Let $\tilde{f}:\underline{dom}\,\tilde{f} \to \underline{B}$ and $f^*:\underline{dom}\,f^* \to \underline{B}$ be the largest closed initial extensions with respect to $\underline{F}(\underline{A},$**TAlg**$(\underline{S}))$ respectively $\underline{F}(\underline{A},K_t)$ -- assuming that f^* exists. Moreover, let $p':\underline{F}(\underline{A},$**TAlg**$(\underline{S})) \to \underline{F}(\underline{A},K_t)$ be the homomorphic extension if id_A, and let p be its restriction to $\underline{dom}\,\tilde{f}$.

Then p induces a homomorphism $p_f:\underline{dom}\,f \to \underline{dom}\,f^*$, and this is a closed and surjective homomorphism satisfying $\tilde{f} = p_f f^*$.

PROOF. Let $i := id_A:\underline{A} \to \underline{dom}\,f^*$, and let $\tilde{i}$ be its closed initial extension within $\underline{F}(\underline{A},$**TAlg**$(\underline{S}))$, then $\tilde{i}f^*$ is a closed initial extension of f, and because of its uniqueness we get $\tilde{i}f^* = \tilde{f}$, and $\tilde{i} = p_f$ immediately follows. ■■■

The preceding observations also lead to a generalization of the Epimorphism Theorem (see 5.8.2 and 5.8.3).

21.1.10 THEOREM (Epimorphism Theorem for Homogeneous E-Varieties). Let **K** be any E-variety, and let $f:\underline{A} \to \underline{B}$ and $g:\underline{A} \to \underline{C}$ be any homomorphisms between **K**-algebras, where f is an epimorphism, and let f^* and g^* be the closed and initial homomorphic extensions of f and g within $\underline{F}(\underline{A},K_t)$, respectively. Then there exists a homomorphism $h:\underline{B} \to \underline{C}$ such that $fh = g$, iff $\ker f^* \subseteq \ker g^*$ and $\underline{dom}\,f^*$ is a weak subalgebra of $\underline{dom}\,g^*$ -- in the case that the domains of f^* and g^* are relative subalgebras of $\underline{F}(\underline{A},K_t)$, the relation between the kernels is sufficient.

PROOF. While the necessity of the condition is trivial, its sufficiency follows from an application of the homomorphism theorem, using the fact that a closed epimorphism is also full and surjective. If the domains of the initial extensions are relative subalgebras, the relation between the domains is implied by the one between the kernels. ■■■

20.1.11 REMARK. The results presented in this subsection show that despite the fact that many thinks work for E-varieties as they do in the whole species, facts become nevertheless to some extent more complicated; and this becomes even worse if we consider ECE-varieties, where in general we do not even know about the existence of relatively free completions, which then have to be replaced by direct systems of relatively free extensions. Mainly because of lack of space we do not go here into more details, but on the other hand we do not yet know about any detailed investigation in this direction. Some hints on how to treat ECE- and QE-varieties can be found in the presentation of [ABN 80].

21.2 On the lattice of initial closed congruence relations -- homogeneous case

21.2.1 REMARK and NOTATION. (i): In the last subsection we have seen that at least in the **homogeneous case** the information about all M-generated **K**-algebras in an E-variety **K** is contained in the pairs $(\underline{E},R)$, where $\underline{E}$ is any M-generated weak **K**-subalgebra of $\underline{F}(M,\mathbf{K}_t)$, and R is a closed congruence relation on $\underline{E}$ (cf. 21.1.6 and 21.1.7) -- and the case, when $\underline{E}$ is not already a relative subalgebra can occur only when $\underline{F}(M,\mathbf{K}_t)$ has only one element. Because of the difficulties which we have encountered in sections 17 and 18 and hence also in 21.1 for the general heterogeneous case, we restrict our considerations in this subsection to the **homogeneous case.**

(ii): Let us introduce the following notation: Let **K** be any (abstract) class of partial $\underline{S}$-algebras, let $\underline{A}$ be any **K**-algebra, and let M be any subset of A. Then we denote by **Wsubalg**$(M,\underline{A},\mathbf{K})$ and **Rsubalg**$(M,\underline{A},\mathbf{K})$ the sets of all M-generated weak respectively relative subalgebras of $\underline{A}$ belonging to **K**. Similarly, we write $\mathrm{Cong}_{c,w}(M,\underline{A},\mathbf{K})$ respectively $\mathrm{Cong}_{c,r}(M,\underline{A},\mathbf{K})$ for the set of all pairs $(\underline{E},R)$, where R is a closed congruence relation on $\underline{E}$, and $\underline{E}$ is any element of **Wsubalg**$(M,\underline{A},\mathbf{K})$ or **Rsubalg**$(M,\underline{A},\mathbf{K})$, respectively, i.e. they are each the union of the sets of closed congruence relations on an M-generated weak respectively relative **K**-subalgebra (say $\underline{E}$) of $\underline{A}$, "labeled" by their domain (i.e. by $\underline{E}$) -- we shall call them M-<u>initial closed</u> **K**-<u>congruence relations</u>. In the case of $\mathrm{Cong}_{c,r}(M,\underline{A},\mathbf{K})$ we shall usually omit mentioning the domain, since then it is uniquely determined by the congruence relation.

21.2.2 LEMMA. Let **K** be any E-variety, let $\underline{A}$ be any **K**-algebra, and M any generating subset of $\underline{A}$. Then $\mathrm{Cong}_{c,w}(M,\underline{A},\mathbf{K})$ and $\mathrm{Cong}_{c,r}(M,\underline{A},\mathbf{K})$ are complete lattices with respect to set theoretical inclusion in each component. The infimum of a family **B** of elements of one of these sets is given as follows:

inf $\mathbf{B} = (\underline{D},R^*)$, where $\underline{D}$ is the subalgebra generated by M in $\cap^*\{\underline{E} \mid (\underline{E},R) \in \mathbf{B}\}$, and R^* is the intersection of $D \times D$ with $\cap\{R \mid (\underline{E},R) \in \mathbf{B}\}$.

PROOF. Since $(\underline{A},\nabla_A)$ is the largest element in both sets of M-initial closed congruence relations, the Lemma is proved once that we have shown the statement about the infima. Since id_D has a homomorphic extension $(id_D)_{\underline{E}}:\underline{F}(D,\mathbf{K}) \to \underline{E}$ for each $(\underline{E},R) \in \mathbf{B}$, and since the graph of all these mappings is always the same, we have a surjective homomorphism $(id_D)':\underline{F}(D,\mathbf{K}) \to \underline{D}$, showing that $\underline{D}$ is indeed a $\mathbf{K}$-algebra -- a similar argument shows that the intersection of weak or relative $\mathbf{K}$-subalgebras of any $\mathbf{K}$-algebra is again a weak or relative $\mathbf{K}$-subalgebra. If we can show that R* is a closed congruence relation on $\underline{D}$, then $(\underline{D},R^*)$ is indeed the infimum of $\mathbf{B}$: Thus assume that $\underline{a}, \underline{b} \in D^{K_\phi}$, that $\underline{a} \in \text{dom}\ \phi^{\underline{D}}$, and that the sequences $\underline{a}$ and $\underline{b}$ are componentwise congruent modulo R*; then the same assumptions apply for every $(\underline{E},R)$ instead of $(\underline{D},R^*)$. But R is a closed congruence relation on $\underline{E}$, therefore $\underline{a}$ and $\underline{b}$ both belong to the domain of $\phi^{\underline{E}}$, and $(\underline{a}\phi^{\underline{E}},\underline{b}\phi^{\underline{E}}) \in R$. But this easily implies that $\phi^{\underline{D}}$ is defined on $\underline{b}$, and that $(\underline{a}\phi^{\underline{D}},\underline{b}\phi^{\underline{D}}) \in R^*$, showing that R* is indeed a closed congruence relation on $\underline{D}$. ∎

21.2.3. Although we do not have counterexamples, we cannot prove in general that inf **B** described above is always already the componentwise set theoretical intersection of the elements of **B**. Nevertheless the results obtained so far show that in E-varieties one can formulate -- at least in the homogeneous case -- a "local model theory" in a similar way as we have done for the whole species in Chapter II. But if we could restrict considerations at least to $\text{Cong}_{c,r}(M,\underline{F}(M,\mathbf{K}_t),\mathbf{K})$ for any given set M, this would make the corresponding results much smoother. Thus we dedicate the rest of this subsection to the study of cases when this special situation arises. We shall heavily need the element basis property in what follows, although we have no idea whether this property is really necessary. If there will really be practical interest in this area of investigations, then there still remains a lot to be done, and at this place we only want to offer an idea of possible methods.

21.2.4 THEOREM. Let M and N be any sets, $M \subseteq N$, let $\mathbf{K}$ be an E-variety, and let $z:U_{\underline{F}(N,\mathbf{K})} \to N$ be an injective mapping such that the z-completion $\underline{A}:= \underline{F}(N,\mathbf{K})^z$ is isomorphic -- over id_M -- to $\underline{F}(M,\mathbf{K}_t)$, the (total) $\mathbf{K}_t$-free $\mathbf{K}_t$-algebra (cf. subsection 17.3). Moreover, let each $x \in F(N,\mathbf{K}) \setminus C(\underline{F}(N,\mathbf{K}))$ have an element basis in N, and assume either $M \neq \emptyset$ or that $\mathbf{C}_{\underline{F}(N,\mathbf{K})}\emptyset = C(\underline{F}(N,\mathbf{K}))$ (cf. section 18.1). Then $\mathbf{Rsubalg}(M,\underline{F}(M,\mathbf{K}_t),\mathbf{K})$ is a closure system on $\underline{F}(M,\mathbf{K}_t)$.

PROOF. Since M generates the whole algebra, the empty intersection of elements of $\mathbf{Rsubalg}(M,\underline{F}(M,\mathbf{K}_t),\mathbf{K})$, namely $\underline{A}$, again belongs to $\mathbf{Rsubalg}(M,\underline{F}(M,\mathbf{K}_t),\mathbf{K})$. Now let $\mathbf{B}:= (\underline{A}_j)_{j\in J}$ be any non-empty family of elements of $\mathbf{Rsubalg}(M,\underline{F}(M,\mathbf{K}_t),\mathbf{K})$. Because of 17.3.6.(i) we have for each $j \in J$ that $A_j = \mathbf{C}_{\underline{F}(N,\mathbf{K})}(N \cap A_j)$. We abbreviate $N_j:= N \cap A_j$ $(j \in J)$ and $\underline{B}:= \bigcap^*\{\underline{A}_j | j \in J\}$. Then $\underline{B}$ is a relative subalgebra of $\underline{A}$, and because of our assumptions it is easy to realize that $C(\underline{F}(N,\mathbf{K})) \subseteq \mathbf{C}_{\underline{F}(N,\mathbf{K})} M \subseteq B$. By possibly introducing new nullary fundamental operations with value x for each $x \in C(\underline{F}(N,\mathbf{K}))$ we change the subalgebra lattice of $\underline{F}(N,\mathbf{K})$ in such a way that $C(\underline{F}(M,\mathbf{K}))$ becomes the smallest closed subset. Then N has the element basis property, and therefore we get for the closure operator $\mathbf{C}^*$ with respect to the possibly extended structure of $\underline{A}$ that

$$B = \bigcap\{A_j | j \in J\} = \bigcap\{\mathbf{C}^* N_j | j \in J\} = \mathbf{C}^*(\bigcap\{N_j | j \in J\}) :$$

We only have to prove the last equality, and the inclusion of the right hand side in the left hand side is trivial; in order to prove the other inclusion let x be any element of $\bigcap\{C^*N_j | j \in J\}$. Since J is non-empty, and since x has an element basis xN in N, we have $xN \subseteq N_j$ for each $j \in J$, and therefore $xN \subseteq \bigcap\{N_j | j \in J\}$, showing that x also belongs to $C^*(\bigcap\{N_j | j \in J\})$. (Because of $C(\underline{F}(N,K)) \subseteq B$ this equality would also have been obtained without changing the similarity type.) Thus we get for $N^* := \bigcap\{N_j | j \in J\}$ that $B = C_{\underline{F}(N,K)} N^*$. We prove indirectly that M generates $\underline{B}$, i.e. we assume that $B' := C_{\underline{B}} M \neq B$. Since $\underline{B}$ belongs to **K** as an intersection of relative subalgebras of $\underline{A}$ belonging to **K**, we also get $\underline{B}' \in K$, where $\underline{B}'$ is the relative subalgebra of $\underline{A}$ on B'. Again 17.3.6.(i) implies that for $N' := N \cap B'$ we get $B' = C_{\underline{F}(N,K)} N'$. Hence the assumption $B \neq B'$ implies $N^* \neq N'$. With δ again designating the ordinal dimension under consideration and $G := G_{\underline{F}(N,K),z}$ as in 17.3 we recall from 17.3.6 that $A = C_{\underline{A}} M = G(\bigcup\{G^{\alpha}M | \alpha < \delta\})$; therefore we have for each $a \in A$ an ordinal $\sigma(a)$ which is the least ordinal ρ such that $a \in G^{\rho}M \setminus \bigcup\{G^{\beta}M | \beta < \rho\}$. By assumption there is $a \in N^* \setminus N'$, and we can choose such an $a \in N^* \setminus N'$, for which $\sigma' := \sigma(a)$ is minimal. It is then easy to realize that we also have $\sigma' < \sigma(b)$ for each $b \in B \setminus B'$ (apply the fact that each $b \in B \setminus B'$ has an element basis bN in N, and that 17.3.6.(v) implies that for $b \in G^{\beta}M$ we also have $bN \subseteq G^{\beta}M$). Because of $M \subseteq B \cap B'$ we have $\sigma' > 0$. Moreover we have

(*) $$B \cap \bigcup\{G^{\beta}M | \beta < \sigma'\} = B' \cap \bigcup\{G^{\beta}M | \beta < \sigma'\} .$$

For the chosen $a \in N^* \setminus N'$ satisfying $\sigma(a) = \sigma'$ there exists exactly one sequence $\underline{a} \in (\bigcup\{G^{\beta}M | \beta < \sigma'\})^{K_\phi}$ for some suitable $\phi \in \Omega$ such that $(\phi, \underline{a}) \in \text{dom } z$ and $a = (\phi, \underline{a})z = \underline{a}\phi^{\underline{A}}$. Since $a \in B = \bigcap\{A_j | j \in J\}$ we get $\underline{a} \in \bigcap\{A_j^{K_\phi} | j \in J\} = B^{K_\phi}$. Because of (*) this implies $\underline{a} \in B'^{K_\phi}$, and since B' is a closed subset of $\underline{B}$, this contradicts the assumption that $a \in B \setminus B'$. Hence $B = B'$, showing that **Rsubalg**$(M, \underline{F}(M, K_t), K)$ is a closure system on $\underline{A} = \underline{F}(M, K_t)$. ■■■

21.2.5 COROLLARY. Let **K** be any E-variety such that for each non-empty set N each $x \in F(N,K) \setminus C(\underline{F}(N,K))$ has an element basis in N. Then **Rsubalg**$(M, \underline{F}(M, K_t), K)$ is a closure system on $\underline{F}(M, K_t)$ for each non-empty set M. ■■■

Combining Corollary 21.2.5 with Theorem 18.2.10 we easily get, when **d** designates the cardinal dimension of the given similarity type, the

21.2.6 COROLLARY. Let **K** be any non-trivial E-variety of homogeneous partial algebras of cardinal dimension $d \leq \#\omega_0$. Then **Rsubalg**$(M, \underline{F}(M, K_t), K)$ is a closure system on $\underline{F}(M, K_t)$ for each non-empty set M -- and if M is empty, then we have additionally to assume that $C_{\underline{F}(N,K)} \emptyset = C(\underline{F}(N,K))$ for some non-empty set N. ■■■

Because of the fact that it is the main consequence of Theorem 21.2.4 we designate the next corollary as

21.2.7 THEOREM. Let **K** be any non-trivial class of homogeneous partial algebras such that for each non-empty set N each $x \in F(N,K) \setminus C(\underline{F}(N,K))$ has an element basis in N. Then $\text{Cong}_{c,r}(M, \underline{F}(M, K_t), K)$ is a closure system on $F(M, K_t) \times F(M, K_t)$ for each non-empty set M -- forgetting about the "first component", which is always induced by the second one. And if M is empty, then we have additionally to assume that

$C_{\underline{F}(N,K)}\ \emptyset = C(\underline{F}(N,K))$ for some non-empty set N. ■■■

21.2.8 REMARKS. (i): Observe that 21.2.7 has a corollary similar to 21.2.6. Moreover, similar results are also true in the inhomogeneous case under additional assumptions derived from earlier observations; nevertheless the formulations are much more complicated, and we do not give them here.

(ii): Observe that for every non-trivial E-variety of finitary homogeneous partial algebras the isomorphism classes (over $M \neq \emptyset$) of M-generated **K**-algebras -- represented as triples $(M,f,\underline{A})$, where $f:M \to A$, and $[M]f$ is a generating subset of $\underline{A}$ -- form a complete lattice, induced by a quasi-ordering, where $(M,f,\underline{A}) \geq (M,f',\underline{A}')$ iff there exists a homomorphism $g:\underline{A} \to \underline{A}'$ such that $fg = f'$;-- this corresponds to the relation $\ker f^* \subseteq \ker f'^*$ (cf. Theorem 21.1.10).

21.3 Relativized algebraic partial operations -- homogeneous case

21.3.1 REMARKS and DEFINITION. In this subsection we want to present some considerations concerning the formulation of "relativized" algebraic partial operations with respect to an E-variety. We recall that it is quite natural to formulate e.g. polynomials for rings and groups and other equationally defined total algebras within the special variety rather than for the whole species. One might be tempted to do the same for partial algebras.

In this subsection we shall only briefly treat the homogeneous case just as an example to show some of the effects arising for partial algebras:

Let **K** be any non-trivial E-variety of homogeneous similar partial algebras, let M be any non-empty set, and $\underline{A}$ be any non-empty **K**-algebra -- we assume additionally that each non-empty subalgebra, say $\underline{B}$, of $\underline{F}(M,\mathbf{K}_t)$ has at least two elements, or $\underline{B}$ has no proper non-empty weak subalgebra in **K**. Then the results in 21.1 imply that each assignment $f:M \to A$ has a largest closed homomorphic extension $f^*:\underline{\mathrm{dom}}\ f^* \to \underline{A}$, where $\underline{\mathrm{dom}}\ f^*$ is an M-generated relative subalgebra of $\underline{F}(M,\mathbf{K}_t)$. As in section 5 we can introduce for each $t \in \underline{F}(M,\mathbf{K}_t)$ an M-ary partial operation, say $t^{*\underline{A}}$, on $\underline{A}$ by defining

$\mathrm{dom}\ t^{*\underline{A}} := \{\underline{a} \in A^M \mid t \in \mathrm{dom}\ \underline{a}^*\}$, and $(\underline{a})t^{*\underline{A}} := (t)\underline{a}^*$ for each $\underline{a} \in \mathrm{dom}\ t^{*\underline{A}}$.

We then call $t^{*\underline{A}}$ a <u>**K**-algebraic M-ary partial operation</u> on $\underline{A}$ or a <u>relativized</u> (to **K**) <u>algebraic partial operation</u> on $\underline{A}$.

Finally we introduce

$APO_K{}^M(\underline{A}) := \{t^{*\underline{A}} \mid t \in \underline{F}(M,\mathbf{K}_t)\}$, the set of all **K**-algebraic M-ary partial operations on $\underline{A}$.

The remainder of this subsection is mainly devoted to the investigation of the relationship between **K**-algebraic partial operations on $\underline{A}$ and algebraic ones as investigated in section 15.

21.3.2 LEMMA. Let **K** be a non-trivial E-variety of homogeneous partial algebras, let $\underline{A} \in \mathbf{K}$ such that each non-empty subalgebra of $\underline{F}(\underline{A},\mathbf{K}_t)$ has at least two elements or has no proper non-empty weak subalgebra in **K**, let M be any non-empty set, $p := id_M$, and let $p':\underline{F}(M,\mathbf{TAlg}(\tau)) \to \underline{F}(M,\mathbf{K}_t)$ be its homomorphic extension. Let $\underline{A}$ be an arbitrary **K**-algebra and $e:M \to E^M(\underline{A})$, defined by $me := e_m{}^M$ for all $m \in M$ (cf. section 15), the canonical mapping which maps M onto the set of all M-ary total projections of $\underline{A}^M$. Let

$\widetilde{e}:\underline{\text{dom}}\ \widetilde{e} \to \underline{F}^M(\underline{A})$ and $e^*:\underline{\text{dom}}\ e^* \to \underline{F}^M(\underline{A})$ be the largest closed homomorphic extensions of e within $\underline{F}(M,\mathbf{TAlg}(\tau))$ and $\underline{F}(M,\mathbf{K}_t)$, respectively; and for $f:M \to A$ let $p_f:\underline{\text{dom}}\ \widetilde{f} \to \underline{\text{dom}}\ f^*$ be the (closed and surjective -- cf. 21.11) restriction of p' to $\underline{\text{dom}}\ \widetilde{f}$.

Then we have the following statements:

(i) For each $q \in \underline{F}(M,\mathbf{K}_t)$:

$\text{graph}(q^{*\underline{A}}) = \bigcup\{\text{graph}(t^{\underline{A}}) \mid t \in (q)p'^{-1}\}$, and $q^{*\underline{A}}$ is a partial operation on $\underline{A}$.

(ii) $\ker \widetilde{f} = [\ker f^*](p_f^{-1} \times p_f^{-1})$.

(iii) $\{t^{\underline{A}} \mid t \in \text{dom}\ \widetilde{e}\} = \{q^{*\underline{A}} \mid q \in \text{dom}\ e^*\} = F^M(\underline{A})$.

(iv) $\ker e^*$ is a fully invariant congruence relation on $\underline{\text{dom}}\ e^*$.

(v) $\underline{\text{dom}}\ e^*$ is freely and $\underline{A}$-freely generated by M.

PROOF. Observing the notation introduced in the lemma we get:

Ad (i): From 21.1.9 we know that for each $f:M \to A$ we have $\widetilde{f} = p_f f^*$, and $p_f:\underline{\text{dom}}\ \widetilde{f} \to \underline{\text{dom}}\ f^*$ is closed and surjective. Now, let $g \in \underline{F}(M,\mathbf{K}_t)$, and let $f \in \text{dom}\ g^{*\underline{A}}$, i.e. $g \in \text{dom}\ f^*$. Then there is $t \in \text{dom}\ p_f$ such that $g = tp_f$, thus $fg^{*\underline{A}} = ft^{\underline{A}}$. Hence we have the inclusion in (i) from the left in the right hand side. Conversely, assume that $f \in \text{dom}\ t^{\underline{A}}$ for some $t \in (g)p_f^{-1}$. Then we have $t \in \text{dom}\ \widetilde{f}$, and therefore $tp_f = g \in \text{dom} f^*$, therefore the other inclusion easily follows. Let us observe, in addition, that for $t,u \in (g)p'^{-1}$ which satisfy $f \in \text{dom}\ t^{\underline{A}} \cap \text{dom}\ u^{\underline{A}}$, we have $ft^{\underline{A}} = t\widetilde{f} = tp_f f^* = gf^* = up_f f^* = u\widetilde{f} = fu^{\underline{A}}$. ∎

(ii) is easily realized. ∎

Ad (iii): We already know from section 15 that the first and the third set are equal. p_e maps $\text{dom}\ \widetilde{e}$ onto $\text{dom}\ e^*$, and (i) implies that all elements of $\text{dom}\ e^*$ also imply total operations on $\underline{A}$, which have to be identical with the ones induced by the corresponding elements of $\text{dom}\ e$. Since p_e is surjective, equality follows. ∎

(iv) and (v) are proved in the same way as this has been done for $\ker \widetilde{e}$ and $\underline{\text{dom}}\ \widetilde{e}$ in section 6. ∎∎∎

21.3.3 EXAMPLE. In order to get a better idea of 21.3.2.(i) we consider the E-variety $\mathbf{R'}$ of all homogeneous partial algebras $\underline{A}:= (A;+,0,-,\cdot,1,h)$ of type $\tau:= (2,0,1,2,0,1)$ such that $(A;+,0,-,\cdot,1)$ is an associative and commutative ring with multiplicative unit (1), and h is an arbitrary additional unary partial operation satisfying $(1)h\cdot 1 \stackrel{e}{=} 1$.

Let $s:= ((x + y) + (xh + (-(xh))) + zh)$

and $t:= ((x + y) + (yh + (-(yh))) + zh)$

be two elements of $\underline{F}(\{x,y,z\},\mathbf{TAlg}(2,0,1,2,0,1))$, and let

$\underline{Z}':= (\mathbf{Z};+,0,-,\cdot,1,h^{\underline{Z}'})$

be the ring of all integers with an additional unary partial operation satisfying $\text{graph}(h'^{\underline{Z}'}):= \{(1,1)\}$. Then we have

$\text{dom}\ s^{\underline{Z}'} = \{((1,b),1) \mid b \in Z'\}$, and

$\text{dom}\ t^{\underline{Z}'} = \{((a,1),1) \mid a \in Z'\}$,

while s and t have the same image in $\underline{F}(\{x,y,z\},\mathbf{R'}_t)$, namely $g:= x + y + zh$, satisfying

$\text{dom}\ g^{\underline{Z}'} = \{((a,b),1) \mid a,b \in Z'\}$.

21.3.4 REMARKS and PROBLEMS. **(i)**: As long as the elements of $\underline{F}(M,\mathbf{K}_t)$ possess a normal form within $\underline{F}(M,\mathbf{TAlg}(\underline{S}))$, each **K**-algebraic partial operation of $\underline{A}$ **K** will also be algebraic, but it is conceivable that this is not the general situation, although we have not yet computed a suitable example.

(ii): We also conjecture, but have not yet checked in detail, that, for each (non-empty) set M, for each non-trivial E-variety of (homogeneous) partial algebras and for each K-algebra $\underline{A}$, $APO_{\mathbf{K}}{}^{M}(\underline{A})$ is the $\mathbf{K}_t$-universal $\mathbf{K}_t$-solution of $APO^{M}(\underline{A})$.

(iii): As it has already been hinted upon earlier, we have also the possibility of relativizing arbitrary X-ary partial polynomial functions for given K-algebras $\underline{A}$ by considering $\underline{F}(\underline{A}\ \dot{\cup}\ X,\mathbf{K})$ and $\underline{F}(\underline{A}\ \dot{\cup}\ X,\mathbf{K}_t)$, where -- at least in the homogeneous case for non-trivial $\underline{A}$ -- the former can be considered as a relative subalgebra of the latter, and each assignment $f:X \to A$ also has a largest closed $(A \cup X)$-initial closed homomorphic extension with respect to the identity mapping on A within $\underline{F}(\underline{A}\ \dot{\cup}\ X,\mathbf{K}_t)$.

(iv): In those cases where each M-generated **K**-algebra (for some E-variety **K**) is representable by a closed congruence relation on an M-generated relative subalgebra of $\underline{F}(M,\mathbf{K}_t)$, the closed M-initial extensions of assignments $f:M \to \underline{A}$ into **K**-algebras $\underline{A}$ provide a tool for formulating a basic model theoretic language within the primitive class. But it should be observed that it is in general coarser than the global one, e.g. it might not be possible to distinguish two globally defined ECE-subvarieties on this local level, meaning that at least one of them might not be locally definable as an ECE-variety.

(v): Most of the trouble which arises for E-varieties of heterogeneous partial algebras -- as discussed earlier -- could be avoided by a more restrictive definition of the concept of E-equations in the heterogeneous case, e.g. by only allowing as E-equations $(M:\ t \stackrel{e}{=} t')$ such expressions which satisfy $[M]v_M = [var(\{t,t'\})]v_{var(\{t,t'\})}$. But this would cause other mostly syntactical difficulties, and we repeat that we have chosen the concept presented so far because of compatibility with the approach of the ADJ-group in computer science, the concept treated in Part II by H.REICHEL, and the category theoretical approach of H.ANDRÉKA, I.NÉMETI and I.SAIN.

APPENDIX

In **section 22** we briefly mention some concepts and results which could not be treated in detail in this book, and we give some further hints to literature concerning applications of the theory presented so far. -- In **section 23** some concepts and results from set theory are presented -- without proof -- which are often used. -- In **section 24** some concepts and results from the theories of partial orders, lattices, closure systems and Galois correspondences are collected, which are often used throughout this book.

22 Survey on some additional concepts and results

22.1 Other concepts of homomorphisms

22.1.1. This book is based on a concept of homomorphisms which corresponds to those in general used for relational systems or total algebras, respectively; we make this more precise for the homogeneous case for any given similarity type $\tau = (K_\phi)_{\phi\in\Omega}$:

The category **TAlg**(τ) of all total algebras of type τ is a full and epireflective subcategory of **Alg**(τ); and in the way described in the introduction, in 1.1.3 and in 13.4.2 even the category **RelS**(τ) of all relational systems of type τ -- i.e. the class of all relational systems $(M,(R_\phi^{\underline{M}})_{\phi\in\Omega})$, where, for each $\phi \in \Omega$, $R_\phi^{\underline{M}}$ is a subset of M^{K_ϕ} (i.e. $R_\phi^{\underline{M}} \subseteq M^{K_\phi}$ is a K_ϕ-ary relation on M) -- and with homomorphisms $f:(M,(R_\phi^{\underline{M}})_{\phi\in\Omega}) \to (N,(R_\phi^{\underline{N}})_{\phi\in\Omega})$ which are mappings from M into N such that, for each $\phi \in \Omega$, $[R_\phi^{\underline{M}}]f^{(K_\phi)} \subseteq R_\phi^{\underline{N}}$ -- can be considered as a full and epireflective subcategory of **Alg**(τ). Moreover, **Alg**(τ) can be considered in a natural way as a full subcategory of **RelS**(τ'), with $\tau' := (K_\phi \cup \{K_\phi\})_{\phi\in\Omega}$, where each K_ϕ-ary partial operation $\phi^{\underline{A}}$ is replaced by the corresponding $(K_\phi \cup \{K_\phi\})$-ary relation graph($\phi^{\underline{A}}$).
However there are other concepts around, and we want to discuss some of them and to show that they can be "modeled" without great difficulties within the presented theory.

22.1.2 DEFINITION. Let $\underline{A},\underline{B} \in$ **Alg**($\underline{S}$) , $h:A \cdot\!\!\to B$ any partial mapping out of A into B:

(i): h is called a <u>quomorphism</u> (short for "<u>quasi-homomorphism</u>") out of $\underline{A}$ into $\underline{B}$ iff one has for every $\phi \in \Omega$ and for every $\underline{a} \in A^{K_\phi}$:

(QH) If $\underline{a}\in\mathrm{dom}\,\phi^{\underline{A}}\cap(\mathrm{dom}\,h)^{K_\phi}$ and $(\underline{a})\phi^{\underline{A}}\in\mathrm{dom}\,h$, then $\underline{a}\cdot h\in\mathrm{dom}\,\phi^{\underline{B}}$ and $(\underline{a}\phi^{\underline{A}})h = (\underline{a}\cdot h)\phi^{\underline{B}}$.

(ii): h is called a <u>conformism</u> (see [Ke 70], or a "p-morphism" in [Po 73]) out of $\underline{A}$ into $\underline{B}$ iff one has for every $\phi \in \Omega$ and for every $\underline{a} \in A^{K_\phi}$:

(C) If $\underline{a}\in(\mathrm{dom}\,h)^{K_\phi}$, and if $\underline{a}\cdot h\in\mathrm{dom}\,\phi^{\underline{B}}$, then $\underline{a}\in\mathrm{dom}\,\phi^{\underline{A}}$ and $(\underline{a}\phi^{\underline{A}})h = (\underline{a}\cdot h)\phi^{\underline{B}}$.

(iii): h is called a <u>partial algebraic mapping</u> out of $\underline{A}$ into $\underline{B}$ iff one has for every $\phi \in \Omega$ and for every $\underline{a} \in A^{K_\phi}$:

(PAM) If $\underline{a}\in\mathrm{dom}\,\phi^{\underline{A}}\cap(\mathrm{dom}\,h)^{K_\phi}$, $\underline{a}\phi^{\underline{A}}\in\mathrm{dom}\,h$, and $\underline{a}\cdot h\in\mathrm{dom}\,\phi^{\underline{B}}$, then $(\underline{a}\phi^{\underline{A}})h = (\underline{a}\cdot h)\phi^{\underline{B}}$.

22.1.3 REMARKS. (i): The above concepts have already been used somewhere in the literature; for instance quomorphisms are investigated by H.-J.HOEHNKE (see [Hoe 72] or [Hoe 76], where they are called partial homomorphisms) and his school, who also study the category **Par** of all sets with partial mappings (see also B.WOJDYŁO [Wo 72]).

(ii): Notice that partial mappings h out of A into B which satisfy (QH) and (C) are sometimes called conformic partial homomorphisms (see [Hoe 72], p.10). The interpretation homomorphisms $v:\underline{F}(M,\mathbf{TAlg}(\underline{S})) \cdot\!\!\to \underline{A}$ for an assignment $v:M \to A$ are always such conformic partial homomorphisms, but we have seen in Chapter II that they can also be expressed in the category **Alg**($\underline{S}$) as the right factor of a (**TAlg**($\underline{S}$)-extendable, **TAlg**($\underline{S}$)-perfect)-factorization of $v:\underline{M}_d \to \underline{A}$.

(iii): It might be useful to observe that, for a partial mapping $h:A \cdot\!\!\to B$, we can characterize its being one of the above morphisms out of $\underline{A}$ into $\underline{B}$ in the following way, where the right hand side always has to be satisfied for all $\phi \in \Omega$:

quomorphism : iff $\mathrm{graph}(\phi^{\underline{A}}\cdot h) \subseteq \mathrm{graph}(h^{(K_\phi)}\cdot\phi^{\underline{B}})$;

conformism : iff $\mathrm{graph}(\phi^{\underline{A}}\cdot h) \supseteq \mathrm{graph}(h^{(K_\phi)}\cdot\phi^{\underline{B}})$;

conformic quomorphism (conformic partial homomorphism)

: iff $\mathrm{graph}(\phi^{\underline{A}}\cdot h) = \mathrm{graph}(h^{(K_\phi)}\cdot\phi^{\underline{B}})$;

partial algebraic mapping : iff $\mathrm{graph}(\phi^{\underline{A}}\cdot h) \cap (\mathrm{dom}(h^{(K_\phi)}\cdot\phi^{\underline{B}}) \times B) \subseteq \mathrm{graph}(h^{(K_\phi)}\cdot\phi^{\underline{B}})$.

(iv): Using a composition of two morphisms, where the target of the first one is discrete, one easily realizes that the smallest category containing the **class Alg**($\underline{S}$) and either (all homomorphisms and all conformisms) or (all quomorphisms and all conformisms) or all partial algebraic mappings, respectively, contains in each case all partial mappings between the carrier sets of the partial algebras involved, i.e. the morphisms have in each of these cases nothing to do with the algebraic structure of the objects.
On the other hand all homomorphisms and all conformic partial homomorphisms are quomorphisms, while the above remark shows that homomorphisms and conformisms cannot be combined in one category which seems to be relevant for partial algebras. Since among conformisms in particular the conformic partial homomorphisms seem to be of greater interest, one might think that the category containing them together with our homomorphisms -- i.e. the category with quomorphisms as morphisms -- might be the most suitable one for partial algebras.
Because of the following reasons we did not take quomorphisms as the basic concept for morphisms:

(a): The reasons already mentioned in 22.1.1: The category (**Alg**($\underline{S}$), homomorphisms) as well as the category (**TAlg**($\underline{S}$), homomorphisms) are neither full nor epireflective subcategories of a category (**Alg**($\underline{S}$), quomorphisms), and all the category theoretical concepts considered in Chapter I would differ very much in the new category from the descriptions given in this book.

(b): However, full and epireflective subcategories inherit a lot of properties from the original one, while an arbitrary subcategory of a given category may have only little connection with the given one. Thus we think it natural that the two categories (**Alg**($\underline{S}$), homomorphisms) and (**Alg**($\underline{S}$), quomorphisms) should be studied separately, and within the scope of this book we could only study one of these categories in detail.

(c): In the category (**Alg**($\underline{S}$), quomorphisms) the description of the basic concepts becomes much more complicated and therefore application of such a theory -- even when it is developed -- seems relatively unlikely. Moreover, there are (category theoretical) tools to handle also quomorphisms, conformisms, etc. within the framework of the category (**Alg**($\underline{S}$), homomorphisms) by considering pairs of homomorphisms with special properties (see 22.1.4 below).

22.1.4 Special categories of pairs. The concepts of morphisms described above can be modeled within the category **Alg**($\underline{S}$) (again we silently assume that the class of morphisms is the class of all homomorphisms) by replacing a partial mapping, say $h:A \dashrightarrow B$, by a suitable pair, say (i,h') of two mappings: $i:A' := \mathrm{dom}\, h \to A$ $(i := id_{A'})$ and $h':A' \to B$ $(\mathrm{graph}(h') := \mathrm{graph}(h))$, see Figure 22.1, and in the case of the above morphisms we only have to choose special properties for each component in the following way:

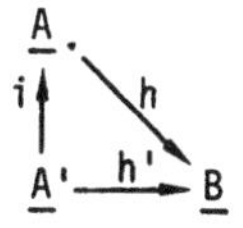

Figure 22.1

h is a quomorphism	iff	i is a full monomorphism, and h' is an arbitrary homomorphism;
h is a conformism	iff	i is a monomorphism, and h' is a closed homomorphism;
h is a conformic quomorphism	iff	i is a full monomorphism, and h' is a closed homomorphism;
h is a partial algebraic mapping	iff	i is a monomorphism, and h' is a homomorphism.

Composition of two partial mappings is then in this "translation" constructed in each case by a suitable pullback as indicated in Figure 22.2, where $(\underline{A}'';p,q)$ is the pullback of $(h',i_g;\underline{B})$. ∎

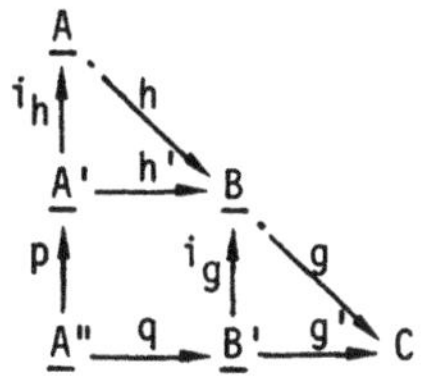

Figure 22.2

22.1.5. We leave the discussion at this place, but we hope that the reader might have gotten the idea, that the concepts presented in this book are not as special as he might have thought.

22.2 Further topics and hints to applications

22.2.1 ECE-varieties. (i): As already mentioned at several other places, there exists another approach to special ECE-varieties -- mainly started by J.SŁOMINSKI in [Sl 68], in some instances further developed by H.HÖFT in [Hö 73] and treated in most generality by R.JOHN in [J 75]. We mean the "notions of validity of equations in partial algebras" (see 8.1.3). Some contributions with respect to "weak validity" (i.e. "weak equations") can also be found in R.KERKHOFF [Ke 70], P.BURMEISTER [B 71], V.S.POYTHRESS [Po 73], and L.RUDAK [Ru 83], where finally the problem of a description of closed sets of "weak equations" has been solved. Applications of this concept can be found in G.A.EDGAR [Ed 73] (to topology) and in S.KOCHEN and E.P.SPECKER [KoSp 68] (to the logic of quantum mechanics). In a diplom thesis [Oe 81] at the Technische Hochschule Darmstadt G.OEHME has combined the results of [Ke 70] and [Po 73] with our concept of ECE-varieties, and he proves that an IS_cP_+-closed class **K** of (homogeneous) partial algebras possesses for each set M a conform-free **K**-algebra iff for every ECE-equation $\Phi \Rightarrow t \overset{e}{=} t'$ which is valid in **K** the "weak equation" $t \overset{e}{=} t \wedge t' \overset{e}{=} t' \Rightarrow t \overset{e}{=} t'$ is also valid in **K**.

(ii): In his doctoral thesis [Si 84] M.SIEGMUND-SCHULTZE has studied the concept of residual smallness for E- and ECE-varieties. Especially for ECE-varieties he gets very satisfactory results -- in particular in connection with eqational compactness by allowing also negations of TE-expressions together with E-equations.

22.2.2 Category theoretical approaches. Besides the approach to a theory of partial algebras presented in this book there are other ones making more use of category theory by using a Lawvere style approach, where partial algebras are functors from some special kind of small categories into the category of all sets with partial mappings. Since space and time limitations do not allow us to go into more details here and to give it due credit, we would just like to mention in this connection the school around H.-J.Hoehnke with papers (among others) from HOEHNKE [Hoe 76], [Hoe 82] (see also [Hoe 84]), J.SCHRECKENBERGER [Sre 80] and H.-J.VOGEL [V 79] and [V 84]. Another approach in this direction is made by A.OBTUŁOWICZ in [O 81] and [O 82]; his results might be quite useful for the study of ECE-varieties, since his concept of equations covers our concept of ECE-equations.

22.2.3 Applications in computer science. We conclude our discussion of the theory of partial algebras by hints -- mainly due to H.Andréka, I.Németi and I.Sain -- to references concerning applications to computer science of the results presented in this book: In a series of papers [AN 79f] H.ANDRÉKA and I.NÉMETI present a survey of applications of universal algebra in computer science, which also indicates at many places main directions of applications of partial algebras -- see also [AN 83].

In [Mr 83] and [Mr 83a] Zs.MARKUSZ thoroughly investigates the problems and difficulties of defining validity for a logic as presented here in section 7 for many-sorted partial algebras (including empty sorts).

The papers [Pa 83] and [Pa 84] of A.PASZTOR apply the theory of partial algebras to obtain results about the concrete categories used in the approach by D.S.Scott and the so-called ADJ-group (S.L.Bloom, J.A.Goguen, J.Meseguer, J.W.Thatcher, E.G.Wagner, J.B.Wright) to computational semantics. E.g. she proves that the structures used there are simply ECE-varieties of partial algebras, thus easily obtaining important results as consequences of the theory presented here. See also the other papers of A.Pasztor.

The papers [T 84] and [T 84a] of A.TARLECKI contain important applications to computer science of the results presented here in sections 12 and 8.2.

In the papers [GuL 82], [GuL 82a], and in particular in [GuL 80] by R.GUITART and C.LAIR mainly the results presented in sections 12 and 8.2 are combined with heavily functorial category theoretical approaches.

Further applications of partial algebras -- mainly in computer science -- can be found among others in the following papers: I.GUESSARIAN and J.MESEGUER [GeMe 84], I.G.ROSENBERG [R 83], B.MAHR and J.A.MAKOWSKY [MhMk 84]and [MhMk 84a], J.A.MAKOWSKY [Mk 84], E.KNUTH and L.RONYAI [KnRo 83], and M.BROY and M.WIRSING [BrWi 81] and [BrWi 81a] (see also [BgBrTuWi 81]).

23 Some definitions and facts from set theory

In addition to subsection 1.1 and the notation scattered throughout the book we want to collect here some remarks on the basic set theoretical notation as well as on the set theoretical background, together with some facts on cardinals and ordinals. For more information about set theory we refer to J.SCHMIDT [Sch 66a], D.MONK [Mo 76], A.LEVY [Lev 79], and with respect to cardinals and ordinals in particular to H.BACHMANN [Bc 67].
From an axiomatic point of view we use the Gödel-Bernays-von Neumann approach. See also Remark 24.4.10.(ii).

23.1 Some remarks on notations

23.1.1 Some notational conventions. From a naive point of view we distinguish proper classes, sets and objects in the following way:
Classes of algebraic structures or sets of subsets (or substructures) of some other set (or structure) are denoted by bold-faced capital Latin letters, as in " $\mathbf{K} \subseteq \mathbf{Alg}(\tau)$ " (to be read as "the class **K** is a subclass of the class $\mathbf{Alg}(\tau)$ of all partial algebras of similarity type τ"). Observe that a **set** is always a class which is an element of some other class as possibly specified by the axioms of axiomatic set theory.
Sets -- sometimes also proper classes -- are usually denoted by capital Latin letters, and their elements are denoted by small Latin letters from the beginning or the end of the alphabet. Exceptions are made when these elements are sequences (see 1.1.1; then the small Latin letters are underscored) or sequences of sequences (see e.g. 4.1.3; then we use small Gothic letters for sequences and underscore them if they designate a sequence of sequences). " $\emptyset$ " always designates the **empty set**, i.e. the set with no elements.
Moreover, closure systems are denoted by capital Gothic letters, and the same holds for categories not further specified (they usually are underscored because of their partial algebraic structure).
Capital letters are underscored when we want to express that we deal with an algebraic or relational structure on the corresponding carrier set.
Ordinals are denoted by small Greek letters, cardinals by bold-faced capital letters.

23.1.2 Some set theoretical notation. (i): " $\in$ " designates the "element relation". Negation: " $\notin$ ": " $a \notin A$ ", if a is not an element of A.
(ii): " $\subseteq$ " (" $\subset$ ") designates (proper) set theoretical inclusion, i.e. " $A \subseteq B$ " means that each element of the class A also belongs to the class B (A and B may be equal), while " $A \subset B$ " means " $A \subseteq B$ and $A \neq B$ (i.e. "not $A = B$ ")". Negation of " $A \subseteq B$ ": " $A \nsubseteq B$ ". (" $\subset$ " is sometimes more explicitely expressed by " $\subsetneqq$ ".)
(iii): For some given property P, $\{x|P(x)\}$ designates the class of all sets x having property P. We sometimes write $\{x \in A|H(x)\}$, if $P(x)$ means " $x \in A$ and $H(x)$ " for some class A and some property H.
(iv): Let A and B be any classes; then:
$A \setminus B := \{a|a \in A$ and $a \notin B\}$ designates the <u>difference class</u> of A without B.
$A \cap B := \{a|a \in A$ and $a \in B\}$ designates the <u>intersection</u> or <u>meet</u> of A and B.
$A \cup B := \{a|a \in A$ or $a \in B\}$ designates the <u>union</u> or <u>join</u> of A and B.

$A \dot{\cup} B := (\{0\} \times A) \cup (\{1\} \times B)$ is called the disjoint union of A and B.

$\cap A := \{b \mid b \in a$ for each $a \in A\} =: \cap\{a \mid a \in A\} =: \cap(a \mid a \in A)$ designates the meet or intersection of the class A (observe that for each class its members are again sets).

$\cup A := \{b \mid$ there exists $a \in A$ such that $b \in a\} =: \{b \mid b \in a$ for some $a \in A\} =: \; =: \cup\{a \mid a \in A\} =: \cup(a \mid a \in A)$ designates the union or join of A.

Similarly, we define for indexed families $(A_t \mid t \in T)$ of sets or classes:

$\cap(A_t \mid t \in T) := \{a \mid a \in A_t$ for each $t \in T\}$,

$\cup(A_t \mid t \in T) := \{a \mid a \in A_t$ for some $t \in T\}$, and as in the case of two sets we define the disjoint union of the family $(A_t \mid t \in T)$ as

$\dot{\cup}(A_t \mid t \in T) := \cup(\{t\} \times A_t \mid t \in T)$ (similarly $\dot{\cup}A := \cup(\{a\} \times a \mid a \in A)$ for every class A).

(v): In connection with the infimum or supremum of partial algebraic (weak) substructures we use " $\cap^*$ " and " $\cup^*$ ", respectively, i.e.:

$\cap^*(\underline{A}_t \mid t \in T)$ means $(\cap(A_t \mid t \in T), (\cap(\phi^{\underline{A}_t} \mid t \in T))_{\phi \in \Omega})$, and $\mathrm{graph}(\cap(\phi^{\underline{A}_t} \mid t \in T)) := \cap(\mathrm{graph}(\phi^{\underline{A}_t}) \mid t \in T)$.

Similarly, $\cup^*(\underline{A}_t \mid t \in T)$ means $(\cup(A_t \mid t \in T), (\cup(\phi^{\underline{A}_t} \mid t \in T))_{\phi \in \Omega})$, where $\mathrm{graph}(\cup(\phi^{\underline{A}_t} \mid t \in T)) := \cup(\mathrm{graph}(\phi^{\underline{A}_t}) \mid t \in T)$, and $\cup^*(\underline{A}_t \mid t \in T)$ is only defined, when for each $\phi \in \Omega$ we again get a partial operation.

(vi): For any two sets a and b the ordered pair (a,b) is a set such that for any two additional sets a' and b':

(*) $(a,b) = (a',b')$ iff $a = a'$ and $b = b'$.

Usually an ordered pair is realized by the so-called Kuratowski-pair $(a,b) := \{\{a\},\{a,b\}\}$. But since we also use this symbolism, when a and b may be proper classes (see e.g. the notation in connection with factorization systems), we want to mention that there exists a set theoretical realization of an ordered pair for -- possibly proper -- classes (then being again a class) having the above property (*): Such a pair is described in detail in [Sch 66a], pp 96ff as

(**) $(a,b) := \{\{\{x\}\} \mid x \in a\} \cup \{\mathbf{Subs}(\{y\}) \mid y \in b\}$ (following an idea of Quine), where $\mathbf{Subs}(A) := M \mid M$ is a set and $M \subseteq A$ designates for any class A the class of all subsets of A, the so-called power-class of A. (**) yields a set iff both a and b are sets.

(vii): The direct or cartesian product of two classes A and B is defined to be the class of all ordered pairs with first component in A and second component in B:

$$A \times B := \{(a,b) \mid a \in A \text{ and } b \in B\}.$$

If $A_1, A_2, \dots, A_n$ are classes, then we define the n-fold cartesian product

$$A_1 \times A_2 \times \dots \times A_n := (\dots(A_1 \times A_2) \times \dots) \times A_n,$$

but usually we do not distinguish this product from the set of all n-ary choice sequences (see 1.1), since there is a canonical bijection between these two sets; and in the case of possibly infinitely many factors, i.e. for $\mathsf{X}(A_i \mid i \in I)$, we only use the representation of its elements by choice functions (I being any set; see also 4.1.3):

$$\mathsf{X}(A_i \mid i \in I) := \{\underline{a} \mid \underline{a}: I \to \cup(A_i \mid i \in I) \text{ and } (i)\underline{a} \in A_i \text{ for every } i \in I\}.$$

23.1.3 Relations.

(i): A binary relation between two classes A and B is simply a subclass $R \subseteq A \times B$ of ordered pairs with first component in A and second component in B. We denote by

$\mathrm{dom}\, R := \{a \in A \mid$ there is $b \in B$ such that $(a,b) \in R\}$, the domain of R, by

$\mathrm{rng}\, R := \{b \in B \mid$ there is $a \in A$ such that $(a,b) \in R\}$, the range of R, and by

$\mathrm{field}\, R := \mathrm{dom}\, R \cup \mathrm{rng}\, R$, the field of R (this is used in particular if $A = B$).

If $A = B$, we call R a binary relation on A.

(ii): Moreover, if $R \subseteq A \times B$, then we define its converse R^{-1} by $R^{-1} := \{(b,a) \mid (a,b) \in R\} \subseteq B \times A$. Thus $\mathrm{rng}\, R = \mathrm{dom}\, R^{-1}$.

(iii): If $R \subseteq A \times B$ is a binary relation, $A' \subseteq A$ and $B' \subseteq B$, then we define $[A']R := \{b \in B \mid \text{there is } a \in A' \text{ such that } (a,b) \in R\} = \mathrm{rng}(R \cap (A' \times B))$, similarly we define $[B']R^{-1} := \mathrm{dom}(R \cap (A \times B'))$.

(iv): If T is any set, and if $(A_t \mid t \in T)$ is a family of sets (or even of classes), then a T-ary relation R on this family is a subclass $R \subseteq \mathbf{X}(A_t \mid t \in T)$.

For some special kinds of relations see 24.1.

23.1.4 Functions and (partial) mappings.

(i): In set theory a function f is just a binary relation on **V** -- the class of all sets -- which is unique in its second component: $(a,b),(a,b') \in f$ implies $b = b'$;

(ii): In this book a function f as above will usually be called the graph of a (partial) mapping (i.e. $f = \mathrm{graph}(g)$) for some (partial) mapping $g: A \multimap B$, since we consider mappings as the morphisms in the category **Set** of all sets, and similarly partial mappings as the morphisms in a category **Par** (again with **V** as class of objects): We define a partial mapping g out of a class A into a class B informally as an assignment assigning to each element a of A at most one element of B -- denoted by ag or (a)g, if it exists, and formally as a triplet $(A, \mathrm{graph}(g), B)$ (sometimes written as (A,f,B), or in symbols as $g: A \multimap B$), where A is called the start object and B is called the target object of g, and $\mathrm{graph}(g) := \{(a,ag) \mid a \in A \text{ and } ag \text{ exists}\} \subseteq A \times B$ is a function in the sense of (i). The fact that $(a,ag) \in \mathrm{graph}(g)$ is also often indicated by writing $a \mapsto ag$ (thinking of the informal definition of g as an assignment).

In analogy with 23.1.3.(i) we define the (exact) domain of a partial mapping g as

$$\mathrm{dom}\, g := \{a \mid a \in A \text{ and there is } b \in B \text{ such that } (a,b) \in \mathrm{graph}(g)\}.$$

In general this is not the category theoretical domain, which is $\mathrm{Dom}(g)$ $(:= A)$.

If $\mathrm{dom}\, g = A$, then we write $g: A \to B$ and by calling g a mapping from A into B.

Instead of "$g: A \multimap B$" we also sometimes write "$g: A \supseteq \mathrm{dom}\, g \to B$" or "$g: \mathrm{dom}\, g \subseteq A \to B$".

If, in particular, A designates a set of variables, we call a mapping $w: A \to B$ from A into some set B an assignment.

(iii): Let A be a class and B a set. By A^B we denote the class of all mappings from B into A: $A^B := \{g \mid g: B \to A\}$.

(iv): A partial mapping $g: A \multimap B$ is called

injective, if $a \neq a'$ implies $ag \neq a'g$ for every $a, a' \in \mathrm{dom}\, g$;

surjective (onto), if for each $b \in B$ there is some $a \in \mathrm{dom}\, g$ such that $b = ag$;

bijective, if g is an injective and surjective mapping **from** A onto B.

(v): Observe that -- at least in abstract situations -- we always apply mappings from the right. As a consequence we then also denote the composition of partial mappings $f: A \multimap B$ and $g: B \multimap C$ by fg $(: A \multimap C)$. In some situations we write $f \cdot g$ rather than fg; and this is done in particular whenever the name of at least one of the mappings under consideration consists -- besides some possible index -- of more than one letter. In particular, id_A designates the identity mapping from some class A onto itself -- or sometimes into some class B containing A as a subclass -- satisfying $(a)id_A := a$ for each $a \in A$. It usually has to be taken from the context, whether or not $A = B$ (at some occasions we shall be more explicit by writing id_{AB} $(: A \to B)$).

Observe, however, that we write composition of mappings in the opposite way when we apply them from the left (as we do in particular with the "operators" defined in section 4 -- "operator" meaning here that the argument as well as the value may be a proper class, and we should extend our set theory by "conglomerates" (see 23.2.1)).

(vi): Observe that, for the composition fg of two partial mappings, we have $\mathrm{dom}\, fg = \{a \in \mathrm{dom}\, f \mid af \in \mathrm{dom}\, g\}$, and $a \in \mathrm{dom}\, fg$ implies $a(fg) := (af)g$.

(vii): Let $f:A \cdot\!\!\rightarrow B$ be a partial mapping, set $R := \mathrm{graph}(f)$, and let $A' \subseteq A$, $B' \subseteq B$; then we define as in 23.1.3: $[A']f := [A']R$, the <u>image</u> of A' with respect to f, and $[B']f^{-1} := [B']R^{-1}$, the <u>preimage</u> of B' with respect to f.
If f is bijective, then the <u>inverse</u> f^{-1} is a mapping $f^{-1}:B \rightarrow A$ defined as the triplet $(B,(\mathrm{graph}(f))^{-1},A)$, i.e. $bf^{-1} = a$ iff $af = b$.

(viii): If $f:A \rightarrow B$, $f':A' \rightarrow B'$ are mappings, then we define $f \times f':A \times A' \rightarrow B \times B'$ by $(a,a')(f \times f') := (af,a'f')$ for each $(a,a') \in A \times A'$.
If $A = A'$, $B = B'$ and $f = f'$, then we write $f^{(2)}$ instead of $f \times f$.
Similarly, if $f:A \rightarrow B$ is a mapping, and if T is any set, then we denote by $f^{(T)}:A^T \rightarrow B^T$ the mapping induced by f and satisfying
$A^T \ni \underline{a} := (a_t \mid t \in T) \mapsto (a_t f \mid t \in T) =: \underline{a}f^{(T)} \in B^T$ (i.e. it is <u>defined componentwise</u> by f).

23.2 Some remarks on an axiomatic approach

23.2.1. On most occasions set theoretical concepts may be used in a naive sense. But at some places some reference is made to set theoretical axioms (e.g. to the Axiom of Choice), and the Axiom of Foundation is tacitly assumed in some set theoretical constructions. For an axiomatic approach to set theory the reader may think of the well-known Zermelo-Fraenkel axioms including the Axiom of Choice; however because of our frequent treatment of classes the Neumann-Bernays approach seems more appropriate as presented -- in part -- in [Sch 66a].
On some occasions we shall even apply operators to proper subclasses of classes, and the reader may think this embedded in an approach using "conglomerates" as a next level, where a <u>conglomerate</u> is a collection of classes (including sets which are just classes which are elements of some other class), but proper conglomerates cannot be elements of other conglomerates (see also the appendix of [HS 73]); conglomerates behave otherwise like classes, but we shall use them only in a very naive way in the text without explicitly mentioning conglomerates there.

23.2.2 Axioms of set theory.
Let **St** be the property of a class to be a <u>set</u>, i.e. to be an element of another class:
St(a): $\exists M(a \in M)$,
and let x,y,z,a,b,M,N be variables for classes (lower case letters usually indicating that they will actually range over sets because of the context). Then we have the following axioms (the set theoretical notation being taken from 23.1, and the logical one from section 7):
Axiom (scheme) of Comprehension: $\exists M \forall x(x \in M \leftrightarrow \mathbf{St}(x) \cap H(x))$
for every first order property H in a language for classes (i.e. having as fundamental non-logical symbol " $\in$ "), where M is not free in H.

Meaning: For every property H there is a class M containing exactly the sets with property H.

Axiom of Extensionality: $\forall y(y \in M \Leftrightarrow y \in N) \Rightarrow M = N$.

Meaning: Two classes are equal iff they contain the same elements.

Axiom of the Null-Set: $\mathbf{St}(\emptyset)$.

Meaning: The empty class is a set (follows from other axioms!).

Axiom of Subsets: $\mathbf{St}(a) \wedge b \subseteq a \Rightarrow \mathbf{St}(b)$.

Meaning: Each subclass of a set is again a set.

Axiom of the Binary Union: $\mathbf{St}(a) \wedge \mathbf{St}(b) \Rightarrow \mathbf{St}(a \cup b)$.

Meaning: The union of two sets is again a set.

Axiom of the One-Element Set: $\mathbf{St}(a) \Rightarrow \mathbf{St}(\{a\})$.

Meaning: For every set a the class containing exactly a as only element is again a set.

Axiom of Union: $\mathbf{St}(a) \Rightarrow \mathbf{St}(\mathbf{U}a)$.

Meaning: For every set its union is again a set.

Inverse of the Axiom of Union: $\mathbf{St}(\mathbf{U}a) \Rightarrow \mathbf{St}(a)$.

Meaning: If the union of a class a is a set, then a is already a set; this axiom implies in particular the "Power Set Axiom": $\mathbf{St}(a) \Rightarrow \mathbf{St}(\mathbf{Subs}(a))$.

Axiom of Replacement: $\mathbf{Function}(a) \wedge \mathbf{St}(\mathrm{dom}\ a) \Rightarrow \mathbf{St}(\mathrm{rng}(a))$.

Meaning: If a is a function (in the sense of 23.1.4.(i)), and if its domain is a set, then its range is a set.

Axiom of Infinity: $\mathbf{St}(\omega)$.

Meaning: There exists an infinite set; as a special item for an infinite set the set ω of all "natural numbers" is chosen, where a set theoretical definition of ω runs as follows: $\omega := \bigcap\{x \mid \emptyset \in x \wedge \forall y(y \in x \Rightarrow y \cup \{y\} \in x)\}$.

Axiom of Foundation (Regularity Axiom): $\forall M(\exists x(x \in M) \Rightarrow \exists y(y \in M \wedge y \cap M = \emptyset))$.

Meaning: Every non-empty class M contains an element which is minimal with respect to the $\in$-relation; in particular there is no class which possesses an infinite descending $\in$-chain, $\in$ is an irreflexive partial order on the class of all sets, and thus no class can be an element of itself.

Axiom of Choice:

$$\forall M(\forall z(z \in M \Rightarrow z \neq \emptyset) \Rightarrow \exists a(\mathbf{Function}(a) \wedge \mathrm{dom}\ a = M \wedge \forall z(z \in M \Rightarrow za \in z))) .$$

Meaning: Each class of non-empty sets possesses a "choice function".

Two well known consequences of the Axiom of Choice are the following ones:

23.2.3 Lemma of Kuratowski-Zorn (Zorn's Lemma). Let $(H;\subseteq)$ be a non-empty inductive subset of some set M. Then H has a maximal element (see 24.1). ∎

23.2.4 Theorem of Well-Ordering. On every set there exists a well-ordering. ∎

We also want to draw some frequently needed consequences from the Axiom of Foundation (see Levy [Lev 79], II, 5, 6, and 7, or Monk [Mo 76], §15):

23.2.5 PROPOSITION. Let **V** designate the class of all sets and **Ord** the class of all ordinals (see 23.3). Then there exists a <u>rank function</u> $\mathbf{rnk}:\mathbf{V} \to \mathbf{Ord}$ such that for every ordinal α the class $\mathbf{V}(\alpha) := \{A \mid A \in \mathbf{V} \text{ and } \mathbf{rnk}(A) < \alpha\}$ is a set, and for any sets a and A, $a \in A$ implies $\mathbf{rnk}(a) < \mathbf{rnk}(A)$. ∎

In what follows we assume a fixed rank function (which can explicitely be specified).

23.2.6 PROPOSITION. Let A be any set, and let $R \subseteq A \times \mathbf{V}$ be any relation. Then there exists a relation $R' \subseteq R$ such that R' is a set, $\mathrm{dom}\, R' = \mathrm{dom}\, R$, and $\mathrm{rng}\, R'$ is a set. In particular we may choose $R' := \{(a,b) \in R \mid \mathbf{rnk}(b) = \bigcap\{\mathbf{rnk}(c) \mid (a,c) \in R\}\}$ having the required properties. ∎

From the following result we often use in particular its corollary (similar results can be found in [Sch 66a] (end of §18) without using the Axiom of Foundation).

23.2.7 PROPOSITION. Let A be any set. Then $B := \{\{x,A\} \mid x \in A\}$ is a set disjoint from A with a canonical bijection $i : B \to A$, $\{x,A\} \mapsto x$ $(x \in A)$.
(Since $\mathbf{rnk}(\{x,A\}) > \mathbf{rnk}(A) > \mathbf{rnk}(x)$, we have $A \cap B = \emptyset$.) ∎∎∎

23.2.8 COROLLARY. Let A and B be any sets, and let $f : A \to B$ be any injective mapping from A into B. Then there exists a set C with a bijection $g : C \to B$ such that $A \subseteq C$ and the restriction $g|A$ of g to A is equal to f.

PROOF. Consider $C' := \{\{x, A \cup B\} \mid x \in B \setminus [A]f\}$, define $i : C' \to B$, $\{x, A \cup B\} \mapsto x$ $(x \in B \setminus [A]f)$, $C := A \cup C'$ and finally $g : C \to B$ for each $c \in C$ by

$$(c)g := \begin{cases} (c)f, & \text{if } c \in A, \\ (c)i, & \text{else.} \end{cases}$$

Obviously, g is a bijective mapping from C onto B, whose restriction to A equals f. ∎∎∎

23.2.9. If, in 23.2.8, $\underline{A}$ and $\underline{B}$ are partial algebras, and if $f : \underline{A} \to \underline{B}$ is a full and injective homomorphism, then the construction in 23.2.8 can be extended to provide C with a partial algebraic structure such that g becomes an isomorphism $g : \underline{C} \to \underline{B}$. In this case $\underline{A}$ becomes a relative subalgebra of $\underline{C}$ (and if f is closed, then $\underline{A}$ is even a subalgebra of $\underline{C}$). ∎

23.2.10. Another consequence of the Axiom of Foundation is the fact that for every set M the set $M \cup \{M\}$ contains besides the elements of M exactly one element -- namely M -- which is not an element of M. This fact is frequently used in this book.

23.3 Ordinals

23.3.1 DEFINITION. A class A is called <u>∈-transitive</u> iff for all classes x: $x \in A \Rightarrow x \subseteq A$ (i.e. iff, for all x, y: $y \in x \in A \Rightarrow y \in A$). -- A class A is an <u>ordinal</u>, iff A is an ∈-transitive **set**, and each element of A is ∈-transitive. (From a more naive point of view an ordinal is a distinguished element in a class of all well-ordered sets isomorphic to some given well-ordered set, as will be made more explicit below.) -- $\mathbf{Ord} := \{x \mid x \text{ is an ordinal}\}$ designates the class of all ordinals.

23.3.2 PROPOSITION and DEFINITION. As an immediate consequence we have that $\emptyset \in \mathbf{Ord}$: $\emptyset =: 0$ is the ∈-smallest ordinal. Moreover, $x \in \mathbf{Ord}$ implies $x \cup \{x\} \in \mathbf{Ord}$; $x \cup \{x\} =: x+1$. ∎

23.3.3 REMARK. This approach leads to a model of ordinals which contains the natural numbers in the following way (see also 23.2.2): $0:=\emptyset$, $1:=\{\emptyset\}$, $2:=\{\emptyset,\{\emptyset\}\}=\{0,1\}$, $3:=\{0,1,2\}$,..., $n:=\{0,1,\dots,n-1\}$, $n+1:=n\cup\{n\}$,.... Finally, $\omega:=\omega_0:=\mathbf{U}\{n|n$ is a natural number$\}$, and this is again an ordinal as well as $\omega+1$, $\omega+2,\dots$; this can be "extended without ending". ω is called the <u>set of natural numbers</u>.

23.3.4 PROPOSITION. **Ord** is totally (irreflexively) ordered by the $\in$-relation: $x,y\in$ **Ord** implies $x=y$ or $x\in y$ or $y\in x$. -- If $A\subseteq$ **Ord** and $A\neq\emptyset$, then $\cap A\in A$ is an ordinal; indeed $\cap A$ is the smallest element of A with respect to the $\in$-relation. -- Define for $x,y\in$ **Ord** : $x\leq y$ iff $x=y$ or $x\in y$; then (**Ord**;$\leq$) and (A;$\leq$) are well-ordered classes for every ordinal A. 0 is the smallest element of **Ord** and of every ordinal $A\neq\emptyset$. ∎

23.3.5 DEFINITION and NOTATION. In what follows we shall use lower case Greek letters $\alpha,\beta,\gamma,\dots$ to denote ordinals. α is called a <u>successor ordinal</u>, if there is an ordinal β such that $\alpha=\beta+1$, otherwise, if $\alpha\neq 0$, then α is called a <u>limit ordinal</u> (we shall often use the letter λ to designate limit ordinals).

23.3.6 PROPOSITION. For ordinals α and β we have:

(i) $\alpha\in\beta$ iff $\alpha\subset\beta$; and $\alpha\in\beta$ implies $\alpha+1\in\beta$ or $\alpha+1=\beta$.

(ii) If $C\subseteq\alpha$, then $\mathbf{U}C\in\alpha$ or $\mathbf{U}C=\alpha$; $\mathbf{U}(\alpha+1)=\alpha$.

(iii) $\alpha=\mathbf{U}\alpha\cup\{\mathbf{U}\alpha\}$ (iff α is a successor cardinal) or $\alpha=\mathbf{U}\alpha$ (iff $\alpha=0$ or α is a limit ordinal).

(iv) $\alpha=\{\beta\in\mathbf{Ord}|\beta<\alpha\}$.

(v) There is no ordinal γ such that $\alpha<\gamma<\alpha+1$.

(vi) For any **set** $A\subseteq$ **Ord** , $\mathbf{U}A\in$ **Ord** is the supremum of A in (**Ord**;$\leq$).

(vii) $\alpha<\beta$ iff $\alpha+1<\beta+1$ iff $\alpha+1\leq\beta$; $\alpha+1=\beta+1$ iff $\alpha=\beta$. ∎

23.3.7 Principles of transfinite induction. Let $A\in$ **Ord** (or $=$ **Ord**) and B be any class.

(1) If $(\forall\alpha\in A)(\forall\beta(\beta<\alpha\Rightarrow\beta\in B)\Rightarrow\alpha\in B)$, then $A\subseteq B$.

(2) If $0\in B$ and $(\forall\alpha\in A)(\alpha\in B\Rightarrow\alpha+1\in B)$ and $(\forall\lambda\in A)(((\lambda$ is a limit ordinal$)\wedge(\beta<\lambda\Rightarrow\beta\in B))\Rightarrow\lambda\in B)$, then $A\subseteq B$. ∎

23.3.8 DEFINITION. A class R is called <u>well-founded</u> iff R is a relation and for every non-empty class $A\subseteq$ field R there is an $x\in A$ such that $A\cap\{y|(y,x)\in R\}=\emptyset$.

23.3.9. Because of the Axiom of Foundation $\in$ is a well-founded relation on the class **V** of all sets, and so is its restriction to the class **Ord** or to any ordinal.

23.3.10 General Recursion Principle. Let R be a well-founded relation such that, for all $x\in$ field R , the class $\{y|(y,x)\in R\}$ is a set, and let $F:(\text{field } R)\times\mathbf{V}\to\mathbf{V}$ be a mapping. Then there is a unique mapping $G:\text{field } R\to\mathbf{V}$ such that, for all $x\in$ field R : $(x)G=(x,G|\{y|(y,x)\in R\})F=(x,\{(y,(y)G)|(y,x)\in R\})F$.
This applies in particular to the $\in$-relation on **Ord**, where it is sometimes formulated as follows:
Let $A\neq\emptyset$ be an ordinal or $A=$ **Ord** , let B be any class, $a\in B$ any set, and let $F:B\to B$, $G:C\to B$ be mappings, where $C:=\{f|f\in B^{\alpha}$ for some $\alpha\in A\}$. Then there is a unique function $H:A\to B$ such that

(1) $(0)H=a$;

(2) $(\alpha+1)H = ((\alpha)H)F$ for every α with $\alpha+1 \in A$;
(3) $(\lambda)H = (H|\lambda)G$ for every limit ordinal $\lambda \in A$. ∎

23.3.11 DEFINITION. **(i)**: The addition of ordinals $+:\mathbf{Ord} \times \mathbf{Ord} \to \mathbf{Ord}$ is the unique mapping defined recursively (with $\alpha,\beta,\lambda \in \mathbf{Ord}$) by:
(1) $\alpha+0 = \alpha$;
(2) $\alpha+(\beta+1) = (\alpha+\beta)+1$;
(3) $\alpha+\lambda = \mathbf{U}(\alpha+\beta|\beta < \lambda)$, if $\lambda = \mathbf{U}\lambda \neq 0$.
(ii): Let $\xi \in \mathbf{Ord}$ and $(\alpha_\zeta|\zeta < \xi) \in \mathbf{Ord}^\xi$, then we define the sum $\Sigma(\alpha_\zeta|\zeta < \xi)$ of the sequence $(\alpha_\zeta|\zeta < \xi)$ recursively (on $\beta \leq \xi$) as follows:
(4) $\Sigma(\alpha_\zeta|\zeta < 0) = 0$ if $\beta = 0$;
(5) $\Sigma(\alpha_\zeta|\zeta < \beta+1) = (\Sigma(\alpha_\zeta|\zeta < \beta))+\alpha_\beta$, if $\beta+1 \leq \xi$;
(6) $\Sigma(\alpha_\zeta|\zeta < \beta) = \mathbf{U}(\Sigma(\alpha_\zeta|\zeta < \gamma)|\gamma < \beta)$, if $0 \neq \beta = \mathbf{U}\beta$.

23.3.12. Observe that there is a unique order isomorphism between the sum $(\Sigma(\alpha_\zeta|\zeta < \xi);\leq)$ of a sequence of ordinals and the poset $(\{(\gamma,\zeta)|\gamma < \alpha_\zeta , \zeta < \xi\};\prec)$, where $(\gamma,\zeta) \prec (\beta,\eta)$ iff $\zeta < \eta$ or $(\zeta = \eta$ and $\gamma \leq \beta$ $(< \alpha_\zeta))$. ∎
This observation together with 23.3.13 below show that for every $\gamma < \Sigma(\alpha_\zeta|\zeta < \xi)$ there exist unique $\nu < \xi$ and $\beta < \alpha_\nu$ such that $\gamma = \Sigma(\alpha_\zeta|\zeta < \nu)+\beta$. ∎

23.3.13 PROPOSITION. $\alpha+(\beta+\gamma) = (\alpha+\beta)+\gamma$. -- $\alpha < \beta$ iff there is $\gamma > 0$ such that $\alpha+\gamma = \beta$. -- If $\alpha < \beta$, then $\alpha+\gamma \leq \beta+\gamma$. -- $0+\alpha = \alpha+0 = \alpha$; $\beta \leq \alpha+\beta$. -- If $\beta < \gamma$, then $\alpha+\beta < \alpha+\gamma$. -- $\omega_0 \leq \alpha$ iff $1+\alpha = \alpha$. -- In general $\alpha+\beta \neq \beta+\alpha$ $(1+\omega = \omega \neq \omega+1)$. ∎

23.3.14 DEFINITION. **(i)**: Let $\underline{P} = (P;\leq)$ be a poset, and let $H \subseteq P$. H is called cofinal in $\underline{P}$ (a cofinal subset of $\underline{P}$), iff for all $p \in P$ there is $h \in H$ such that $p \leq h$ (iff $\downarrow H = P$).
(ii): Let $\alpha,\beta \in \mathbf{Ord}$ be any ordinals. α is called cofinal with β iff there exists a strictly increasing order preserving mapping $f:(\beta;\leq) \to (\alpha;\leq)$ such that $[\beta]f$ is cofinal in $(\alpha;\leq)$ (cf. 11.1.1).
(iii): An ordinal α is called regular iff α is not cofinal with any ordinal $\beta < \alpha$. In every other case α is called singular.

23.3.15 PROPOSITION. If α is cofinal with β, then $\beta \leq \alpha$. -- If $\alpha = \beta+1$, then α is cofinal with 1. -- 0, 1, ω are the first three regular ordinals; in particular 0 and 1 are the only regular ordinals which are no limit ordinals. -- α is a regular ordinal, iff for every $\gamma < \alpha$ and for every mapping $f:\gamma \to \alpha$ there is $\beta < \alpha$ such that $[\gamma]f \subseteq \beta$. ∎

In connection with the ordinal dimension we mainly need the

23.3.16 COROLLARY. Let α be a regular ordinal, and let **M** be a set allowing a bijective mapping, say $g:\gamma \to \mathbf{M}$, from some $\gamma < \alpha$ onto **M**. Then there exists, for every mapping $f:\mathbf{M} \to \alpha$, an ordinal $\beta < \alpha$ such that $[\mathbf{M}]f \subseteq \beta$, i.e. $[\mathbf{M}]f$ is bounded in α. ∎

23.4 Cardinals

23.4.1. While ordinals generalize the types of well-orderings from finite to infinite sets, cardinals generalize the process of determining (by counting) the number of elements of sets from finite to infinite sets (i.e. of comparing magnitudes by numbers).

23.4.2 DEFINITION. A set A is called <u>equipotent</u> with a set B iff there is a bijective mapping from A onto B. -- A set A is called a <u>cardinal number</u> or simply a <u>cardinal</u> iff $A \in \mathbf{Ord}$, and A is not equipotent with any $\alpha \in A$. -- $\mathbf{Card} := \{\alpha | \alpha \text{ is a cardinal}\}$ designates the class of all cardinals. -- Bold-faced small Latin letters like **m, n, p, q, c, d,** etc. are used in this book as variables or names for cardinals (mainly since cardinal arithmetic is quite different from ordinal arithmetic).

23.4.3 PROPOSITION. The relation of equipotency is an equivalence relation on the class **V** of all sets. Every equivalence class except that of the empty set is a proper class; and every equivalence class contains exactly one cardinal (this is a consequence of the Axiom of Choice). ∎

23.4.4 DEFINITION. For any set A let #A designate the unique cardinal **m** equipotent with A ($\mathbf{m} =: \#A$). #A is called the <u>power</u> or <u>cardinality</u> of A (sometimes also, in analogy to the finite case, #A is called the "number of elements of A").

23.4.5 LEMMA. **Card** is a system of unique representatives for the system of equipotency classes (this is no class!), and if A is any member of any such equipotency class, then #A is the unique cardinal representing this class. -- If $A \subseteq B$, then $\#A \leq \#B$. ∎

23.4.6 PROPOSITION. Let α be any ordinal, and A be any set. Then $\#\alpha \leq \alpha$; and if α and A are equipotent, then $\#A \leq \alpha$. -- α is a cardinal iff $\#\alpha = \alpha$. -- If $\omega \leq \alpha$, then $\#(\alpha+1) = \#\alpha$. ∎

23.4.7 PROPOSITION. For any sets A and B the following statements are equivalent:

(i) $\#A \leq \#B$.

(ii) There exists an injective mapping from A into B.

(iii) $A = \emptyset$, or there is a surjective mapping from B onto A. ∎

23.4.8 COROLLARY (Cantor-Bernstein-Theorem). If A and B are sets such that each is equipotent with a subset of the other set, then A and B are equipotent. (This can be proved without using the Axiom of Choice). ∎

23.4.9 PROPOSITION. Each natural number is a cardinal: $\omega \subseteq \mathbf{Card}$. -- $\omega \in \mathbf{Card}$; ω is the smallest infinite cardinal. -- Every set has a smaller cardinal than its power set: $\#A < \#\mathbf{Subs}(A)$. -- For every $\alpha \in \mathbf{Ord}$ there is $\mathbf{m} \in \mathbf{Card}$ such that $\alpha < \mathbf{m}$. -- If $G \subseteq \mathbf{Card}$ is a set, then $\mathbf{U}G$ is a cardinal. ∎

23.4.10 DEFINITION. **(i):** For any ordinal α let α^+ designate the smallest cardinal greater than α (observe 23.4.9.(iv) and $\mathbf{Card} \subseteq \mathbf{Ord}$).

(ii): The mapping $\#\omega : \mathbf{Ord} \to \mathbf{Card}$ (usually denoted by $\aleph$) is defined recursively (the argument being written as an index!): $\#\omega_0 := \omega$; $\#\omega_{\alpha+1} := (\#\omega_\alpha)^+$; $\#\omega_\lambda := \mathbf{U}(\#\omega_\beta | \beta < \lambda)$, if $\lambda = \mathbf{U}\lambda \neq 0$ is a limit ordinal.

23.4.11 PROPOSITION. $\#\omega$ is a strictly increasing mapping which maps **Ord** onto $\{\alpha \in \mathbf{Card} | \alpha \geq \omega\}$. Thus, in particular, **Card** is a proper class. ∎

23.4.12 LEMMA. Let $\mathbf{m}$, $\mathbf{n}$ be cardinals. Then: $\mathbf{m} < \mathbf{n}^+$ iff $\mathbf{m} \leq \mathbf{n}$, and $\mathbf{m} < \mathbf{n}$ iff $\mathbf{m}^+ \leq \mathbf{n}$. -- $\mathbf{m}^+ \leq \#\mathbf{Subs}(\mathbf{m})$ (equality for each infinite $\mathbf{m}$ would be the statement of the Generalized Continuum Hypothesis, which we do not need in this book). -- If R is an equivalence relation on some set A, then $\#(A/R) \leq \#A$. -- $\mathbf{m}^+ = \{\alpha | \#\alpha \leq \mathbf{m}\}$. ∎

23.4.13 DEFINITION. Let A be any set. A is called: <u>finite</u> iff $\#A < \#\omega_0$, <u>infinite</u> iff $\#A \geq \#\omega_0$, <u>denumerable</u> iff $\#A = \#\omega_0$, <u>countable</u> iff $\#A \leq \#\omega_0$, <u>uncountable</u> iff $\#A > \#\omega_0$. -- The ordinals representing infinite cardinals are called <u>initial ordinals</u>; and the initial ordinal representing $\mathbf{n}$ is denoted by $\omega(\mathbf{n})$.

23.4.14 PROPOSITION. If A is a finite set and $B \subset A$ is a proper subset, then $\#B < \#A$; and a set A is infinite iff A is equipotent with a proper subset of A. ∎

23.4.15 DEFINITION (Cardinal arithmetic). Let $\mathbf{m},\mathbf{n} \in \mathbf{Card}$.

(i): Then $\mathbf{m}+\mathbf{n} := \#(\mathbf{m} \dot{\cup} \mathbf{n})$ is called the <u>sum</u> of these two cardinals and $\mathbf{mn} := \#(\mathbf{m} \times \mathbf{n})$ is called their <u>product</u>. And if T is any set and $(\mathbf{m}_t | t \in T) \in \mathbf{Card}^T$, then

$\Sigma(\mathbf{m}_t | t \in T) := \#(\dot{\bigcup}(\mathbf{m}_t | t \in T))$ is called the <u>sum</u> and

$\Pi(\mathbf{m}_t | t \in T) := \#(\mathbf{X}(\mathbf{m}_t | t \in T))$ is called the <u>product</u> of this family of cardinals.

(ii): $\mathbf{m}^{\mathbf{n}} := \#(\mathbf{m}^{\mathbf{n}})$ is called the $\mathbf{n}$-th <u>power</u> of $\mathbf{m}$ (the <u>exponentiation</u> of $\mathbf{m}$ with $\mathbf{n}$) -- we hope that the same notation for the power of cardinals in cardinal arithmetic and of their power "just as sets" will cause no confusion.

23.4.16. As an immediate consequence of the definitions in 23.4.15 addition and multiplication of (arbitrarily many) cardinals are "commutative" and "associative" (i.e. they do not depend on any bijective change of the index set, nor do they depend on any subdivision of this index set and first adding (respectively multiplying) with respect to the subsets and then adding (multiplying) the results). ∎

23.4.17 PROPOSITION. Let $\mathbf{m}$ and $\mathbf{n}$ be cardinals, and let at least one of them be infinite; then: $\mathbf{m}+\mathbf{n} = \max(\mathbf{m},\mathbf{n})$; and if $\mathbf{m} \neq 0 \neq \mathbf{n}$, then $\mathbf{mn} = \max(\mathbf{m},\mathbf{n})$. ∎

23.4.18 PROPOSITION. Let $(\mathbf{m}_t | t \in T)$, $(\mathbf{n}_t | t \in T) \in \mathbf{Card}^T$.

(i) If $\mathbf{m}_{t'} = 0$ for some $t' \in T$, then $\Pi(\mathbf{m}_t | t \in T) = 0$.

(ii) If there is some $t' \in T$ such that $\mathbf{m}_{t'}$ is infinite, or if $\{t \in T | \mathbf{m}_t \neq 0\}$ is infinite, then $\Sigma(\mathbf{m}_t | t \in T) = \sup(\#\{t \in T | \mathbf{m}_t \neq 0\}, \mathbf{m}_t | t \in T)$.

(iii) If $\mathbf{m}_t \leq \mathbf{n}_t$ for each $t \in T$, then
$\Sigma(\mathbf{m}_t | t \in T) \leq \Sigma(\mathbf{n}_t | t \in T)$ and $\Pi(\mathbf{m}_t | t \in T) \leq \Pi(\mathbf{n}_t | t \in T)$.

(iv) If $\mathbf{m}_t \leq \mathbf{n}_t$ and $\mathbf{n}_t \geq 2$ for each $t \in T$, then $\Sigma(\mathbf{m}_t | t \in T) \leq \Pi(\mathbf{n}_t | t \in T)$. ∎

23.4.19 DEFINITION. A cardinal $\mathbf{m}$ is called <u>additively inaccessible</u> or <u>regular</u> iff for every set T with $\#T < \mathbf{m}$ and for every family $(\mathbf{m}_t | t \in T) \in \mathbf{Card}^T$ satisfying $\mathbf{m}_t < \mathbf{m}$ for each $t \in T$ one has $\Sigma(\mathbf{m}_t | t \in T) < \mathbf{m}$; else $\mathbf{m}$ is called <u>additively accessible</u> or <u>singular</u>.

23.4.20 PROPOSITION. 0, 1, 2, and ω are regular cardinals. -- If α is a regular ordinal, then $\#\alpha$ $(=\alpha)$ is a regular cardinal. -- A cardinal $\mathbf{m} \neq 2$ is regular iff its corresponding initial ordinal is a regular ordinal (but observe that 2 is regular as a cardinal, but singular as an ordinal). -- For each cardinal $\mathfrak{m}$ its successor cardinal $\mathfrak{m}^+$ is regular. -- $\#\omega_\omega$ is additively accessible (i.e. a singular cardinal). ∎

24 Ordered sets, algebraic lattices and Galois connections

24.1 Partially ordered sets and lattices

24.1.1 (Partial) Orders. Let A be any set or class. Recall that a binary relation R on A (i.e. $R \subseteq A \times A$) is

reflexive	iff $(x,x) \in R$ for all $x \in A$,
irreflexive	iff $(x,x) \notin R$ for all $x \in A$,
symmetric	iff $(y,x) \in R$, whenever $(x,y) \in R$,
antisymmetric	iff $(x,y) \in R$ and $(y,x) \in R$ implies $x = y$,
transitive	iff $(x,y) \in R$ and $(y,z) \in R$ implies $(x,z) \in R$.

A reflexive, symmetric and transitive binary relation R on A is called an equivalence relation on A; (if A is a set) each equivalence relation R on A induces a partition $\{[a]R \mid a \in A\}$ of A, where $[a]R := \{b \in A \mid (a,b) \in R\}$ is the equivalence class of a with respect to R and two equivalence classes are either equal or disjoint; conversely, if $Z := \{A_i \mid i \in I\}$ is a partition of A, then

$$R_Z := \{(a,b) \mid a,b \in A \text{ , a and b belong to the same class of } Z\}$$

is an equivalence relation on A.

A reflexive and transitive relation R on A is called a quasi-order on A, and if R is in addition antisymmetric, then it is called a partial order (sometimes also: partial ordering or simply order) on A, and (A;R) is called a partially ordered set or briefly a poset -- sometimes irreflexivity rather than reflexivity is required, and in this book we shall usually indicate a reflexive partial order by a symbol like " $\leqq$ ", " $\preccurlyeq$ ", " $\subseteq$ " (set theoretic inclusion) or similar symbols, and an irreflexive partial order by a symbol like " $<$ ", " $\prec$ ", " $\subset$ ", etc., and we shall write " a R b " (infix notation) rather than " $(a,b) \in R$ ".

If R is a partial order on A which additionally satisfies

for all $x,y \in A$ one has $(x,y) \in R$ or $(y,x) \in R$

(or $x = y$ -- in the case of irreflexivity),

then R is called a linear or total order on A, and (A;R) is called a linearly or totally ordered set or simply a chain.

If " $\leqq$ " is a (quasi-) order on A, then the relation " $\geqq$ " defined by $a \geqq b$ iff $b \leqq a$ (for all $a,b \in A$) is called the dual (quasi-) order on A.

24.1.2 LEMMA. Let A be any set, $R \subseteq A \times A$.

(i) If R is a reflexive quasi-order on A, then the relation

$$\equiv := \{(a,b) \in A \times A \mid (a,b) \in R \text{ and } (b,a) \in R\}$$

is an equivalence relation on A, and the quotient structure $(A/\equiv; R/\equiv)$ is a poset (the universal solution of (A;R) in the subcategory of all posets with (quasi-) order preserving mappings), where

$$R/{\equiv} := \{([a]_{\equiv},[b]_{\equiv}) \mid (a,b) \in R\} =$$
$$= \{([a]_{\equiv},[b]_{\equiv}) \mid a,b \in A \,,\ (a',b') \in R \ \text{for some} \ a' \in [a]_{\equiv} \,,\ b' \in [b]_{\equiv}\} \,.$$

(ii) If $(A;R)$ is a poset (respectively a chain), and if $M \subseteq A$, then $(M;\, R \cap (M \times M))$ is a poset (respectively a chain). ∎

24.1.3. **(i)**: Let $(A;\leq)$ be a poset, $M \subseteq A$ any subset.
$u \in A$ is called an *upper* (respectively *lower*) *bound* of M, if $m \leq u$ (respectively $u \leq m$) for every $m \in M$.
$s \in A$ is called a *supremum* or *least upper bound* (respectively *infimum* or *greatest lower bound*) of M (and denoted by sup(M) (inf(M)) iff $m \leq s$ $(s \leq m)$ for all $m \in M$, and if $m \leq s'$ $(s' \leq m)$ for all $m \in M$ and some $s' \in A$, then $s \leq s'$ (respectively $s' \leq s$) -- if it exists.
Notice that in a poset an infimum respectively supremum of some subset M is uniquely determined by M, whenever it exists (this need not be true in a quasi-order), and it is denoted by inf(M) (respectively sup(M)). If sup(M) (inf(M)) exists and belongs to M for some $M \subseteq A$, then sup(M) is called *the largest element* or *maximum* max(M) of M (respectively *the smallest element* or *minimum* min(M) of M). If $\sup(\emptyset)$ exists in $\underline{A}$, then it is the smallest element of A, if $\inf(\emptyset)$ exists, then it is the largest element of $\underline{A}$.
An element $a \in A$ is called *maximal* (*minimal*) if, for all $b \in A$, $a \leq b$ implies $a = b$ ($b \leq a$ implies $a = b$).
If, for two -- not necessarily distinct -- elements $a,b \in A$, $\sup(\{a,b\})$ $(\inf(\{a,b\}))$ exists, then it is often denoted by $a \vee b$, $\sup(a,b)$ or $\sup\{a,b\}$ (respectively $a \wedge b$, $\inf(a,b)$ or $\inf\{a,b\}$). We always have $\sup(a,a) = a = \inf(a,a)$, and if it exists, then $\sup(a,b) = \sup(b,a)$, $\inf(a,b) = \inf(b,a)$.
(ii): Let $(P;\leq)$ be a poset. If each non-empty subset of P has a smallest element, then $\leq$ is called a *well-ordering* of P, and $(P;\leq)$ is called a *well-ordered set*, *well-ordered by* $\leq$. Observe that every well-ordered set is in particular a chain. More information about well-ordered sets can be obtained in 23.3 in connection with ordinals.

24.1.4. A poset $\underline{A} := (A;\leq)$ in which for any two elements a, b of A sup(a,b) exists (inf(a,b) exists) is called an *upper* or *join-* (*lower* or *meet-*) *semilattice*; and if $\underline{A}$ is both an upper and a lower semilattice, then it is called a *lattice*. If, for every $M \subseteq A$, sup(M) (inf(M)) exists, then $\underline{A}$ is called a *complete upper semilattice* or *complete sup-semilattice* (*complete lower semilattice* or *complete inf-semilattice*), and if $\underline{A}$ is both a complete upper and complete lower semilattice, then it is called a *complete lattice*.
If for each subset $M \subseteq A$ one has: if M has an upper bound, then M also has a supremum, and if M has a lower bound, then M has an infimum, then $\underline{A}$ is called *conditionally complete*; in particular a *conditionally complete upper semilattice* is an upper semilattice which is conditionally complete.

24.1.5. It is well-known that every total algebra $(A;\circ)$ with a binary idempotent (i.e. $a\circ a = a$), commutative (i.e. $a \circ b = b \circ a$) and associative (i.e. $(a\circ b)\circ c = a\circ(b\circ c)$) operation " $\circ$ " is an **upper semilattice**, when one defines, for all $a,b \in A$: $a \leq b$ iff $a \circ b = b$ (or dually a lower semilattice by defining: $a \leq b$ iff $a \circ b = a$), in which $\sup(a,b) = a \circ b$ ($\inf(a,b) = a \circ b$ in the dual case).

Conversely, for every, say, upper semilattice $(A;\leq)$ the partial algebra $(A;\sup(.,.))$ is a total algebra of type (2), in which $\sup(.,.)$ is idempotent, commutative and associative and satisfies $a \leq b$ iff $\sup(a,b) = b$.
Similarly, $(A;\leq)$ is a lattice, iff $(A;\vee,\wedge)$ is a total algebra of type (2,2) such that both $\vee$ and $\wedge$ are idempotent, commutative and associative, and we have the absorption laws $a\vee(a\wedge b) = a$ and $a\wedge(a\vee b) = a$; and moreover: $a \leq b$ iff $a\vee b = b$ (iff $a\wedge b = a$). ∎

24.2 k-directed posets, k-inductivity, k-algebraic lattices

24.2.1 DEFINITION. Let **k** be any cardinal and $\underline{A} := (A;\leq)$ be any poset. $\underline{A}$ is called (upwards) **k**-directed -- observe that we shall not use "downwards **k**-directed -- iff every subset M of A with cardinality less than **k** ($\#M < \mathbf{k}$) has an upper bound in $\underline{A}$.
$\underline{A}$ is simply called directed iff every finite subset has an upper bound, i.e. iff it is $\#\omega_0$-directed.

24.2.2 Special cases. Let $\underline{A}$ be any poset.
"$\underline{A}$ is 0-directed" does not mean anything for $\underline{A}$.
$\underline{A}$ is 1-directed iff the empty subset has an upper bound, i.e. iff $A \neq \emptyset$.
$\underline{A}$ is 2-directed iff $\underline{A}$ is 1-directed (iff $A \neq \emptyset$).
$\underline{A}$ is 3-directed iff $\underline{A}$ is **k**-directed for any cardinal with $3 \leq \mathbf{k} \leq \#\omega_0$, iff every finite subset of A has an upper bound in A (iff $A \neq \emptyset$ and any two elements of A have an upper bound in A, iff $\underline{A}$ is directed). ∎

24.2.3. If $\mathbf{k}' \leq \mathbf{k}$ and $\underline{A}$ is a **k**-directed poset, then $\underline{A}$ is **k'**-directed.
If **k'** is an additively accessible cardinal, if **k** is the least additively inaccessible cardinal greater than **k'**, and if $\underline{A}$ is **k'**-directed, then $\underline{A}$ is **k**-directed. ∎

24.2.4 DEFINITION. Let **k** be any cardinal, and let $(L;\leq) = \underline{L}$ be a **complete lattice.**
(i): A subset M of L is called **k**-inductive iff for any **k**-directed subset T of M the supremum $\sup_{\underline{L}}(T)$ of T formed in $\underline{L}$ belongs to M: $\sup_{\underline{L}}(T) \in M$.
(ii): $a \in \underline{L}$ is called **k**-compact iff one has for every **k**-directed subset D of L:

$$a \leq \sup_{\underline{L}}(D) \text{ iff there is } d \in D \text{ such that } a \leq d ;$$

(iff, for every $T \subseteq \underline{L}$, $a \leq \sup_{\underline{L}}(T)$ implies $a \leq \sup_{\underline{L}}(T')$ for some $T' \subseteq L$, $\#T' < \mathbf{k}$).
(iii): $\underline{L}$ is called **k**-compactly generated or **k**-algebraic iff every $l \in L$ is the supremum of a set of **k**-compact elements of $\underline{L}$, i.e. in particular

$$l = \sup_{\underline{L}}(\{c \in L | c \leq l \text{ and } c \text{ is } \mathbf{k}\text{-compact in } \underline{L}\}) .$$

(iv): If $\mathbf{k} = \#\omega_0$, then a **k**-algebraic lattice $\underline{L}$ is simply called algebraic, and **k**-compact elements are called compact.

24.2.5 LEMMA. Let $\mathbf{k} \leq \mathbf{k}'$ be any cardinals and $\underline{L}$ be any complete lattice. Then:
(i) The fact that $a \in L$ is **k**-compact implies that a is **k'**-compact.
(ii) The fact that $M \subseteq L$ is **k**-inductive implies that M is **k'**-inductive.
(Since every **k'**-directed set is **k**-directed.) ∎∎∎

24.2.6 Special cases. Let $(L;\leq) = \underline{L}$ be a complete lattice, $M \subseteq L$, $a \in L$:

(i): $\mathbf{k} = 0$: M is 0-inductive iff, for every $T \subseteq M$, $\sup_{\underline{L}}(T) \in M$, i.e. iff $(M;\leq)$ is a complete $\sup_{\underline{L}}$-subsemilattice of $\underline{L}$ -- in this case $\underline{M} := (M;\leq)$ is also a complete lattice, since $0_{\underline{L}} := \sup_{\underline{L}}(\emptyset) \in M$, but $\inf_{\underline{M}}$ need not coincide with $\inf_{\underline{L}}$, since for $T \subseteq M$, $\inf_{\underline{M}}(T) = \sup_{\underline{L}}(\{m \in M | m \text{ is a lower bound of } T\})$,

and this set of lower bounds of T within $\underline{M}$ may have a supremum t which is strictly smaller than that of the set of all lower bounds of T within $\underline{L}$.

$\inf_{\underline{L}}(L)$ is not 0-compact. The 0-compact elements of $\underline{L}$ are sometimes called the completely supremum-prime elements of $\underline{L}$ (see (ii)). If $\underline{L}$ is the four-element lattice given in Figure 24.1 by its Hasse-diagram (there is a line upward from a to b if $a < b$ and if there is no other element between a and b), then the two "atoms" (the elements joined by a line with the smallest element) are 0-compact; however the lattice in Figure 24.2 does not have any 0-compact elements.

Figure 24.1.

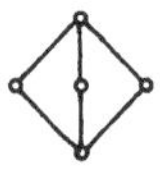

Figure 24.2.

(ii): $\mathbf{k} = 1$ or $\mathbf{k} = 2$: M is $\mathbf{k}$-inductive iff, for every non-empty subset T of M, $\sup_{\underline{L}}(T)$ belongs to M: $T \subseteq M$ and $T \neq \emptyset$ implies $\sup_{\underline{L}}(T) \in M$. Hence M need not be a complete sublattice. -- $a \in L$ is $\mathbf{k}$-compact iff $a = \inf_{\underline{L}}(L)$ or a is completely supremum prime: if $a \leq \sup_{\underline{L}}(D)$ for some non-empty set D, then $a \leq d$ for some $d \in D$.

(iii): $3 \leq \mathbf{k} \leq \mathbf{l} \leq \#\omega_0$: Then, because of 24.2.1, M is $\mathbf{k}$-inductive iff it is $\mathbf{l}$-inductive, and $a \in L$ is $\mathbf{k}$-compact iff a is $\mathbf{l}$-compact.

(For $\mathbf{k} > 0$, $\mathbf{k}$-directedness of M implies $M \neq \emptyset$.) ∎

24.2.7 LEMMA. Let $\underline{L}$ be a complete lattice, $\mathbf{k} \geq 2$ a regular cardinal. Then the supremum of less than $\mathbf{k}$ $\mathbf{k}$-compact elements of L is $\mathbf{k}$-compact; hence the subset $L_{\mathbf{k}}$ of L of all $\mathbf{k}$-compact elements of $\underline{L}$ is $\mathbf{k}$-directed. ∎

24.3 Closure systems and closure operators

Although we keep the following survey quite general, we do not claim that all assumptions are presented in greatest generality.

24.3.1 DEFINITION. Let $\underline{L} = (L;\leq)$ be a complete lattice (in most applications $\underline{L}$ will be the complete lattice $(\mathbf{Subs}(E);\subseteq)$ of all subsets of some set E with set theoretical inclusion as partial order).

(i): A subset $\mathcal{H} \subseteq L$ is called a closure system in $\underline{L}$ iff for every subset $\mathcal{B} \subseteq \mathcal{H}$ also its infimum with respect to $\underline{L}$ belongs to $\mathcal{H}$:

(CS) $\quad \mathcal{B} \subseteq \mathcal{H}$ implies $\inf_{\underline{L}}(\mathcal{B}) \in \mathcal{H}$

(if $\underline{L} = (\mathbf{Subs}(E);\subseteq)$, then this means $\bigcap\mathcal{B} \in \mathcal{H}$).

(ii): A mapping $\mathbf{C}: L \to L$ is called a closure operator in $\underline{L}$ iff $\mathbf{C}$ has the following properties for all $a,b \in L$:

(C1) $a \leq b$ implies $\mathbf{C}a \leq \mathbf{C}b$ (C is monotone);

(C2) $a \leq \mathbf{C}a$ (C is extensive);

(C3) $\mathbf{CC}a = \mathbf{C}a$ (C is idempotent).

(iii): If $\mathbf{C}$ is a closure operator in $\underline{L}$, and if $b = \mathbf{C}a$, then we say that a generates b with respect to $\mathbf{C}$ (a $\mathbf{C}$-generates b).

24.3.2. (i): In particular we have, if $\underline{L} = (\mathbf{Subs}(E);\subseteq)$, that $\mathcal{H}$ is a closure system on E and $\mathbf{C}$ is a closure operator on E, respectively, iff
$\mathcal{H} \subseteq \mathbf{Subs}(E)$, $\mathbf{C}:\mathbf{Subs}(E) \to \mathbf{Subs}(E)$, and for $\mathcal{B} \subseteq \mathcal{H}$ and $M,N \subseteq E$:

(CS) $\quad \bigcap\mathcal{B} \in \mathcal{H}$;

(C1) $\quad M \subseteq N$ implies $\mathbf{C}M \subseteq \mathbf{C}N$;

(C2) $\quad M \subseteq \mathbf{C}M$;

(C3) $\quad \mathbf{CC}M = \mathbf{C}M$.

(ii): In 24.3.1 we have introduced the more general concepts of a closure system and a closure operator in a complete lattice, since in applications we often take into consideration the algebraic structure on a set and consider complete lattices with closure systems and closure operators on $\mathbf{Wsub}(\underline{A})$ with the inclusion of partial algebraic structures for any given partial $\underline{A}$ (see subsections 3.1, 3.2 and 23.1).

23.3.3 LEMMA. Let $\underline{L} = (L;\leq)$ be a complete lattice, and let $\mathbf{C}$ be a closure operator on $\underline{L}$. Then one has for every family $(a_t \mid t \in T) \in L^T$:
$\mathbf{C}(\inf_{\underline{L}}(\mathbf{C}a_t \mid t \in T)) = \inf_{\underline{L}}(\mathbf{C}a_t \mid t \in T)$ and $\mathbf{C}(\sup_{\underline{L}}(\mathbf{C}a_t \mid t \in T)) = \mathbf{C}\sup_{\underline{L}}(a_t \mid t \in T)$. ∎

24.3.4 THEOREM. Let $\underline{L}$ be any complete lattice.

(i) Let $\mathcal{H}$ be a closure system in $\underline{L}$, and define $\mathbf{C}_{\mathcal{H}}:L \to L$ by
$\mathbf{C}_{\mathcal{H}}a := \inf_{\underline{L}}(h \in \mathcal{H} \mid a \leq h)$;
then $\mathbf{C}_{\mathcal{H}}$ is a closure operator in $\underline{L}$ (called the closure operator corresponding to the closure system $\mathcal{H}$), and one has $\mathcal{H} = \{a \in L \mid \mathbf{C}_{\mathcal{H}}a = a\}$.

(ii) Let $\mathbf{C}:L \to L$ be a closure operator in $\underline{L}$. Then the set $\mathcal{H}_{\mathbf{C}} := [L]\mathbf{C} = \{\mathbf{C}a \mid a \in L\}$ is a closure system in $\underline{L}$ (called the closure system corresponding to the closure operator $\mathbf{C}$), and one has for each $a \in L$: $\mathbf{C}a = \inf_{\underline{L}}(b \in \mathcal{H}_{\mathbf{C}} \mid a \leq b)$.

(iii) With the assumptions and notation from above one has $\mathcal{H}_{(\mathbf{C}_{\mathcal{H}})} = \mathcal{H}$ and $\mathbf{C}_{(\mathcal{H}_{\mathbf{C}})} = \mathbf{C}$; i.e. there is a one-to-one correspondence between the closure operators and the closure systems in $\underline{L}$ described by (i) and (ii). ∎

24.3.5 THEOREM. Let $\mathbf{k} \geq 2$ be any (regular) cardinal, let $\underline{L} = (L;\leq)$ be a complete $\mathbf{k}$-algebraic lattice, let $\mathbf{C}$ be a closure operator in $\underline{L}$, and let $\mathcal{H}$ be its corresponding closure system. Then the following statements are equivalent:

(i) $\mathcal{H}$ is $\mathbf{k}$-inductive in $\underline{L}$.

(ii) For every $\mathbf{k}$-directed system $\mathcal{B} \subseteq L$ one has $\mathbf{C}(\sup_{\underline{L}}(\mathcal{B})) = \sup_{\underline{L}}(\mathbf{C}b \mid b \in \mathcal{B})$.

(iii) For every $a \in L$ one has $\mathbf{C}a = \sup_{\underline{L}}(\mathbf{C}b \mid b \leq a$, b is $\mathbf{k}$-compact in $\underline{L})$.

Observe that (i) and (ii) are equivalent without $\underline{L}$ being $\mathbf{k}$-algebraic (or $\mathbf{k}$ being regular). ∎

24.3.6 LEMMA. For every cardinal $\mathbf{k} \geq 2$ and for every set E, $(\mathbf{Subs}(E);\subseteq)$ is a $\mathbf{k}$-algebraic complete lattice. And if, in addition, $\mathbf{k}$ is regular, then the $\mathbf{k}$-compact elements of $(\mathbf{Subs}(E);\subseteq)$ are exactly the subsets of E of cardinality less than $\mathbf{k}$. ∎

24.3.7 LEMMA. For each closure system $\mathcal{H}$ in a complete lattice $(L;\leq)$ the subposet

$\underline{H} := (H;\leq)$ is itself a complete lattice such that

$$\inf_{\underline{H}} = \inf_{\underline{L}}, \text{ in particular } 1_{\underline{L}} := \inf_{\underline{L}}(\emptyset) \text{ belongs to } H.$$

Moreover, for each $B \subseteq H$, $\sup_{\underline{H}}(B) = C_H(\sup_{\underline{L}}(B))$;

and if $B \subseteq H$ is **k**-directed, then (because of 24.3.5.(ii)): $\sup_{\underline{H}}(B) = \sup_{\underline{L}}(B)$. ∎

24.3.8 PROPOSITION. Let $\mathbf{k} \geq \#\omega_0$ be a regular cardinal, let $\underline{L}$ be a **k**-algebraic complete lattice, and let H be a closure system in $\underline{L}$ with closure operator **C**.

(i) If H is **k**-inductive, then $\underline{H} := (H;\leq)$ is a **k**-compactly generated (i.e. **k**-algebraic) complete lattice.

(ii) Let $H_{\mathbf{k}}$ designate the set of all **k**-compact elements of H; then

$$H_{\mathbf{k}} = \{h \in H \mid \text{there is a } \mathbf{k}\text{-compact } l \in L \text{ such that } Cl = h\},$$

i.e. the **k**-compact elements of $\underline{H}$ are exactly the closures of **k**-compact elements of $\underline{L}$. ∎

Moreover, we have the following connection between abstract **k**-algebraic lattices and subalgebra lattices of (total) algebras:

24.3.9 THEOREM. Let $\underline{L} = (L;\leq)$ be a complete lattice, and let $\mathbf{k} \geq 2$ be a regular cardinal. Then $\underline{L}$ is **k**-algebraic iff there is a partial (or even a total) algebra $\underline{A}$ with cardinal dimension **k** such that $\underline{L} \cong (\{\mathbf{C}_{\underline{A}}M \mid M \subseteq A\};\subseteq)$.

SKETCH of PROOF. (ii) implies (i) because of 3.3.2.

"(i) ⇒ (ii)": Let $A := L_{\mathbf{k}}$ be the set of all **k**-compact elements of $\underline{L}$ including $0_{\underline{L}} := \inf_{\underline{L}}(L)$, and let $c_0^{\underline{A}} := 0_{\underline{L}}$ be the nullary constant of $\underline{A}$.

For each $a \in L_{\mathbf{k}}$ define a unary (total) operation $\phi_a^{\underline{A}}$ mapping $b \in L_{\mathbf{k}}$ onto a, if $a \leq b$, and onto $0_{\underline{L}}$, otherwise. -- For each cardinal **c** with $2 \leq \mathbf{c} < \mathbf{k}$ define a total c-ary operation $\phi_{\mathbf{c}}^{\underline{A}}$ by $(c_i \mid i \in \mathbf{c})\phi_{\mathbf{c}}^{\underline{A}} := \sup_{\underline{L}}(c_i \mid i \in \mathbf{c})$; then 24.2.7 implies that the value is **k**-compact and therefore belongs to $A = L_{\mathbf{k}}$. Then $\underline{L} \cong (\mathbf{Csubs}(\underline{A});\subseteq)$. ∎∎∎

Since every complete lattice $\underline{L}$ is $(\#L)^+$-algebraic, we get the

24.3.10 COROLLARY. Each complete lattice is isomorphic to the lattice of all closed subsets of a suitable (possibly infinitary) total algebra. ∎∎∎

In a similar way one shows:

24.3.11 PROPOSITION. Let $\mathbf{k} \geq 2$ be a regular cardinal, and let H be a **k**-inductive closure system on some set A (i.e. in $(\mathbf{Subs}(A);\subseteq)$). Then there is a total algebraic structure $(\phi^{\underline{A}})_{\phi \in \Omega}$ of cardinal dimension **k** on A such that $(H;\subseteq) = (\mathbf{Csubs}(\underline{A});\subseteq)$.

Namely, let $f:\mathbf{Subs}(A) \setminus \{\emptyset\} \to A$ be a choice function ($Mf \in M$ for each non-empty $M \subseteq A$). For each set $M \subseteq A$ with $\#M < \mathbf{k}$, and for each $a \in C_H M$ define for $M = \emptyset$: $\phi_{\emptyset,a}^{\underline{A}} := a$, and else:

$$(a_m \mid m \in M)\phi_{M,a}^{\underline{A}} := \begin{cases} a & \text{, if } a_m = m \ (m \in M); \\ a_{Mf} & \text{, else.} \end{cases}$$ ∎∎∎

24.4 Galois connections

24.4.1. The concept of Galois connections (sometimes also called Galois correspondences) originates from Galois theory in connection with the solution of polynomial equations with coefficients in some field, but similar facts can be observed quite often in mathematics, in particular in connection with logics, and in category theory in connection with pairs of adjoint functors and with factorization systems. We restrict this survey to posets and in particular to power sets. Up to minor changes we take this survey from [Sch 68a].

24.4.2 DEFINITION. Let $\underline{P}:= (P;\leq)$ and $\underline{Q}:= (Q;\leq)$ be two posets, and let $':P \to Q$, $°:Q \to P$ be two mappings such that for each $p \in P$ and for each $q \in Q$ one has: $q \leq p'$ iff $p \leq q°$; then the pair $(',°)$ is called a <u>Galois connection</u> (<u>Galois correspondence</u>) between $\underline{P}$ and $\underline{Q}$. -- If, in particular, $\underline{P} = (\mathbf{Subs}(A);\subseteq)$ and $\underline{Q} = (\mathbf{Subs}(B);\subseteq)$ (A and B sets), then $(',°)$ is called a <u>Galois connection on</u> A and B.

24.4.3 PROPOSITION. Let $(':P \to Q\ ,\ °:Q \to P)$ be a Galois connection between two posets $(P;\leq)$ and $(Q;\leq)$, and $p,p_1,p_2 \in P$, $q,q_1,q_2 \in Q$. Then:

(G1) $p_1 \leq p_2$ implies $p_2' \leq p_1'$ and $q_1 \leq q_2$ implies $q_2° \leq q_1°$ (i.e. $'$ and $°$ are <u>antimonotone</u> mappings).

(G2) $p \leq p'°$ and $q \leq q°'$.

(G3) $p'°' = p'$ and $q°'° = q°$. ∎

We now restrict considerations to Galois connections on some sets, say A and B.

24.4.4 COROLLARY 1. Let A and B be any sets and let $':\mathbf{Subs}(A) \to \mathbf{Subs}(B)$ and $°:\mathbf{Subs}(B) \to \mathbf{Subs}(A)$ be two mappings. Then $(',°)$ is a Galois connection on A and B iff $'$ and $°$ are antimonotone with respect to set theoretical inclusion, and $'°$ and $°'$ are extensive and idempotent (see 24.3.1). ∎

24.4.5 COROLLARY 2. With the assumptions of 24.4.4 the operators $'°$ and $°'$ are closure operators on A and B, respectively.
The closed sets of the corresponding closure systems, say $\mathcal{L}$ on A and $\mathcal{R}$ on B, are sometimes called <u>left-closed</u> and <u>right-closed</u>, respectively. ∎

24.4.6 COROLLARY 3. The mapping $\varphi:= '|\mathcal{L}:\mathcal{L} \to \mathcal{R}$ is an anti-isomorphism (i.e. a bijection which is antimonotone in both directions, here with respect to set theoretical inclusion) with inverse $\psi:= °|\mathcal{R}$.
In particular the Galois connection $(',°)$ on A and B is uniquely determined by the quadruple $('°,°',\varphi,\psi)$, since $' = '°\varphi$ and $° = °'\psi$.
Moreover, each quadruple $(\mathbf{C},\overline{\mathbf{C}},\varphi,\psi)$ determines a Galois connection on A and B iff $\mathbf{C}$ and $\overline{\mathbf{C}}$ are closure operators on A and B, respectively, and φ is an anti-isomorphism from the closure system $\mathcal{H}_{\mathbf{C}}$ on A onto the closure system $\mathcal{H}_{\overline{\mathbf{C}}}$ on B with inverse ψ. ∎

24.4.7 THEOREM. A mapping $':\mathbf{Subs}(A) \to \mathbf{Subs}(B)$ is one part of a Galois connection on the two sets A and B iff $(\bigcup(M_t|t \in T))' = \bigcap(M_t'|t \in T)$ for all families $(M_t|t \in T)$ of subsets of A (in particular $\emptyset' = B$). ∎

Of great importance for the theory of Galois connections on two sets is the following

24.4.8 THEOREM. Let $R \subseteq A \times B$ be a binary relation between any two sets A and B.

(i) Define for every $M \subseteq A$ and $N \subseteq B$:

$M' := \{b \in B \mid (a,b) \in R$ for all $a \in M\}$, $N^\circ := \{a \in A \mid (a,b) \in R$ for all $b \in N\}$.

Then $(',^\circ)$ is a Galois connection on A and B.

(ii) Conversely, if $(',^\circ)$ is a Galois connection on A and B, then

$R := \{(a,b) \in A \times B \mid a \in \{b\}^\circ\} = \{(a,b) \in A \times B \mid b \in \{a\}'\}$

is a binary relation on $A \times B$ inducing $(',^\circ)$ according to (i). ∎

24.4.9 COROLLARY. The Galois connections $(',^\circ)$ on given sets A and B are in one-to-one correspondence to the binary relations $R \subseteq A \times B$. ∎

24.4.10 REMARKS. (i): Thus the characterization of a certain binary relation $R \subseteq A \times B$ is essentially the task to determine the left-closed and the right-closed sets of the corresponding Galois connection, in particular to describe explicitly the corresponding closure operators and closure systems.

(ii): If A and B are proper classes (or at least one of them is), then the above concepts and observations can be transferred to this situation with the necessary care with respect to the set theoretic concepts. In particular one now gets for a given relation $R \subseteq A \times B$ operators $'$ and $^\circ$ defined as in 24.4.8.(i) such that their compositions $'^\circ$ and $^\circ{}'$ behave like closure operators on A and B, respectively. (One may introduce as an additional level "conglomerates" (see 23.2.1)).

(iii): Let us observe, too, that Galois correspondences have become quite popular now -- and have many applications in different sciences -- in the form of "concept lattices" (see R.WILLE [Wi 82] and [Wi 84]).

References

ABN 80 H.Andréka, P. Burmeister, I.Németi. Quasivarieties of partial algebras - A unifying approach towards a two-valued model theory for partial algebras. Preprint Nr. 557, TH Darmstadt, 1980; appeared in: Studia Sci. Math. Hungar. 16, 1981, 325-372.

AN 76 H.Andréka, I.Németi. Generalization of the concept of variety and quasivariety to partial algebras through category theory. Math. Inst. Hung. Acad. Sci. Preprint No. 5/1976; appeared in Dissertationes Mathematicae (Rozprawy Mat.) No. 204, Warszawa, 1983.

AN 78 H.Andréka, I.Németi. Łos'-lemma holds in every category. Studia Scientiarum Math. Hung. 13, 1978, pp.361-376.

AN 79 H.Andréka, I.Németi. Formulas and ultraproducts in categories. Beiträge zur Algebra und Geometrie 8, 1979, pp. 133-151.

AN 79a H.Andréka, I.Németi. Injectivity in categories to represent all first order formulas. I. Demonstratio Math. 12, 1979, pp. 717-732.

AN 79f H.Andréka, I.Németi. Applications of universal algebra, model theory and categories in computer science (Survey and bibliography). Parts I and II in CL & CL (Comp. Linguist. Comp. Lang.), Vol. 13, 1979, pp. 152-282, and Vol. 14, 1980, pp. 43-65; Part III in Fundamentals of Computation Theory '81, Lect. Notes Comp. Sci., Vol. 117, 1981, pp. 281-290; Part IV: Preprint Math. Inst. Hung. Acad. Sci., 1984.

AN 82 H.Andréka, I.Németi. A general axiomatizability theorem formulated in terms of cone-injective subcategories. In: Universal Algebra (Proc. Coll. Esztergom 1977), Colloq. Math. Soc. J. Bolyai, Vol. 29, North-Holland Publ. Co., Amsterdam, 1982, pp. 13-35.

AN 83 H.Andréka, I.Németi. Importance of universal algebra for computer science. In: Universal Algebra and its links with logic, algebra, combinatorics and computer science (Proceedings of the "25. Arbeitstagung Allgemeine Algebra", Darmstadt, 1983, eds.: P.Burmeister, B.Ganter, C.Herrmann, K.Keimel, W.Poguntke, R.Wille), Heldermann Verlag, Berlin, 1984, pp. 204-215.

ANSa 82 H.Andréka, I.Németi, I.Sain. A complete logic for reasoning about programs via non-standard model theory. Theoretical Computer Sci. 17, 1982, pp. 193-212, 259-278.

APa 84 H.Andréka, A.Pasztor. Relative epis need not be surjective. Preprint, 1984. To appear in Algebra Universalis.

ASa 81 H.Andréka, I.Sain. Connections between algebraic logic and initial algebra semantics of CF languages. I. In: Mathematical Logic in Computer Science (Proc. Salgótarján 1978) Colloq. Math. Soc. J. Bolyai, Vol. 26, North-Holland Publ. Co., Amsterdam, 1981, pp. 25-83.

Bc 67 H.Bachmann. Transfinite Zahlen (2.Aufl.) Springer-Verlag, Berlin, 1967.

BaH 76 B.Banaschewski, H. Herrlich. Subcategories defined by implications. Houston J. Math. 2, 1976, pp. 149-171.

BtRSk 75 W.Bartol, Z.Raś, A.Skowron. Theory of Computing Systems. Banach Center Publications, Vol. 2, Warszaw 1975, pp. 101-165.

BtNiRu W.Bartol, D.Niwiński, L.Rudak. Completion varieties. Preprint, Math. Inst., Univ. of Warsaw.

BeKaRe 81 K.Benecke, H.Kaphengst, H.Reichel. Partial algebras defined by generators and relations. Manuscript, 1981, submitted to Algebra Universalis.

BeRe 81 K.Benecke, H.Reichel. Equational partiality. Manuscript, 1981; Algebra Universalis, 16, 1983, pp. 219-232.

BgBrTuWi 81 J.A.Bergstra, M.Broy, J.V.Tucker, M.Wirsing. On the power of algebraic specifications. Mathematical Foundations of Computer Science '81 (Proc. Conf. Strbske Pleso 1981), Lect. Notes Comp. Sci. 118, Springer Verlag, Berlin, 1981, pp. 193-204.

Bm 71 J.Berman. Strong congruence lattices of finite partial algebras. Algebra Universalis 1, 1971, 133-135.

BoMuMi A.Bertoni, G.Mauri, P.A.Miglioli. Model theoretic aspects of abstract data specification. Manuscript, U. Milano.

Bi 35 G.Birkhoff. On the structure of abstract algebras. Proc. Cambridge Philos. Soc. 31, 1935, pp. 433-454.

BiLi 70 G.Birkhoff, J.D.Lipson. Heterogeneous algebras. J. Combinatorial Th. 8, 1970, pp. 115-133.

BrWi 81 M.Broy, M.Wirsing. On the algebraic extensions of abstract data types. In: Proc. Peniscola, Spain, April 1981 (J.Diaz, I.Ramos, eds.), Lect. Notes Comp. Sci. 107, Springer-Verlag, Berlin, 1981, pp. 244-251.

BrWi 81a M.Broy, M.Wirsing. An analysis of semantic models for algebraic specifications. In: Proc. of Internat. Summer School on Theoret. Found. of Progr. Methodology (München), 1981.

B 68 P.Burmeister. Über die Mächtigkeiten und Unabhängigkeitsgrade der Basen freier Algebren. Fund. Math. LXII, 1968, 165-189.

B 70 P.Burmeister. Free partial algebras. J. reine und angewandte Math. 241, 1970, pp. 75-86.

B 71 P.Burmeister. Primitive Klassen partieller Algebren. Habilitationsschrift, Bonn, 1971.

B 71a P.Burmeister. On problem 12 in G.Grätzer's book "Universal Algebra". Abstract, Notices Amer. Math. Soc. 18, 1971, p. 401

B 73 P.Burmeister. An embedding theorem for partial algebras and the free completion of a free partial algebra within a primitive class. Algebra Universalis 3, 1973, pp. 271-279.

B 82 P.Burmeister. Partial algebras - survey of a unifying approach towards a two-valued model theory for partial algebras. Algebra Universalis 15, 1982, pp. 306-358.

BJPa 78 P.Burmeister, R.John, A.Pasztor. On closed morphisms in the category of partial algebras. Contributions to General Algebra, Proc. Conf. Klagenfurt, 1978 (Ed. by H. Kautschitsch, W.B. Müller, W. Nöbauer), Verlag Joh. Heyn, 1978, pp. 69-76.

BSch 67 P.Burmeister, J.Schmidt. On the completion of partial algebras. Coll. Math. 17, 1967, pp. 235-245.

BSi 84 P.Burmeister, M.Siegmund-Schultze. Subdirect representations of partial algebras. Preprint Nr. 867, TH Darmstadt, 1984; to appear in Contributions to General Algebra 3 (Proceedings of the Vienna Conference, June 21-24, 1984; eds.: G.Eigenthaler, H.K.Kaiser, W.B.Müller,W.Nöbauer), Verlag B.G.Teubner, Stuttgart, 1985, pp. 75-98.

BWo 82 P.Burmeister, B.Wojdyło. A category theoretical background for homomorphism theorems. Preprint Nr.699, TH Darmstadt, 1982. Submitted to Coll. Math.

BWo 82a P.Burmeister, B.Wojdyło. Properties of homomorphisms and quomorphisms between partial algebras. Preprint Nr.657, TH Darmstadt, 1982.

ChKs 73 C.C.Chang, H.J.Keisler. Model Theory. North-Holland Publ. Co., 1973 (2nd ed. 1977).

ClKr 81 D.M.Clark, P.H.Krauss. Topological quasivarieties. Preprint, Math. Dept. State Univ. of New York, New York, USA, 1981.

C 65 P.M.Cohn. Universal Algebra. Harper and Row, 1965 (2nd rev. ed. D.Reidel Publ. Co., Dordrecht, 1981).

E 69 H.-D.Ebbinghaus. Über eine Prädikatenlogik mit partiell definierten Prädikaten und Funktionen. Arch. math. Logik 12, 1969, pp. 39-53.

Ed 73 G.A.Edgar. The class of topological spaces is equationally definable. Algebra Universalis 3, 1973, pp. 139-146.

Eh 82 C.Ehresmann. Oevres complètes et commentées. Parte IV-1. Esquisses et complètions. Supplément des Cahiers Top. Géom. Diff., 1982.

Ev 51 T.Evans. Embeddability and the word problem. J. London Math. Soc. 26, 1951, pp. 64-71.

Fl 81 I.Fleischer. Equational classes of partial algebras. Manuscript, 1981.

FR 80 N.Francez and M.Rodeh. A distributed abstract data type implemented by a probabilistic communication scheme. Preprint 1980. TECHNION Comp. Sci. Dept. Haifa.

GlZJu 80 W.M.Gluschkow, G.J.Zeitlin, J.L.Justschenko. Algebra, Sprachen, Programmierung. Akademie-Verlag, Berlin, 1980.

Go 78 J.A.Goguen. Abstract errors for abstract data types. Proc. IFIP Working Conference on Formal Description of Programming Concepts, St.Andrews, 1977.

GoMe 81 J.A.Goguen, J.Meseguer. Completeness of many-sorted equational logic. Manuscript, 1981; appeared in SIGPLAN Notices.

GoTWa 78 J.A.Goguen, J.W.Thatcher, E.G.Wagner. An initial algebra approach to the specification, correctness, and implementation of abstract data types. IBM Research Report RC-6487, 1976; and: Current Trends in Programming Methodology, IV: Data Structuring (R. Yeh, ed.), Prentice Hall, 1978, pp. 80-144.

G 68 G.Grätzer. Universal Algebra. D.van Nostrand Co., 1968 (2nd ed., Springer Verlag, 1979)).

G 68a G.Grätzer. On polynomial algebras and free algebras. Can. J. Math. 20, 1968, pp. 575-581.

G 78 G.Grätzer. General Lattice Theory. Akademie-Verlag, Berlin, 1978.

G 85 G.Grätzer. Axiom of Choice is equivalent with Birkhoff's subdirect irreducible representation. Preprint, Univ. of Manitoba, Canada, 1985.

GeMe 84 I.Guessarian, J.Meseguer. Axiomatization of "if..then..else" revisited. Laboratoire Inf. Theor. et Programmation, Preprint 84-18, April, 1984.

Gu 81 R.Guitart. The theory of sketches (revisited). Journées Faisceaux et Logique (19th P.S.S.L.) U. Paris-Nord & S.M.F., Paris, 23 et 24 mai 1981.

Gu 82 R.Guitart. Qu'est-ce que la logique dans une categorie? Cahiers de Top. Géom. Diff. 23, 1982, pp. 115-148.

GuL 80 R.Guitart, C.Lair. Calcul syntaxique des modèles et calcul des formules internes. Diagrammes, Vol. 4, Dec. 1980.

GuL 82 R.Guitart, C.Lair. Existence de diagrammes localement libres II. Diagrammes, Vol.7, 1982.

GuL 82a R.Guitart, C.Lair. Limites et colimites pour présenter les formules. Diagrammes, Vol.7, 1982.

GuL 82b R.Guitart, C.Lair. La continuité pour représenter les formules. Diagrammes, Vol.7, 1982.

Hy 72 F.Harary. Graph Theory. Addison Wesley Publ. Comp., 1972.

Ha 77 R.Hartwig. Pseudogleichungen in formalen Kalkülen über partiellen Algebren. Vorträge zu Grundlagen der Informatik, Heft 27/77, TU Dresden, 1977, pp. 13-22.

HnMoT 85 L.Henkin, D.Monk, A.Tarski. Cylindic Algebras, Part II. North-Holland Publ. Co., 1985.

He 63 H.Hermes. Einführung in die mathematische Logik. Teubner Verlag, Stuttgart, 1963 (English translation: Introduction to Mathematical Logic. Unitext, Springer-Verlag, 1973).

H 72 H.Herrlich. Perfect subcategories and factorizations. Coll. Math. Soc. Janos Bolyai 8. Topics in Topology, Keszthely (Hungary), 1972, pp. 387-403.

HS 73 H.Herrlich, G.E.Strecker. Category Theory - An Introduction. Allyn and Bacon, 1973 (2nd ed.: Heldermann-Verlag).

Hi 83 B.H.Hien. On a problem of algebraic model theory. Bull. of Section in Logic 11, Warszaw-Lodz, 1983, pp. 103-108.

Hi 84 B.H.Hien. In which categories are first order axiomatizable hulls characterizable by ultraproducts? Periodica Math. Hungar. 15, 1984, to appear.

HiSa 83 B.H.Hien, I.Sain. In which categories are first order axiomatizable hulls characterizable by ultraproducts? Cahier Top. Géo. Diff. XXIV-2, 1983, pp. 215-222.

Hg 63 P.J.Higgins. Algebras with a scheme of operators. Math. Nachr. 27, 1963, pp. 115-132.

Hö 72 H.Höft. Operators on classes of partial algebras. Algebra Universalis 2, 1972, pp. 118-127.

Hö 73 H.Höft. Weak and strong equations in partial algebras. Algebra Universalis 3, 1973, pp. 203-215.

Hoe 72 H.-J.Hoehnke. Superposition partieller Funktionen. Studien zur Algebra und ihre Anwendungen (Hoehnke, ed.), Akademie Verlag Berlin, 1972, pp. 7-26.

Hoe 76 H.-J.Hoehnke. On partial algebras. Preprint Akad. d. Wiss. d. DDR, Berlin, 1976. Published in: Universal algebra (Proc. Coll. Esztergom 1977) Colloq. Math. Soc. J. Bolyai, Vol. 29, North-Holland Publ. Co., Amsterdam, 1982, pp. 373-412.

Hoe 82 H.-J.Hoehnke. Fully invariant algebraic closure systems of congruences and quasivarieties of algebras. Manuscript, 1982.

Hoe 84 H.-J.Hoehnke. On the Lemma of Yoneda-Schreckenberger in certain monoidal categories. Preprint, 1984.

Hoe 84a H.-J.Hoehnke. On Yoneda-Schreckenberger's embedding of a class of monoidal categories. Submitted to Beiträge zur Algebra und Geometrie, 1984

Ja 83 T.M.V.Janssen. Foundations and applications of Montague grammar. Mathematisch Centrum, Amsterdam, 1983.

J 75 R.John. Gültigkeitsbegriffe für Gleichungen in partiellen Algebren. Dissertation, TH Darmstadt, 1975.

J 75a R.John. A note on implicational subcategories. Contributions to Universal Algebra (Proc. Coll. Szeged, 1975); Eds.: B.Csakany, J.Schmidt; Colloq. Math. Soc. J. Bolyai, Vol. 17, North-Holland Publ. Co., Amsterdam, 1977. pp. 213-222.

J 78 R.John. Gültigkeitsbegriffe für Gleichungen in partiellen Algebren. Math. Zeitschrift 159, 1978, pp. 25-35.

JoSe 67 J.Johnson, R.L.Seifert. A survey of multi-unary algebras (report). Univ. of California, Berkeley, 1967.

Ka 81 H.Kaphengst. What is computable for abstract data types? Fundamentals of Computation Theory (Proc. Conf. Szeged, 1981), Lect. Notes Comp. Sci. 117. Springer-Verlag, Berlin, 1981, pp. 173-181.

Ka 82 H.Kaphengst. Zum Aufbau einer mehrsortigen Logik. Manuscript, Dresden, 1982.

KaRe 77 H.Kaphengst, H.Reichel. Initial algebra semantics for non-context-free languages. Fundamentals of Comp. Sci., Lect. Notes Comp. Sci. 56, Springer-Verlag, 1977, pp. 120-127.

K 64 C.R.Karp. Languages with expressions of infinite length. North-Holland Publ. Co., 1964.

Ky D.Kelly. Complete rules of inference for universal sentences. Studia Sci. Math. Hungar.

Ke 65 R.Kerkhoff. Eine Konstruktion absolut freier Algebren. Math. Annalen 158, 1965, pp. 109-112.

Ke 70 R.Kerkhoff. Gleichungsdefinierbare Klassen partieller Algebren. Math. Annalen 185, 1970, pp. 112-133.

Kl 52 S.C.Kleene. Introduction to Metamathematics. North-Holland Publ. Co., 1952 (5th reprint 1967).

KnRo 83 E.Knuth, L.Rónyai. Closed convex reference schemes. (A junction between computer science, cylindric and partial algebras.) Preprint, MTA-SZTAKI, March 1982. II/32. (Proc. IFIP Conf., 1983, Hungary).

KoSp 68 S.Kochen and E.P.Specker. The problem of hidden variables in Quantum Mechanics. J. Math. Mech. 17, 1968, pp. 59-87.

Ku 77 I.Kupka. Partial algebras for representing semantics of information processing. Habilitationsschrift, Hamburg, 1977 (und Preprint am Fachbereich Informatik, Hamburg, 1980).

LePa 80 D.Lehmann, A.Pasztor. On a conjecture of Meseguer. Preprint Techn. Dept. of Comp. Sci., Haifa 1980; also under the title "Epis need not be dense" in Theoretical Computer Science 17, 1982, pp. 151-162.

Lev 79 A.Levy. Basic Set Theory. Springer-Verlag, 1979.

Lj 76 E.S.Ljapin. Abstract characterization of partial groupoids of words with synonyms. Algebraic Theory of Semigroups (Proc. Coll. Szeged, 1976); Colloq. Math. soc. J. Bolyai, Vol. 20, 1979, pp. 341-356.

Lu 76 H.Lugowski. Grundzüge der Universellen Algebra. Teubner, Leipzig, 1976.

Ml 71 S.MacLane. Categories for the Working Mathematician. Springer-Verlag, 1971.

MhMk 84 B.Mahr, J.A.Makowski. An axiomatic approach to semantics of specification languages. In: Theor. Computer Science (6th GI Conference, Dortmund, 1983; A.B.Cremers, H.P.Kriegel, eds.), Lect. Notes Comp. Sci. 145, Springer-Verlag, Berlin, 1984, pp. 211-219.

MhMk 84a B.Mahr, J.A.Makowski. Characterizing specification languages which admit initial semantics. Theor. Computer Science, 31, 1984, pp. 49-59.

Mk 84 J.A.Makowski. Model theoretic issues in theoretical computer science. Part I. Relational data bases and abstract data types. Logic Coll. '82 (G.Lolli, G.Longo, A.Marcja, eds.), North-Holland Publ. Co. 1984, pp. 303-345.

Mal 73 A.I.Mal'cev. Algebraic Systems. Springer-Verlag, 1973.

Mal 74 A.I.Mal'cev. Algorithmen und rekursive Funktionen. Vieweg, 1974.

Mar 58 E.Marczewski. A general scheme of independence in mathematics. Bull. de l'Acad. Polonaise des Sci., Sér. Math. Astr. et Phys. 6, 1958, pp. 731-736.

Mr 83 Zs.Markusz. On first-order many-sorted logic. MTA SZTAKI Tanulmányok, 151/1983, pp. 1-85.

Mr 83a Zs.Markusz. Different validity concepts in many-sorted logic. Proc Conf. Algebra, Combinatorics and Logic in Computer Science (Györ, Hungary, 1983), to appear.

Mo 76 J.D.Monk. Mathematical Logic. Springer-Verlag, 1976.

Ms 51 A.Mostowski. On the rules of proof in the pure functional calculus of the first order. J. Symbolic Logic 16, 1951, pp. 107-111.

N 75 I.Németi. Partial algebra theory via categories. Dissertation, Hungar. Acad. Sci. 1975 (in Hungarian); appeared as "Partial Algebras" (in Hungarian) as booklet by SZAMALK, 1976.

N 78 I.Németi. From hereditary classes to varieties in abstract model theory and partial algebras. Beiträge zur Algebra und Geometrie 7, 1978, pp. 69-78.

N 81 I.Németi. Connections between cylindric algebras and initial algebra semantics of CF languages. II. In: Mathematical Logic in computer Science (Proc. Salgótarján 1978), Colloq. Math. Soc. J. Bolyai, Vol. 26, North-Holland Publ. Co., Amsterdam, 1981, pp. 561-605.

N 82 I.Németi. On notions of factorization systems and their applications to cone-injective subcategories. Periodica Math. Hungar. 13/3, 1982, pp.229-235.

NSa 77 I.Németi, I.Sain. Cone-implicational subcategories and some Birkhoff-type theorems. In: Universal Algebra (Proc. Coll. Esztergom 1977), Colloq. Math. Soc. J. Bolyai, Vol. 29, North-Holland Publ. Co., Amsterdam, 1981, pp. 535-578.

O 81 A.Obtułowicz. Theories classifying partial algebras. submitted to: Proceedings Conf. on Category Theory, Gummersbach, 1981.

O 82 A.Obtułowicz. The logic of categories of partial functions and its applications. Preprint, 1982.

Oe 81 G.Oehme. Konformfreie Algebren und ihr Zusammenhang mit speziellen ECE-Varietäten. Diplom thesis, Fachbereich Math., TH Darmstadt, 1981.

Pa 79 A.Pasztor. Faktorisierungssysteme in der Kategorie der partiellen Algebren - Kennzeichnung von (Homo)-Morphismen. Dissertation, TH Darmstadt, 1979 appeared in:Hochschulverlag, Freiburg, 1979.

Pa 82 A.Pasztor. Ordered algebras are monoreflective in the category of ordered partial algebras. Proc. 7th Coll. on Trees in Algebra and Programming (March 4-6, 1982), Lille, France.

Pa 83 A.Pasztor. Chain-continuous algebras - a variety of partial algebras. Fund. Inform. 6, 1983, pp. 275-288.

Pa 83a A.Pasztor. The epis of the category of ordered Z-continuous homomorphisms. Cahiers Top. Géom. Diff. 26, 1983, pp. 203-214.

Pa 83b A.Pasztor. Epis of some categories of Z-continuous partial algebras. Acta Cybernet. 6, 1983, pp. 111-123.

Pa 84 A.Pasztor. On the variety concept for ω-continuous algebras. Applications of a general approach. (Dept. Math., Carnegie Mellon Univ., Pittsburgh, USA, Research report 84-2, 1984); Proc. Conf. on the mathematical foundations of programming semantics (Kansas State Univ., Manhatten, April 11-12, 1985), Lect. Notes Comp. Sci., Springer-Verlag, to appear.

Pc 64 H.E.Pickett. Subdirect representations of relational systems. Fund. Math. 56, 1964, pp. 223-240.

Pi 68 R.S.Pierce. An Introduction to Abstract Algebras. Holt, Rinehart and Winston, 1968.

Pig 66 D.Pigozzi. On some operations on classes of algebras. Notices Amer. Math. Soc. 13, 1966, p. 829.

Po 73 V.S.Poythress. Partial morphisms on partial algebras. Algebra Universalis 3, 1973, pp. 182-202.

PuTr 80 A.Pultr, V.Trnková. Combinatorial, Algebraic and Topological Representations of Groups, Semigroups and Categories. Academia, Praha 1980.

Re 78a H.Reichel. Limit-colimit doctrines in computer science. Algebra and its Applications, Banach Center Semester, Warszawa, 1978. To appear in Banach Center Publications.

Re 78b H.Reichel. Algebraic specifications of abstract data types. Algebra and its Applications. Banach Center Semester, Warszawa, 1978. To appear in Banach Center Publications, Vol. 9, 1980.

Re 79 H.Reichel. Theorie der Äquoide. Dissertation B, Humboldt-Universität, Berlin, 1979.

Re 81 H.Reichel. Homomorphism theorem for equationally partial heterogeneous algebras. Manuscript, submitted to Algebra Universalis.

Re 84 H.Reichel. Structural Induction on Partial Algebras (Part II of: Introduction to Theory and Applications of Partial Algebras; Part I: this volume). Mathematical Research, Vol. 18, Akademie-Verlag, Berlin, 1984.

Ri 82 H.Riedel. Category theory and universal algebra. Preprint, Univ. of Waterloo, Canada, 1982.

R 83 I.G.Rosenberg. Galois theory for partial algebras. Preprint, 1983.

Ru 83 L.Rudak. Completeness theorem for partial algebras with weak satisfaction. Algebra Universalis, 16, 1983, pp. 331-337.

Sa 74 I.Sain. Category theoretical investigations in order to generalize identities and quasi-identities. Master Thesis (in Hungarian), Eötvös Lórand Univ., Budapest, 1974.

Sa 77 I.Sain. On classes of algebraic systems closed with respect to quotients. Preprint Math. Inst. Hung. Acad. Sci. 1977; appeared in: Algebra and its applications, (Banach Center Semester Warsaw 1978), Banach Center Publ. Vol. 9, PNW - Polish Scientific Publishers, Waszawa, 1980, pp. 127-131.

Sche 76 B.Schepull. Über Quasivarietäten von partiellen Algebren. Dissertation, Akad. D.Wiss. d. DDR, Berlin, 1976.

Sch 62 J.Schmidt. Die Charakteristik einer Allgemeinen Algebra. I. Arch. Math. XIII, 1962, pp. 457-470.

Sch 64 J.Schmidt. Die Charakteristik einer Allgemeinen Algebra. II. Arch. Math. XV, 1964, pp. 286-301.

Sch 64a J.Schmidt. Some properties of algebraically independent sets in algebras with infinitary operations. Fund. Math. LV, 1964, pp. 123-137.

Sch 66 J.Schmidt. A general existence theorem on partial algebras and its special cases. Coll. Math. 14, 1966, pp. 73-87.

Sch 66a J.Schmidt. Mengenlehre I (Einführung in die axiomatische Mengenlehre). Bibliogr. Institut, HTB 56/56a, 1966.

Sch 66b J.Schmidt. Allgemeine Algebra. Lecture notes at the U. of Bonn, 1965/66.

Sch 68 J.Schmidt, Seminar on Universal Algebra. Lecture notes at the U. of Athens, Ga., 1967/68.

Sch 70 J.Schmidt. A homomorphism theorem for partial algebras. Coll. Math. 21, 1970, pp. 5-21.

Sre 80 J.Schreckenberger. Über die Einbettung von dht-symmetrischen Kategorien in die Kategorie der partiellen Abbildungen zwischen Mengen. Preprint der Akad. d. Wiss. d. DDR, ZAMM, 1980.

Sr 77 P.Schreiber. Grundlagen der Mathematik. VEB Deutscher Verlag der Wissenschaften, Berlin 1977.

Sc 77 D.Scott. Identity and existence in intuitionistic logic. Durham Proceedings, 1977, Applications of Sheaves (ed. M.P. Fourman); Lect. Notes Math. 753, Springer-Verlag 1979, pp. 660-696.

Si 82 M.Siegmund-Schultze. Subdirectly irreducible and equationally compact monounary partial algebras. Preprint Nr. 695, TH Darmstadt, 1982.

Si 84 M.Siegmund-Schultze. Residual kleine Varietäten partieller Algebren. Dissertation, Fachbereich Mathematik, TH Darmstadt, 1984.

Sl 64 J.Słomiński. A theory of extensions of quasi-algebras to algebras. Rozprawy Matematyczne 40, 1964.

Sl 68 J.Słomiński. Peano-algebras and quasi-algebras. Dissertationes Mathematicae (Rozprawy Mat.) 62, 1968.

S 72 G.E.Strecker. Epireflection operators vs perfect morphisms and closed classes of epimorphisms. Bull. Austral. Math. Soc. 7, 1972, pp. 359-366.

T 84 A.Tarlecki. On the existence of free models in abstract algebraic institutions. Preprint, Univ. of Edinbourgh, 1984.

T 84a A.Tarlecki. Note: Quasi-varieties in abstract algebraic institutions. Preprint, Univ. of Edinbourgh, 1984.

Th 66 H.Thiele. Wissenschaftstheoretische Untersuchungen in algorithmischen Sprachen I (Theorie der Graphschemata-Kalküle). VEB Deutscher Verlag der Wissenschaften, Berlin 1966.

Ti 81 J.Tiuryn. A survey of the logic of effective definitions. Preprint MIT/LCS/TR-246, MIT Laboratoy for Computer Science, Cambridge, Massachusetts, 1981.

vdW 66 B.L.van der Waerden. Algebra, Erster Teil. Springer-Verlag, 1966.

V 79 H.-J.Vogel. On Birkhoff-algebras in dht-symmetric categories. Coll. Math. Soc. J.Bolyai 28 (Proceedings of "Finite Algebras and Multiple-Valued Logic", Szeged, Hungary, 1979), North-Holland Publ. Co., 1981.

V 84 H.-J.Vogel. Eine kategorientheoretische Sprache zur Beschreibung von Birkhoff-Algebren. Report R-Math-06/84 der Akad. d. Wiss. d. DDR, 1984.

We 70 G.H.Wenzel. Subdirect irreducibility and equational compactness in unary algebras <A;f>. Arch. Math. 21, 1970, pp. 256-264.

W 78 H.Werner. Einführung in die allgemeine Algebra. B.I.-Hochschultaschenbücher, Band 120, 1978.

Wi 70 R.Wille. Subdirekte Produkte und konjunkte Summen. J. für Math., Bd. 239/240, 1970, pp. 333-338.

Wi 82 R.Wille. Restructuring Lattice Theory: An Approach Based on Hierarchies of Concepts. In: Ordered Sets (I.Rival, ed.); D.Reidel Publ. Co., Dordrecht, 1982, pp. 445-470.

Wi 84 R.Wille. Line diagrams of hierarchical concept systems. Internat. Classification 11, 1984, pp. 77-86.

Wo 72 B.Wojdyło. Categories of quasi-algebras. N. Copernicus University, Torun, Preprint No. 2, 1972.

LIST OF SYMBOLS

Symbolism consisting only in a special kind of writing the parameters can be found at the end; otherwise, the symbol is followed by an example with parameters. Symbols only used at some particular place are usually not listed here.

Fixed abbreviations starting with a letter

Alg(τ), 19
Alg(τ)$\downarrow\underline{S}$, 41
Alg($\underline{S}$), 41
AO, AO2($\underline{A};X;\underline{S}$), 206
APF, APF($\underline{A};\underline{S}$), $\mathrm{APF}_S(\underline{A};\underline{S})$, 247
APO, APO1($\underline{A};\omega(\mathbf{d}^{\&});\underline{S}$), $\mathrm{APO}^M(\underline{A};\underline{S})$, APO2($\underline{A};X;\underline{S}$), 203
$\underline{\mathrm{APO1}}(\underline{A};\omega(\mathbf{d}^{\&});\underline{S})$, $\underline{\mathrm{APO}^M}(\underline{A};\underline{S})$, $\underline{\mathrm{APO2}}(\underline{A};X;\underline{S})$, 203
$\mathrm{APO3}^M(\underline{A};\underline{S})$, APO3($\underline{A};X;\underline{S}$), $\underline{\mathrm{APO3}^M}(\underline{A};\underline{S})$, $\underline{\mathrm{APO3}}(\underline{A};X;\underline{S})$, 207
$\mathrm{APO}_K{}^M(\underline{A})$, 274
Aut, Aut($\underline{A}$), **27**, 43
$\underline{\mathrm{Aut}}(\underline{A})$:= (Aut($\underline{A}$);$\cdot$,$\mathrm{id}_A$) , **27**, 43
Aut($\underline{A}$\`), $\underline{\mathrm{Aut}}(\underline{A}$\`), 43

B, **BM**, $\mathbf{B}_{\underline{A}}{}^n\mathbf{M}$, 50

c, 112
C, $\mathbf{C}_{\underline{A}}\mathbf{M}$, 49
$C_{\bar{\mathcal{H}}}$, 193
C, $C^M(\underline{A};\underline{S})$, 203
C($\underline{A}$), 240
¢, ¢$_a$, 214
Card, 289
Cat, **44**, 120
Cat*, **14**, 120
char, char($\underline{A}$;;**K**), 93
CMpI, 185
Co, $\underline{\mathrm{Co}}$(M:Φ), 123
Cod, Cod(f), 21
Cod*, Cod*(f), 14
Cong, Cong($\underline{A}$), $\underline{\mathrm{Cong}}(\underline{A})$, **30**, **43**
Cong($\underline{A}$\`), **Cong**($\underline{A}$\`), 43
Cong, $\mathbf{Cong}_A$(M), **31**, **43**
$\mathbf{Cong}_{A}$\`($\bar{R}$), 43
Cong_C, $\overline{\mathrm{Cong}}_C(\underline{A})$, $\underline{\mathrm{Cong}_C}(\underline{A})$, **30**, **43**
$\mathrm{Cong}_C(\underline{A}$\`), 43
$\mathrm{Cong}_C(A,\underline{B})$, $\underline{\mathrm{Cong}_C}(A,\underline{B})$, 91
$\mathrm{Cong}_C(\underline{A}$;;**K**), 93
$\mathrm{Cong}_C(\subseteq Y,\underline{F}(Y,$**TAlg**($\underline{S}$))), 193
$\mathrm{Cong}_{cfi}(\underline{A})$, 103
$\mathrm{Cong}_{c,r}(M,\underline{A},$**K**), 271
$\mathrm{Cong}_{c,w}(M,\underline{A},$**K**), 271
$\mathrm{Cong}_E(\underline{A})$, 256
$\mathrm{Cong}_{fi}(\underline{A})$, 103
$\mathrm{Cong}_p(\underline{A})$, 252
Csubs, **Csubs**($\underline{A}$), 45

d, $\underline{A}_d$, 19
d:= **d**(τ) , 23
d*:= max(#ω_0,**d**) , **23**, 95
$\mathbf{d}^{\&}$:= max(**d***,$(\#S)^+$) , **23**, 95, 112
deg, deg(**K**), 259
D, **DM**, $\mathbf{D}_{\underline{A}}{}^n\mathbf{M}$, 50
Difip, $\overline{\mathrm{Difip}}$(e,f), 128
dom, dom ϕ, 18, **283**
dom R (for a relation), 104, **282**
$\underline{\mathrm{dom}}$, $\underline{\mathrm{dom}}$ f, **58**, 269
$\underline{\mathrm{d}}$om, $\underline{\mathrm{d}}$om f*, 269
Dom, Dom(f), 21, **283**
Dom*, Dom*(f), 14

e, **e**(**K**), 72
e, $e_m{}^M$, $(e_m{}^M)^{\underline{A}}$, 87, **199**, 213
E, E^M(A;S), E($\underline{A};X;\underline{S}$), 199
$E^{n,v}(A;\underline{S})$, 200
E($\underline{A}$;;$\underline{S}$) = E($\underline{A};\omega(\mathbf{d}^{\&});\underline{S}$), 200
e(.:.), e(M:Φ), 123
E, $\mathbf{E_S{}^H}$, $\mathbf{E_S{}^{H,k}}$, 187
e_F, 124
$\mathbf{E_k}$, $\mathbf{E_k}$(**K**), 188
ECEeq, **ECEeq**, **ECEeq**(**K**), 121
ECH, **ECH**, **ECH**(**K**), 121
ECHEeq, ECHEeq(**K**), 121
Eeq, Eeq_M, Eeq^*_X, **96**, 121
Eeq, $\mathbf{Eeq_M}$(**K**), **Eeq**(**K**), **96**, 121
End, End($\underline{A}$), **27**, 43
$\underline{\mathrm{End}}(\underline{A})$:= (End($\underline{A}$);$\cdot$,$\mathrm{id}_A$) , **27**, 43
End($\underline{A}$\`), $\underline{\mathrm{End}}(\underline{A}$\`), 43
Epi, Epi($\mathfrak{C}$), 39
eval, eval_M, $\mathrm{eval}_M{}^{\underline{A}}(t)$, **203**, 213
Ext, Ext(**K**), 144

F, **162**, 186
$\mathbf{F_C}$, 69
F, $\underline{F}$, $\underline{F}$(.,.), $\underline{F}(\underline{A}$,**K**), $\underline{F}$(A,**K**), **76**, 224 (F($\underline{A}$,**K**) for the carrier set)
$F^M(\underline{A};\underline{S})$, 211
$\underline{F}$((M,G),Q), $\underline{F}$((M,G),**K**), 234
FALSE, 13
field, field R, 282
FS, (FS1)-(FS4), 140
(FS5), (FS6), 141
fvar, fvar(Φ), 113

$\bar{G}$, 106
G, $\mathbf{G}_{\underline{A},z}{}^n$, $\mathbf{G}_{\underline{A},z}{}^n\mathbf{M}$, 228
graph, graph(ϕ), 18, **283**

$\mathbf{H}$, $\mathbf{H}(K)$, 71, redefined 162
$\mathbf{H}_b$, 151
$\mathbf{H}_c$, **151**, 258
$\mathbf{H}_c$, $\mathbf{H}_c(\dot{K})$, 71
$\mathbf{H}_f$, **149**, 258, 263
$\mathbf{H}_f$, $\mathbf{H}_f(K)$, 71
$\mathbf{H}_{fp}$, **151**, 258
$\mathbf{H}_F$, 172
$\mathbf{H}_i$, 151
$\mathbf{H}_p$, 149
$\mathbf{H}_{p-i}$, 173
$\mathbf{H}_w$, 148
$\mathbf{H}_w$ (= $\mathbf{H}$), $\mathbf{H}_w(K)$, 71
Heq, **Heq**, **Heq**(K), 121
Hom(**Alg**(τ)), 24
Hom(**Alg**($\underline{S}$)), 43
Horn, **Horn**, **Horn**(K), 121

i, **i**(K), 72
I, $I_{a,b}$, 70
I, I^W, 115
I, 172
I, **I**(K), 71
$I_n := I_n(\underline{A},M)$ $(0 \le n \le 3)$, 67
I_4, 69
id, id_A, 21, **283**
id_{AB}, 283
Imp, Imp_{Prem}, 190
$\mathbf{Imp}_{Prem}$, 191
Impl, **Impl**, **Impl**(K), 121
ind, $ind_{\underline{A}}(M)$, 76
inf, $inf(\overline{M})$, $inf_{\underline{L}}(M)$, $inf(a,b)$, 292
Inj, Inj(**E**), 76
Inj(K), 163
INPUT $(:\Omega \to S)$, (ϕ)INPUT, 22
Iso, Iso($\underline{\mathfrak{C}}$), 25

K, K_X, K_X(**K**), 163
K_e, K_H, $K_{pr,k}$, $K_{P_+^{r,k}}$, K_{P_u}, K_S, 162
K_ϕ, 19
ker, ker f, 30

$L := L_{c,q,d,v}(\tau,\underline{S},(X,v_X)) = {:}L_{c,q}(\underline{S},X^{`})$, 112
$\underrightarrow{\lim}$, $\underrightarrow{\lim}\ \underline{A}_I$, 65

$\mathbf{M}_c$, 147
$\mathbf{M}_c$, $\mathbf{M}_c(K)$, 126
$\mathbf{M}_{f_i}$, 150
$\mathbf{M}_i$, 149
$\mathbf{M}_{ip}$, 173
$\mathbf{M}_n$, 150
$\mathbf{M}_p$, 147
$\mathbf{M}_{p-i}$, 173
max, max(M), max(a,b), 292
min, min(M), min(a,b), 292
Mod, **Mod**(G), 96
Mod(F), 119
$\mathbf{Mod}^t(Q)$, $\mathbf{Mod}^t_\theta(Q)$, 191
Mono, Mono($\underline{\mathfrak{C}}$), 39
Mor, Mor($\underline{\mathfrak{C}}$), 14, 21
Mor(A,B), 21
Mor, **Mor**(P), 254

$\mathbf{N}_k$, $\mathbf{N}_k(K)$, 187
nat, $nat_R: A \to A/R$, $nat_R: \underline{A} \to \underline{A}/R$, 34
$nat_c(M:\Phi)$, 123
$nat_p(M:\Phi)$, 123
NEeq, **NEeq**, **NEeq**(K), 121
Neg(Q'), 196
NTE, **NTE**, **NTE**(K), 121

o, **o**(K), 72
O, $\underline{\mathbf{O}}$, 72
O, $O^M(A;S)$, $O^M(A)$, $O(\underline{A};X;\underline{S})$, 199
$O(\underline{A};;\underline{S}) = O(\underline{A};\omega(\mathbf{d}^{\&});\underline{S})$, 200
$O^{n,v}(\underline{A};\underline{S})$, 200
Ob($\underline{\mathfrak{C}}$), 21
Ord, 286
OUTPUT $(:\Omega \to S)$, (ϕ)OUTPUT, 22

P, **P**(K), 71
$\mathbf{P}^r$, $\mathbf{P}^r(K)$, 72
$\mathbf{P}^{r,k}$, $\mathbf{P}^{r,k}(K)$, 125
$\mathbf{P}_u$, $\mathbf{P}_u(K)$, 72
$\mathbf{P}_+$, $\mathbf{P}_+(K)$, 72
$\mathbf{P}_+{}^r$, $\mathbf{P}_+{}^r(K)$, 72
$\mathbf{P}_+{}^{r,k}$, $\mathbf{P}_+{}^{r,k}(K)$, 125
P, (P1)-(P3), 77
P_1, P_2, 188
$P(X,\underline{S})$, $\underline{P}(X,\underline{S})$, 78
Par, **$\underline{\text{Par}}$**, **278**, 283
Pj, Pj(**H**), 128
PO, $PO^M(A;S)$, $PO^M(A)$, $PO(\underline{A};X;\underline{S})$, $PO(\underline{A};;\underline{S})$, 199
$\underline{PO}^M(A;S)$, $\underline{PO}^M(A)$, $\underline{PO}(\underline{A},X;\underline{S})$, 201
$PO(\underline{A};;\underline{S}) = PO(\underline{A};\omega(\mathbf{d}^{\&});\underline{S})$, 200
$\underline{PO}(\underline{A};\omega(\mathbf{d}^{\&});\underline{S})$, 201
$PO^{n,v}(A;S)$, 200
POP, 12
POS, 183
POSω, 184
POS, (POS1)-(POS4), 183
(POSω1)-(POSω6), 184
(POS$\bot$7), (POS$\bot$8), 184
pr, $pr_j{}^A$, $pr_j: \mathsf{X}(A_i \mid i \in I) \to A_j$, 56
$\underline{Pr}$, $\underline{Pr}(M:\Phi)$, 123
PRED, 12
Prem, 190
Ps, $\mathbf{Ps}^{(E,M)}(K)$, 259
PUSH, 12

q, 112
Q, $(\Phi)Q$, 190
Q_H, 194
Q*, 197
QEeq, **QEeq**, **QEeq**(K), 121

$r_{.,.}$, $r_{\underline{A},K}: \underline{A} \to \underline{F}(\underline{A},K)$, **76**, 224
$(r_{\underline{A},K}, \underline{F}(\underline{A},K))$, **77**, 95
(r_P, F_P), 187
$r_{M,G}$, 234
r$:= \sup(c \mid c < d)$, 23
R, $R_{(a,b)}$, 248
$R_E(a,b)$, 257

$R_{E,f,\underline{a},a}$, $R_{E,c,\underline{a}}$, 261
$R_c := R_c(\underline{A})$, **32**, 247, 248
R_f, 30
R_F, 70
rng, rng R, 282
rnk, **rnk**(M), 285
Rsuba, **Rsuba**($\underline{A}$), 45
Rsuba(A,$\underline{B}$), 91
Rsuba($\underline{A}$;;K), 93
Rsuba($\subset$Y,$\underline{F}$(Y,**TAlg**($\underline{S}$))), 193
Rsubalg(M,$\underline{A}$,K), **271**, 272

s, s', s_1,..., 22
S, 22, **40**
$\underline{S}$, 40
$\underline{S}_M$, 236
$\underline{S}'$, K($\underline{S}'$-t), $\underline{A}|_{S'}$, $\Omega(S')$, $\tau(S')$, 228
S, 71; redefined 162
S_c, **149**, 258, 262
S_c (= **S**), **S_c**(K), 71
S_f, **S_f**(K), 71
S_f = **S_i**, 262
S_i, 148
S_{ip}, 149
S_p, **149**, 258
S_{p-i}, 173
S_r, **S_r**(K), 71
S_w, **149**, 258, 262
S_w, **S_w**(K), 71
SECH, **SECH**, **SECH**(K), 121
Set, 256
$\underline{\text{Set}}$, 283
SI, (SI), 256
(SR), 261
(SRc), 261
(SRf), 261
si, **si**$^{(E,M)}$(K), 259
St, **St**(a), 284
STACK(NAT), 101
Suba, **Suba**($\underline{A}$), 45
subf, subf(Φ), 113
Subs, **Subs**(M), 92
Subs_k(M), 190
Subs_k*(Δ_f), 190
Subs(S,Q*), 197
Subst, 114
Subst((N:Φ),f,(M:(Φ)f)), 114
SUCC, 12
sup, sup(M), $\text{sup}_{\underline{L}}$(M), sup(a,b), 14, 292

t, K_t, 224
$\underline{T}$, $\underline{T}$(X,**TAlg**($\underline{S}$)), 95
$\underline{T}_c$, $\underline{T}_c$(M:Φ), 123
T_{loc}, 96
$\underline{T}_p$, $\underline{T}_p$(M:Φ), 123
T_γ(G), 106
TAlg(τ), 19
TAlg($\underline{S}$), 41
TE, **TE**, **TE**(K), 121
Term, **Term**(H), 254
Th, **Th**(K), 118
TOP, 12
Tr, Tr(M$\cup\Omega$,**d***), 85
Trees, Trees(M,τ), 86
TRUE, 13

U, $U_{\underline{A}}$, 224

v, v_A, 40
v, 112
V, 283
var, var(Φ), 113

Wsuba, **Wsuba**($\underline{A}$), 45
WRsubalg(M,$\underline{A}$,K), 271

X, 95, 112

... starting with another symbol

1, 21
$\cup$, A $\cup$ B, 281
$\dot\cup$, A $\dot\cup$ B, 12, 282
$\dot\cup_M L_m$, 200
$\cup$, **$\dot\cup$**, **$\cup$**(A_t|t $\in$ T), 282
$\cup$*, **$\cup$***($\underline{A}_t$|t $\in$ T), 48, 282
$\cap$, A $\cap$ B, 281
$\cap$, **$\cap$**(A_t|t $\in$ T), 282
$\cap$*, $\cap$*($\underline{A}_t$|t $\in$ T), 48, 282
$\subset$, $\subseteq$, $\not\subseteq$, A $\subseteq$ B, 281
$\setminus$, A $\setminus$ B, 281
x, A x B, A x...x A, 18, 42, **56**, **282**
A\` x B\`, 42
f x f', 284
$f^{(2)}$, $f^{(T)}$, 284
X, **X**(A_i|i $\in$ I), **X**($\underline{A}_i$|i $\in$ I), **56**, 282
X, **X**(f_i|i $\in$ I), 57
X^k, **X^k**($\underline{A}_i$|i $\in$ I)/**F**, 70
X($\underline{A}_i$|i $\in$ I)/**F**, 70
/, A/R, $\underline{A}$/R, 34
/, $\underline{A}$/F, a/F, $\underline{a}$/F, 70
\`, A\`:= $(A_s)_{s\in S}$, 22; redefined: A\`:= (A,v_A) , 40
^, **46**, 105, 214
^, $\underline{A}$^:= $\underline{F}$($\underline{A}$;**TAlg**($\underline{S}$)), 217
_\`, $\underline{A}$\`:= (A\`,$(\phi^{\underline{A}`})_{\phi\in\Omega}$) , 22; redefined: $\underline{A}$\`:= ($\underline{A}$,v_A) , 40
_, $\underline{a}$, (k$\underline{a}$|k $\in$ K), ((k)$\underline{a}$|k $\in$ K), $(a_k)_{k\in K}$ 18
_, $\underline{A}$:= (A,$(\phi^{\underline{A}})_{\phi\in\Omega}$), 19
$\underline{c}$, 21
_, $\underline{a}$:= (a_k|k $\in$ K_ϕ) , 56
(.,.), (a,b), 282
(.|.), (a_k|k $\in$ K), 18
$(a_k)_{k\in K}$, (a_0,...,a_{n-1}), 18
(.:.), (M:t), (v:t), **84**, 86
(M:t $\triangleq$ t'), 96
$(M:t)^{\underline{A}}$, 86
(.:.|.), (M:t|D), 146
$(M:t|D)^{\underline{A}}$, 146
(.,(.|.)), (m,(m_i|i $\in$ I)), 152
(_,.), ($\underline{E}$,R), 271
$\langle . | . \rangle$, $\langle g_i | i \in I \rangle$, 57
$\langle .,. \rangle$, $\langle \underline{A}, P \rangle$, 146
$\triangleq$, (M:t $\triangleq$ t'), **96**, 112

$\models$, $\underline{A} \models (M:t \stackrel{e}{=} t')[w]$,
$\underline{A} \models (M:t \stackrel{e}{=} t')$,
$\underline{A} \models G$, 96
$\underline{A} \models (M:\Phi)[w]$,
$\underline{A} \models (M:\Phi)$, 115
$K \models F$, 118
$\underline{A} \models c$ (for cones), 163
$\vdash$, $G \vdash (M:t \stackrel{e}{=} t')$, 107
$\neg$, $\neg\Phi$, $(M:\neg\Phi)$, 112, **115**
$\wedge$, $(N:(\Phi \wedge \Psi))$, 112, **113**
$\wedge^*$, $(N:\wedge^*\{\Phi|(N:\Phi) \in F\})$, 112, **115**
$\vee$, $(N:(\Phi \vee \Psi))$, 112, **113**
$\vee^*$, $(N:\vee^*\{\Phi|(N:\Phi) \in F\})$, 112, **113**
$\Rightarrow$, $(N:(\Phi \Rightarrow \Psi))$, 112
$\Leftrightarrow$, $(N:(\Phi \Leftrightarrow \Psi))$, 112
$\forall$, $(M\setminus\{x\}:\forall x\Phi)$, 112
$\forall^*$, $(M \setminus Y:\forall^* Y\Phi)$, 112, **115**
$\exists$, $(M\setminus\{x\}:\exists x\Phi)$, 112, **113**
$\exists^*$, $(M \setminus Y:\exists^* Y\Phi)$, 112, **113**
$[.].$, $[a]R$, 33,
$[B]R$, 283
$[B]f$, 26, **284**
$[B]R$ (saturation), 54
$\{.|.\}$, $\{x|P(x)\}$, $\{x\in A|H(x)\}$, 281
$\#$, $\#A$, 18, **289**
$\cdot$, $f\cdot g = fg$, 21, 25, **283**
$\underline{g}\cdot h$, 200
$\circ$, $R\circ R'$, 32
$\underline{g}\circ h$, 200
$\circ$, $\underline{A}^\circ$, 219
$*$, $\underline{g}*h$, 200
$*$, $R*E$, 261
$*$, f^*, 267, **269**
$t^{*\underline{A}}$, 274
$+$, $\underline{A} + \underline{B}$, 63
$^+$, n^+, $(\#S)^+$, 23, **289**, 290
$\overset{k}{\blacksquare}$, 162
$-t$, $K(S'-t)$, 228
$\cong$, $\underline{A} \cong \underline{B}$, 25
∇, $\nabla_A:= \{(a,b)|a,b \in A\}$, 30; redefined: 43 $(:= \ker \nu_A)$
$\sim$, $\tilde{f}$, $(id_A)^\sim$, $(\tilde{f})_{\underline{B}}$, **82**, 269
$(\tilde{f})_{\underline{A},\underline{B}}$, $(g\cdot h)^\sim$, 83
$\sim$, $\underline{B} \sim \underline{C}$, 217
$\rightarrow$, $\underline{a}:K \rightarrow A$, 18
$f:A \rightarrow B$, 19, **283**
$f:\underline{A} \rightarrow \underline{B}$, 24
$f:\underline{A}` \rightarrow \underline{B}`$, 24
-1, $f^{-1}:B \rightarrow A$, 26, **284**
R^{-1} (converse of a relation), 283
$\dashrightarrow$, $\phi:A^K \dashrightarrow A$, 19
$g:A \dashrightarrow B$, 283
$\mapsto$, $f:a \mapsto b$, 283
$\uparrow$, $\uparrow M$, 69
$\downarrow$, $\downarrow M$, $\downarrow m$, $\underline{\downarrow M}$, 83
$\downarrow G$ (G a set of equations), 106
$\downarrow\Phi$ (for a formula), **113** 114
∞, **162**, **265**
$<$, $\leq$, $\prec$, $\preccurlyeq$, $\geq$, 291
$\lhd$, $\lhd_{\underline{A}}$, $a \lhd_{\underline{A}} a'$, 83
$<$, $<_{\underline{A}}$, $a <_{\underline{A}} a'$, 83
$\leq$, $\leq_{\underline{A}}$, $a \leq_{\underline{A}} a'$, 83
$>$, $>_{\underline{A}}$, $\underline{B} >_{\underline{A}} \underline{C}$, 217
$\prec$, $\overline{(\bar{M},f,\underline{A})} \prec (M,f',\underline{A}')$, 274

... starting with Gothic letters

$\mathfrak{a}$ (see also $\underline{a}$), 56
$\mathfrak{A}$, $\mathfrak{A}_I$, 64
$\mathfrak{A}_F$, 70

... starting with Greek letters

$\delta:= \delta(\tau)$, 23
Δ, $\Delta_A:= \{(a,a)|a \in A\}$, 30
Δ_T, Δ_M, Δ_f, 190
$\in$, $\notin$, 112, **281**
Λ, $\Lambda(M)$, Λ^{op}, $\Lambda^{op}(M)$, 128
$\Lambda^{op}(S)^k_k$, 169
σ, $\sigma_{\underline{A}}(a,M)$, $\sigma(a,M)$, 50
$\Sigma = (\tau,\mathrm{INPUT},\mathrm{OUTPUT})$, 22
Σ, $\Sigma(\underline{A}_i|i \in I)$, 63
Σ, $\Sigma(L_m|m \in M)$, $\Sigma(\alpha_\zeta|\zeta < \xi)$, 288
$\Sigma_M L_m$, 200
$\tau = (K_\phi)_{\phi\in\Omega}$, 19, 22
τ_M, 236
$\tau(S')$, 228
$\tau = (K_1,\ldots,K_n)$, 20
$(\tau,\mathrm{INPUT},\mathrm{OUTPUT})$, 24
ϕ, 18, 112
$\phi^{\underline{A}}$, **19**, 41
$\phi^{\underline{A}`}$, **22**, 41
ϕ_R, 18, **183**
$\phi^{\underline{A}}:A^{K_\phi} \dashrightarrow A$, 19
$\emptyset$, **18**, **285**
Ω, 19, 22
$\Omega = (\phi_1,\ldots,\phi_n)$, 20
(S'), 228
$\omega = \omega_o$, 287
$\omega(n)$, $\omega(d^{\&})$, 200, **290**

Further notation

$\underline{A}_A$, $\underline{A}_M$, 236, 247
$\underline{A}_d$, 19
A^n, A^K, $A^{\{k\}}$, 18, **42**, **56**, 283
A_s, 40
$\underline{A}_F$, 70
$\underline{A}_X:= \underline{A} + \underline{X}_d$, 88
F, (e.g. F_c), 69
F (redefined), 162
F_P, 133
$f^{(K)}$, $f^{(K \cup \{K\})}$, 25, 284
G_M, 104
$h_{p,N}$, 204
M_t, 84
p, $p_{FG}:\underline{A}_F \rightarrow \underline{A}_G$, 70
p_f, 270
R_B, 54
xM, 239
$.^z$, $\underline{A}^z$, 224

INDEX

Compound expressions are not always listed by each possible keyword. If there is more than one reference to some notion, then the bold-faced one refers to the place of definition (see also the list of Contents for reference to the application of some item). Page numbers in brackets mean that this page contains an important contribution to this item without containing the keyword explicitly.

absorption laws, 293
ADA, 5
additively accessible cardinal, 23, **290**
additively inaccessible cardinal, 23, **290**
ADJ-group, 13, 216, 276, **280**
admissible (mapping) w.r.t. substitution, 114
algebra, 19 (see also "partial algebra")
- of global partial polynomials, 88
canonical term -, 236
closed sub-, 45
complete -, 19
conclusion -, 123
conform-free **K**--, 280
error -, 221
factor -, 34
free **K**-- presented by, 234
global term -, 96
heterogeneous -, 12, **22**, **41**
heterogeneous (partial) $\underline{S}$--, 40
heterogeneous sort -, 40
homogeneous partial -, 19
homogeneous total -, 19
initial **K**--, 95
K-free **K**--, 76
K-freely generated **K**--, 76
k-small -, 156
many-sorted -, 12
ω-continuous -, 15, 184
ω-continuously ordered -, 184
partial -, 12, **19**, **43**
partial $\underline{S}$--, 40
partially ordered -, 184
partial Peano - on X, 77
Peano -, 77
premise -, 123
quotient -, 34
relative sub-, 45
Σ--, 22
sort -, 40
strongly **k**-small -, 156
sub-, 45
term -, 77
canonical --, 236
total -, **19**, 26, 47, (53), 183, 277
total $\underline{S}$--, **41**, 183
universal -, 19
weak (relative) sub-, 45
weak sub- induced by a monomorphism, 45
word -, 77
algebraic closure operator, 31, 293, 296
algebraic independence, **222**, 239, 242
algebraic induction, 51
algebraic lattice, 49, **293**
algebraic operator, 123
algebraic partial order, 83
algebraic quasi-order, 83
algebroidal
k-dual -, 159
strongly **k**--, 156
amalgamation, 216
antimonotone, 297
arity, 18
assignment, 86, **96**, 115, **283**
local -, 96
associativity, 292
- of composition, 21
atomic formula, 112
automaton
Mealy(-type) -, 12, 33, 146
stack -, 101
automorphism, 26, **27**
- group, 28
identity -, 21
Axiom
- of Choice, 256, **285**
- of Comprehension, 284
- of Extensionality, 285
- of Foundation, **285**, 286
- of Induction, 77
- of Infinity, 285
- of Replacement, 285
- of Subsets, 285
- of the Binary Union, 285
- of the Nullset, 285
- of the One Element Set, 285
- of Union, 285
generalized Peano -, 77
Inverse of the - of Union, 285
Regularity -, 285
axiomatizable class, 119

Baire class, 50
Baire operator, 50
basic formula, 113
basis, 76
- (of a free partial algebra), 76
- of the induction, 51
element -, 239
-- property, **239**, 267, 272
Peano -, **77**, 239
bicategory, 140
bimorphism, **39**, 145
body of a formula, 113
Borel class, 50
Borel operator, 50
bound
greatest lower -, 292
least upper -, 292
lower -, 292
upper -, 292

cancellable, left -, 138
right -, 138
canonical term, 81
-- algebra, 236
cardinal, 281, **289**
additively accessible -, 23, **290**
additively inaccessible -, 23, **290**
- dimension, 23
exponentiation of -s, 290
power of -s, 290
product of -s, 290
regular -, 290
singular -, 290
sum of -s, 290
cardinality, 18, **289**
carrier (set), 19
cartesian product, 56
category, 14, 21
abstract -, 21
bi-, 140
- of pairs, 279
comma -, 41
concrete -, 256
co-(well-powered) -, 91
full epireflective sub-, 137, 277
small -, 14, **21**, 44, 102, **120**
strongly **k**-algebroidal -, 156
sub-
epireflective --, **145**, 277
full --, 187,
full epireflective --, 137, 277
monoreflective --, 145
reflective --, 145
well-powered -, 142
chain, 291
characteristic, (A--), 93
choice function, 56
class, 281
axiomatizable -, 119
compatible pair of -es of morphisms, 171
difference -, 281
equationally definable -, 98
independence -, 76
injectivity -, 76
model -, 96, 119
power--, 282
primitive -, **72**, 76
quasi-primitive -, **72**, 76
reflective sub-, 145
transitive -, 286
well-founded -, 287
clone, **207**, 209
closed
A-initial - congruence, 91
- congruence, **30**, 42
- homomorphic image, 37
- homomorphism, **28**, 29, 35, **42**, 89, 90, 130, 215
- initial extension, 82
- S-congruence, 42
- subalgebra, 45
- subset, **45**, 46, 49
- w.r.t. pullbacks, 138
- w.r.t. pushouts, 138
- w.r.t. superposition, 207
left--, 297
P-- congruence, 252,
weakly ----, 253
relatively - homomorphism, 133
right--, 297
closure operator, 31, (49), 128, **294**, 295
(**k**-) algebraic --, 31, (**293**), **296**
(**k**-) inductive --, 31, (**293**), **296**
many-step --, 50
closure system, 31, 49, **294**, 295
c-inductive -, 49, (**293**)
cocone, colimiting -, 65
codomain of a morphism, 21
cofinal morphism between directed systems, 153
cofinal subset of a poset, 153, **288**
colimit, 62
directed -, 62, **64**
colimiting cocone, 65
commutative, 292
compact (**k**--), 293
equationally -, 260, 280
compactification, Stone-Čech--, 138
compactly generated (complete) lattice, 49, **293**
compatible
- classes of morphisms, 171
- family of morphisms, 65
complete
- (K-ary) operation, 18
- lattice, 292
- (universal) algebra, 19
k-- filter, 69
completeness degree, **93**, 259
completeness of a system of rules, 109
completion, 13, **217**
(absolutely) free -, 80
- greater than, 217
(heterogeneous) normal one-point -, 219
homogeneous -, 217
inner -, 218, **224**
inner one-point -, 27
minimal -, 80, **217**
normal -, **217**, 220
weak -, 217
z--, **224**, 272
component, connected -, 264
composition, 18
- of functions, 18
- of homomorphisms, 25
- of morphisms, 21
- of partial mappings, 283
computational semantics, 280
computer, stored program -, 266
conditionally complete, 292
cone, 162
(e--, H--, $P^{r,k}$--, $P_+^{r,k}$--, P_u--, S--, 162)
conform-free, 280
conformic partial homomorphism, 278
conformic quomorphism, **278**, 279
conformism, 270, **277**, 279
conformism-freely generated, 270
conglomerate, **284**, 298
congruence (relation), **30**, 42, 46
(A-) initial closed -, 91
closed -, **30**, 42
closed S--, 42
- generated by a set, 31
fully invariant -, **103**, 275

congruence (continued)
(M-) initial closed **K**--, 271
P-closed -, 252
principal -, 248
S--, 42
semigroup -, 34
weakly P-closed -, 253
conjunction
binary -, 113
c-restricted -, 113, **115**
connected component, 264
connective, sentential -, 112
constant, 18
- partial operation, 240
extended fundamental -, 203
individual -, 113
nullary -, 12, **18**
converse (of a relation), 283
coproduct, **61**, 139, 216
- of partial algebras, 62
co-(well-powered), 91, **142**

data type, 216
abstract --, 216
dense, 52
dependence (on a variable), 240
derivable (E-equation), 107
derivation, rules of -, 107
diagonal-fill-in, 127
- property, 127
diagram
commutative -, 24
Hasse -, 34
positive -, 238
difference class, 281
dimension,
cardinal -, 23
infinite --, 23
heterogeneous ---, 23
ordinal -, **23**, 50
directed
- colimit, 62, **65**
- partially ordered set (- poset), 64, **293**, (**k**-----, 293)
- system, (**k**---), 64, **65**
direct limit, (**k**---), 62, **65**
direct product, 56
discrete (K-ary) partial operation, 18
discrete partial algebra, 19
discrete partial operation, 18
disjunction
binary -, 113
c-restricted -, 113
domain, 282
category theoretical -, 21, **283**
- of a morphism, 21
- of a partial operation, 18
- of a relation, 282
exact - of a partial mapping, 283

edge, 264
directed -, 85
element
completely supremum prime -, 294
(**k**-) compact -, 293
- basis, 239
-- property, **239**, 267, 272
largest -, 292
maximal -, 292
minimal -, 292
smallest -, 292
elementary embedding, 135
elementary implication, **97**, 121
embeddability, 186
- problem, 216
embedding
elementary -, 135
(L-) pure -, 135
- morphism, 62
empty (K-ary) partial operation, 18
empty partial algebra, 62
empty sequence, 18
empty set, 281
end, principal -, 38
endomorphism, 27
- monoid, 27
epimorphism, **39**, 52, 67, 89, 154, 155, 263
encoding -, 123
- in the sense of R.S.Pierce, 40
extendable -, **75**, 90, 268, 278
strong -, **142**, 143
equalizer, **52**, 70
equation, 12, **96**
E-- (existence--), 13, **96**, 276
ECE-- (existentially conditioned E--), 14, 102, **121**
Evans' -, 122
existence--, E--, 13, **96**, 121, 276
small ---, 96
NE-- (negation of an E--), 121
QE-- (quasi-existence -), 121
regular -, **121**, 225
strong -, 122
weak -, **122**, 279
equationally definable class, 98
equational compactness, 260
equational theory, **96**, 97
equipotent, 289
equivalence, (logical -) 113
- class, 291
- of completions, 217
- of partial operations, 201
- of sequences, 261
- of subobjects, 45
- relation, 291
equoid, hierarchical -, 39
error algebra, 221
error message, 216
error value, 13, 216, 221
evaluable term, **86**, 87
evaluation, **96**
E-variety, 98
existential quantification, 113
expression
NTE-- (negation of a TE--), 121
TE-- (term existence -), 116, **121**
extendable epimorphism, **90**, 268
extendable morphism, **K**--, **75**, 268
extension
(A-) initial, 82
inner -, 224
normal -, 134
z--, 224
extensive, 294

factor
-algebra, 34
left -, 140
right -, 140
factorization, 140
- system, 128, **140**
family
natural -, 189
strongly --, 189
separating -, 58
field, 15, 20 **186**
- (of a binary relation), 282
filter, 69
generalized Fréchet -, 69
k-complete -, 69
principal -, 69
proper -, 69
ultra-, 69
final homomorphism, **132**, 136, 171
finitary partial algebra, 19
finitary (similarity) type, 19
fixed point, 52
forest, split -, 253
formula, 112
atomic -, 112
basic -, 113
body of a -, 113
ECH-- (existentially conditioned universal Horn--), 121
Horn-- (H--), 121
open -, 119
quantifier free -, 114
SECH-- (special ECH--), 121
universal -, 119
Fréchet-Urysohn limit space, 246
free, **K**--, 76
absolutely - completion, 80
- completion, 80
- generation, 76
- group, 75
- partial algebra, **K**----, 76
relatively ---, 76
- sum, 61
full (epireflective) subcategory, 137
full homomorphic image, 37
full homomorphism, **28**, 29, **42**, 136, 157
full (K-ary) operation, 18
fully invariant closure system, 193
fully invariant congruence, **103**, 275
function, 283
algebraic (partial) -, **214**, 248
unary ---, 248
choice -, 56, 285
output -, 12
partial polynomial -, 276
polynomial -, 214
rank -, 83 (set theoretical --, 285)
transition -, 12
functor, 14, **44**
faithful -, 256
full -, 256
inclusion -, 187
left adjoint -, **145**, 187
fundamental constant, 18
extended --, 203
fundamental (partial) operation, 19
Galois connection, 119, 128, 163, 254, **297**
- correspondence, 128, 163, **297**
generalization
- of one variable, 113
q-restricted -, 113, **116**
generating subset, 44, **49**
generation (w.r.t. a closure operator), 295
free -, 76
- of a subalgebra, 49
- of a closed subset, 49
- of a congruence, 31
generator, 233
set of -s, 49
global, 221
global term, 95
-- algebra, 95
graph, 13, 18, **283**
group, 20, 46, 233, 241, 274
abelian -, 63
free -, 75

Hasse diagram, 34
hereditaey, 127
heterogeneous, 22
- infinite cardinal dimension, 23
- partial algebra, 22
- similarity type, 22
- term, 84
- total algebra, 22
hierarchical equoid, 39
hierarchy condition, 184
homogeneous, 19
- partial algebra, 19
- total algebra, 19
homomorphic extension, initial --, 268
homomorphic image, 37
closed --, **37**, 71
full --, **37**, 71
strong --, 37
weak --, **37**, 71
homomorphism, 13, 15, **24**, 129, 277
bijective -, 26
closed -, **28**, 29, 35, **42**, 53, 67, 89, 90, 130, 215
composition of -s, 25
conformic partial -, 278
final -, **132**, 136, 150, 171
full -, **28**, 29, 35, 42, 67, 89, 136, 154, 155, 157
group -, 35
- (from A into B), 24
- between heterogeneous partial algebras, 24
- between (partial) S-algebras, 41
- between total algebras, 25
initial -, 132
injective -, 67, 89, 135
K-perfect -, **145**, 255, 268, 278
monoid -, 35
partial -, 278
conformic --, 278
P-closed -, 133
P-initial -, **134**, 150
P-normal -, **134**, 150
P-reflecting -, 133

homomorphism (continued)
  quasi--, 277
   conformic ---, 278
  relatively closed -, **133**, 157
  relatively normal -, **134**, 150
  relatively P-reflecting -, 157
  semiclosed -, 53
  semigroup -, 34
  S--, 41
  strong -, 28
  surjective -, 67, 89, 154, 155, 263

ideal, principal -, 251
idempotent, **292**, 294
identity, 12
  - automorphism, 21
  - mapping, 25
iff = if and only if
image, 284
  closed homomorphic -, **37**, 71
  full homomorphic -, **37**, 71
  pre-, 284
  weak homomorphic -, **37**, 71
implication, 13, 113
  elementary -, 13, **97**, 121, 123
independence of an operation from a set
   of variables, 240
independence class, 76
independent, **K**--, (B--), **76**, 210
  algebraically -, **222**, 239, 242
induced weak subalgebra, 45
induction
  algebraic -, 51
  basis of the -, 51
  - hypothesis, 51
  - step, 51
  principles of transfinite -, 287
  transfinite -, 51
inductive, **K**--, 293
  c-- closure system, 31, 49, (**293**)
  - limit, 65
  - set, 256
  k-- class of morphisms, 159
infimum, 292
  - of a set of M-initial congruences,
   271
  - of a set of substructures, 48
infinitary partial algebra, 14, 19
infinitary (similarity) type, 19
infix notation, 18, **20**
initial
  closed - homomorphic extension, 82
  - closed congruence, 91
  - closed **K**-congruence, 271
  - extension, **82**, 268
  - homomorphism, 132
  - **K**-algebra, 95
  - object, 12, 62, **235**
  - ordinal, 290
  - segment, 31, **83**, **91**
  - structure, 57
injection, 62
injective homomorphism, 67, 89, 135
injective mapping, 130
injective w.r.t. a cone, 163
injective w.r.t. a morphism, 75
injectivity class, 76
input element, 12
input set, 33
integer, 73, 275
interpretation, 96, **115**
  - corresponding to a valuation, 115
intersection, 281
  - of a set, 282
  - of weak substructures, 48
inverse (morphism), 25
irreducible
  completely meet -, 256
  subdirectly -, 255
isomorphic, 25
  - copy, 71
isomorphism, **25**, 26, 35, 67, 89, 158
  canonical -, 21
  identity -, 21
  - between partial algebras, 26
  - type, 27

join, 281
  - of a set, 282

kernel, 30
  J.Schmidt-- (S--), 91
Konformismus, 270

label, 85
lattice, 34, 251, **292**
  algebraic -, 251
  c-algebraic -, 49, **293**
  c-compactly generated -, 49, **293**
  complete -, 193, **292**
   (**k**-) algebraic --, 293
   k-compactly generated --, 293
leaf, 85
left zero, 62
Lemma
  Diagram Completion -, 35
   --- for Epimorphisms, 89
  - of Kuratowski-Zorn, 285
  Łos' -, 119, 161
  Tacking -, 68
  Zorn's -, 285
limit, 62
  direct -, 65
  inductive -, 65
  - space, 246
local assignment, 96
local term, 96
local valuation, 96

many-sorted algebra, 22
many-sorted partial algebra, 22
many-sorted similarity type, 22
mapping
  admissible -, 114
  antimonotone -, 297
  bijective -, 283
  composition of partial -s, 283
  identity -, 25,**283**
  injective -, 130
  injective partial -, 283
  inverse of a -, 284
  linear -, 14
  - from, 283
  - onto, 283
  partial - out of, 283
  partial algebraic -, **277**, 279

mapping (continued)
  restriction -, 70
  sort -, 40
  surjective partial -, 283
  underlying -, 256
matrix, 14
maximal, 292
maximum, 292
Mealy-type automaton, 12, 33, 146
meet, **281**, 282
minimal, 292
  - completion, 80
minimum, 292
model (of a system of E-equations), 96
  local - theory, 272
  - class, 96
monoid
  commutative - with partial inverse, 185
  - homomorphism, 35
monomorphism, **39**, 89
  strong -, **142**, 143
monotone, 294
morphism (of a category), 14, **21**
  bi-, 145
  cofinal -, 153
  colimit -, 65
  compatible classes of -s, 171
  composition of -s, 21
    --- between directed systems, 153
  embedding -, 62
  epi-, 39 (see "epimorphism")
  inclusion -, 153
  induced colimit -, 65
  injectivity w.r.t. a -, 75
  k-dual algebroidal class of -s, 159
  K-extendable -, 75
  k-inductive class of -s, 159
  machine -, 135
  mono-, 39 (see "monomorphism")
  - between **k**-directed systems, 153
    cofinal ----, 153
    inclusion ----, 153
    special ----, 153
  p--, 270, **277**
  product -, 56
multiple pullback, 59
  -- object, 61
multiple pushout, 64

natural family, 188
  strongly --, 188
natural number, 78, 101, **287**
natural projection, 34
negation, 113, **115**
normal
  - extension, 134
  - form, 276
  - one point completion, 219
  - relative subalgebra, 80
  - subgroup, 39
  relatively - homomorphism, **134**, 150
nullary constant, 12, **18**
nullary partial operation, 18
number
  natural -, 78, 101, **287**
  rational -, 73
  step -, 50

object (of a category), 21
  enough **H**-projective -s, 163
  initial -, 12, 62, **235**
  k-small -, 155
    strongly ---, 155
  left zero -, 62
  multiple pullback -, 61
  right zero -, 56
  start -, 21, **283**
  subdirectly irreducible -, 255/6
  sub-, 45
  target -, 21, **283**
  terminal -, 56
operation (see also "partial operation")
  algebraic - of the first kind, 203
    -- of the second kind, 203
  algebraic total -, **211**, 215
  binary -,14
  complete -, 18
  conditioned term - (c-term -), 146
  full algebraic -, 211
  full -, 18
  fundamental (partial) -, 19
  global term -, 86
  K-ary partial -, 18
  - symbol, 19
  partial - of type (arity), 18
  partial term -, 86
  term -, 253
  total -, 18
  total term -, 102
  unary -, 14, 20
operator, 283
  Baire -, 50
  Borel -, 50
  closure -, 51, 128, **294**, 295
  many-step closure -, 50
  mixed step -, 228
  preserved by some algebraic -, 122
  step--, 50
order, 291
  dual -, 291
  linear -, 291
  partial -, 15, **291**
  quasi--, 291
  total -, 291
  well--, 292
ordinal, 20, 200, 281, **286**
  addition of -s, 288
  cofinal -, 288
  initial -, 290
  limit -, 287
   - dimension, **23**, 50
  regular -, 23, **288**
  singular -, 288
  successor -, 287
  sum of -s, 288
output, 12
  - element, 12
  - function, 12
  - set, 33

pair
  category of -s, 279
  Kuratowski -, 282
  ordered -, 282
  - of homomorphisms, 279
Part II, 5, 304

partial algebra, 12, **19**, 43
  discrete --, 19
  empty --, 19
  finitary --, 19
  free --, 76
  homogeneous --, 19
  heterogeneous --, 22
  homogeneous --, 19
  infinitary --, 14, 19
  k-small, 155
  many-sorted --, 22, (40)
  -- of premise terms, 123
  relatively free --, 76
  strongly **k**-small --, 155
partial function, algebraic --, **214**, 248
  unary ---, 248
partially ordered set, see "poset"
partial mapping, 18
  -- out of, 18, **283**
partial operation, **18**, 198
  algebraic -- of the first kind, 203
  algebraic -- of the second kind, 203
  algebraic -- of the third kind, 207
  constant --, 240
  discrete --, 18
  empty --, 18
  eqivalence of --s, 201
  fundamental --, 19
    heterogeneous ---, 22
  heterogeneous --, 22
  K-algebraic --, 274
  K-ary --, 18
    --- with output sort s, 199
  nullary --, 18
  -- of type (arity) K, 18
  relativized algebraic --, 274
  unary --, 19
partial M-ary term operation, 86
partial order, 15
  algebraic --, 83
  k-directed -, 64
partial Peano algebra, 77
partial S-algebra, 40
PASCAL, 5, 209, 216
path, 85
Peano algebra, 77
  partial --, 77
  - basis, **77**, 239
Peano axioms, generalized --, 77
perfect homomorphism (**K**---), **145**, 255, 268
phylum, **22**, 40
polynomial (in rings), 89
  (global partial) - with coefficients in A, 88
  M-ary - function, 214
  partial - function, 276
  relativized - function, 276
poorer structure, 48
poset, 183, **291**
  directed -, 293
  k-directed -, 293
  ω-continuous -, 183
  ω--, 14
positive part of a set of Horn-formulas, 196
power, n-th -, 290
power-class, 282
predecessor (of an element), 83
  immediate -, 83
preimage, 284
premise algebra, 123
premise term, 123
presentation of a **K**-free partial algebra, 234
preservation of a formula by a mapping, 129
preservation of a set of formulas by an operator, 123
prime, completely supremum -, 294
primitive class, **72**, 76
  quasi---, **72**, 76
principal ideal, 251
problem, (119), 134, 150, 156, 174, 176, 180, 186(2), 190, 213, 216, 236, 238, 243, 260, (272), 276
product, **55**, 139
  cartesian -, 56, **282**
  co-, 61
  direct -, 56, **282**
  free -, 216
  k-reduced -, 70
  - morphism, 56
  - of cardinals, 290
  - of partial algebras, 56
  - of partial S-algebras, 61
  - of relations, 32
  ultra-, 70
projection, 37, **56**, 87
  i-th -, 56
  M-ary -, **87**, 213
  natural -, 34
  partial K-ary -, 18
projective (**M**--), 127, **128**
  enough **H**-- objects, 163
pullback, **59**, 138, 279
  multiple -, **59**, 138
    -- object, 61
pure embedding, 135
pushout, **64**, 138
  multiple -, **64**, 139

quantification
  existential - with one variable, 113
  q-restricted existential -, 113
quantifier, 112
  hidden existential -, 116
  --free, 114
quantum mechanics, 5, 279
quasi-homomorphism, 277
quasi-order, algebraic --, 83
quasi-primitive class, 72
quomorphism, **277**, 279
quotient algebra, 34
quotient set, 34

range (of a relation), 282
rank (of a similarity type), **23**, 92
  - function, **50**, 83
    (set theoretical) --, 285
  - with respect to a subset, 50
recursion, general - principle, 287
reduced product, **k**---, 70
reduct
  Ω'--, 62
  S'--, 228

reduction, 33
S'--, 228
reflection of a formula by a mapping, 129
weak -------, 136
reflective subclass, 145
regular cardinal, 23, **290**
regular equation, **221**, 225
regular ordinal, 23, **288**
relation, 13, 15, 19, **282/3**, 291
- (with different meaning), 233
antisymmetric -, 291
binary -, **282**, 283
congruence -, 30
closed --, 30
equivalence -, 291
irreflexive -, 291
symmetric -, 291
T-ary -, 283
well-founded -, 287
relational system, 15, 19, 183, **277**
relative subalgebra, 45
normal --, 80
residually small variety, **255**, 280
restriction mapping, 70
retract, 158
retractable morphism, 127
retraction, 158
- of a formula by a mapping, 136
richer structure, 48
ring, 73, 75, 89, 274
commutative -, 275
polynomial -, 75, 89
root, 85
rules of derivation, 107

satisfaction (of an E-equation), 96
- of a formula, 115
saturation w.r.t. a relation, 54
section, 158
segment, initial -, 31, **83**
A---, 91
semigroup congruence, 34
semigroup homomorphism, 34
semilattice, 34
complete inf--, 292
complete sup--, 292
complete upper -, 292
conditionally complete -, 219, **292**
join -, 292
lower -, 292
meet -, 292
upper -, 292
sentential connective, 112
separating family, 58
separation of a set, 58
sequence, **18**, 281
corresponding -, 70
empty -, 18
K-ary -, 18
- of (arity) type K, 18
undefined -, 224
set, 281, **284**
algebraically independent -, **222**, 239, 242
carrier - (of a partial algebra), 19
countable -, 290
denumerable -, 290
empty -, 281
finite -, 290
inductive - of congruences, 256
infinite -, 290
left-closed -, 297
partially ordered -, 291
quotient -, 33
- of all partial operations, 200
- of all total operations, 200
- of all total projection operations, 200
- of free variables, 113
- of generators, 49
- of natural numbers, 287
- of operation symbols, 19
- of **k**-small subsets, 190
- of sorts, 22
- of subformulas, 113
- of subterms, 113
- of terms, 112
- of variables, 86, **113**
small - (of variables), 96
totally ordered -, 291
uncountable -, 290
well-ordered -, 292
S-congruence, 42
closed -, 42
S-homomorphism, 41
closed -, 42
full -, 42
similar, 19
similarity type, **19**, 46
finitary --, 19
hetrogeneous --, 22
infinitary --, 19
many-sorted --, 22
sink, 59
small
k-- partial algebra, 155
strongly -----, 155
k-- total algebra, 156
strongly -----, 156
residually -, 255, 280
- category, **14**, 21, 44, 102, **120**
- E-equation, 96
- set of variables, 96
- subset, 190
solution (of an E-equation), 96
sort, 22
phylum of - s, 22
- algebra, 40
heterogeneous --, 40
- homomorphism, 40
- mapping, 40
soundness, 109
source, 55
species (of partial algebras of type), 19
split forest, 253
stack automaton, 101
stack of integers, 12
start object, 21, **283**
state, 33
internal -, 12
step-number with respect to a subset, 50
step-operator with respect to a subset, 50
Stone-Čech compactification, 138
strong epimorphism, **142**,143

strong homomorphism, 28
strong monomorphism, 142
structure, initial-, 57
subalgebra, **45**, 49
closed -, 45
relative -, 45
normal --, 80
relatively normal --, 211
- generated by a set, 49
weak (relative) -, 45
subclass, reflective -, 145
subdirectly irreducible, 255
subdirect product, 255
subdirect representation, 255
subset,
algebraically independent -, 242
closed -, **45**, 46, 49
cofinal -, 288
generating -, 49
(**k**-) directed -, 293
K-free -, 76
K-independent -, 76
(**k**-) inductive -, 293
k-small -, 190
substitution, 105, **114**
generalized -, 114
simultaneous -, 114
uncritical -, 114
substructure
poorer -, 48
richer -, 48
weaker -, 48
subterm, 83
M--, 86
sum
free -, 61
- of cardinals, 290
- of ordinals, 288
superposition, 87
- of the first kind, **200**, 206
ordinal -----, 200
set theoretical -----, 200
- of the second kind, **200**, 206
supremum, 14, **292**
- of substructures, 48
symbol, operation -, 19
system
closure -, 193, **294**, 295
fully invariant --, 193
factorization -, 128, **140**
iterative -, 266
relational -, 19, 183, 277

target (object), 21, **283**
TE-expression, 97, 116, **121**, 129
term, 12, **83**, 112
canonical -, 81
canonical - algebra, 236
conditioned - (c-term), 146
-- operation, 146
evaluable -, 86
global -, 95
global - algebra, 95
global - operation, 86
heterogeneous -, 84
local -, 96
M-sub-, 86
partial (M-ary) - operation, 86
sub-, 83
- algebra, 77
- existence expression (TE-expression), 97, 116, **121**, 129
negation of a ---, 121
- operation, 253
total (M-ary) - operation, 102
TE-statement, see "TE-expression"
Theorem
Cantor-Bernstein -, 289
Characterization - of Free Partial Algebras, 231
Coincidence -, 117
Compactness -, 119
Epimorphism -, 90
-- for Homogeneous E-Varieties, 270
Existence - for **K**-Universal **K**-Solutions, 94
First Birkhoff -, 100
First Homomorphism -, 36
First Isomorphism -, 54
General Homomorphism -, 90
Meta Birkhoff -, 164
Recursion -, **79**, 266
Generalized --, 82
Recursion - for Free Completions, 81
Second Birkhoff -, 104
Second Isomorphism -, 38
Skolem's -s, 119
- of Keisler and Shelah, 119
- of Soundness and Completeness, 109
- of Substitution, 117
--- for Validity, 118
- of Well-Ordering, 285
Ultraproduct -, 119, 161
theory, 119
equational -, 96
local model -, 272
topology, 137
total (universal) algebra, **19**, 47, (53), 277
total (K-ary) operation, 18
total $\underline{S}$-algebra, **41**, 183
transformation of variables, 204
transition function, 12
tree, 85
rooted, labeled, directed -, 85
tuple, n--, 18
type, 19
finitary -, 19
heterogeneous (similarity) -, 22
infinitary -, 19
isomorphism -, 27
many-sorted -, 22
similarity -, **19**, 46
heterogeneous --, 22
many-sorted --, 22
- of a sequence, 18

ultrafilter, 69
ultraproduct, 70
ultraroot, 119
unary partial operation, 20
union, 281
disjoint -, 282
universal, (**K**--), 75
K-- **K**-solution, 75

universal (continued)
- algebra, 19
- solution, K---, 75

validity, 13, **122**, 213, 280
Evans' -, 122
notion of -, 122, 279
strong -, **122**, 213
- of a cone, 163
- of a formula, 115
- of a morphism, 75
- of an E-equation, 96
weak -, **122**, 279
valuation, 86, **96**, **115**
admissible - (for substitution), 114
local -, 96
variable, 86, **95**
bound -, 114
global set of -s, 95
set of -s, 86
transformation of -s, 204
variety
E--, **98**, **121**, 126, 259
ECE--, 102, **121**, 126, 259, 260, 279
ECH--, **121**, 126
hep--, **121**, 184
Horn--, **121**, 126, 235
NE--, 121
NTE--, 121
residually small -, **255**, 280
SECH--, **121**, 126
TE--, **121**
vector space, 21, 50
vertex, 85, 264
end -, 85

weak homomorphic image, **37**, 71
weak reflection of a formula by a (homo)morphism, 136
weak (relative) subalgebra, 45
weaker structure, 34, **48**
well-ordered, 292
well-powered category, 142
word, 63, 79
- algebra, 77
- problem, 216
w.l.o.g. (without loss of generality)
w.r.t. (with respect to)

zero
left--, 62
right--, 56
z-completion, **224**,272
z-extension, 224

Band 1

Inverse and Improperly Posed Problems in Differential Equations

Proceedings of the Conference on Mathematical and Numerical Methods, held in Halle Saale (GDR) from May 29 to June 2, 1979

Edited by Gottfried Anger
296 Seiten, 1979 (vergriffen)

Band 2

Fundamentals of Computation Theory FCT '79

Proceedings of the Conference on Algebraic, Arithmetic and Categorial Methods in Computation Theory, held in Berlin/Wendisch-Rietz (GDR) September 17–21, 1979

Edited by Lothar Budach
575 Seiten, 1979

Band 3

Decidability and Generalized Quantifiers

By Andreas Baudisch, Detlef Seese, Hans-Peter Tuschik, and Martin Weese
XII, 235 Seiten, 1980

Band 4

Zur Elastizitäts- und Thermoelastizitätstheorie

Potentialtheoretische Methoden in Anwendung auf verschiedene Modelle

Von Lothar Jentsch und Johannes Maul
157 Seiten, 1980

Band 5

Systems Analysis and Simulation 1980

Proceedings of the International Symposium held at Berlin (GDR) September 1–5, 1980

Edited by Achim Sydow
496 Seiten, 1980 (vergriffen)

Band 6

Nichtlineare diskrete Optimierung

von Eberhard Girlich und Michail M. Kowaljow
218 Seiten, 1981

Band 7

Completeness Problems in the Structural Theory of Automata

By Jürgen Dassow
148 Seiten, 1981

Band 8

Darstellung von Lösungen linearer elliptischer Differentialgleichungen

Von Günther Wildenhain
92 Seiten, 1981

Band 9

Introduction to the Spectral Theory of Operators in Spaces with an Indefinite Metric

By I. S. Iohvidov, M. G. Krein, and H. Langer
120 Seiten, 1982 (vergriffen)

Band 10

Nonlinear Time-discrete Systems

A General Approach by Nonlinear Superposition

By Michael Gössel
115 Seiten, 1982 (vergriffen)

Band 11

Preprimal Algebras

By Klaus Denecke
162 Seiten, 1982

Band 12

Ergodic Theory and Related Topics

Proceedings of the Conference held in Vitte/Hiddensee (GDR) October 19–23, 1981

Edited by Horst Michel
232 Seiten, 1982

Band 13

Singularly Perturbed Differential Equations

By H. Goering, A. Felgenhauer, G. Lube, H.-G. Roos, and L. Tobiska
176 Seiten, 1983

Band 14

Leben wir in einer Volterra-Welt?

Ein ökologischer Zugang zur angewandten Systemanalyse

Von Manfred Peschel und Werner Mende
233 Seiten, 1983

Band 15

Markov Decision Problems with Countable State Spaces

Optimality Criteria – Algorithms – Clustering

By Hans Michael Dietz and Volker Nollau
174 Seiten, 1983

Band 16

Stochastische Verfahren zur Suche nach dem Minimum einer Funktion

Von Ryszard Zieliński und Peter Neumann
133 Seiten, 1983

Band 17

Invertibility and Asymptotics of Toeplitz Matrices

By Albrecht Böttcher and Bernd Silbermann
200 Seiten, 1983

Band 18

Structural Induction on Partial Algebras

Introduction to Theory and Application of Partial Algebras – Part II

By Horst Reichel
205 pages, 1984

Band 19

Algebraic Methods for Toeplitz-like Matrices and Operators

By Georg Heinig and Karla Rost
212 pages, 1984

Band 20

Frege Conference 1984

Proceedings of the International Conference held at Schwerin (GDR), September 10–14, 1984
Edited by Gerd Wechsung
408 pages, 1984

Band 21

Numerical Analysis of Selected Semilinear Differential Equations

Edited by Christian Großmann, Thomas Riedrich, and Hartmut Schönheinz
203 pages, 1984

Band 22

Singular Ordinary Differential Operators and Pseudodifferential Equations

By Johannes Elschner
200 pages, 1985

Band 23

Lotka-Volterra-Approach to Cooperation and Competition in Dynamic Systems

Proceedings of the 5th Meeting of UNESCO's Working Group on Systems Theory held on the Wartburg, Eisenach (GDR) at March 5–9, 1984
Edited by Werner Ebeling and Manfred Peschel
271 pages, 1985

Band 24

Convergence Structures 1984

Proceedings of the Conference on Convergence held in Bechyně (Czechoslovakia); September 24–28, 1984
Edited by Josef Novák, Werner Gähler, Horst Herrlich, and Jan Mikusiński
254 pages, 1985

Band 25

Parcella '84

Proceedings of the II. International Workshop on Parallel Processing by Cellular Automata and Arrays held in Berlin (GDR), September 25–27, 1984
Edited by Wolfgang Händler, Tamás Legendi, and Gottfried Wolf
239 pages, 1985

Band 26

Multiobjective and Stochastic Optimization Based on Parametric Optimization

By Jürgen Guddat, Francisco Guerra Vasquez, Klaus Tammer, and Klaus Wendler
175 pages, 1985

Band 27/28

Systems Analysis and Simulation 1985
I: Theory and Foundations
II: Applications

Proceedings of the International Symposium held in Berlin (GDR), August 26–31, 1985
Edited by Achim Sydow, Manfred Thoma, and Robert Vichnevetsky
per volume 495 pages, 1985

Band 29

Parcella '86

Proceedings of the III. International Workshop on Parallel Processing by Cellular Automata and Arrays held in Berlin (GDR), September 9–11, 1986
Edited by Tamás Legendi, Dennis Parkinson, Roland Vollmar, and Gottfried Wolf
about 272 pages, 1986

Band 30

Fuzzy Sets Applications, Methodological Approaches, and Results

Proceedings of the International Workshop on Fuzzy Sets Applications cosponsored by IIASA and convened in cooperation with IAEA held on the Wartburg, Eisenach (GDR), at March 3–8, 1985
Edited by Steffen Bocklisch, Sergei Orlovski, Manfred Peschel, and Yasushi Nishiwaki
233 pages, 1986

Band 31

Mathematical Methods of Specification and Synthesis of Software Systems '85

Proceedings of the International Spring School, held in Wendisch-Rietz (GDR), April 22–26, 1985
Edited by Wolfgang Bibel and Klaus P. Jantke
245 pages, 1986

Band 32

A Model Theoretic Oriented Approach to Partial Algebras

Introduction to Theory and Application of Partial Algebras – Part I
By Peter Burmeister
319 pages, 1986

Band 33

Finite-difference Methods on Irregular Networks

A Generalized Approach to Elliptic Problems of Second Order
By Bernd Heinrich
about 200 pages, 1986

Bei Fragen zur Produktsicherheit wenden Sie sich bitte an:
If you have any questions regarding product safety,
please contact:

Walter de Gruyter GmbH
Genthiner Straße 13
10785 Berlin
productsafety@degruyterbrill.com